Methods in Molecular Biology™

Series Editor
John M. Walker
School of Life Sciences
University of Hertfordshire
Hatfield, Hertfordshire, AL10 9AB, UK

For further volumes:
http://www.springer.com/series/7651

Calcium-Binding Proteins and RAGE

From Structural Basics to Clinical Applications

Edited by

Claus W. Heizmann

Department of Pediatrics, Division of Clinical Chemistry and Biochemistry, University of Zürich, Zürich, Switzerland

Humana Press

Editor
Claus W. Heizmann
Department of Pediatrics
Division of Clinical Chemistry and Biochemistry
University of Zürich
Zürich, Switzerland

ISSN 1064-3745 ISSN 1940-6029 (electronic)
ISBN 978-1-4939-5976-1 ISBN 978-1-62703-230-8 (eBook)
DOI 10.1007/978-1-62703-230-8
Springer New York Heidelberg Dordrecht London

Printed on acid-free paper

Humana Press is a brand of Springer
Springer is part of Springer Science+Business Media (www.springer.com)

Preface

A major direction in medical research leading to clinical applications targets the regulation of intracellular calcium and the various human diseases associated with an altered homeostasis of this global second messenger. These diseases include, for example, cardiomyopathy, inflammation, brain disorders, diabetes, and cancer.

After entering the cell, Ca^{2+} binds reversibly to specific Ca^{2+}-binding and sensor proteins that fulfill multiple cellular functions. The human genome contains over 200 genes coding for this large protein family, characterized by a common helix–loop–helix structural motif, the EF-hand.

In the 1960s troponin was discovered by Ebashi [1] as the first intracellular Ca^{2+}-sensor protein, regulating muscle contraction. Today highly sensitive cardiac-specific troponin assays are routinely used in most hospitals to diagnose patients with acute coronary syndrome (ACS).

Calmodulin interacts with more than 300 variable target sequences, regulating many biological functions. New strategies are described in this book to quantify calmodulin–target interactions. These techniques can be applied to analyze the interaction between any calcium-binding protein with its binding partner.

In 1971 muscle parvalbumin was the first Ca^{2+}-binding protein to have its amino acid sequence and 3D structure resolved. On this basis, Kretsinger developed the concept of the EF-hand structural motif [2]. Today parvalbumin is known as a marker of a subpopulation of GABAergic neurons in the brain and as a major food allergen sharing the allergenic potential of many other Ca^{2+}-binding proteins. Parvalbumin and S100A1, when introduced by gene therapy, are able to restore heart function in animal models.

These and other examples, included in this book, underline the diagnostic and clinical importance of this family of proteins in human diseases and as drug targets. For example, longistatin—a plasminogen activator from vector ticks (blood-sucking ectoparasites affecting humans and animals)—is a candidate for the development of anti-tick vaccines.

S100 proteins (first described by Moore;[3]) constitute the largest family within this EF-hand superfamily [4]. Most S100 genes are clustered on a region of human chromosome 1q21 (for nomenclature see: [5] and update [6]) that is prone to chromosomal rearrangements. Several S100 proteins are secreted, and they exert the role of cytokines through the activation of various cell surface receptors, e.g., the Receptor for Advanced Glycation Endproducts (RAGE), an immunoglobulin-like multi-ligand receptor for AGEs, amyloid beta peptides, HMGB1, TTR, and several S100 proteins. This receptor and its ligands are associated with Alzheimer's disease, inflammation and cancer. The interaction of RAGE with its ligands is described, using the Surface Plasmon Resonance (SPR) technology. Site-specific blocking of RAGE by specific inhibitors, antibodies, or selected peptides that target RAGE or its ligands is also described, using phage display technology. This approach is presently discussed as a therapeutic strategy to attenuate cell toxicity.

This volume is a collection of chapters written by leading experts in the field, containing state-of-the-art, lab-based methods and easy-to-follow protocols for daily use. These methods and techniques are generally applicable—after modifications—and are not restricted to Ca^{2+}-binding proteins and their targets.

Methods and protocols include:

Calcium-measurements, structural analysis, target binding, and localization: Intracellular Ca^{2+}-measurement in single cells using ratiometric calcium dyes; protein expression systems; equilibrium and flow dialysis and isothermal calorimetry to determine metal binding; isotope labeling of proteins for NMR analysis; crystallization and X-ray analysis using a synchroton source; fluorescence anisotropy; Surface Plasmon Resonance (SPR); Phage display selection; super-resolution light microscopy, and human cell line authentication.

Screening methods: HPLC-ES-MS analysis of body fluids; GST-pulldown to screen for novel Ca^{2+}-binding proteins; grafting approach with luminescence resonance energy transfer to search for viral Ca-binding motifs; primer extension microarray for the analysis of genes associated with disease ; in vivo screening of S100B inhibitors in melanoma.

Clinical Chemistry: Diagnostic and prognostic use of cardiac troponin; S100B as a biomarker for traumatic brain injury, certain neurodegenerative disorders, and malignant melanoma.

Therapy: In vivo S100A1 gene delivery to correct heart failure; Nesfatin and diagnosis and treatment of obesity.

I am very grateful to all contributors for making this volume interesting for basic and medical researchers, cell biologists, clinicians, clinical chemists, and the diagnostic industry. I am also thankful to my wife Erika Heizmann for her patience and understanding during the process of editing this book.

Zürich, Switzerland ***Claus W. Heizmann***

References

1. Ebashi S. (1980) Regulation of muscle contraction. Proc R Soc Lond B207:259–286
2. Kretsinger RH, and Nockolds CE (1973) Carp muscle calcium-binding protein II. Structure determination and general description. J Biol Chem 248:3313–3326
3. Moore BW (1965) A soluble protein characteristic of the nervous system. Biochem Biophys Res Commun 189:739–744
4. Leclerc E, and Heizmann CW (2011) The importance of Ca^{2+}/Zn^{2+} signaling S100 proteins and RAGE in translational medicine. Frontiers Biosci S3:1232–1262
5. Schäfer BW, Wicki R, Engelkamp D, Mattei MG, and Heizmann CW (1995) Isolation of a YAC clone covering a cluster of nine S100 genes on human chromosome 1q21: rationale for a new nomenclature of the S100 calcium-binding protein family. Genomics 25:638–643
6. Kizawa K, Takahara H, Unno M, and Heizmann CW (2011) S100 and fused-type protein families in epidermal maturation with special focus on S100A3 in mammalian hair cuticles. Biochimie 93:2038–2047

Contents

Part III Interaction of Ca^{2+}-Binding Proteins with Targets and Receptors/RAGE

Part IV RAGE-Meditated Cell Signaling and Pathology

Part V Ca^{2+}-Binding Proteins in Human Diseases and As Drug Targets

Part VI Cell Line Authentication

Part VII Biomarkers, Diagnostic, Clinical Chemistry, and Gene Therapy

Contributors

M. Abdul Alim • *National Institute of Animal Health, National Agricultural and Food Research Organization, Tsukuba, Ibaraki, Japan*
Carlos Alonso • *Centro de Biologia Molecular Severo Ochoa, Consejo Superior de Investigaciones Cientificas, Universidad Autonoma de Madrid, Madrid, Spain*
Anisuzzaman • *Department of Global Agricultural Sciences, Graduate School of Agricultural and Life Sciences, The University of Tokyo, Tokyo, Japan*
National Institute of Animal Health, National Agricultural and Food Research Organization, Tsukuba, Ibaraki, Japan
Ramona Astrand • *Department of Neurosurgery, Rigshospitalet, Copenhagen University Hospital, Copenhagen, Denmark*
Emilie Audran • *Faculté de Pharmacie, Laboratoire d'Innovation Thérapeutique, URM 7200, Université de Strasbourg, F-Illkirch, France*
Tiziana Cabras • *Dipartimento di Scienze della Vita e dell 'Ambiente, Università di Cagliari, Monserrato, CA, Italy*
Massimo Castagnola • *Facoltà di Medicina, Istituto di Biochimica e Biochimica Clinica, Università Cattolica, Rome, Italy*
Benjamin Chagot • *Departments of Biochemistry and Chemistry, Vanderbilt University, Nashville, TN, USA*
Center for Structural Biology, Vanderbilt University, Nashville, TN, USA
CNRS/Université de Lorraine, UMR 7214, Vandoeuvre-lès-Nancy, France
Walter J. Chazin • *Departments of Biochemistry and Chemistry, Vanderbilt University, Nashville, TN, USA*
Center for Structural Biology, Vanderbilt University, Nashville, TN, USA
Yanyi Chen • *Department of Chemistry, Georgia State University, Atlanta, GA, USA*
Jos A. Cox • *Department of Biochemistry, University of Geneva, Geneva, Switzerland*
Rania Dagher • *Physiopathologie et Epidémiologie de l'Insuffisance Respiratoire, Faculté de Médicine, Inserm U700, Université Paris Diderot-Paris 7, Paris, France*
Steven M. Damo • *Departments of Biochemistry and Chemistry, Vanderbilt University, Nashville, TN, USA*
Center for Structural Biology, Vanderbilt University, Nashville, TN, USA
Yago Pico de Coaña • *Centro de Biologia Molecular Severo Ochoa, Consejo Superior de Investigaciones Cientificas, Universidad Autonoma de Madrid, Madrid, Spain*
Hugo Murua Escobar • *Small Animal Clinic, University of Veterinary Medicine, Hannover, Germany*
Chiara Fanali • *Facoltà di Medicina, Istituto di Biochimica e Biochimica Clinica, Università Cattolica, Rome, Italy*

MICHAEL D. FELDKAMP • Departments of Biochemistry and Chemistry, Vanderbilt University, Nashville, TN, USA

Center for Structural Biology, Vanderbilt University, Nashville, TN, USA

GÜNTER FRITZ • Department of Neuropathology, Neurozentrum, University of Freiburg, Freiburg, Germany

MIGUEL ÁNGEL FUERTES • Centro de Biologia Molecular Severo Ochoa, Consejo Superior de Investigaciones Cientificas, Universidad Autonoma de Madrid, Madrid, Spain

SOPHIE GIORIA • PCBIS, UMS 3286, Bd Sébastien Brand, Illkirch, France

YOUSANG GWACK • Department of Physiology, David Geffen School of Medicine at UCLA, Los Angeles, CA, USA

JACQUES HAIECH • Laboratoire d'Innovation Thèrapeutique, Facultè de Pharmacie, URM 7200, Université de Strasbourg, Illkirch, France

JOERG HEINEKE • Klinik für Kardiologie und Angiologie, Rebirth-Cluster of Excellence, Medizinische Hochschule Hannover, Hannover, Germany

CLAUS W. HEIZMANN • Division of Clinical Chemistry and Biochemistry, Department of Pediatrics, University of Zürich, Zürich, Switzerland

DANIELLE HOF • Institute for Clinical Chemistry, University Hospital Zürich, Zürich, Switzerland

JOHANNES HRADSKY • Research Group, Neuroplasticity, Leibniz Institute for Neurobiology (LIN), Magdeburg, Germany

HAO HUANG • Samuel Lunenfeld Research Institute, Mt. Sinai Hospital, Toronto, Canada

FEDERICA IAVARONE • Facoltà di Medicine, Istituto Biochimica e Biochimica Clinica, Università Cattolica, Rome, Italy

M. KHYRUL ISLAM • National Institute of Animal Health, National Agricultural and Food Research Organization, Tsukuba, Ibaraki, Japan

Department of Veterinary Science, The University of Melbourne, Parkville, VIC, Australia

ANNA KARPOVA • Research Group, "Neuroplasticity", Leibniz Institute for Neurobiology, Magdeburg, Germany

HUGO A. KATUS • Department of Internal Medicine III, University of Heidelberg, Heidelberg, Germany

MARIE-CLAUDE KILHOFFER • Laboratoire d'Innovation Thérapeutique, Faculté de Pharmacie, URM 7200, Université de Strasbourg, Illkirch, France

KENJI KIZAWA • Skin Science Research Group, Innovative Beauty Science Laboratory, Kanebo Cosmetics Inc., Odawara, Kanagawa, Japan

ROLAND KLINGENBERG • Department of Cardiology/Cardiovascular Research, University Hospital Zürich, Zürich, Switzerland

Zürich Center of Integrated Physiology, University of Zürich, Zürich, Switzerland

MICHAEL R. KREUTZ • Research Group, "Neuroplasticity", Leibniz Institute for Neurobiology (LIN), Magdeburg, Germany

JUHA KUJA-PANULA • Neuroscience Center, University of Helsinki, Helsinki, Finland

RENA G. LAPIDUS • Center for Biomolecular Therapeutics, Marlene and Stewart Greenebaum Cancer Center, University of Maryland, Baltimore, MD, USA

ESTELLE LECLERC • Department of Pharmaceutical Sciences, North Dakota State University, Fargo, ND, USA

MASATOSHI MAKI • Department of Applied Molecular Biosciences, Graduate School of Bioagricultural Sciences, Nagoya University, Nagoya, Japan

ROBERT S. MCLAREN • Promega Corporation, Madison, WI, USA

Irene Messana • *Dipartimento di Scienze della Vita e dell' Ambiente, Università di Cagliari, Cittadella Universitaria di Monserrato, Monserrato, CA, Italy*
Marina Mikhaylova • *Research Group, "Neuroplasticity", Leibniz Institute for Neurobiology, Magdeburg, Germany*
Masatomo Mori • *Department of Medicine and Molecular Science, Graduate School of Medicine, Gunma University, Maebashi, Gunma, Japan*
Patrick Most • *Department of Internal Medicine III, Center for Molecular and Translational Cardiology, University of Heidelberg, Heidelberg, Germany*
Ingo Nolte • *Small Animal Clinic, University forVeterinary Medicine, Hannover, Germany*
Eneli Oitmaa • *Asper Biotech Ltd, Tartu, Estonia*
Kanae Osugi • *Department of Applied Molecular Biosciences, Graduate School of Bioagricultural Sciences, Nagoya University, Nagoya, Japan*
Nuria Parody • *Centro de Biologia Molecular Severo Ochoa, Consejo Superior de Investigaciones Cientificas, Universidad Autonoma de Madrid, Madrid, Spain*
R&D Department, Laboratorios LETI S.L, Madrid, Spain
Sven T. Pleger • *Department of Internal Medicine III, Center for Molecular and Translational Cardiology, University of Heidelberg, Heidelberg, Germany*
Philip Raake • *Department of Internal Medicine III, University of Heidelberg, Heidelberg, Germany*
Heikki Rauvala • *Neuroscience Center, University of Helsinki, Helsinki, Finland*
Yvonne Reid • *American Type Culture Collection (ATCC™), Manassas, VA, USA*
Bertil Romner • *Department of Neurosurgery, Rigshospitalet, Copenhagen University Hospital, Copenhagen, Denmark*
Ari Rouhiainen • *Neuroscience Center, University of Helsinki, Helsinki, Finland*
Hideki Shibata • *Department of Applied Molecular Biosciences, Graduate School of Bioagricultural Sciences, Nagoya University, Nagoya, Japan*
Hiroyuki Shimizu • *Department of Health and Nutrition, Faculty of Health and Welfare, Takasaki University of Health and Welfare, Takasaki, Gunma, Japan*
Sonal Srikanth • *Department of Physiology, David Geffen School of Medicine at UCLA, Los Angeles, CA, USA*
Katharina Anna Sterenczak • *Small Animal Clinic, University for Veterinary Medicine, Hannover, Germany*
Douglas R. Storts • *Promega Corporation, Madison, WI, USA*
Hidenari Takahara • *Laboratory of Biochemistry and Molecular Biology, Department of Applied Biological Resource Sciences, Ibaraki University, Inashiki, Ibaraki, Japan*
Naotoshi Tsuji • *Department of Global Agricultural Sciences, Graduate School of Agricultural and Life Sciences, The University of Tokyo, Tokyo, Japan*
National Institute of Animal Health, National Agricultural and Food Research Organization, Tsukuba, Ibaraki, Japan
Sarka Tumova • *Faculty of Biological Sciences, Multidisciplinary Cardiovascular Research Center and the Institute of Membrane & Systems Biology, University of Leeds, Leeds, UK*
Janne Üksti • *Asper Biotech Ltd, Tartu, Estonia*
Johan Undén • *Department of Anesthesiology and Intensive Care, Skane University Hospital, Malmoe, Sweden*
Masaki Unno • *Frontier Research Center for Applied Atomic Sciences, Ibaraki University, Naka, Ibaraki, Japan*

KADRI VAIDLA • *Asper Biotech Ltd, Tartu, Estonia*
STEFAN W. VETTER • *Department of Pharmaceutical Sciences, North Dakota State University, Fargo, ND, USA*
HANS J. VOGEL • *Department of Biological Sciences, University of Calgary, Calgary, AB, Canada*
ARNOLD VON ECKARDSTEIN • *Institute for Clinical Chemistry, University Hospital Zürich, Zürich, Switzerland*

Zürich Center of Integrated Human Physiology, University of Zürich, Zürich, Switzerland
CHRISTOPHE WEBER • *Department of Internal Medicine III, Center for Molecular and Translational Cardiology, University of Heidelberg, Heidelberg, Germany*
DAVID J. WEBER • *Center for Biomolecular Therapeutics, The University of Maryland, School of Medicine, The University of Maryland, Baltimore, MD, USA*

Marlene and Stewart Greenebaum Cancer Center, The University of Maryland, Baltimore, MD, USA
SHENGHUI XUE • *Department of Biology, Georgia State University, Atlanta, GA, USA*
JENNY J. YANG • *Department of Chemistry, Georgia State University, Atlanta, GA, USA*
CHRISTINA ZEITZ • *INSERM, U968, Paris, France*

Institute de la Vision, UPMC University of Paris 06, UMR_S 968, Paris, France

CNRS, UMR_7210, Paris, France
YUBIN ZHOU • *Department of Chemistry, Georgia State University, Atlanta, GA, USA*

Division of Signaling and Gene Expression, La Jolla Institute for Allergy and Immunology, La Jolla, CA, USA
DANNA B. ZIMMER • *Department of Biochemistry and Molecular Biology, Center for Biomolecular Therapeutics, The University of Maryland School of Medicine, Baltimore, MD, USA*
WERNER ZUSCHRATTER • *Special Laboratory Electron-and Laserscanning Microscopy, Leibniz Institute for Neurobiology (LIN), Magdeburg, Germany*

Part I

Intracellular Ca^{2+}-Measurements and Cation Interactions with EF-Hand Proteins

Chapter 1

Measurement of Intracellular Ca^{2+} Concentration in Single Cells Using Ratiometric Calcium Dyes

Sonal Srikanth and Yousang Gwack

Abstract

Measurement of intracellular Ca^{2+} concentration ($[Ca^{2+}]_i$) is useful to study the upstream and downstream events of Ca^{2+} signaling. Ca^{2+}-binding proteins including EF-hand-containing proteins are important downstream effector molecules after an increase of $[Ca^{2+}]_i$. Conversely, these proteins can also act as key modulators for regulation of $[Ca^{2+}]_i$ by sensing the Ca^{2+} levels in the intracellular organelles and cytoplasm. Here we describe a single-cell Ca^{2+} imaging technique that was used to measure the intracellular Ca^{2+} levels to examine the function of Ca^{2+}-binding proteins, STIM1 and Calcium release-activated Calcium channel regulator 2A (CRACR2A), using ratiometric Ca^{2+} dye Fura-2 in adherent and non-adherent cells.

Key words: Store-operated Ca^{2+} entry, Calcium release-activated Calcium (CRAC) channel, Nuclear factor of activated T cells (NFAT), Ratiometric Ca^{2+} measurements, Fura-2, Intracellular Ca^{2+} concentration, EF-hand-containing proteins, STIM1, Orai1, CRAC channel regulator (CRACR) 2A

1. Introduction

In non-excitable cells, Ca^{2+} entry via store-operated Ca^{2+} (SOC) channels is a predominant mechanism to raise intracellular Ca^{2+} concentrations ($[Ca^{2+}]_i$) (1–4). Under resting conditions, cytoplasmic $[Ca^{2+}]$ is in the range of ~100 nM while that in the endoplasmic reticulum, which serves as an intracellular Ca^{2+} store, is much higher (~400 μM). SOC channels were so named because they are activated by depletion of intracellular Ca^{2+} stores (2). The Ca^{2+} release-activated Ca^{2+} (CRAC) channel is a specialized class of SOC channel and is well characterized in immune cells. Ca^{2+} influx via the CRAC channel activates many downstream Ca^{2+}-binding proteins including calmodulin (1, 4, 5). Ca^{2+}-bound calmodulin activates further downstream signaling molecules like

Claus W. Heizmann (ed.), *Calcium-Binding Proteins and RAGE: From Structural Basics to Clinical Applications*, Methods in Molecular Biology, vol. 963, DOI 10.1007/978-1-62703-230-8_1, © Springer Science+Business Media New York 2013

calmodulin-dependent kinases, the protein phosphatase calcineurin, and others. Other EF-hand-containing Ca^{2+}-binding proteins also play an important role in activation and modulation of CRAC channel activity, by sensing the Ca^{2+} concentrations in specific cellular compartments including cytoplasm (e.g., CRACR2A) (6) and endoplasmic reticulum (e.g., STIM1) (7–9).

Recent studies have identified STIM1, a Ca^{2+}-binding protein localized predominantly in the endoplasmic reticulum (ER) as an important component of CRAC channel-mediated Ca^{2+} entry (7–9). STIM1 contains an N-terminal EF-hand that detects luminal ER [Ca^{2+}], a single transmembrane domain, and a long C-terminal cytoplasmic region. Subsequent studies have identified Orai1 (also known as CRACM1 or TMEM142A) as a pore subunit of the CRAC channels (10–13). Upon ER Ca^{2+} depletion, STIM1 becomes Ca^{2+}-unbound, multimerizes, translocates to PM proximal ER, mediates clustering of Orai proteins on the PM, and stimulates Ca^{2+} entry (7–9, 14). Amplified CRAC currents have been observed upon co-expression of Orai1 and STIM1, suggesting that these are the limiting and essential components of CRAC channels (15–17).

A direct protein interaction between STIM1 and Orai1 is crucial for activation of CRAC channels. Recently an important domain in the cytoplasmic region of STIM1, termed the CRAC activation domain (CAD)/STIM1 Orai1 activating region (SOAR) was identified by truncation analysis (1–4). The CAD/SOAR fragment of STIM1 was shown to play a pivotal role in activation of Orai1 by direct binding to cytosolic N- and C-termini of Orai1. These studies indicate that CRAC channel activation involves multiple steps including STIM1 oligomerization, co-clustering of Orai1 and STIM1 at the PM-ER junctions, and gating of Orai1 (18–23). While the role of Orai1-STIM1 interaction is emphasized in the multiple activation steps of CRAC channels, the cellular machinery modulating these processes is currently under investigation.

In a previous study, we identified a novel cytoplasmic EF-hand-containing protein, CRACR2A, as an important regulator of Orai1–STIM1 interaction (6). We showed that CRACR2A directly interacts with Orai1 and STIM1, forming a ternary complex. Interestingly, CRACR2A dissociates from Orai1 and STIM1 upon increase of $[Ca^{2+}]_i$. An EF-hand mutant of CRACR2A elevated cytoplasmic [Ca^{2+}], enhanced STIM1 clustering, and induced cell death. These observations suggest a role for CRACR2A as a cytoplasmic Ca^{2+} sensor that modulates multiple steps of CRAC channel activation including oligomerization, translocation, and cluster formation of Orai1 and STIM1 by direct protein interaction. Recently, CaM binding to the N terminus of Orai1 has been shown to induce inactivation of CRAC channels (24). CaM-binding sites on Orai1 and STIM1 partially overlap with those of CRACR2A.

Interestingly, CaM was found to co-immunoprecipitate with Orai1 only in the presence of Ca^{2+}, contrary to CRACR2A that dissociates from Orai1 and STIM1 at elevated $[Ca^{2+}]_i$. These results suggest a scenario where CRACR2A binding favors an active CRAC channel complex upon store depletion (at low $[Ca^{2+}]_i$) while CaM facilitates inactivation of CRAC channels at high $[Ca^{2+}]_i$. In summary, several Ca^{2+}-binding proteins are key players in regulation of $[Ca^{2+}]_i$ and to study their role, accurate measurement of a change in $[Ca^{2+}]_i$ is very important.

Two different methods such as current measurements using electrophysiological recordings and intracellular Ca^{2+} measurements using Ca^{2+}-binding dyes are widely used to examine CRAC channel function. Both of them have their benefits and limitations and should be used based on desired sensitivity, specificity, and feasibility of the techniques. Whole-cell patch clamp recording is a specialized and accurate technique to measure CRAC currents. Because CRAC channel-mediated currents are defined based on their characteristic inwardly rectifying current–voltage relationship, whole-cell patch clamp recording remains an indisputable method of choice for accurate measurement of CRAC channel function.

Although electrophysiological recording is an excellent tool to study biophysical properties of CRAC channels, it also has limitations. Because CRAC currents are very small with an estimated single-channel conductance of <1 pS (25, 26), it is not possible to directly measure single-channel currents, a hallmark of electrophysiological analyses, with the currently available tools. In addition, endogenous CRAC currents have been measured only from immune cells including T cells and mast cells. Because these cells express relatively few ion channels, it is easy to isolate the small CRAC currents. Whole-cell patch clamp recording is an invasive method involving exchange of intracellular content of a cell with a synthetic solution, which may cause diffusion of intracellular factors regulating channel activity. Lastly, it is very difficult to measure small changes in basal (resting) Ca^{2+} levels or Ca^{2+} oscillations using electrophysiological tools.

On the other hand, Ca^{2+} imaging can determine $[Ca^{2+}]_i$ in 20–100 cells simultaneously and allows to detect even a small change in $[Ca^{2+}]_i$. However, Ca^{2+} imaging lacks specificity of detecting a change in $[Ca^{2+}]_i$ resulting from the activity of a single type of channels. One often relies on agonists or antagonists of specific channels in combination with Ca^{2+} imaging to measure changes in $[Ca^{2+}]_i$ specifically derived from a single type of channels (e.g., La^{3+}, Gd^{3+}, or 2-aminoethoxydiphenyl borate for CRAC channels). While a wide range of currents (e.g., ranging from pA to nA) can be easily measured using whole-cell patch-clamp techniques without saturation, for ratiometric Ca^{2+} imaging, one relies on dyes with different affinities for Ca^{2+} to avoid saturation of signals. In addition, it is not feasible to study the detailed channel properties such as gating and inactivation using Ca^{2+} imaging. On the positive side, SOCE from hundreds of Jurkat or primary T cells can be measured simultaneously using this technique, allowing for population analysis. With combinations of surface staining and fluorescence marker expression (e.g., GFP or mCherry), data can be analyzed in specific subsets of cells. Single-cell Ca^{2+} imaging is relatively noninvasive, requiring loading of cells with a Ca^{2+}-sensitive dye, leaving intracellular milieu intact for regulation of channel activity. Hence, the method of choice for any experiment would depend on the purpose of the experiment. In this report, we focus on intracellular Ca^{2+}-measurement using noninvasive ratiometric Ca^{2+}-imaging technique. This method can be used for analysis of a population of cells after exposure to various conditions including knockdown and overexpression of specific genes, after co-expression of compatible fluorescence markers, etc. This method can be widely applicable to determine the role of possible modulators for $[Ca^{2+}]_i$ in agonist-dependent, store-dependent, or store-independent manners.

2. Materials

2.1. Transfection of Plasmids and siRNAs and Quantification of Knockdown Efficiency Using Quantitative RT-PCR

1. Lipofectamine 2000 and Lipofectamine RNAimax (Invitrogen, Carlsbad, CA).
2. siRNAs (Dharmacon).
3. Complete Medium—DMEM (Mediatech) + 10% Fetal Bovine Serum (Hyclone) + Penecillin–Streptomycin (1%, Mediatech) + HEPES (1 mM, Mediatech) + Glutamine (1 mM, Mediatech).
4. ECM-830 Electro Square Porator (Harvard apparatus).
5. Gene pulser 0.4 cm electroporation cuvettes (Biorad).
6. OptiMEM (Invitrogen) or plain DMEM (without additives or serum).
7. Reporter plasmids (pEGFPN1 or pN1-mCherry, Clontech).
8. TRIzol LS reagent and oligo(dT)-primed first-strand cDNA synthesis kit (Superscript III kit; Invitrogen).

2.2. Reagents for Ratiometric Ca^{2+} Imaging

1. Fura 2-acetoxymethyl ester (AM) and Fura 4F-AM (Invitrogen, Carlsbad, CA) resuspended in dimethyl sulfoxide (DMSO) to generate a 1 mM stock solution.
2. Ca^{2+}-free Ringer's solution (in mM): 145 NaCl, 4.5 KCl, 3 $MgCl_2$, 10 D-glucose, 0.05 EGTA, and 10 HEPES (pH 7.35).
3. 2 mM Ca^{2+}-containing Ringer's solution (in mM): 145 NaCl, 4.5 KCl, 1 $MgCl_2$, 2 $CaCl_2$, 10 D-glucose, and 10 HEPES (pH 7.35).

4. Thapsigargin (from EMD Biosciences), final concentration of 1 μM in Ca^{2+}-free Ringer's solution.
5. Fura-2 Ca^{2+} Imaging Calibration kit (Invitrogen).

2.3. Software for Data Acquisition and Analysis

1. Slidebook (Intelligent Imaging Innovations, Inc.).
2. Origin Pro (Origin, Northampton, USA).

2.4. Equipment for Imaging

1. Olympus 1X-51 inverted microscope (Olympus Inc.) connected to a Lambda LB-LS/30 (Sutter instruments) light source housing a 300 W xenon-arc lamp (Perkin-Elmer) and an excitation filter-wheel, regulated by an external controller (Fig. 1a). Image acquisition is done using Fura-2 (z340/15 and z380/15 excitations, T400LP dichroic and D510/40 emission) filter cube (Chroma Technology Corp.).
2. Microscope is attached to a deep-cooled CCD camera (ORCA-ER-AG; Hamamatsu). Filter switching and image acquisition are

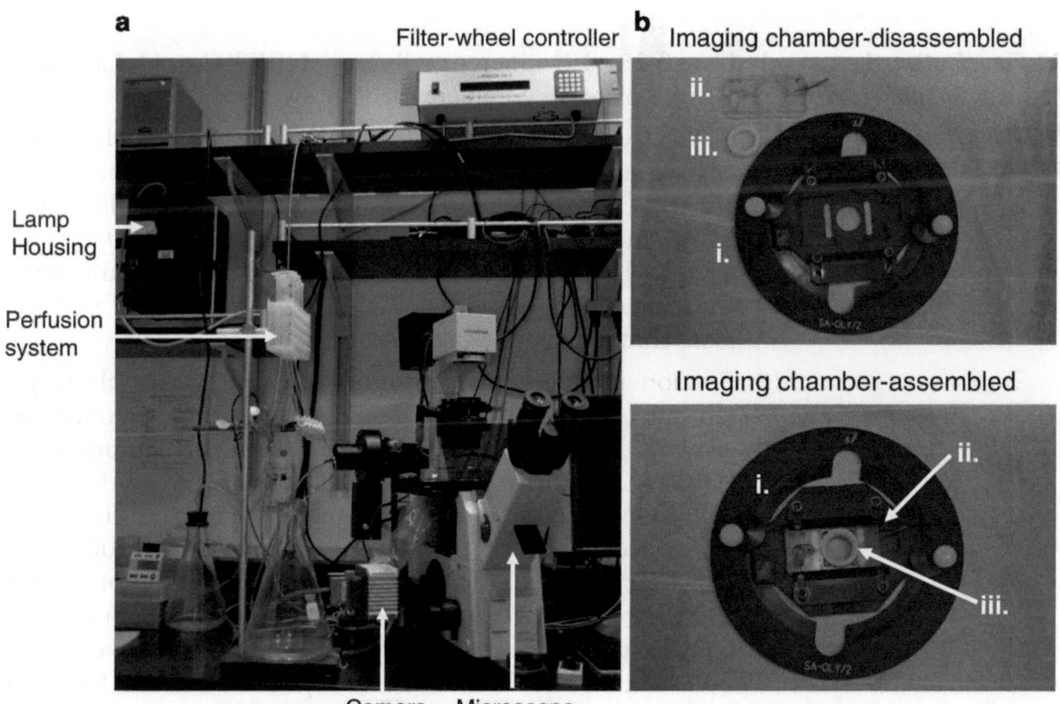

Fig. 1. Pictures of setup for measurement of SOCE using ratiometric Ca^{2+} imaging. (**a**) Microscope setup with microscope, lamp housing, filter-wheel controller, perfusion apparatus, and camera. (**b**) Perfusion chamber setup. *Top image* depicts the disassembled chamber showing a series 20 chamber platform attached to Olympus stage adaptor (i) and the imaging chamber RC-20 H (ii) with the white-ring (iii). *Bottom image* shows the RC20H chamber assembled on the platform with the white-ring on top.

controlled by scripts written in Slidebook (Intelligent Imaging Innovations, Inc.). All fluorescence image acquisition is done with 2 × 2 pixel binning at 5-s frame intervals and all the experiments are performed at room temperature (22–25°C, Fig. 1a).

3. 15 mm round poly-D-Lysine coated coverslips (Warner Instruments).
4. SA-OLY/2 (Warner Instruments), stage adapter for series 20 chamber platform.
5. Series 20 chamber platform P-5 (Warner instruments).
6. Cell culture/Imaging Chamber RC-20 H (Warner Instruments).
7. Perfusion system (6-channel multi-valve perfusion system, Warner Instruments, Hamden, CT).
8. A 3-ml syringe filled with petroleum jelly and a 26-gauge needle attached to the tip of the syringe.
9. Philips #1 screwdriver.

3. Methods

3.1. Transfection of Plasmids or siRNA into Adherent and Non-adherent Cells

1. Plate 0.25×10^6 adherent cells (e.g., HEK293 or HeLa cells) in 12-well plates in complete medium overnight.
2. Next day, transfect with control or experimental siRNAs (Dharmacon) using lipofectamine RNAi Max reagent. We co-transfect pEGFPN1 or pN1-mCherry reporter plasmid to select transfected cells for analysis. We use 20–50 nM siRNA + 100 ng reporter plasmid (pEGFPN1 or pN1-mCherry) with 2.5 μg of lipofectamine RNAimax for one well of a 12-well plate containing adherent cells.
3. For non-adherent cells such as Jurkat T cells, we use electroporation to introduce siRNA. For electroporation, exponentially growing Jurkat T cells are washed with media lacking serum (OptiMEM [Invitrogen, Carlsbad] or DMEM without any additives) twice and resuspended in the same serum-free media at a concentration of 10×10^6 cells/ml. The cells (400 μl) are transferred into a 0.4 cm Gene pulser electroporation cuvette (Biorad), incubated with 100–200 nM of siRNA + 2 μg of reporter plasmid (or alternatively 15 μg DNA for overexpression analysis) for 15 min and electroporated at 200 V for 40 ms on an ECM-830 Electro Square Porator (Harvard Apparatus). After electroporating, cells are incubated for further 10 min, before transferring into 6-well plates (Greiner, Bio-one) containing 3 ml of pre-warmed complete medium and incubated in a 37°C incubator with 5% CO_2. Experiments are performed 72–96 h after electroporation.

4. For adherent cells, the protocol of transfection is repeated 24 h after the first transfection and cells are harvested for reverse transcription-polymerase chain reaction (RT-PCR) or $[Ca^{2+}]_i$ measurement 3 days after the first transfection.
5. Knockdown efficiency is quantified by quantitative RT-PCR analysis. RNA is purified using TRIzol LS reagent (Invitrogen) and oligo(dT)-primed for first-strand cDNA synthesis (Superscript III kit; Invitrogen) using manufacturer's protocols. Quantitative real-time PCR is performed using an iCycler IQ5 (Bio-Rad).
6. 72 h after the first transfection, adherent cells are plated onto UV-sterilized 15 mm coverslips (individual coverslips added onto 12-well plates) and used for imaging on the next day.

3.2. Fura-2 Imaging

1. Adherent cells plated on coverslips overnight are loaded with 1 μM Fura 2-AM in complete medium for 45 min at 25°C in the dark.
2. Non-adherent cells are loaded at 1×10^6 cells/ml with 1 μM Fura 2-AM for 30 min at 25°C and attached to poly-D-lysine-coated 15 mm coverslips for 15 min.
3. The P-5 platform is inserted onto the stage adapter (SA-OLY/2) and a thin ring of petroleum jelly is applied around the edges for placement of coverslip on top. Once the coverslip is placed on the platform, the imaging chamber (RC-20H) pretreated with petroleum jelly (as mentioned for the platform above) on both the sides is placed on top of the coverslip. Then a blank coverslip (without any cells) is attached on top of the imaging chamber. The imaging chamber is held onto the platform by four tiny screws. Once assembled, it forms the imaging area (Fig. 1b). Application of petroleum jelly ensures a leak-free imaging chamber.
4. After chamber assembly, cells are visualized using a 20× (NA 0.5) objective and a field containing a good number of cells and some blank area for background selection is selected. The cells are first perfused with Ringer's solution containing 2 mM $CaCl_2$ to get a baseline reading for ~1 min (in the absence of any stimulus, Fig. 2, i). Slidebook 5.0 is used for image acquisition, using settings for time-lapse ratiometric Fura-2 imaging, allowing for visualization of real-time pseudocolored images of intracellular Ca^{2+} during the acquisition.
5. After acquiring the baseline, the chamber is perfused with Ca^{2+}-free Ringer's solution and measurements are continued for further 1 min.
6. For depletion of ER Ca^{2+}-stores, the chamber is perfused with 1 μM thapsigargin in Ca^{2+}-free Ringer's solution for 5 min to observe an increase in $[Ca^{2+}]_i$ derived from depletion of ER

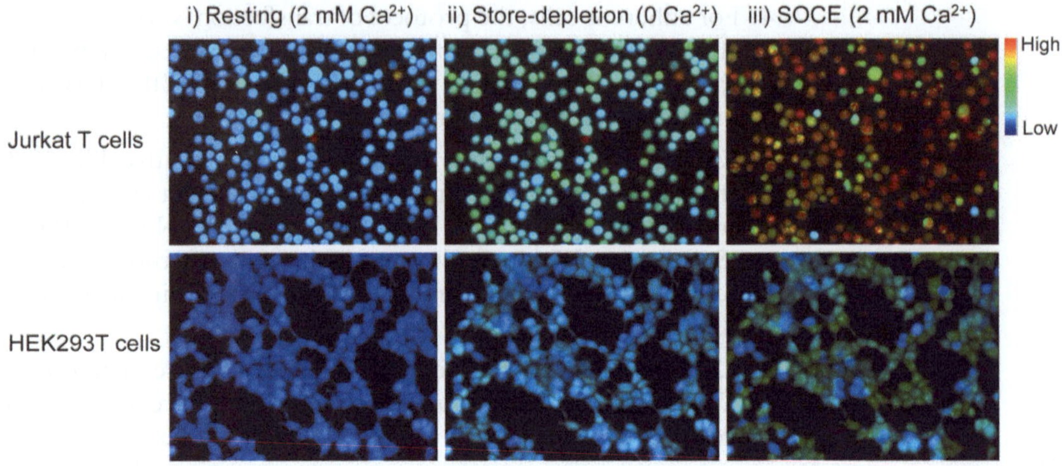

Fig. 2. Measurement of the intracellular Ca^{2+} concentrations using Fura-2 in adherent and non-adherent cells. Images of Jurkat T cells and HEK293 cells at different stages of SOCE measurement: (i) resting, (ii) peak of store depletion after thapsigargin treatment, and (iii) peak of SOCE after exchanging the Ca^{2+}-free extracellular solution with that containing 2 mM Ca^{2+}.

Ca^{2+} store (Fig. 2, ii). Complete depletion of the store is ensured when $[Ca^{2+}]_i$ fluorescence levels return to the baseline.

7. SOCE is measured by perfusing the chamber with 2 mM Ca^{2+}-containing Ringer's solution (Fig. 2, iii). A sharp increase in $[Ca^{2+}]_i$ is observed upon addition of 2 mM Ca^{2+}-containing Ringer's solution and depending on the cell type, it is either sustained (in T cells) or slowly decays (in HEK293 and HeLa cells) (Fig. 3a, b).
8. eGFP or mCherry images are acquired using appropriate filter cubes, at the start and end of the experiment, to allow for selection of cells expressing the fluorescence marker (Fig. 3a).
9. Measurement of Ca^{2+} oscillations in T cells: Jurkat T cells are treated with 10 nM thapsigargin in 2 mM Ca^{2+}-containing Ringer's solution to induce asynchronous $[Ca^{2+}]_i$ oscillations.

Fig. 3. Method for single cell analysis. (**a**) Images on the left depict cells co-expressing GFP fluorescence marker while those on the right show individual regions of interest (ROIs) drawn around GFP^+ cells (represented as *circles* in the *top panel* and *irregular shape* in the *bottom panel*) using Slidebook. (**b**) Raw graphs of individual ROIs (cells) selected in panel (**a**). *Green line* shows the time point of ratio images depicted in panel (**a**). (**c**) Graphs of average ± s.e.m from several ROIs. 30–60 cells (ROIs) were selected and their 340/380 ratio was averaged. Averaged ratio values of individual cells (as seen in panel (**b**)) ± s.e.m are plotted versus time using graph options of Origin Pro.

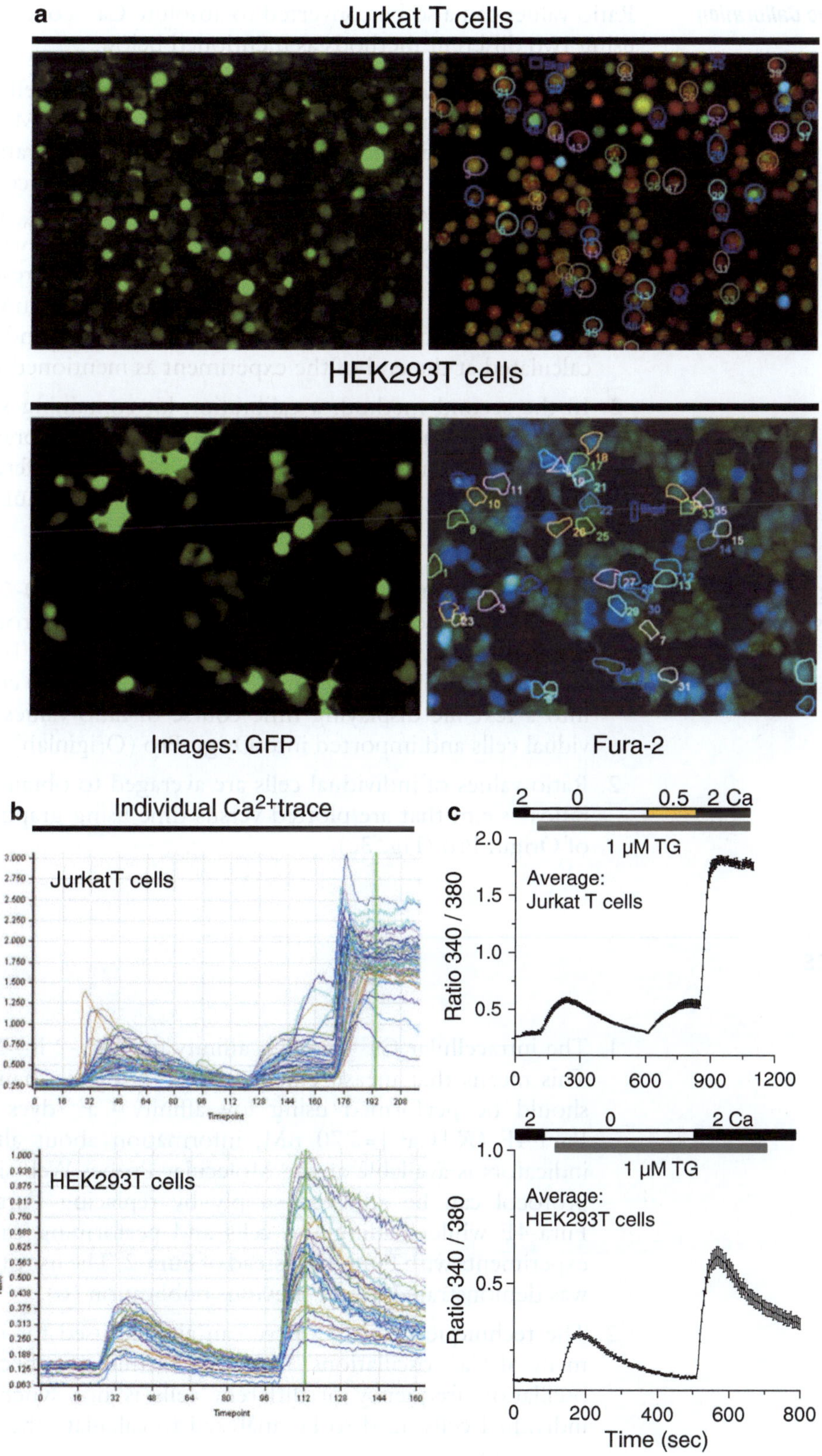
a
Jurkat T cells
HEK293T cells
Images: GFP
Fura-2
b
Individual Ca2+trace
JurkatT cells
HEK293T cells
Ratio
Timepoint
c
2
0
0.5
2 Ca
1 μM TG
Average:
Jurkat T cells
Ratio 340 / 380
2
0
2 Ca
1 μM TG
Average:
HEK293T cells
Ratio 340 / 380
Time (sec)

3.3. Ratio Calibration

Ratio values can also be converted to absolute Ca^{2+} concentrations using two different methods as mentioned below.

1. In the first method, at the end of the experiment, cells are first perfused with a Ca^{2+}-free solution containing 2 mM ethylene glycol tetraacetic acid (EGTA, to obtain minimum ratio [R_{min}] value) and then with 1 μM Ionomycin in 2 mM Ca^{2+}-containing Ringer's solution (to obtain maximum ratio [R_{max}] value). These values are then incorporated into the Grynkiewicz equation $[Ca^{2+}]_i = K \times (R - R_{min})/(R_{max} - R)$ (27). K represents the Ca^{2+} affinity of Fura-2 which is calculated as ~224 nM within cells (and 145 nM in extracellular solutions). R_{min} and R_{max} are calculated at the end of the experiment as mentioned above.
2. In the second method, a calibration kit containing solutions with known amounts of free Ca^{2+} (Fura-2 Ca^{2+} Calibration Kit, Invitrogen) is used and fluorescence values at different Ca^{2+} concentrations are obtained as described in the manufacturer's protocol.

3.4. Single Cell Analysis

1. For each experiment, 25–35 HEK293/HeLa and 50–70 T cells are selected by drawing regions of interest (ROIs) around cells using the analysis options of Slidebook (Fig. 3a). Individual cell traces are observed (Fig. 3b) and the results are converted into a text file displaying time course of ratio values of individual cells and imported into OriginPro (Originlab).
2. Ratio values of individual cells are averaged to obtain average ratios ± s.e.m that are plotted versus time using graph options of Origin Pro (Fig. 3c).

4. Notes

1. The intracellular Ca^{2+}-binding affinity for Fura-2 is ~224 nM. This means that measurement of $[Ca^{2+}]_i$ higher than ~1 μM should be performed using low-affinity Ca^{2+} dyes such as Fura-4F ($K_d[Ca^{2+}] = 770$ nM, information about alternative indicators is available at the Molecular Probes Web site). The protocol can be adjusted simply by replacing Fura-2 with Fura-4F while loading the cells and performing calibration experiments with Fura-4F instead of Fura-2. The use of Fura-4F was demonstrated in our previous publication (6).
2. The technique described here can also be used for measurement of Ca^{2+} oscillations. However, in this case, because the oscillation frequency of different cells is not synchronized, individual cells need to be analyzed to calculate their oscillation frequency.

3. When GFP is coexpressed as a fluorescence marker, high GFP levels can interfere with Fura-2 signals. In these cases, select only cells with mild GFP expression. It is advisable to use m-Cherry as a reporter with Fura-2 imaging to avoid any interference.

Acknowledgments

This work was supported by the National Institute of Health grants AI-083432 and AI-088393 to YG.

References

1. Lewis RS (2011) Store-operated calcium channels: new perspectives on mechanism and function. Cold Spring Harb Perspect Biol. doi: 10.1101/cshperspect.a003970
2. Putney JW (2009) Capacitative calcium entry: from concept to molecules. Immunol Rev 231:10–22
3. Cahalan MD, Chandy KG (2009) The functional network of ion channels in T lymphocytes. Immunol Rev 231:59–87
4. Hogan PG, Lewis RS, Rao A (2010) Molecular basis of calcium signaling in lymphocytes: STIM and ORAI. Annu Rev Immunol 28: 491–533
5. Hogan PG, Chen L, Nardone J et al (2003) Transcriptional regulation by calcium, calcineurin, and NFAT. Genes Dev 17:2205–2232
6. Srikanth S, Jung HJ, Kim KD et al (2010) A novel EF-hand protein, CRACR2A, is a cytosolic Ca2+ sensor that stabilizes CRAC channels in T cells. Nat Cell Biol 12:436–446
7. Liou J, Kim ML, Heo WD et al (2005) STIM is a Ca2+ sensor essential for Ca2+-store-depletion-triggered Ca2+ influx. Curr Biol 15:1235–1241
8. Roos J, Digregorio PJ, Yeromin AV et al (2005) STIM1, an essential and conserved component of store-operated Ca2+ channel function. J Cell Biol 169:435–445
9. Zhang SL, Yu Y, Roos J et al (2005) STIM1 is a Ca2+ sensor that activates CRAC channels and migrates from the Ca2+ store to the plasma membrane. Nature 437:902–905
10. Feske S, Gwack Y, Prakriya M et al (2006) A mutation in Orai1 causes immune deficiency by abrogating CRAC channel function. Nature 441:179–185
11. Vig M, Peinelt C, Beck A et al (2006) CRACM1 is a plasma membrane protein essential for store-operated Ca2+ entry. Science 312: 1220–1223
12. Zhang SL, Yeromin AV, Zhang XH et al (2006) Genome-wide RNAi screen of Ca(2+) influx identifies genes that regulate Ca(2+) release-activated Ca(2+) channel activity. Proc Natl Acad Sci USA 103:9357–9362
13. Gwack Y, Srikanth S, Feske S et al (2007) Biochemical and functional characterization of Orai proteins. J Biol Chem 282:16232–16243
14. Carrasco S, Meyer T (2011) STIM proteins and the endoplasmic reticulum-plasma membrane junctions. Annu Rev Biochem 80: 33.31–33.28
15. Mercer JC, Dehaven WI, Smyth JT et al (2006) Large store-operated calcium selective currents due to co-expression of Orai1 or Orai2 with the intracellular calcium sensor, Stim1. J Biol Chem 281:24979–24990
16. Peinelt C, Vig M, Koomoa DL et al (2006) Amplification of CRAC current by STIM1 and CRACM1 (Orai1). Nat Cell Biol 8:771–773
17. Soboloff J, Spassova MA, Tang XD et al (2006) Orai1 and STIM reconstitute store-operated calcium channel function. J Biol Chem 281: 20661–20665
18. Muik M, Frischauf I, Derler I et al (2008) Dynamic coupling of the putative coiled-coil domain of ORAI1 with STIM1 mediates ORAI1 channel activation. J Biol Chem 283: 8014–8022
19. Park CY, Hoover PJ, Mullins FM et al (2009) STIM1 clusters and activates CRAC channels via direct binding of a cytosolic domain to Orai1. Cell 136:876–890
20. Yuan JP, Zeng W, Dorwart MR et al (2009) SOAR and the polybasic STIM1 domains gate and regulate Orai channels. Nat Cell Biol 11:337–343

21. Navarro-Borelly L, Somasundaram A, Yamashita M et al (2008) STIM1-Orai1 interactions and Orai1 conformational changes revealed by live-cell FRET microscopy. J Physiol 586:5383–5401
22. Muik M, Fahrner M, Derler I et al (2009) A cytosolic homomerization and a modulatory domain within STIM1 C terminus determine coupling to ORAI1 channels. J Biol Chem 284:8421–8426
23. Liou J, Fivaz M, Inoue T et al (2007) Live-cell imaging reveals sequential oligomerization and local plasma membrane targeting of stromal interaction molecule 1 after Ca2+ store depletion. Proc Natl Acad Sci USA 104:9301–9306
24. Mullins FM, Park CY, Dolmetsch RE et al (2009) STIM1 and calmodulin interact with Orai1 to induce Ca2+-dependent inactivation of CRAC channels. Proc Natl Acad Sci USA 106:15495–15500
25. Zweifach A, Lewis RS (1993) Mitogen-regulated Ca2+ current of T lymphocytes is activated by depletion of intracellular Ca2+ stores. Proc Natl Acad Sci USA 90: 6295–6299
26. Prakriya M, Lewis RS (2002) Separation and characterization of currents through store-operated CRAC channels and Mg^{2+}-inhibited cation (MIC) channels. J Gen Physiol 119: 487–507
27. Grynkiewicz G, Poenie M, Tsien RY (1985) A new generation of Ca2+ indicators with greatly improved fluorescence properties. J Biol Chem 260:3440–3450

Chapter 2

Divers Models of Divalent Cation Interaction to Calcium-Binding Proteins: Techniques and Anthology

Jos A. Cox

Abstract

Intracellular Ca^{2+}-binding proteins (CaBPs) are sensors of the calcium signal and several of them even shape the signal. Most of them are equipped with at least two EF-hand motifs designed to bind Ca^{2+}. Their affinities are very variable, can display cooperative effects, and can be modulated by physiological Mg^{2+} concentrations. These binding phenomena are monitored by four major techniques: equilibrium dialysis, fluorimetry with fluorescent Ca^{2+} indicators, flow dialysis, and isothermal titration calorimetry. In the last quarter of the twentieth century reports on the ion-binding characteristics of several abundant wild-type CaBPs were published. With the advent of recombinant CaBPs it became possible to determine these properties on previously inaccessible proteins. Here I report on studies by our group carried out in the last decade on eight families of recombinant CaBPs, their mutants, or truncated domains. Moreover this chapter deals with the currently used methods for quantifying the binding of Ca^{2+} and Mg^{2+} to CaBPs.

Key words: Equilibrium dialysis, Fluorescent Ca^{2+} indicators, Flow dialysis, Isothermal titration calorimetry, Penta-EF-hand proteins, S100 proteins, Invertebrate-specific CaBPs, Centrins, NOX5

1. Introduction

A rise in cytosolic Ca^{2+} concentration is used as a universal signaling mechanism to control very diverse biological processes, such as exocytosis, contraction, cell growth, and cell death. It consists of an oscillating and transient increase of the mean Ca^{2+} concentration from 100 nM to 0.5–2 μM during stimulation (1). It is noteworthy that at the surface of different membranes cytosolic microdomains build up in which Ca^{2+} reaches a few hundred μM (2). The intracellular Ca^{2+}-binding proteins (CaBPs) have affinities which are fine-tuned to handle the intracellular Ca^{2+} signal. The

Claus W. Heizmann (ed.), *Calcium-Binding Proteins and RAGE: From Structural Basics to Clinical Applications*, Methods in Molecular Biology, vol. 963, DOI 10.1007/978-1-62703-230-8_2, © Springer Science+Business Media New York 2013

affinities of CaBPs for Ca^{2+} vary considerably; moreover for a given protein the affinity for Ca^{2+} can be strongly affected by Mg^{2+}; for instance in the absence of Mg^{2+} parvalbumins display affinity constants for Ca^{2+} (K_{Ca}) of 10^9 to 10^7 M^{-1} which are decreased to 10^6 M^{-1} in the presence of millimolar concentrations of Mg^{2+} (3). On the other hand calmodulin shows affinity constants of the order of 10^5 M^{-1} and these values are nearly not affected by Mg^{2+} (4). Some CaBPs show much lower affinities: for instance several S100 proteins display K_{Ca} values of 3 to 8×10^3 M^{-1} (5). For such proteins the importance of the microdomains becomes evident. Also the interaction of CaBPs with target proteins may dramatically alter their affinity for Ca^{2+} (6). Moreover in the case of calmodulin the rather low affinity is compensated by its high cellular concentration and huge excess over the target enzymes (7).

Anthology of Ca^{2+}-binding to CaBPs: A first anthology has been reported in 1996 (8) and comprised calmodulin, troponin C, calbindin D-9k, parvalbumin and oncomodulin, sarcoplasmic Ca^{2+}-binding proteins (SCPs), and neuron-specific Ca^{2+}-binding proteins (NCS). Below are summarized more recently studied (in the present century) families of CaBPs, with emphasis on what they have added to our knowledge on divalent cation interaction to EF-hand motifs. By convention the term K_{dCa} (dissociation constant for Ca^{2+}) will be used when the sites in a CaBP do not display cooperative binding, while $[Ca^{2+}]_{0.5}$ will be used in cooperative systems.

1.1. Apoptosis-Linked Protein ALG2

ALG-2 encodes a dimeric CaBP of the penta EF-hand family that promotes Ca^{2+}-dependent apoptosis (9). We reported that cloned ALG-2 cDNA from mouse liver mRNA contained two types of sequences (10). One (named ALG-2,5) corresponded to the sequence reported before; the second (named ALG-2,1) was six nucleotides shorter, and the corresponding protein lacked the amino acid residues Gly121 and Phe122. Both transcripts are present in mouse tissues in the same 2:1 molar ratio. The amino acid sequence shows 5 EF-hands but only EF-I and EF-III are canonical Ca^{2+}-binding sites (9). Contrary to the metal-free form, in the presence of Ca^{2+} ALG-2 is completely insoluble and 0.5% Tween must be added in order to perform the Ca^{2+}-binding experiments. Both isoforms possess two sites with high affinity for Ca^{2+} ($K_{dCa} = 1.2$ μM for ALG-2,5 and 3.1 μM for ALG-2,1) and one site of much lower affinity (K_{dCa} ca 300 μM). All the sites are Ca^{2+} specific, i.e., 2 mM Mg^{2+} does not affect the binding isotherms. The crystal structure of ALG-2,5 shows bound Ca^{2+} in EF-I, EF-III, and EF-V (11). EF-V is also a dimerization domain, as in most other penta EF-hand proteins. One target of ALG-2,5 is Alix, but the shorter isoform ALG-2,1 does not interact with this target. Thus the minor sequence difference, which occurs in the linker between EF-III and EF-IV, affects the function of the proteins. In human uveal melanoma ALG-2,5 is down regulated and mutations in the EF-hand revealed that EF-I and EF-III are the high-affinity sites and the binding of Ca^{2+} to both EF-I and EF-III is necessary for interaction with Alix (12).

1.2. Grancalcin

Grancalcin is a penta-EF-hand protein especially abundant in human neutrophils. Native and recombinant grancalcin always exist as a dimer regardless of the Ca^{2+} load (13). Metal-free grancalcin is fully soluble and in Ca^{2+}-containing buffers the protein is partly soluble, especially at micromolar concentrations; in the presence of 25 mM octylglycoside the protein is soluble even at 10 μM Ca^{2+}. From the amino acid sequence one can predict that, as for ALG-2, EF-I and EF-III are canonical Ca^{2+}-binding sites, whereas EF-II, EF-IV, and EF-V are abortive sites. As in other penta-EF-hand proteins EF-V is the dimerization module. The protein binds two Ca^{2+} with moderate affinity, with a K_{dCa} of 25 μM in the presence of octylglycoside and 83 μM in the absence of the detergent. The first Ca^{2+} seems to bind to EF-III without inducing any conformational change (14), whereas the binding of Ca^{2+} to the first EF-hand leads to pronounced changes in the Trp environment. In neutrophils the target of grancalcin is L-plastin, a leukocyte-specific actin-bundling protein, a mosaic protein containing two EF-hands, one calmodulin-binding domain, and two actin-binding domains. Ca^{2+} regulates the interaction in a negative manner, i.e., in the presence of EGTA the proteins interact and upon addition of an excess of Ca^{2+} the complex dissociates.

1.3. Calmodulin-Like Skin Protein

Calmodulin-like skin protein (CLSP), a 15.9 kDa CaBP specifically expressed in differentiated human keratinocytes (15), displays 52% sequence identity to calmodulin and possesses four EF-hands. The protein binds four Ca^{2+} ions at two pairs of sites with $[Ca^{2+}]_{0.5}$ of 1.2 (with very strong positive cooperativity: $n_H = 1.99$) and 150 μM (without cooperativity). Mg^{2+} at millimolar concentrations strongly decreases the affinity for Ca^{2+} of the two high-affinity sites, but has no effect on the low-affinity sites. The protein can also bind two Mg^{2+} ($K_{dMg} = 57$ μM) at the sites of high Ca^{2+} affinity (16). Thus, as in fast skeletal muscle troponin C, CLSP possesses two high-affinity Ca^{2+}/Mg^{2+} mixed sites and two low-affinity Ca^{2+}-specific sites. Studies on the isolated N- and C-terminal half domains of CLSP revealed that, in contrast to the case of TNC, the high-affinity Ca^{2+}–Mg^{2+} mixed sites reside in the N-terminal half. The binding of Ca^{2+} to the pairs of sites in each half, respectively, leads to the exposure of two hydrophobic patches, which are of equal quality as probed with the hydrophobic fluorescent TNS (see Note 5) and are directly governed by the Ca^{2+} affinity of that domain.

1.4. Calretinin

Calretinin is an abundant neuronal CaBP containing six EF-hand motifs and closely related to calbindin-D28k. In the cochlea it likely provides a strong, fast, and mobile buffer to cope with the unusually high, localized Ca^{2+} spikes. Calretinin-22k is a splice product of calretinin (17 kDa) found specifically in cancer cells, and possesses four EF-hands (17). Calretinin binds four Ca^{2+} ions with high affinity ($[Ca^{2+}]_{0.5}$ = 2 μM; n_H = 1.3) and one Ca^{2+} with much lower affinity ($[Ca^{2+}]_{0.5}$ = 50 μM). Calretinin-22k binds three Ca^{2+} with high affinity ($[Ca^{2+}]_{0.5}$ = 1.2 μM; n_H = 1.3) and one Ca^{2+} with low affinity (K_{dCa} above 1 mM). Since EF-VI shows a noncanonical amino acid sequence, it is likely that EF-V, which is paired to EF-VI, is the low-affinity site in both proteins. All the sites are of the Ca^{2+}-specific type. The secondary structure is not significantly altered upon the binding of Ca^{2+}, but Ca^{2+} induces changes in the microenvironment of Trp and leads to the exposure of a hydrophobic patch on the surface of both proteins (18).

1.5. S100 Proteins

S100 proteins (soluble in 100% ammonium sulfate) are dimeric proteins of ca 20 kDa. Each monomer contains 2 EF-hands: the C-terminal EF-hand displays the canonical sequence, whereas the N-terminal EF-hand, consisting of 14 amino acids, is specific for S100 proteins. To date, some 21 different proteins, which display various degrees of amino acid sequence homology, have been assigned to the S100 protein family (19). The protein contains a dimerization domain, an X-type four-helix bundle consisting of the N-terminal helix of EF-I and the C-terminal helix of EF-II. The conformation of this socle is insensitive to Ca^{2+}. Binding of Ca^{2+} induces small changes in the loop of the second EF-hand and a reorientation of the third helix. This swing-out movement on both sides of the protein dimer leads to unmasking of hydrophobic residues, thus forming two symmetric hydrophobic patches. The different S100 proteins interact in a Ca^{2+}-dependent manner with a large number of target proteins. Since a previous review covered the studies of the last quarter of the twentieth century (5), this overview deals with the ion-binding properties of S100 proteins reported in the last decade: S100A5, S100A13, and S100A16.

S100A5 is expressed in some restricted regions containing neurons of the brain (20). The four Ca^{2+} in the homodimer bind with high affinity and positive cooperativity ($[Ca^{2+}]_{0.5}$ = 6 μM; n_H = 1.9) and the sites are of the Ca^{2+}-specific type. The dimer of S100A5 also binds 2 Zn^{2+} ($[Zn^{2+}]_{0.5}$ = 2 μM) and 4 Cu^{2+} ($[Cu^{2+}]_{0.5}$ = 3 μM). In the monomer Zn^{2+} binds to a site different from the Ca^{2+} site, likely opposite to the two Ca^{2+} sites. In contrast Cu^{2+} seems to use the Ca^{2+} ligands. The secondary structure does not change upon binding of metal ions, but the microenvironment of the Tyr changes upon binding of Ca^{2+}, but not of Zn^{2+}.

S100A13 is expressed in different tissues, but often in a cell type-specific manner in these tissues; it seems to function in exocytosis since it is a target of anti-allergic drugs (21, 22). The protein dimer binds 4 Ca^{2+}, two with $[Ca^{2+}]_{0.5}$ of 8 μM (n_H = 1.3) and two with $[Ca^{2+}]_{0.5}$ of 400 μM (n_H = 1.3). Mg^{2+} (2 mM) has no effect on the Ca^{2+}-binding isotherm. As for S100A5 Ca^{2+}-binding does not change the secondary structure, but it has a strong effect on the microenvironment of the single Trp. Surprisingly, S100A13 does not bind to phenyl-Sepharose, as do all other S100 proteins, and the Ca^{2+}-loaded protein does not influence the hydrophobic probe TNS, thus strongly suggesting that this Ca^{2+}-S100A13 isoform has no hydrophobic patch on its surface.

S100A16 is predominantly expressed in astrocytes in adult brain where it is present in the nucleus (concentrated in the nucleolus) and cytoplasm in vivo (23); in fact the intracellular free Ca^{2+} regulates the nuclear translocation of S100A16 in the sense that high intracellular Ca^{2+} levels induce nucleolar exit and nucleocytoplasmic transport and lowering the intracellular Ca^{2+} concentration leads to nuclear translocation. S100A16 binds only 2 Ca^{2+} ions per dimer with a K_{dCa} of 750 μM ($n_H = 1$) for the human protein and 430 μM for that of mouse. From the amino acid sequence it can be inferred that Ca^{2+} binds to the second EF-hand in each monomer. The mouse protein binds also 2 Zn^{2+} per dimer with an estimated K_{dZn} of 25 μM and, as for S100A5, likely at a different site than that for Ca^{2+}. Surprisingly Ca^{2+}-binding does not induce a hydrophobic patch on the surface of the mouse protein, whereas Zn^{2+} does so.

1.6. Invertebrate-Specific Ca^{2+}-Binding Proteins

A family of intracellular CaBPs was found to be spread over all forms of life, from bacteria to protochordates, in invertebrate muscle and brain, except in vertebrates. We called it the SCP family since for a long time only the proteins present in the sarcoplasm, i.e., in the supernatant after centrifugation of a muscle homogenate extracted in physiological ion salt concentrations, were known. They have all a very similar three-dimensional structure, which is characteristic, and unique for this family, and are involved in a variety of functions including intracellular Ca^{2+} buffering, bioluminescence, enzymatic activity, and associated learning. Members of this family are SCP, calerythrin (a Ca^{2+}-buffering protein in the gram-positive bacterium *Saccharopolyspora erythraea*), aequorin (a bioluminescent protein in the jellyfish *Aequorea aequorea*), and calexcitin found in the nervous tissues of insects and mollusks. The Ca^{2+}-binding properties of SCPs (considered to function only as intracellular Ca^{2+}-buffering proteins) of arthropods, annelids, protochordates mollusks, and insects have been described in the period 1975–1990 and have been reviewed (24).

Calerythrin possesses 4 EF-hands, but EF-II does not possess a canonical Ca^{2+}-binding loop. It displays three high-affinity Ca^{2+}-binding sites, two of which are strongly cooperative. The NMR solution structure (25) shows much similarity with sandworm SCP. Aequorin consists of a 22 kDa apoprotein, the hydrophobic prosthetic coelenterazine moiety, and an oxygen molecule. Aequorin binds three Ca^{2+} with high positive cooperativity and subsequently undergoes a series of autocatalyzed reactions leading to the emission of a photon. The crystal structure (26) shows an overall conformation very similar to sandworm SCP, but, contrary to the latter, in the center of aequorin there is a cavity with a volume of 600 Å^3. The cavity is entirely filled with the substrate 2-hydroperoxycoelenterazine which is oxidatively decarboxylated to coelenteramide.

Calexcitin, the most recent member of the SCP family (27), plays a role in associative learning through inhibition of K^+ channels and activation of the ryanodine receptor, both in a Ca^{2+}-dependent manner (28). Calexcitin from the optic lobe of squid *Loligo pealei* has four EF-hands but EF-IV is clearly inactive. The sequence identity with amphioxus SCP is 31%. Calexcitin is monomeric and binds three Ca^{2+} ions with a $[Ca^{2+}]_{0.5}$ of 0.8 μM and n_H of 1.6, meaning pronounced cooperativity. Similar to be case of SCPs the relation between Ca^{2+} and Mg^{2+} is complex: the data can best be interpreted as follows: calexcitin possesses one Ca^{2+}-specific site and two Ca^{2+}/Mg^{2+} mixed sites. Of the mixed sites one is very sensitive to Mg^{2+}, whereas the other displays a weak Mg^{2+} effect (29). The thiol reactivity, circular dichroism, Trp fluorescence, thermal stability, and urea denaturation, all indicate that the protein displays three distinct structures based on the type of ion that is bound, as do the genuine SCPs. From hydrophobic probe binding it can be inferred that the metal-free protein has molten globule characteristics, the Ca^{2+}-form displays a hydrophobic grove, likely involved in protein–protein interaction, and the Mg^{2+} form is very compact without a hydrophobic patch. The three-dimensional structure at 1.8 Å resolution of calexcitin from *Loligo paelei* (30) displays a compact molecule with a very pronounced hydrophobic core and a scaffold similar to that of amphioxus SCP.

1.7. Centrins

Centrins are ubiquitous Ca^{2+}-binding proteins in higher plants, yeast, algae, invertebrates, and vertebrates, concentrated in microtubule-organizing centers (MTOCs), which are named spindle pole bodies in yeast, basal bodies in algae, and centrosomes in higher eukaryotes. Centrins are essential for cell division, for mRNA nuclear transport, and for nucleotide excision repair. Moreover in green algae centrin participates in a network of Ca^{2+}-sensitive fibers forming connections between them and with the nucleus. These fibers contract in a Ca^{2+}-dependent, ATP-independent manner and centrins seem to be the Ca^{2+} sensor for contraction. The molecular mechanism of centrin-mediated motility is actually not known, but a clear sign of progress was the discovery of a protein called Sfi1, containing a large number of similar sequence motifs, each capable to bind independently one centrin molecule (31). Salisbury (32) postulated a novel mechanism of MTOC duplication based on a "unit of duplication" requiring either C-terminal tail-to-tail homodimerization of Sfi1, thus forming the spindle pole body in yeast, or oligomerization (a 9-mer) in green algae and likely in vertebrates, thus forming a cartwheel. Ca^{2+}-binding to the multiple centrins aligned on the Sfi1 fibers may cause changes in the length of these fibers. All these functions are centrin specific and not supported by calmodulin.

Centrins (ca 18 kDa) are structurally closely related to calmodulin with four EF-hands and structural independence of the two EF-hand domains. But, whereas calmodulin displays four Ca^{2+}-specific sites of almost the same affinity and no, or nearly no, effect of Mg^{2+}, the Ca^{2+}- and Mg^{2+}-binding properties are different, and even differ widely within the centrin family. Comparative sequence analysis allowed the establishment of a phylogenetic three that has four main subfamilies. *Chlamydomonas reinhardtii* (CrCen) is the prototype of the largest subfamily which comprises also human centrin 1 (HsCen1) and HsCen2. Yeast centrin (Cdc31) and HsCen3 form the next subfamily, while plant and *Paramecium* centrins define the two other subfamilies.

Human centrin 2 (HsCen2) is a homodimer and binds only one Ca^{2+} in each monomer with rather low affinity ($[Ca^{2+}]_{0.5} = 30\ \mu M$) and negative cooperativity ($n_H = 0.68$) (33). Mg^{2+} does not affect the Ca^{2+}-binding curve. In the amino acid sequence EF-IV is the only EF-hand with a canonical Ca^{2+}-binding loop. But in the presence of a model peptide (33) or in the isolated C-terminal half which possesses in addition helix D (F86-T94) at its N-terminus, called LC-HsCen2 (34), two bound Ca^{2+} ions are present per monomer; thus under certain conditions EF-III, which contains Thr in the +Z and Asn in the −Z position, can bind Ca^{2+}. The binding of Ca^{2+} induces a hydrophobic patch on the surface of HsCen2, only in the C-terminal half, consistent with the observation that the protein binds one mole of melittin, a model peptide for calmodulin interactions. HsCen1 is in several of these characteristics very similar to HsCen2 (not published observations).

Human centrin 3 (HsCen3) contains one high-affinity Ca^{2+}/Mg^{2+} mixed site ($[Ca^{2+}]_{0.5} = 3\ \mu M$, $[Mg^{2+}]_{0.5} = 10\ \mu M$) and two Ca^{2+}-specific sites of low affinity ($[Ca^{2+}]_{0.5} = 140\ \mu M$). The high-affinity site could be assigned to the N-terminal half of HsCen3, very likely to EF-I which contains an acid pair (35). Temperature denaturation, circular dichroism, and utilization of fluorescent hydrophobic probes allowed us to propose that the metal-free protein has molten globule characteristics and that the dication-bound forms are compact with a polar surface for the Mg^{2+} form and a hydrophobic exposed surface for the Ca^{2+} form.

Clamydomonas reinhardtii centrin (CrCen, also called caltractin) is a major component of the basal body connector and of the distal striated fiber that connects the basal bodies of contractile fibers in green algae and plays an essential role in flagellar excision. It has four functional EF-hands in two distinct and independent domains and may be the most ancestral of centrins since it resembles the most to calmodulin. But, whereas calmodulin has four Ca^{2+}-specific sites of nearly identical affinity, CrCen possesses two high-affinity EF-hands K_{dCa} = (1.2 μM) in the N-terminal domain and two low-affinity sites ($[Ca^{2+}]_{0.5}$ = 160 μM) in the C-terminal domain (36). The effect of Mg^{2+} on these Ca^{2+}-binding properties has not been studied. A target peptide excised from the yeast protein Kar1 binds to each of the CrCen domains in a Ca^{2+}-dependent manner, with the C-terminal domain binding stronger than the N-terminal domain. Thus CrCen binds two Kar1p and can act as a bridging factor in MTOCs.

Scherffelia dubia centrin (SdCen) is, as CrCen, a component of basal body-associated contracting fibers in this green alga, and it resembles this latter centrin also in Ca^{2+}-binding (37). SdCen possesses two high-affinity Ca^{2+} sites ($[Ca^{2+}]_{0.5}$ = 3 μM) and one low-affinity site ($[Ca^{2+}]_{0.5}$ = 20 μM). In the presence of Mg^{2+} one of the high-affinity sites displays a fourfold lower affinity. Experiments with the isolated halves showed that the two high-affinity sites correspond to EF-I and EF-II and the low-affinity site likely is EF-IV. Ca^{2+}-binding to each half induces a hydrophobic patch with a sensitivity toward Ca^{2+} which is dictated by the affinity of Ca^{2+} for that domain. Each of the hydrophobic patches can bind a target peptide corresponding to one repeat of Sfi1 or of an identified consensus motif in the XPC protein which is involved in nuclear excision repair. As in the case of CrCen, the peptide interaction is stronger for the C-terminal domain. Thus SdCen is in many aspects similar to CrCen, except that it binds only one Ca^{2+} in the C-terminal domain.

Saccharomyces cerevisiae centrin (Cdc31) binds to yeast Sfi1, which contains 21 conserved repeats, each able to bind one molecule of centrin. The crystal structure of a Cdc31 complex with two or three repeats of Sfi1 shows Sfi1 as a long α-helix with the centrins wrapped around (38). As calmodulin, the centrin molecules are organized in two domains; the N-terminal domain is in a closed conformation (antiparallel EF-hand α-helices), whereas the C-terminal domain is in an open conformation (perpendicular EF-hand α-helices). Cdc31 binds Sfi1 via a deep hydrophobic cavity of the C-terminal domain, while its N-terminal domain has only limited contacts with Sfl1. Adjacent centrin molecules are rotated by 65° along the Sfi1 α-helix. Cdc31 contains one high-affinity Ca^{2+}/Mg^{2+} mixed site (K_{dCa} = 0.3 μM; K_{dMg} = 44 μM) and two low-affinity Ca^{2+}-specific sites ($[Ca^{2+}]_{0.5}$ = 28 and 41 μM) (39). The high-affinity mixed site could be attributed to the N-terminal domain. Both Ca^{2+} and Mg^{2+} increase the α-helix content and stabilize the protein. In apo Cdc31 the N- and C-terminal domains are fused together, whereas in the Mg^{2+} state they are well separated as shown by the groove in the middle of the molecule. The affinity of Cdc31 for five different peptides excised from natural target proteins shows a Ca^{2+} and Mg^{2+} dependence and the peptides occupy the central groove. Contrary to the green algae centrins, but similar to the human proteins, Cdc31 interacts with peptides in a 1:1 molecular ratio.

1.8. NADPH Oxidase 5

NADPH Oxidase 5 (NOX5) is a monomeric transmembrane Ca^{2+}-dependent oxidase with multiple domains (40): two heme groups attached to two of the six transmembrane α-helices, a long intracellular C-terminus with NADP- and FAD-binding sites, called *c*atalytic *deh*ydrogenase *d*omain (CDHD), and an intracellular N-terminus with four EF-hands called NOX5-EF. When NOX5 is activated it transports electrons from intracellular NADPH by means of FAD and the two heme groups to extracellular oxygen resulting in the formation of superoxide radicals. Contrary to the other NOX isoenzymes, which need regulatory subunits, NOX5 activation is initiated by the binding of Ca^{2+} to the N-terminal regulatory domain which interacts then directly with the catalytic C-terminal domain (41). Two short segments that represent the binding site of NOX5-EF have been identified (42); they are intrinsically autoinhibitory and the inhibition can be relieved by Ca^{2+}-dependent binding to NOX5-EF. In addition, in the presence of Ca^{2+}, calmodulin binds to the enzyme at a well-identified 19 residue long segment in the CDHD and this represents an additional sophistication in the regulation of superoxide production by NOX5 (43).

NOX5-EF possesses two pairs of EF-hands with a higher affinity (K_{dCa} = 3.8 μM) for the C-terminal pair than the N-terminal pair (K_{dCa} = 17 μM) and these affinities are not affected by Mg^{2+} (41); thus as in calmodulin they are of the Ca^{2+}-specific type. Interestingly whereas sites II, III, and IV have canonical sequences, site I displays a sequence typical of some sites in the PEF proteins sorcin, grancalcin, and the heavy and light chains of calpains, with Ala in the +X position providing its carbonyl oxygen and a water molecule oxygen in +Y. Upon binding of Ca^{2+} a hydrophobic patch is exposed on the surface of NOX5-EF, but only in the C-terminal pair. However conformational changes that affect the intrinsic fluorescence of Trp residues occur in both pairs of EF-hands.

2. Materials

2.1. Solutions

All solutions are prepared with ultrapure water bidistilled in a quartz apparatus to attain a conductivity of 10 mΩ cm. The standard working buffer in our studies is 50 mM Tris–HCl, pH 7.5, 150 mM KCl. Buffer solutions are decalcified by passage over a 2–15 cm column of Chelex-100 (Bio-Rad Laboratories).

Trichloroacetic acid is from Calbiochem. The contaminating Ca^{2+}, Mg^{2+}, or Zn^{2+} was measured by atomic absorption spectrophotometry (Perkin Elmer 2380).

2.2. Equilibrium Dialysis

Equlibrium dialysis of 1 ml protein solution was performed for 48 h in the cold room in Spectra/Por dialysis tubes (Spectrum Europe B. V.) versus 200 ml buffer in polypropylene bottles on a lab shaker with one buffer change after 24 h of dialysis.

2.3. Fluorescent Ion Indicators

Benzothiazole coumarin (BTC) and Mag-fura-2 are from Molecular Probes, Eugen, Oregon.

2.4. Flow Dialysis

Flow dialysis was performed in homemade flow dialysis cells with volumes of the upper and lower compartment of 500 μl and 50 μl, respectively (see Note 1). The flow rate was 1 ml/min, the volume of a fraction 0.4 ml. As tracer, 5 μM $^{45}Ca^{2+}$ (950 mBq/mg) were added to the upper compartment. The titration, done by a Microlab-2200 (Hamilton), consisted of additions every 2 min of 1–4 μl of the following stock solutions of Ca^{2+}: 0.63, 3.05, 9.58, 82.5 mM and the final chase was done with 1 μl of 1 M Ca^{2+}.

2.5. Isothermal Titration Calorimetry

Isothermal titrations were carried out at 30°C on a VP-ITC instrument (Microcal Inc Northampton, MA) in 50 mM Mops buffer, pH 7.4, 100 mM NaCl. A solution of 30 μM protein was titrated by automatic injections of 5–10 μl of 1 mM Ca^{2+}.

3. Methods

Since most methods require removal of the metal ions from the protein, a short first section is devoted to this topic.

3.1. Metal Removal from CaBPs

All the methods for metal binding described below require a strict control of the Ca^{2+} and Mg^{2+} levels and of the contamination of the buffers and protein solutions. Therefore it is nearly indispensable to have access to an atomic absorption spectrometer (AAS), called also flame-spectrometer. The natural contamination of solutions by Mg^{2+} is negligible (i.e., below 0.03 μM), but the Ca^{2+} contamination is several μM. Therefore all solutions used must be rendered “calcium-free” by treatment with Chelex 100 resin (Bio-Rad Laboratories). To reduce Ca^{2+} contamination all solutions must be kept in plastic containers, which have been soaked in 1 N HCl and rinsed with bidistilled water. In the second step Ca^{2+} must be removed from the CaBPs by methods which do not alter the binding properties. The softest protein decalcifications are done at neutral pH under non-denaturating conditions, but one can take advantage of the fact that CaBPs are surprisingly stable and easily accommodate reversible denaturation. A panoply of fast (taking 1 h or less) methods is listed below.

1. After addition of 1–2 mM EDTA 1 ml of the concentrated protein solution is chromatographed on a 0.8–40 cm column of Sephadex G-25 equilibrated in the "Ca-free" working buffer (see Note 2). There is a risk that complexes of weak affinity are formed between the protein and EDTA, but likely they dissociate during the gel filtration.
2. Passage of the concentrated protein solution over a column of Chelex-100 resin or of EDTA-agarose equilibrated in the working buffer. It must be noted that the Ca^{2+} binding leads to proton exchange, thus necessitating strong pH buffers. In this method the protein is in fact not perfectly equilibrated in the working buffer.
3. Acidification of the protein solution to pH 2 followed by Sephadex G-25 chromatography in the working buffer.
4. Denaturation of the protein in 6 M guanidine-HCl at pH 8.0 followed by Sephadex G-25 chromatography in the working buffer.
5. Protein precipitation by addition of 3% trichloroacetic acid (TCA) and resolubilization at pH 8.0. The contaminating Ca^{2+} is monitored by AAS and eventually the precipitation is repeated At the end the dissolved protein is put on the Sephadex-25 column.

The last method is very efficient and has been applied successfully in my laboratory on dozens of proteins. In several cases we compared the Ca^{2+}-binding data for samples obtained by either the TCA method (numbered 5, above) or methods 1 and 2: no significant differences were detected. In contrast the Ca^{2+}-binding properties of some SCPs, CaVP, and certain mutants of parvalbumins are damaged by the TCA method.

3.2. Ca^{2+} Binding

The first question to be solved after one realizes that a given new protein contains EF-hand motifs is if it really binds Ca^{2+}. For this purpose qualitative methods exist such as the Ca^{2+}-dependent mobility shift in polyacrylamide gel electrophoreses (PAGE). Different optical methods such as fluorimetry on the intrinsic Trp or Tyr residues can be performed in most laboratories and lead to a valid diagnosis provided ion-binding leads to signal changes. But quantitative information, leading to the evaluation of the physiological significance of Ca^{2+}-binding to a protein, cannot be gathered by these methods. This section deals with the four most common methods to determine Ca^{2+}-binding isotherms.

3.2.1. Equilibrium Dialysis

In this method an equilibrium is established on both sides of a solute-permeable, but protein-impermeable, membrane (Spectra/

Por 3). In practice, aliquots of a protein solution (20–100 μM of binding sites) are equilibrated against a series of buffers of increasing ion (Ca^{2+}) concentrations. At the end the concentrations of protein and of total ligand are measured in both compartments. The difference in ligand concentration between both compartments corresponds to the bound ligand, which can be directly related to the protein concentration. Ion quantification can be done with AAS. A major complication is that it is impossible to decontaminate buffers of physiological ionic strength to free Ca^{2+} values lower than 0.2 μM. In order to attain free Ca^{2+} levels of 10^{-9} to 10^{-5} M, it is thus necessary to "clamp" the free Ca^{2+} by including 1 mM EGTA in the dialysis experiment. The free Ca^{2+} is then calculated using a computer program (WEBMAXC STANDARD version 2.10) of "speciation" taking into account the affinity constants of EGTA + Ca^{2+} and EGTA + H^+ and the total concentrations of EGTA and Ca^{2+}. Even with this complication the method is quite simple and needs minimal equipment. But three negative points must be mentioned: (1) The length (24 h) precludes determination on labile, sticky, or hydrolysis-susceptible proteins. (2) In the EGTA buffering system the free Ca^{2+} versus total Ca^{2+} relationship is not linear and lacks precision in the 10^{-7} to 2×10^{-5} M range, i.e., in the zone one usually wants to cover. (3) The workload is quite heavy and the number of manipulations requires maximal attention of the experimenter.

3.2.2. Fluorescent Ion Indicators

The principle of this method is competition for Ca^{2+} of the CaBP and an indicator whose binding parameters are well known and whose degree of saturation can be quantified exactly (44). To the mixture of protein and indicator increasing Ca^{2+} concentrations are added until the indicator is saturated (see Note 3). If the physical and optical properties of the indicator are the same in the presence and absence of protein, one can determine the free Ca^{2+} concentration by Eq. 1:

$$[\mathrm{Ca}^{2+}] = \Delta s_i / (\Delta s_{max} - \Delta s_i) K_{indicator} \quad (1)$$

where $[Ca^{2+}]$ is the concentration of free Ca^{2+}, Δs_i and Δs_{max} are the optical signal change at titration point i and at the end (maximal saturation of the chelator), respectively, and $K_{indicator}$ is the affinity constant of the indicator. The value of $K_{indicator}$ under the experimental conditions can be determined by a separate titration of the indicator alone. Protein-bound Ca^{2+} (Ca_{bound}) can then be calculated as follows:

$$\mathrm{Ca}_{bound} = \mathrm{Ca}_{Tot} - [\mathrm{Ca}^{2+}] - \{K_{indicator} C_{indicator} [\mathrm{Ca}^{2+}] / (1 + K_{indicator} [\mathrm{Ca}^{2+}])\}, \quad (2)$$

where Ca_{Tot} and $C_{indicator}$ are the concentration of added Ca^{2+} and of the indicator, respectively. If the protein concentration is in large excess over that of the indicator the last factor in Eq. 2 can be neglected.

Fluorescent probes are used at submicromolar concentrations while the protein concentration is 2–3 μM. In fluorescent titrations the spectra of the indicator often show an isosbestic point, a wavelength where no ligand-induced changes occur during the titration. The precision is much better when the signal at the wavelength of maximal fluorescence change is rationed to the signal at the isosbestic wavelength. An important consideration is the choice of the indicator with respect to its affinity for Ca^{2+}. Ideally $K_{indicator}$ should be slightly lower than that of the lowest affinity sites of the protein. Fortunately, many indicators have been developed by Molecular Probes with different affinities for Ca^{2+}, ranging from 10^7 to 10^4 M^{-1}. If a CaBP shows pronounced negative cooperativity or displays more than one site with markedly different affinities, the indicator method is not recommended. A last caveat is the selectivity of the indicator, especially if one wants to determine the Ca^{2+}-binding properties in the presence of physiological concentrations of Mg^{2+}. Ideally the Ca^{2+}/Mg^{2+} selectivity ratio should be higher than 10,000. Indicators with a low selectivity ratio can in fact be used to determine the Mg^{2+}-binding properties of a CaBP by similar Mg^{2+} titration experiments in the absence of Ca^{2+} (i.e., in the presence of EGTA).

3.2.3. Flow Dialysis

In this method an immobile solution (upper compartment) is separated from the lower compartment by a semipermeable membrane which retains the protein but not the small ligands; the lower chamber is continuously flushed with the working buffer (45). Flow dialysis is based in the principle that the amount of Ca^{2+} diffusing per unit of time from the upper compartment is proportional to the free Ca^{2+} in that compartment. If the Ca^{2+} pool in the upper compartment is tagged with $^{45}Ca^{2+}$, this rate can be measured by determining the concentration of the isotope in the effluent of the lower chamber (for commodity called here cpm_i), which is collected in a fraction collector. After perfusion with four times the volume of the lower chamber a steady state is reached for which the rates (dcpm/dt) of isotope entering and leaving the lower chamber are practically equal, as indicated by Eq. 3:

$$\mathrm{dcpm_i} / \mathrm{dt} = D[\mathrm{Ca}^{2+}] - \mathrm{cpm_i}(f / V) = 0, \quad (3)$$

hence $[Ca^{2+}]_i = cpm_i\ (f/V.D)$,

where D is a constant depending on both the diffusion properties of Ca^{2+} and the geometry of the dialysis apparatus, f is the flow rate of buffer through the lower chamber, and V is the volume of this chamber. In essence, Eq. 3 indicates that after an equilibrium is reached, the concentration of isotope (cpm_i) in the effluent becomes a true measure of free $[Ca^{2+}]$ in the upper compartment. After each addition of unlabeled Ca^{2+} to the upper chamber, a new equilibrium is established and a new steady state of cpm_i is reached.

The parameters $(f/V.D)$ are not determined directly; instead a large excess of cold Ca^{2+} is added at the end of the titration so that all Ca^{2+} and $^{45}Ca^{2+}$ can be considered as free. The corresponding cpm_{end} value is an index standing for 100% free $[Ca^{2+}]$ and the ratio of cpm_i/cpm_{end} represents thus the fraction of free Ca^{2+} in the upper compartment. From the known total Ca^{2+} concentration at each increment the concentration of free and bound Ca^{2+} in the upper chamber is calculated. This data treatment is not complete enough since three factors are not negligible: (1) The dilution ($Fdil_i$, or dilution factor) in the upper compartment by the increments; (2) the losses ($Floss_i$, or loss factor); and (3) the Ca^{2+} (and tracer $^{45}Ca^{2+}$) contamination in the upper chamber at the start of the experiment. For this purpose the following additional parameters must be measured:

1. The initial Ca^{2+} contamination, $[Ca^{2+}]_{contam}$: After the cell is mounted, the upper chamber is filled with the protein solution and tracer $^{45}Ca^{2+}$ is added; then 250 μl is withdrawn to measure the Ca^{2+} concentration by AAS.
2. The initial (CPM_{init}) and final $^{45}Ca^{2+}$ counts (CPM_{end}) in the upper compartment: The ΔCPM factor, defined as $(CPM_{init} - CPM_{end})/CPM_{init}$, determines the total loss of Ca^{2+} in the upper chamber, which can vary substantially (from 10 to 20%) depending on the profile of free Ca^{2+} change in this chamber during the experiment. The fraction of free Ca^{2+}, $f_{[Ca2+]i}$, after increment i is as follows:

$$f_{[Ca2+]i} = (cpm_i Fdil_i Floss_i)/(cpm_{end} Fdil_{end} Floss_{end}), \tag{4}$$

with $Fdil_i = Vol_i/Vol_{init}$ and $Fdil_{end} = Vol_{end}/Vol_{init}$ and $Floss_i = 1 + \Delta CPM \{\Sigma_{(i=1 \text{ to } n)} cpm_i/\Sigma_{(i=1 \text{ to } n)} cpm_i\}$.

The total Ca^{2+} concentration, $Catot_i$, after increment i is

$$\begin{aligned} Catot_i &= ([Ca^{2+}]_{contam.} + S[Ca^{2+}]_{increm.i})/Fdil_i Catot_i \\ &= ([Ca^{2+}]_{contam.} + S[Ca^{2+}]_{increm.i})/Fdil_i. \end{aligned} \tag{5}$$

The free Ca^{2+} concentration corresponds to $f_{[Ca2+]i} \cdot Catot_i$ and the protein-bound Ca^{2+} are then calculated by $Ca_{bound}^{2+} = Catot_i - [Ca^{2+}]$.

In practice the following experimental conditions are used for most of the CaBPs. The flow device (see Note 1) was mounted using Spectra/Por membranes, which were boiled twice: once in EDTA-alkalinized water and once in distilled water. The upper chamber contained 750 μl of a protein solution of 60 μM Ca^{2+}-binding sites. From this chamber 250 μl is withdrawn just after the addition of tracer $^{45}Ca^{2+}$. The lower chamber is perfused at a flow rate of 1 ml/min and the effluent is collected in fractions of 0.4 ml. Every 2-min increments of Ca^{2+} are added so that the total Ca^{2+} concentrations in the upper compartment vary according to an exponential scale until a value of 2–3 mM is attained. The samples of the fraction collector are supplemented with scintillation liquid and counted to 98% precision. This method has allowed us to determine quite easily and precisely the Ca^{2+} affinity of a great number of proteins with K_{Ca} values varying from 5×10^7 to 2×10^3 M^{-1}. The precision is greatly improved by automation (robot) of the pipetting of the increments.

3.2.4. Isothermal Calorimetry

If Ca^{2+} or Mg^{2+}-binding leads to an enthalpy change (ΔH), either exothermic or endothermic, their affinity constants can be determined by isothermal calorimetry (ITC) using a MicroCal VP-ITC Microcalorimeter (Microcal Inc). This method, known since 1989 (46), has been applied for Ca^{2+}- and Mg^{2+}-binding experiments only recently and only in a few cases (37, 39, 47), but is expected to become a very important tool in the next future since it is fast (circa 2 h), nearly completely automated, and yields not only the affinity constants but also the enthalpy changes of each binding step. As for the true ΔH, it should be stressed that divalent cation binding to CaBPs implies exchange of protons and that these proteins in turn will be buffered by the working buffer leading to a positive or a negative heat release, depending on which type of proton buffer is used (see Note 4).

In practice circa 50 μM of the metal-free protein is titrated with 50 successive automatic injections of 5–10 μl each of a stock solution of 0.5–2 mM divalent cation. The additions were 4 min apart to allow heat accompanying each increment to return to the baseline prior to the next addition. For each addition the molar heat (Q_t) was measured as a function of the total ligand concentration (X_t):

$$Q_t = nP_t\Delta H{\cdot}V\left\{\begin{matrix}1 + X_t/(nP_t) + 1/(nK_aP_t) \\ -\left[(1 + X_t/(nP_t) + 1/(nK_aP_t)^2 - 4X_t/(nP_t)\right]^{0.5}\end{matrix}\right\}/2, \quad (6)$$

where n is the number of sites, Pt is the total protein concentration, K_a is association constant, and V is the cell volume. The differential heat (dQt) is then fitted to various models by using a nonlinear least squares minimization method using Microcal Origin ITC software. The free energy of binding (ΔG) and entropy (ΔS) is obtained using the classical thermodynamic formulas $\Delta G = -RT \ln Ka$ and $\Delta G = \Delta H - T\Delta S$, where R is the gas constant and T is the absolute temperature in kelvin.

3.3. Mg^{2+} Binding

Under physiological ionic conditions at low Ca^{2+} concentrations many EF-hand-containing CaBPs interact with Mg^{2+} with affinities ranging from 10^2 to 10^5 M^{-1}. The binding of Mg^{2+} can be measured by equilibrium dialysis or by the fluorescent indicator method, but unfortunately not by flow dialysis since there is no stable radioactive isotope of Mg^{2+} with a long enough half-life time. A method well suited for Mg^{2+}-binding is equilibrium gel filtration, first proposed by Hummel and Dryer (48). This method takes advantage of the fact that on a gel filtration column the protein is rapidly and efficiently brought to equilibrium with the buffer in which the gel matrix is equilibrated. In practice 0.5–1 ml of a protein solution is applied to a 0.8 × 40 cm column of Sephadex G-25 equilibrated in the working buffer plus a given concentration of free Mg^{2+}, and eluted at 0.4 ml/min. In the eluent fractions the protein concentration is monitored by its UV absorbance and Mg^{2+} by AAS. Analysis of three protein-containing fractions across the protein peak allows us to verify if the equilibrium is reached, which is the case when the bound Mg^{2+}-to-protein ratio is the same in the three fractions. In the method originally proposed the protein sample was loaded without any addition of the ligand. However, in Mg^{2+}-binding experiments it is preferable for kinetic reasons that the loaded protein already contains Mg^{2+}. A practical and rapid way of generating a Mg^{2+}-binding isotherm is to chromatograph first the CaBP in buffer containing 1 mM Mg^{2+} and, after analysis, rechromatograph each of the protein-containing fractions on columns equilibrated in lower Mg^{2+} concentrations (from 50 to 500 μM). Excess Mg^{2+} dissociates and a new equilibrium is rapidly reached. When performing Hummel–Dryer experiments in buffer containing less than 50 μM Mg^{2+}, the flow rate must be reduced to 0.1 ml/min. Since generating a correct isotherm takes 3–5 days, the determination of the Mg^{2+}-binding isotherms now constitutes the rate-limiting step in the unraveling of the ion-binding parameters of a given CaBP.

3.4. Analysis of the Binding Data

3.4.1. Stoichiometric, Microscopic, and Intrinsic Constants

The above methods yield model-independent isotherms which describe the binding of the ion to the protein. They provide macroscopic, also named stoichiometric binding constants, K_n, which refer to the binding of the first, second, etc. Ca^{2+} to the protein (49); for an excellent review see ref. (50). The most general description of the equilibria in terms of stoichiometric binding is provided by the Adair-Klotz model (51), which analyzes the binding data in terms of a set of constants that describe the binding of the first, second, ... nth Ca^{2+} ion to the protein. The binding then obeys the following equation:

$$r = \frac{\left(K_1[\mathrm{Ca}^{2+}] + 2K_1K_2[\mathrm{Ca}^{2+}]^2 + \ldots \mathrm{n}K_1K_2\ldots K_\mathrm{n}[\mathrm{Ca}^{2+}]^\mathrm{n}\right)}{\left(1 + K_1[\mathrm{Ca}^{2+}] + K_1K_2[\mathrm{Ca}^{2+}]^2 + \ldots K_1K_2\ldots K_\mathrm{n}[\mathrm{Ca}^{2+}]^\mathrm{n}\right)}, \quad (7)$$

where r is the ratio of bound Ca^{2+} per protein and n is the number of Ca^{2+}-binding sites. These stoichiometric constants yield information on the pace of the successive binding steps, but not on the affinity constant of any particular site. In a CaBP with four functional Ca^{2+}-binding sites I, II, III, and IV and the affinity constants

k_I, k_{II}, k_{III}, and k_{IV} (called microscopic or site-binding constants), the relationship between K_n and these constants is as follows (49):

$$\begin{aligned} K_1 &= k_I + k_{II} + k_{III} + k_{IV} \\ K_1K_2 &= k_Ik_{II} + k_Ik_{III} + k_Ik_{IV} + k_{II}k_{III} + k_{II}k_{IV} + k_{III}k_{IV} \\ K_1K_2K_3 &= k_Ik_{II}k_{III} + k_Ik_{II}\mathrm{k}_{IV} + k_{II}k_{III}k_{IV} \\ K_1K_2K_3K_4 &= k_Ik_{II}k_{III}k_{IV}. \end{aligned} \tag{8}$$

It should be noticed that the microscopic k values may change during the binding process, e.g., the affinity of site II for Ca^{2+} may be different depending on the presence or not of Ca^{2+} in site I. Cornish-Bowden and Koshland (49) analyzed the binding isotherm in terms of the intrinsic affinity constant, K'_n, the meaning of which is intuitively the easiest understood since it represents the mean affinity constant of site n in a protein. If all the microscopic constants are identical there is only one intrinsic constant and the relation between stoichiometric and intrinsic constant is as follows: $K_1 = 4\,K'$, $K_2 = 3/2\,K'$, $K_3 = 2/3\,K'$, and $K_4 = ¼\,K'$. If the k values do not change during the titration (i.e., in a noncooperative system), but are different, the four different intrinsic constants, K'_1, K'_2, K'_3, and K'_4, describe the binding isotherm. In many instances, as in the first part of this chapter, the binding parameters are expressed as dissociation constants K_d, since these are closest to the notion of the Ca^{2+} concentration which induces half maximal binding or response: in most (simplified) cases $K_d = 1/K'$.

In the case of positive cooperativity in a two-site system the relationship between the intrinsic constants is as follows: $K'_1 < K'_2$ and the degree of allostery is given either by the ratio $k_{I,II}/k_I$ or by $\Delta G = \Delta G_{I,II} - \Delta G_I$, i.e., by the difference between the change in free energy when Ca^{2+} binds to a given site in the presence ($\Delta G_{I,II}$) or absence (ΔG_I) of bound Ca^{2+} in the second site. However a much more popular cooperativity index is given by the empirical Hill model, with

$$r = \mathrm{n}K^{\mathrm{nH}}[\mathrm{Ca}^{2+}]/(1 + K^{\mathrm{nH}}[\mathrm{Ca}^{2+}]), \tag{9}$$

where n is the number of sites, n_H the Hill coefficient, and K an apparent binding constant, not related to the stoichiometric, intrinsic, or macroscopic constants (49). Linearization of Eq. 9 yields the widely used Hill function:

$$\log\{\mathrm{r}/(\mathrm{n}-\mathrm{r})\} = \log K + \mathrm{n_H}\log[\mathrm{Ca}^{2+}]. \tag{10}$$

3.4.2. The Ca^{2+}/Mg^{2+} Antagonism

In most cases the affinities of CaBPs for Ca^{2+} are modulated by millimolar concentrations of Mg^{2+}, either by direct competition at the so-called Ca^{2+}/Mg^{2+} mixed sites or by indirect antagonism in proteins with Ca^{2+}-specific sites. A primordial task after the

Ca^{2+}-binding isotherms have been collected at different Mg^{2+} concentrations is to determine which model of antagonism is functioning.

1. In the model of *direct competition* the EF-hand site binds either Ca^{2+} or Mg^{2+} and the shift of the isotherms to higher $[Ca^{2+}]$ is unlimited. The competition model obeys the following equation:

$$K_{Ca} / K_{Ca.app} = 1 + K_{Mg}[Mg^{2+}], \tag{11}$$

where K_{Ca} and $K_{Ca.app}$ are the Ca^{2+}-binding constants in the absence and presence of Mg^{2+}, respectively, and K_{Mg} the Mg^{2+}-binding constant in the absence of Ca^{2+}. Examples of the analysis of simple competition models have been presented for troponin C (52) and parvalbumin (53); complex, but still direct, competition models have been described for sarcoplasmic Ca^{2+}-binding proteins (3, 54, 55).

2. *Indirect Ca^{2+}–Mg^{2+} antagonism* is exemplified by calmodulin (4). In this protein Mg^{2+} does not bind to the same sites as Ca^{2+}, but to auxiliary sites (56), physically different from the four EF-hands. Calmodulin can bind 4 Ca^{2+} and 4 Mg^{2+} simultaneously. The Ca^{2+}- and Mg^{2+}-binding sites influence each other's affinity due to negative free energy coupling between the sites. The amount of bound ligand per mole of site is given by (see ref. 4)

$$\text{Bound Ca}^{2+} / \text{site} = \frac{1 + (K_{Mg(Ca)}[Mg^{2+}])^{-1}}{1 + (K_{Ca(Mg)}[Ca^{2+}])^{-1} + (1 + K_{Ca}[Ca^{2+}])^{-1}(K_{Mg(Ca)}[Mg^{2+}])^{-1}}; \tag{12}$$

similarly:

$$\text{Bound Mg}^{2+} / \text{site} = \frac{1 + (K_{Ca(Mg)}[Ca^{2+}])^{-1}}{1 + (K_{Mg(Ca)}[Mg^{2+}])^{-1} + (1 + K_{Mg}[Mg^{2+}])^{-1}(K_{Ca(Mg)}[Ca^{2+}])^{-1}},$$

where K_{Ca} and K_{Mg} are the association constants for Ca^{2+} and Mg^{2+}, respectively, in the absence of the other ligand, and $K_{Ca(Mg)}$ and $K_{Mg(Ca)}$ the corresponding constants in the presence of infinite concentrations of the second ligand. As in the model of straight competition, upon increasing the Mg^{2+} concentrations the Ca^{2+}-binding isotherms shift to higher free Ca^{2+} concentrations, but in this model the shift is not unlimited and levels off at the isotherm dictated by $K_{Ca(Mg)}$.

3.4.3. Conclusion

The examples mentioned above show the great diversity of the thermodynamic behavior of different CaBP families, diversity necessary to cope with the complexity of the Ca^{2+} signal. The kinetics of Ca^{2+} and Mg^{2+}-binding, not treated here and in general rarely studied, adds a new dimension to the action of CaBPs in the cell.

4. Notes

1. Occasionally, we used flow dialysis cells with a spiraled (Feldman cell) lower compartment instead of the cubic lower chamber. Its smaller volume assures a faster installation of a steady state between the two compartments.
2. In our studies the working buffer for nearly all methods is 50 mM Tris–HCl, pH 7.5 150 mM KCl, except for ITC experiments where a Mops buffer, pH 7.4, 100 mM NaCl is used (its protonation enthalpy is negligible).
3. To a solution (500 μl) containing 15 μM of protein and 1 μM fluorescent Ca^{2+} indicator (BTC or Mag-fura-2) 20 μM increments of Ca^{2+} were added and the fluorescence excitation spectra were monitored after each addition. For BTC the fluorescence intensities at 401 (increasing) and 469 nm (decreasing) were ratioed with respect to that at the isosbestic point at 433 nm to eliminate interferences or lamp intensity changes. For example, for Mag-fura-2 the data were extracted at 332 (increasing), 352 (isosbestic point) and 385 nm (decreasing). The signal changes were normalized from 0 (nominaly zero [Ca^{2+}]) to +1 (saturating [Ca^{2+}]), yielding ΔS at the respective wavelengths which were averaged to yield ΔS_i. The concentration of free Ca^{2+} was then calculated by extrapolation of these ΔS_i values at each increment to the Ca^{2+}-binding isotherms of BTC or Mag-fura-2. The latter were estimated independently and yielded affinity constants of 1.4×10^5 and 1.8×10^4 M^{-1} for BTC and Mag-fura-2, respectively. Protein-bound Ca^{2+} was then calculated by subtracting the free [Ca^{2+}] from the total amount of added Ca^{2+} *plus* the initial contamination determined by atomic absorption.
4. The heat production or absorption (Q) as a function of the degree of saturation of a site (α) is given by $Q = \alpha(\Delta H_{Ca} + n\Delta H_{proton})$, where ΔH_{Ca} is the enthalpy of the Ca^{2+}-binding process and n and ΔH_{proton} are the number of protons liberated and the protonation enthalpy of a given buffer, respectively. When the same reaction is studied in two buffers of different protonation enthalpy, ΔH_{p1} and ΔH_{p2}, the value of n can be formulated by $n = (Q_1 - Q_2)/[\alpha \times (\Delta H_{p1} - \Delta H_{p2})]$. Thus, by calorimetric monitoring of a reaction in two different buffers one can calculate the proton release and determine subsequently the real ΔH_{Ca} of the process.
5. The Ca^{2+}-dependent exposure of a hydrophobic patch on the surface of a CaBP I is measured by fluorescent indicators such as 8-naphtalene-1-sulfonate (ANS), 2-p-toluidinyl-naphtalene-6-sulfonate (TNS), and 4,4′-dianilino-1,1′-binaphtyl-

5,5′-disulphonate (BISANS). In our studies nearly always TNS (Sigma) was used: a 400 μl solution of 40 mM TNS and 1 μM protein was excited at 322 nm and the emission spectra measured from 380 to 540 nm.

References

1. Parekh AB (2011) Decoding cytosolic Ca^{2+} oscillations. Trends Biochem Sci 36:78–87
2. Rizzuto R, Pozzan T (2006) Microdomains of intracellular calcium: molecular determinants and functional consequences. Physiol Rev 86:369–408
3. Wnuk W, Cox JA, Stein EA (1982) Parvalbumins and other soluble sarcoplasmic Ca-binding proteins. In: Cheung WY (ed) Calcium and cell function, vol II. Academic, New York, pp 243–278
4. Milos M, Schaer J-J, Comte M, Cox JA (1986) Calcium-proton and calcium-magnesium antagonism in calmodulin. Biochemistry 25:6279–6287
5. Heizmann CW, Cox JA (1998) New perspectives on S100 proteins: a multifunctional Ca^{2+}-, Zn^{2+}- and Cu^{2+}-binding protein family. Biometals 11:383–397
6. Mamar-Bachi A, Cox JA (1987) Quantitative analysis of the free energy coupling in the system calmodulin, calcium, smooth muscle myosin light chain kinase. Cell Calcium 8: 473–482
7. Cox JA (1988) Interactive properties of calmodulin. Biochem J 249:621–629
8. Cox JA (1996) Techniques for measuring the binding of Ca^{2+} and Mg^{2+} to calcium-binding proteins. In: Celio MR, Pauls T, Schwaller B (eds) Guidebook to the calcium-binding proteins. Oxford University Press, Oxford, pp 1–12
9. Vito P, Lacana E, D'Adamio L (1996) Interfering with apoptosis: Ca^{2+}-binding protein ALG-2 and Alzheimers' disease gene. Science 271:521–525
10. Tarabykina S et al (2000) Two forms of the apoptosis-linked protein ALG-2 with different Ca^{2+} affinities and target recognition. J Biol Chem 275:10514–10518
11. Jia J et al (2001) Structure of apoptosis-linked gene product ALG-2: insights into Ca^{2+}-induced changes in penta-EF-hand proteins. Structure 9:267–275
12. Subramanian L et al (2004) Ca^{2+} binding to EF-hands 1 and 3 is essential for the interaction of apoptosis-linked gene-2 with Alix/AIP1 in ocular melanoma. Biochemistry 43:11175–11186
13. Lollike K et al (2001) Biochemical characterization of the penta-EF-hand protein grancalcin and identification of L-plastin as a binding partner. J Biol Chem 276:17762–17769
14. Jia J et al (2000) Crystal structure of human grancalcin, a member of the penta-EF-hand protein family. J Mol Biol 300:1271–1281
15. Méhul B, Bernard D, Simonetti L, Bernard MA, Schmidt R (2000) Identification and cloning of a new calmodulin-like protein from human epidermis. J Biol Chem 275: 12841–12847
16. Durussel I et al (2001) Cation- and peptide-binding properties of human calmodulin-like skin protein. Biochemistry 41:5439–5448
17. Schwaller B, Celio MR, Hunziker W (1995) Alternative splicing of calretinin mRNA leads to different forms of calretinin. Eur J Biochem 230:424–430
18. Schwaller B et al (1997) Comparison of the Ca^{2+}-binding properties of human recombinant calretinin-22k and calretinin. J Biol Chem 272:29663–29671
19. Leclerc E, Heizmann CW (2011) The importance of the Ca^{2+}/Zn^{2+} signaling proteins and RAGE in translational medicine. Front Biosci (Schol Ed) S3:1232–1262
20. Schäfer BW et al (2000) Brain S100A5 is a novel calcium-, zinc-, and copper-binding protein of the EF-hand superfamily. J Biol Chem 275:30623–30630
21. Ridinger K et al (2000) S100A3 Biochemical characterisation and subcellular localization in different cell lines. J Biol Chem 275: 8686–8694
22. Hsieh H-L, Schäfer BW, Cox JA, Heizmann CW (2002) S100A13 and S100A6 exhibit distinct translocation pathways in endothelial cells. J Cell Sci 115:3249–3258
23. Sturchler E et al (2006) S100A16, a novel calcium-binding protein of the EF-hand superfamily. J Biol Chem 281:38905–38917
24. Hermann A, Cox JA (1995) Sarcoplasmic calcium-binding proteins. Comp Biochem Physiol 111B:337–345
25. Tossavainen H et al (2003) NMR solution structure of calerythrin, an EF-hand calcium-binding protein of Saccharopolyspora erythraea. Eur J Biochem 270:2505–2512
26. Head JF, Inouye S, Teranishi K, Shimomura O (2000) The crystal structure of the photoprotein

aequorin at 2.3 Å resolution. Nature 405: 372–376

27. Gombos Z et al (2001) Calexcitin B is a new member of the sarcoplasmic calcium-binding protein family. J Biol Chem 276:22529–22536
28. Nelson TJ et al (1999) Calexcitin interacts with neuronal ryanodine receptors. Biochem J 341:423–433
29. Gombos Z et al (2003) Conformational coupling of Mg^{2+} and Ca^{2+} on the three-state folding of calexcitin B. Biochemistry 42: 5531–5539
30. Erskine PT et al (2006) Structure of the neuronal protein calexcitin suggests a mode of interaction in signalling pathways of learning and memory. J Mol Biol 357:1536–1547
31. Kilmartin JV (2003) Sfip has conserved centrin-binding sites and an essential function in budding yeast spindle body duplication. J Cell Biol 162:1211–1221
32. Salisbury JL (2004) Sfip and centrin unravel a structural riddle. Curr Biol 14:R27–R29
33. Durussel I et al (2000) Cation- and peptide-binding properties of human centrin 2. FEBS Lett 472:208–212
34. Matei E et al (2003) The C-terminal half of human centrin 2 behaves like a regulatory EF-hand domain. Biochemistry 42:1439–1450
35. Cox JA et al (2005) Calcium and magnesium binding to human centrin 3 and interaction with target peptides. Biochemistry 44:840–850
36. Veeraghavan S et al (2002) Structural independence of the two EF-hands of caltractin. J Biol Chem 277:28564–28571
37. Radu L et al (2010) Scherffelia dubia centrin exhibits a specific mechanism for Ca^{2+}-controlled target binding. Biochemistry 49:4383–4394
38. Li S et al (2006) Structural role of Sfip-centrin filaments in budding yeast spindle pole body duplication. J Cell Biol 173:867–877
39. Miron S et al (2011) Binding of calcium, magnesium, and target peptides to Cdc31, the centrin of yeast Saccharomyces cerevisiae. Biochemistry 50:6409–6422
40. Banfi B et al (2001) A calcium-activated NADPH oxidase in testis, spleen, and lymph nodes. J Biol Chem 276:37594–37601
41. Banfi B et al (2004) Mechanism of Ca^{2+} activation of NADPH oxidase 5 (NOX5). J Biol Chem 279:18383–18591
42. Tirone F, Radu L, Craescu CT, Cox JA (2010) Identification of the binding site for the regulatory calcium-binding domain in the catalytic domain of NOX5. Biochemistry 49:761–771
43. Tirone F, Cox JA (2007) NADPH oxidase 5 (NOX5) interacts with and is regulated by calmodulin. FEBS Lett 581:1202–1208
44. Linse S, Thulin E, Sellers P (1993) Disulfide bonds in homo- and heterodimers of EF-hand subdomains of calbindin D9k, stability, calcium binding and NMR studies. Protein Sci 2:985–1000
45. Colowick SP, Womack FC (1969) Binding of diffusible molecules by macromolecules: rapid measurement by rate of dialysis. J Biol Chem 244:774–777
46. Wiseman T, Williston S, Brands JF, Lin L-N (1989) Rapid measurement of binding constants and heats of binding using a new titration calorimeter. Anal Biochem 179:131–137
47. Osawa M et al (2005) Mg^{2+} and Ca^{2+} differentially regulated DNA binding and dimerization of DREAM. J Biol Chem 280:18008–18014
48. Hummel JP, Dryer WJ (1962) Measurement of protein-binding phenomena by gel filtration. Biochim Biophys Acta 63:530–532
49. Cornish-Bowden A, Koshland DE (1975) Diagnostic use of the hill (Logit and Nernst) plots. J Mol Biol 95:201–212
50. Linse S, Forsen S (1995) Determinants that govern high-affinity calcium binding. Adv Second Messenger Phosphoprotein Res 30:89–151
51. Klotz IM, Hunston DL (1979) Protein affinities for small molecules: conceptions and misconceptions. Arch Biochem Biophys 193: 314–428
52. Potter JD, Gergely J (1975) The calcium and magnesium binding on troponin and their role in the regulation of myofibrillar ATPase. J Biol Chem 250:4628–4633
53. Moeschler H, Schaer J-J, Cox JA (1980) A thermodynamic analysis of the binding of calcium and magnesium ions to parvalbumin. Eur J Biochem 111:73–78
54. Engelborghs Y et al (1990) Kinetics of conformational changes in Nereis calcium-binding protein upon calcium and magnesium binding. J Biol Chem 265:18801–18815
55. Luan-Rilliet Y, Milos M, Cox JA (1992) Thermodynamics of cation binding to Nereis sarcoplasmic calcium-binding protein. Direct binding studies, microcalorimetry and conformational changes. Eur J Biochem 208:133–138
56. Milos M, Comte M, Schaer J-J, Cox JA (1989) Evidence for four capital and six auxiliary cation-binding sites of calmodulin: divalent cation interactions monitored by direct binding and microcalorimetry. J Inorg Biochem 25: 6279–6287

Chapter 3

Probing Ca^{2+}-Binding Capability of Viral Proteins with the EF-Hand Motif by Grafting Approach

Yubin Zhou, Shenghui Xue, Yanyi Chen, and Jenny J. Yang

Abstract

Ca^{2+} is implicated in almost every step of the life cycle of viruses, including virus entry into host cells, virus replication, virion assembly, maturation, and release. However, due to the lack of prediction algorithms and rigorous validation methods, only limited cases of viral Ca^{2+}-binding sites are reported. Here, we introduce a method to predict continuous EF-hand or EF-hand-like motifs in the viral genomes based on their primary sequences. We then introduce a grafting approach, and the use of luminescence resonance energy transfer and Ca^{2+} competition assay for experimental verification of predicted Ca^{2+}-binding sites. This protocol will be valuable for the prediction and identification of unknown Ca^{2+}-binding sites in virus.

Key words: Calcium, CD2, EF-hand, Luminescence resonance energy transfer, Protein engineering, Protein–ligand interaction, Terbium, Virus

1. Introduction

Ca^{2+} is a universal signal for "life and death" that regulates almost every aspect of cellular activities. Ca^{2+} is evolutionally chosen as a second messenger to modulate a myriad of essential biological events, including gene transcription, fertilization, cell proliferation, secretion, cell motility, and apoptosis. Under physiological conditions, Ca^{2+} homeostasis is maintained by the exquisite choreography of a variety of signaling components, including ion channels, pumps, exchangers, receptors, Ca^{2+} buffers, Ca^{2+}-responsive proteins (e.g., Ca^{2+}-sensitive enzymes and transcription factors) in cytoplasm, ER, Golgi complex, mitochondria, nucleus, and the extracellular matrix (1). Errors in any step of the calcium signaling are likely to cause abnormal cellular activities.

Claus W. Heizmann (ed.), *Calcium-Binding Proteins and RAGE: From Structural Basics to Clinical Applications*, Methods in Molecular Biology, vol. 963, DOI 10.1007/978-1-62703-230-8_3, © Springer Science+Business Media New York 2013

Viruses are adept at hijacking Ca^{2+} over other biological metal ions (Na^+, K^+, Zn^{2+}, Fe^{3+}, etc.) for their own benefit (2). Ca^{2+} is closely involved in viral infection, replication, virion assembling, maturation, and release. In general, the interplay between Ca^{2+} and a virus falls into three categories: (1) the virus perturbs the Ca^{2+} homeostasis of the host cell by altering the membrane permeability or disrupting key component(s) of the calcium signaling network; (2) viral proteins directly bind Ca^{2+} to maintain structural integrity and their optimal function; and (3) the virus interacts with cellular Ca^{2+}-binding proteins or Ca^{2+}-signaling proteins (2, 3).

Based on the arrangement of the Ca^{2+}-binding ligand residues in space, viral Ca^{2+}-binding proteins can be divided as EF-hand and non-EF-hand proteins. EF-hand Ca^{2+}-binding proteins with a continuous helix-loop-helix topology were originally described in the structure of parvalbumin by Kretsinger (4). A typical EF-hand Ca^{2+}-binding motif contains 12 residues in the loop region responsible for coordinating Ca^{2+} (Fig. 1a, b). Among these residues, the side chain oxygen atoms in positions 1, 3, 5, and 12, the main chain oxygen in position 7, and a bridged water molecular in position 9 bind to Ca^{2+} to form a pentagonal bipyramidal geometry. Until now, only very limited cases of viral calcium-binding proteins were reported in the literature, possibly due to challenges in the prediction and investigation of Ca^{2+}-binding sites. First, we lack robust prediction algorithms and effective validating methods. Second, given the physiological abundance and spectroscopic silence nature of Ca^{2+}, identification and analyzing Ca^{2+}-binding sites in proteins is not an easy task. Most experimental methods to determine Ca^{2+}-binding are sensitive to the total Ca^{2+} concentration. The background Ca^{2+} concentration, however, is usually hard to control and evaluate. Finally, the binding of Ca^{2+} to EF-hands usually induces conformational change of the EF-hands. Therefore, evaluating site-specific Ca^{2+}-binding affinity is often complicated by the conformational cooperativity in the case of proteins possessing multiple Ca^{2+}-binding sites (e.g., calmodulin and troponin C) (5).

To overcome the first limitation, we have generated a handful of motif signatures (Table 1) for prediction of continuous EF-hand or EF-hand-like motifs (6). More importantly, we have developed a grafting approach coupled with luminescence resonance energy transfer (LRET) to evaluate isolated Ca^{2+}-binding sites from EF-hand Ca^{2+}-binding proteins (7–9). To circumvent the spectroscopically silent nature of Ca^{2+}, we use lanthanides to probe Ca^{2+}-binding because of their similar ionic radii (Ca^{2+}, 1.00 Å; Tb^{3+}, 0.92 Å) and metal coordination chemistry (10, 11). As shown in Fig. 2, a putative Ca^{2+}-binding sequence can be inserted in a flexible loop in the domain 1 of CD2 (CD2.D1) with minimal perturbations to the overall structure. The Trp residue (W32) within CD2. D1 is capable of transferring energy to Tb^{3+} that binds to the inserted Ca^{2+}-binding loop. We optimized the insertion position

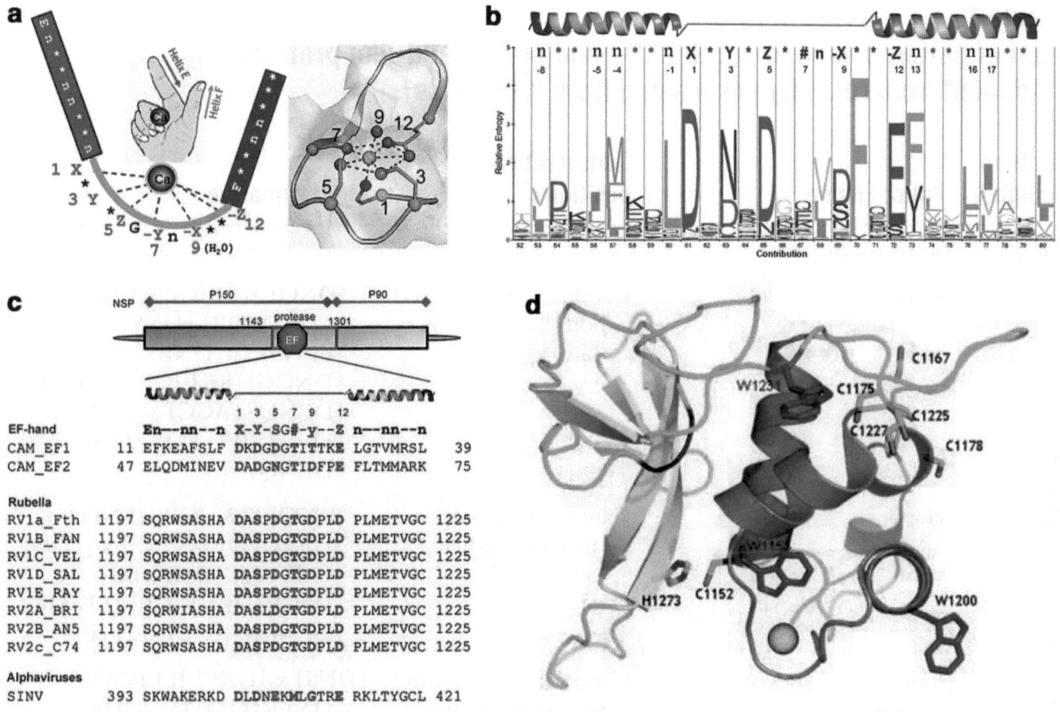

Fig. 1. Examples of predicted EF-hand calcium-binding sites in the *Togaviridae* family of virus. (**a**) Cartoon representation of a typical EF-hand Ca^{2+}-binding motif (adapted from ref. 2 with permission from Elsevier). The EF-hand motif contains a helix-loop-helix topology with residues from the loop responsible for coordinating Ca^{2+}. The oxygen atoms from the side chains at positions 1, 3, 5, and 12, backbone in position 7, and one bridged water molecule (via position 9) directly chelate Ca^{2+} to form a pentagonal bipyramidal geometry (12). (**b**) The sequence consensus for a classical EF-hand motif. The conservation of amino acids at several positions makes it possible to predict EF-hand motifs from genomic sequences (adapted from ref. 6 with permission from Wiley-Liss). (**c**) The predicted EF-hand motifs in Sindbis virus (SINV) and eight genotypes of rubella virus (RV). The Ca^{2+}-binding sites of EF-hand motifs are compared to the classical EF-hands 1 and 2 in calmodulin, a well-studied eukaryotic Ca^{2+}-binding protein with four EF-hand motifs. (**d**) The EF-hand Ca^{2+}-binding motif (highlighted in blue with Ca^{2+} ion shown as cyan sphere) in the protease domain of rubella virus nonstructural protein p200 (adapted from reference (7) with permission from American Society of Microbiology).

and the length of the glycine linkers to optimize LRET efficiency, and meanwhile, to minimize the perturbation to the host structure (9). Insertion of the putative Ca^{2+}-binding sequence between residues Ser52 and Gly53 flanked by triple glycine linkers proves to be most effective. The native structure of host protein CD2.D1 after EF-hand insertion is essential for the accuracy of the metal binding assay. Next, the correct protein folding of engineered proteins is confirmed by both intrinsic tryptophan fluorescence and circular dichroism spectroscopy. Finally, the site-specific Tb^{3+}-binding affinity is obtained by titrating Tb^{3+} into the sample containing engineered proteins, whereas the site-specific Ca^{2+}-binding affinity is subsequently obtained with competition assay after obtaining site-specific Tb^{3+}-binding affinity.

Table 1
List of patterns used to predict EF-hand or EF-hand-like proteins (adapted from ref. 6 by permission from Wiley-Liss)

Pattern number	Pattern name	Conservation positions	Motif signature
To predict canonical EF-hand			
1	eloopf 	Canonical EF-loop and both flanking helices	x-{DNQ}-x(2)-{GP}-{ENPQS}-x(2)-{DPQR}-[DNS]-x-[DNS]-{FLIVWY}-[DNESTG]-[DNQGHRK]-{GP}-[LIVMC]-[DENQSTAGC]-x(2)-[ED]-[FLYMVIW]-x(2)-{NPS}-{DENQ}-x(3)
2	eloop 	Canonical EF-loop and the entering helix	x-{DNQ}-x(2)-{GP}-{ENPQS}-x(2)-{DPQR}-[DNS]-x-[DNS]-{FLIVWY}-[DNESTG]-[DNQGHRK]-{GP}-[LIVMC]-[DENQSTAGC]-x(2)-[ED]
3	loopf 	Canonical EF-loop and the exiting helix	[DNS]-x-[DNS]-{FLIVWY}-[DNESTG]-[DNQGHRK]-{GP}-[LIVMC]-[DENQSTAGC]-x(2)-[ED]-[FLYMVIW]-x(2)-{NPS}-{DENQ}-x(3)
4	loop 	Canonical EF-loop	[DNS]-x-[DNS]-{FLIVWY}-[DNESTG]-[DNQGHRK]-{GP}-[LIVMC]-[DENQSTAGC]-x(2)-[ED]
5	PS00018 	Canonical EF-loop	D-x-[DNS]-{ILVFYW}-[DENSTG]-[DNQGHRK]-{GP}-[LIVMC]-[DENQSTAGC]-x(2)-[DE]-[LIVMFYW]
To predict pseudo EF-hand			
6	PC 	Both helices of the pseudo EF-motif, both helices and the loop of the paired canonical EF-motif	[LMVITNF]-[FY]-x(2)-[YHIVF]-[SAITV]-x(5,9)-[LIVM]-x(3)-[EDS]-[LFM]-[KRQLE]-x(20,28)-[LQKF]-[DNG]-x-[DNSC]-x-[DNK]-x(4)-[FY]-x-[EKS]
7	Pseudo 	Both helices of the pseudo EF-motif	[LMVITNF]-[FY]-x(2)-[YHIVF]-[SAITV]-x(5,9)-[LIVM]-x(3)-[EDS]-[LFM]-[KRQLE]

(continued)

Table 1 (continued)

Pattern number	Pattern name	Conservation positions	Motif signature
8	PS00303 	Both helices and the loop of the paired canonical EF-motif	[LIVMFYW](2)-x(2)-[LK]-D-x(3)-[DN]-x(3)-[DNSG]-[FY]-x-[ES]-[FYVC]-x(2)-[LIVMFS]-[LIVMF]
To predict EF-hand-like			
9	Excalibur 	The 10-residue loop	D-x-D-x-D-G-x(2)-C-E
10	EF-hand-like 	The loop	D-x-[DNS]-{ILVFYW}-[DEN]-G-{GP}-x(5, 6)-[DE]

In this chapter, we describe a general approach to probe the Ca^{2+}-binding capability of viral EF-hand proteins as outlined in Fig. 3. First, putative viral calcium-binding proteins can be predicted using our developed calcium-binding site prediction algorithm (http://chemistry.gsu.edu/faculty/Yang/Calciomics.htm) or using the pattern search server available at ScanProsite (http://ca.expasy.org/prosite/PDOC00018). A total of 93 putative EF-hands or EF-hand-like motifs have been found in the viral genomes, including 80 different viruses that spread in the majority of virus families (2). This protocol will further summarize our methods for the identification of putative calcium-binding sites in viral proteins, which include the construction, expression and purification, biophysical characterization of engineered proteins, Tb^{3+}-LRET assay, and Ca^{2+} competition assay. The experimental flowchart is shown in Fig. 3 with the predicted EF-hand Ca^{2+}-binding sites in the *Togaviridae* family (e.g., rubella virus and Sindbis virus) as examples (Figs. 1 and 2).

2. Materials

2.1. Bioinformatic Resources

1. Protein sequence search from NCBI: http://www.ncbi.nlm.nih.gov/protein/.
2. Pattern search for EF-hands: http://chemistry.gsu.edu/faculty/Yang/Calciomics.htm.

   Input user-defined patterns at ScanProsite (http://prosite.expasy.org/scanprosite/).

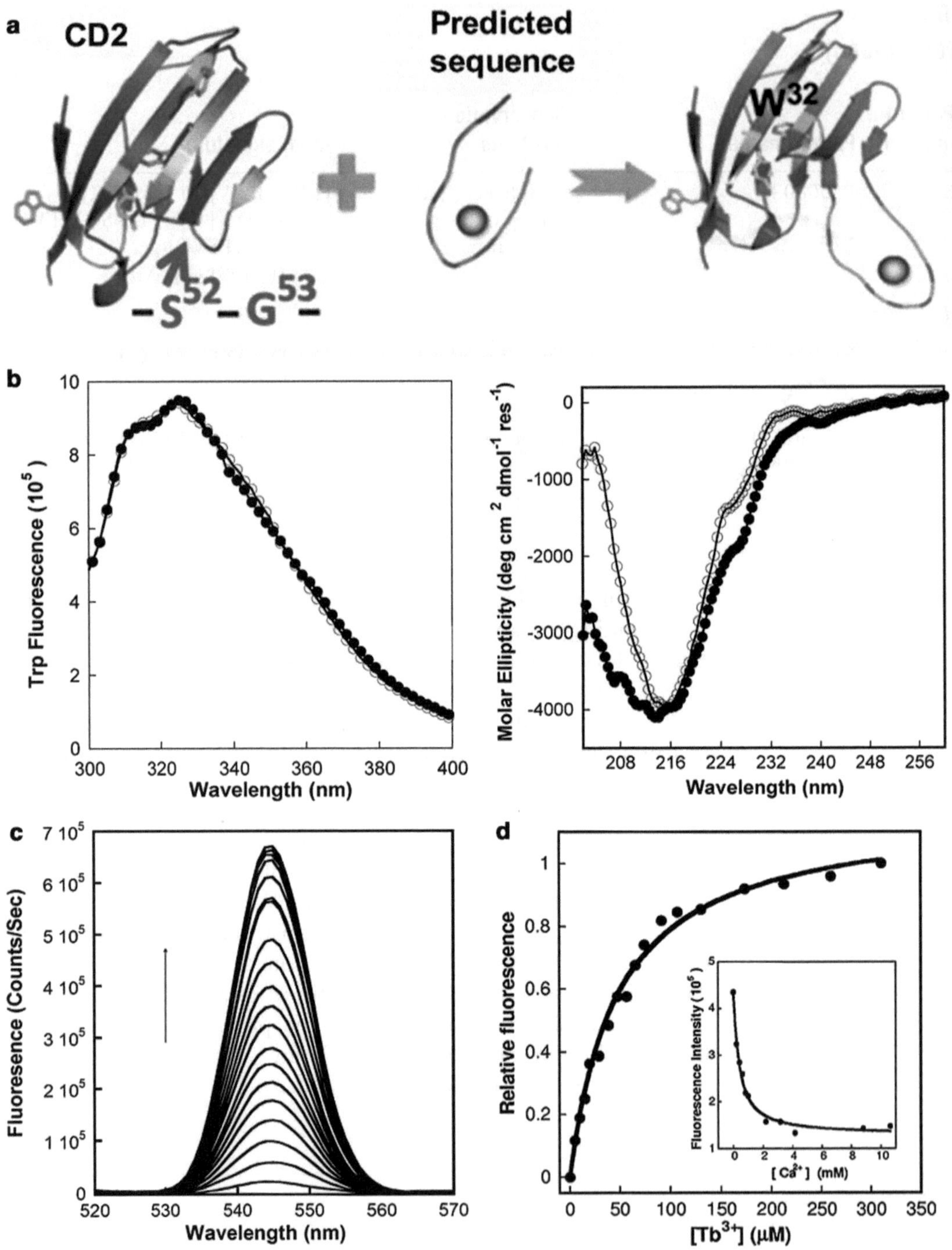

Fig. 2. Use of the grafting approach to verify the predicted viral EF-hand calcium-binding sites. (**a**) Schematics of the grafting approach. CD2.D1 is used as the host scaffold. The putative EF-hand Ca^{2+}-binding site is inserted between residues 52 and 53, flanked by three glycines on each side. The grafting approach allows one to conveniently monitor the metal binding process with fluorescence spectroscopy by taking advantage of the aromatic residue (W32 in CD2.D1)-sensitized Tb^{3+} luminescence resonance transfer (Tb^{3+}-LRET) (adapted from ref. 2 with permission from Elsevier). (**b**) Characterization of the engineered protein by intrinsic Trp fluorescence (*left*) and circular dichroism spectroscopy (*right*).

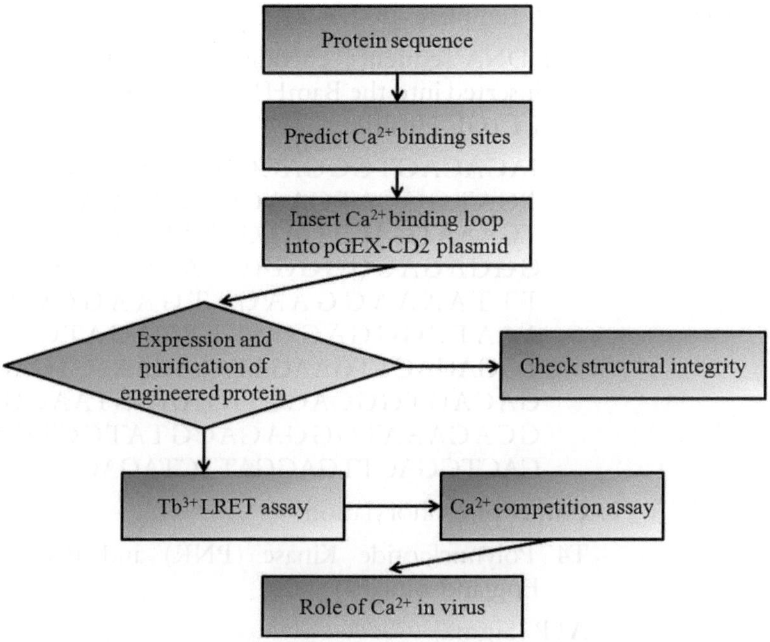

Fig. 3. Experimental approaches to validate predicted continuous Ca^{2+}-binding motifs. Continuous Ca^{2+}-binding sites can be predicted from the primary sequence by using the program CaPS (available at http://chemistry.gsu.edu/faculty/Yang/Calciomics.htm). The predicted Ca^{2+}-binding sequence is subsequently grafted into a scaffold protein CD2.D1. After expression and purification of the engineered protein, one can perform Tb^{3+} titration assay (Tb^{3+}-LRET) to confirm the metal binding capability. Ca^{2+} competition assay can be further performed based on Tb^{3+}-LRET to obtain Ca^{2+}-binding affinity. If metal binds to the engineered protein, mutations on the proposed ligand residues can be subsequently introduced to double confirm the binding event. If the mutagenesis results in a decrease in binding, the predicted site has a high chance of binding Ca^{2+}. Possible Ca^{2+}-induced conformational changes will be examined by expressing the protein of interest. In addition, the functional correlation of Ca^{2+}-binding can be followed by comparing the phenotypes of WT and mutant viruses.

Fig. 2. (continued) The structural integrity of the engineered protein is confirmed by (1) the fluorescence emission peaks at 315 nm and 330 nm and (2) a negative trough at 215 nm in the CD spectrum. *Solid circle*: engineered CD2.D1 with the insertion of EF-hand loop from rubella virus; *Open circle*: wild-type CD2.D1. *C*, Tb^{3+} titration assay (reproduced from ref. 7 with permission from American Society for Microbiology). (**c**) Tb^{3+} titration spectra of the engineered protein. If Tb^{3+} binds to the predicted sequence, the luminescence signal at 545 nm will increase until it reaches saturation. (**d**) Typical Tb^{3+} titration and Ca^{2+} competition curves (*inset*) (reproduced from ref. 7 with permission from American Society for Microbiology). The dissociation constants of Tb^{3+} and Ca^{2+} can be calculated from by fitting the curves using Eqs. 2 and 3, respectively.

2.2. Molecular Cloning Reagents

1. PCR Template: pGEX-2T-CD2:

   The DNA sequence encoding CD2.D1 (shown below) was inserted into the BamHI-EcoRI sites in the plasmid pGEX-2T (GE Healthcare):

   "AGAGACAGTGGACCGTCTGGGGTGCCCTGGG TCATGGCATCAACCTGAACATCCCTAACTT TCAAATGACTGATGATATTGATGAGGTGCGAT GGGAGAGGGGGAGCACCCTGGTTGCCGAG TTTAAAAGGAAGATGAAGCCTTTTTTG AAATCGGGAGCATTTGAGATCTTAGCAAA TGGAGACTTGAAGATAAAGAATCTGACAAGAGAT GACAGTGGCACCTATAATGTAACGGTATACA GCACAAATGGGACACGTATCCTGAACAAGG CACTGGACTTGAGGATTCTAGAG".

2. Primer phosphorylation:

   T4 Polynucleotide Kinase (PNK) and PNK buffer (New England Biolabs).

   ATP (Sigma).

3. PCR:

   KOD hot start DNA polymerase and buffer (Novagen).

   dNTP (Novagen).

   $MgSO_4$ (Novagen).

4. Ligation:

   T4 DNA ligase and ligation buffer (New England Biolabs).

   ATP (Sigma).

5. Transformation:

   Competent cells: DH5α and BL21(DE3) (Invitrogen).

   LB broth media and LB agar (EMD Chemicals).

   Ampicillin (Inalco S.p.A, Milano, Italy).

6. Molecular cloning kits:

   QIAprep spin miniprep kit and QIAquick Gel Extraction Kit (Qiagen).

2.3. Protein Expression and Purification

1. Protein expression:

   Isopropyl β-D-1-thiogalactopyranoside (IPTG) stock solution: 1 M IPTG in ddH_2O, sterilized by passing through a 0.2 μm syringe filter.

   UV-VIS spectrophotometer (Shimadzu).

2. Protein purification and preparation:

   Sonicator.

   Lysis buffer: 1× PBS supplemented with 5 mM DTT.

GS4B beads (GE Healthcare).

Bio-Rad Protein Assay Dye Reagent Concentrate (Bio-Rad).

Lysate buffer: 1% sarcosyl in 1× PBS.

Dithiothreitol (DTT, Inalco S.p.A, Milano, Italy).

Thrombin (Sigma).

FPLC (GE Healthcare).

HiTrap SP column (GE Healthcare).

2.4. Instruments for Biophysical Characterization of Engineered Proteins

QuantaMaster™40 fluorometer (PTI, Birmingham, NJ).

Jasco-810 circular dichroism spectropolarimeter (JASCO, Easton, MD).

2.5. LRET-Based Metal Binding Assay

1. Tb^{3+} LRET assay:

   PIPES buffer: 50 mM PIPES (EMD Chemicals), pH 6.8, 10–100 mM KCl.

   $TbCl_3$ (Sigma).

2. Ca^{2+} competition assay:

   PIPES buffer: 50 mM PIPES (EMD Chemicals), pH 6.8, 10–100 mM KCl.

   Ca^{2+} standard solution (Thermo scientific): 100 mM $CaCl_2$.

3. Methods

3.1. Prediction of EF-Hand Motifs from Viral Genomes by Pattern Search

The primary amino acid sequence is required to predict the EF-hand or EF-hand-like motifs in virus genome. Here, we use the nonstructural protein 1 (nsp1) of Sindbis virus (NCBI Reference Sequence: NP_740670.1) as an example.

1. Search viral proteins of interest at NCBI (http://www.ncbi.nlm.nih.gov/protein/). Here we input NP_740670.1 and selected the amino acid sequence for nsp1 of Sindbis virus.
2. Enter the calcium-binding pattern search Web site (http://chemistry.gsu.edu/faculty/Yang/Calciomics.htm) and choose "CaPS (Calcium Pattern Search)" (see Note 1). Input the selected protein sequence of interest. The Web site will immediately tell the number(s) of predicted Ca^{2+}-binding sites and the binding pattern for each binding site. For example, one putative Ca^{2+}-binding site was detected in nsp 1 of Sindbis virus with the sequence "DLDNEKMLGTRE." Likewise, a putative Ca^{2+}-binding site (sequence "DASPDGTGDPLD") within the rubella virus nonstructural protein p200 was detected.

3.2. Grafting the Predicted Ca^{2+}-Binding Site into Host Protein CD2.D1

The putative calcium-binding sequence is inserted into the CD2 by PCR using procedures illustrated in Fig. 4. To simplify the protein purification, pGEX-2T vector is used, which encodes a GST tag, which entails the use of affinity purification (see Note 2). This section contains information on primer design for PCR, PCR setup, ligation, transformation, plasmid extraction, and sequencing.

1. Primer design for PCR:

   The design of primers is shown in Fig. 4b. The forward primer starts with the DNA sequence encoding putative calcium-binding sites and a flexible linker (i.e., DNA sequences encoding protein residues GGG). The 3′ end of the forward primer is paired with the antisense strand of the template DNA corresponding to the protein sequences after Gly53 from CD2. The reverse primer starts with 9 bp that encodes the flexible linker GGG, followed by sequences pairing with the sense strand of the template corresponding to the region of CD2 protein before G53. The paired bases ("→" as shown in Fig. 4b) should have a calculated annealing temperature (Tm) of 64–68°C (Tm is estimated by the equation shown in Note 3). By this way, we can ensure that the putative Ca^{2+}-binding site, flanked by three glycines on each side as linkers, is inserted between Ser52 and Gly53 of CD2.D1.

   As an example, the forward primer can be designed as "XXX…XXXGGTGGCGGAGCATTTGAGATCTTAGCAAATGGA," and the reverse primer can be designed as "GCCGCCGCCCGATTTCAAAAAGGCTTCATCTTC," where

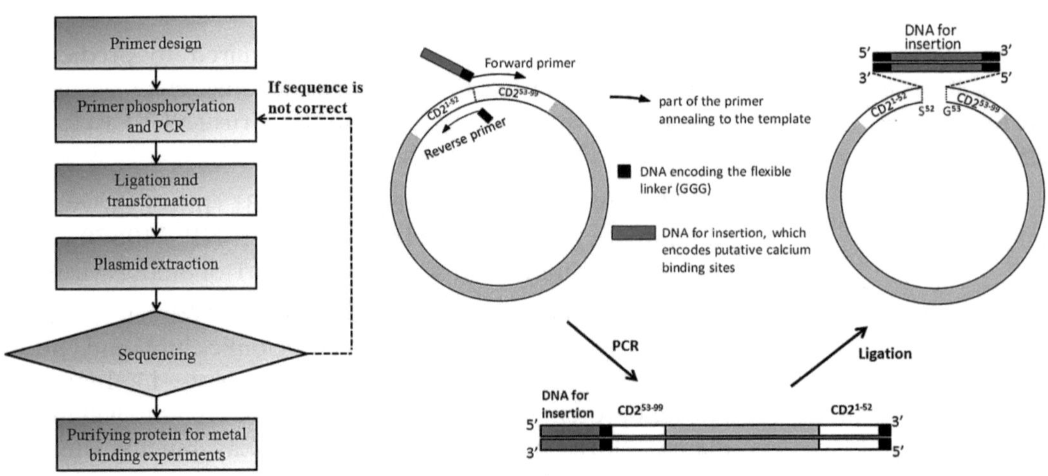

Fig. 4. Instructions on inserting predicted Ca^{2+}-binding sites into the host protein by one single PCR reaction. Primers are designed as described in text (Subheading 3.2) and phosphorylated. The DNA sequences encoding for the calcium-binding site and linkers are inserted into the template plasmid by PCR. The linear PCR products are subsequently circularized by DNA ligase. The final construct contains the predicted Ca^{2+}-binding site with triple-glycine on each side between residues Ser52 and Gly53 of CD2.D1.

"XXX…XXX" standards for the DNA sequence encoding the putative Ca^{2+}-binding sites and the underlined sequences ("GGTGGCGGA" in forward primer and "GCCGCCGCC" in reverse primer) encode the triple-glycine linker region of the engineered protein.

2. Primer phosphorylation:
   (a) Dilute the primer in ddH_2O with a final concentration of 100 μM.
   (b) Add 6 μL primer, 2 μL PNK buffer (10×), 1 μL ATP (10 mM), and 1 μL PNK in 10 μL ddH_2O.
   (c) Incubate at 37°C for 30 min and then incubate at 65°C for 15 min to inactivate PNK activity.
3. PCR:
   (a) PCR mixture includes 100–200 ng pGEX-2T-CD2 plasmid, 1 μL primer prepared in Subheading 3.2, step 2, 5 μL KOD buffer, 5 μL dNTP (2 mM), 3 μL $MgSO_4$ (25 mM), and 1 μL KOD enzyme in 33 μL H_2O. PCR was performed by following the manufacturer's instructions.
   (b) Add 2 μL of DpnI restriction enzyme to the amplified PCR reaction and then incubate at 37°C for 1–2 h to digest the parental plasmid (see Note 4).
   (c) Add all the PCR products in the sample well of 0.8% agarose gel and then run the gel at 80–120 V for 30 min. Recover the band at ~5.3 kbp in a total volume of 30 μL with Qiaquick gel extraction kit.
4. Ligation and transformation:
   (a) Mix 30 μL recovered DNA in Subheading 3.2, step 3, with 1 μL T-4 ligase, 5 μL ligase buffer, 1 μL ATP (100 mM stock), and 13 μL ddH_2O. The reaction mixture is incubated at 16°C for 2 h.
   (b) Transformation of DH5α is carried out according to the manufacturer's instructions.
5. Verifying the cloning products:
   (a) The next day after transformation, check if the plate has colonies. If it does, pick out single colonies to 10 mL LB medium supplemented with 100 μg/mL ampicillin (LB-amp medium).
   (b) Shake the culture overnight at 200 rpm at 37°C.
   (c) Harvest the bacteria and extract the plasmid using QIAprep spin miniprep kit.
   (d) Sequence the extracted plasmid to confirm that it contains correctly inserted sequence.

3.3. Protein Expression and Purification

1. Protein expression:
   (a) Transform the correct plasmid encoding the engineered protein into BL21(DE3) competent cells for protein expression. Positive colonies were selected on ampicillin-resistant LB agar plates.
   (b) Inoculate a single colony into 10 mL LB-amp medium and shake the culture overnight with 200–250 rpm at 37°C.
   (c) Transfer the overnight culture to 1 L fresh LB-amp medium and allow the culture to grow until the OD_{600} reaches 0.6–0.8.
   (d) Induce protein expression with 0.3 mM IPTG and allow the bacteria to grow for another 3–4 h at 200 rpm, 37°C.
   (e) Harvest cells by centrifugation at 5,000 × *g* for 20 min and collect the cell pellets in a 50-mL falcon tube.
   (f) Store the cell pellets at −80°C for future use (see Note 5).
2. Protein purification and preparation:
   (a) Thaw the cell pellets.
   (b) Add at least 25 mL (per liter culture) of lysis buffer and thoroughly resuspend cell pellets with Vortexer.
   (c) Sonicate the bacteria suspension for 20 s, six times. Wait for 2–5 min between each cycle to cool down the tip.
   (d) Centrifuge the sonicated solution at 35,000 × *g* for 30 min, and then collect the clarified supernatant.
   (e) While the solution is centrifuging, add 4 mL GS4B beads to a gravity-flow column. Wash the GS4B beads with ~4 column volumes of 1× PBS (pH 7.30). Take 20 μL aliquot of the equilibrated column beads for SDS-PAGE (beads before binding; Note 6).
   (f) Filter the clarified supernatant through a 0.45 μM syringe filter. Apply the supernatant to the equilibrated GS4B column. Take 20 μL aliquot of this supernatant for SDS-PAGE analysis (supernatant).
   (g) Collect the flow-through solution and rebind it to the column once. Take a 20 μL aliquot of the flow-through for SDS-PAGE analysis (flow-through).
   (h) Wash the column with at least 10 column volumes of 1× PBS. Take 20 μL aliquot of the beads for SDS-PAGE (beads after binding; Note 7).
   (i) On column cleavage of GST fusion proteins: Cap column bottom, and add 2.5–3 mL of 1× PBS and 20–30 μL (20–30 units) of thrombin. Cap the column top and leave the column at 4°C overnight with gentle agitation.

(j) Open the column top and bottom; collect the flow-through. Wash columns with ~10 mL of 1× PBS to rinse off residual cleaved products, pool together all the flow-through, and concentrate to ~1 mL (see Note 8).

(k) Filter the flow-through by passing through a 0.45 μm syringe filter. Inject 1 mL cleaved product to FPLC using the Superdex 75 column pre-equilibrated with 10 mM Tris-HCl at pH 7.4.

(l) Collect the proper peak fractions (corresponding to a size of ~11 kDa). Take 20 μL sample from each peak for SDS-PAGE analysis.

(m) Determine the concentration of purified proteins by measuring absorbance at 280 nm ($\varepsilon_{280\ nm} = 11{,}700\ M^{-1}\ cm^{-1}$). The purified protein can be kept at −20°C for up to 1 month.

3.4. Structural Analyses on Engineered Proteins

1. Intrinsic Trp fluorescence measurement:

   Fluorescence emission spectra are recorded using a 1-cm-path-length cell at ambient temperature using a QuantaMaster™40 fluorometer.

   (a) Prepare 800 μL of protein sample with a final concentration of 2–5 μM in a quartz fluorescence cuvette in a buffer consisting of 10 mM Tris-HCl at pH 7.4.

   (b) Turn on the fluorometer and set the excitation at 280 nm, and the emission spectrum is collected from 300 nm to 400 nm. The integration time was set as 1 s.

   (c) Adjust the excitation and emission slit width until fluorescence signal reaches within the linear range of 10^5–10^6.

   (d) The emission peaks are expected at 315 nm and 330 nm for a well-folded engineered protein (Fig. 2b).

2. Secondary structure analysis by circular dichroism spectroscopy:

   The circular dichroism spectrum of an engineered protein is acquired with a Jasco-810 spectropolarimeter at ambient temperature using a quartz cell of 1-mm path length. All spectra are obtained as the average of at least eight scans with a scan rate of 50 nm/min.

   (a) Prepare 300 μL of the engineered protein sample with a final concentration of 10–20 μM in 10 mM Tris-HCl at pH 7.4. Transfer ~250 μL to a 1-mm-path-length quartz cuvette.

   (b) CD spectrum is recorded in the far-UV (190–260 nm) range. Adjust the protein concentration until the CD signals are between −10 and −20 mdeg.

   (c) The CD signals from the buffer are also recorded and used for background subtraction.

(d) Compare the spectrum of CD2.D1 and the engineered protein. The secondary structure of CD2 is characteristic of β sheet with a major negative trough at 215 nm. For the engineered protein, the majority of secondary structure is still β sheet, along with ~10% contribution of random coils from inserted EF-hand Ca^{2+}-binding motif (Fig. 2b).

3.5. LRET-Based Metal Binding Assay

1. LRET assay:

LRET experiments are carried out with a QuantaMaster™40 fluorometer with the slit width for excitation and emission set at 6–12 nm. A glass filter is used to avoid secondary Raleigh scattering.

(a) Turn on the fluorometer and log in the FELIX software.

(b) Set the excitation wavelength at 280 nm, and emission wavelength between 500 nm and 600 nm.

(c) Rinse the quartz fluorescence cuvette thoroughly with ddH_2O and air-dry the cuvette completely (see Note 9).

(d) Prepare 2 μM of protein in 1,000 μL PIPES buffer. Add 800 μL into the cuvette (defined as Sample 1).

(e) Add 800 μL PIPES buffer without the protein in another cuvette (defined as Sample 2), which will be used as Tb^{3+} background signal during titration.

(f) Scan Sample 1 in fluorometer using the Emission mode. Adjust the slit width for both excitation and emission to make sure that the initial fluorescence signal is about 1/100 of the maximal intensity of the instrument (see Note 10).

(g) Titrate increasing amounts of $TbCl_3$ into Sample 1 until the luminescence signal is saturated. For each titration point, make sure to mix the solution well and wait for at least 10 min. For the grafted EF-hand Ca^{2+}-binding loop from rubella virus, the luminescence signal reaches saturation when the total Tb^{3+} concentration is over 300 μM (see Note 11). A typical titration spectra and curve are shown in Fig. 2c, d.

(h) Repeat the same titration procedure (steps f, g) for Sample 2, which is used for background subtraction.

(i) Export the fluorescence spectra for data analysis.

(j) Calculate the luminescence intensity (I_{LRET}) using Eq. 1:

$$I_{LRET} = I_{sample1} - I_{sample2} \quad (1)$$

where $I_{sample1}$ is the intensity of the sample 1 at each titration point and $I_{sample2}$ is the intensity of buffer at each titration point.

(k) Plot the total Tb^{3+} concentration as X-axis and I_{LERT} at different Tb^{3+} concentration as Y-axis.

(l) Fit the titration curve by assuming a 1:1 binding process with Eq. 2:

$$f = \frac{([P]_T + [M]_T + K_d) - \sqrt{([P]_T + [M]_T + K_d)^2 - 4[P]_T[M]_T}}{2[P]_T}, \quad (2)$$

where f is the fractional change of luminescence intensity, K_d is the dissociation constant for Tb^{3+}, and $[P]_T$ and $[M]_T$ are the total concentrations of the protein and Tb^{3+}, respectively.

2. Ca^{2+} competition assay:

(a) Add 40 μM Tb^{3+}, 1–2 μM protein in PIPES buffer with a total volume of 800 μL (see Note 12).

(b) Before adding $CaCl_2$, scan the emission spectrum with the excitation wavelength at 280 nm. The emission spectrum is recorded between 500 nm and 600 nm.

(c) Titrate increasing amounts of Ca^{2+} to the Tb^{3+}–protein mixture in the cuvette until the luminescence signal stops decreasing. Each time after adding Ca^{2+}, turn the cuvette up and down, and then wait for at least 15 min before the next addition (see Note 13).

(d) Plot the total Ca^{2+} concentration as X-axis and luminescence intensity at each Ca^{2+} titration point as Y-axis (Fig. 2d). The dissociation constant of Ca^{2+} for the engineered protein can be calculated by Eq. 3:

$$K_{d,Ca} = K_{app} \times \frac{K_{d,Tb}}{K_{d,Tb} + [M_{Tb}]}, \quad (3)$$

where $K_{d,Ca}$ and $K_{d,Tb}$ are the dissociation constants of Ca^{2+} and Tb^{3+}, respectively. K_{app} is the apparent dissociation constant obtained by fitting the competition curve using Eq. 2.

4. Notes

1. ScanProsite (http://prosite.expasy.org/scanprosite/) can also be freely accessed to perform user-defined pattern search. Users are advised to use either our optimized patterns (Table 1) (6) or PS00018.
2. The engineered protein was expressed as GST fusion protein, which makes the protein expression more efficient by enhancing

protein solubility and improving protein folding. However, GST Tag may influence the subsequent LRET measurements. GST tags have to be removed by enzymatic cleavage with thrombin.

3. Tm = 64.9°C + 41°C × (number of G's and C's in the primer—16.4)/N, where N is the length of the primer.
4. This step aims to digest parental methylated plasmid pGEX-2T-CD2 with DpnI. If a high frequency of template DNA is observed in sequenced colonies, one is advised to either prolong the incubation time or increase the amount of DpnI.
5. Snap-freeze the cell pellets with liquid nitrogen and store at −80°C for long-term use. It is strongly suggested to proceed to purification steps in the same day to minimize degradation and/or unfolding of engineered proteins.
6. To monitor the protein expression and purification process, one is advised to collect samples at each step and analyze them with SDS-PAGE. For example, we always keep nine samples for SDS-PAGE analysis: cells before induction, cells after induction (1.5 and 3 h), cell pellets after sonication, supernatant before binding, beads before binding, beads after binding, flow-through, and eluent.
7. To ensure that the GST fusion proteins bind to the beads, one can take 20 μL of beads and mix it with 100 μL of 1× Bio-Rad Protein Assay solution. If the solution turns blue, the beads should have bound proteins.
8. To monitor the cleavage and elution efficiency, we usually keep appropriate samples (before cleavage, after cleavage (1 and 16 h), eluent, and beads after cleavage) for SDS-PAGE.
9. The quartz cuvette is stored in 2% HNO_3 solution to remove residual Tb^{3+} bound to the wall. Make sure that the cuvette is completely immersed into the storage solution.
10. It is hard to adjust the excitation and emission slit width without knowing the saturated fluorescence signal. Based on experience, the fluorescence may increase less than 100-fold of the initial intensity after Tb^{3+} signal is saturated. The safest way is to do a pilot experiment to find out the optimal excitation and emission slit widths.
11. It is important to ensure that pH is maintained at ~6.8. Tb^{3+} tends to precipitate if the pH is over 7.0. The precipitation of Tb^{3+} will cause significant shift of background signal and obscure the explanation of titration data.
12. For the Ca^{2+} competition assay, the ratio of Tb^{3+} over protein has to be optimized for each protein. As a rule of thumb, the ideal Tb^{3+} concentration should be at a level of reaching ~80% of the maximal signal in the Tb^{3+} titration curve.

13. The 15-min interval between each titration point is essential for reaching full equilibrium between metals and protein. Due to competitive binding of Ca^{2+} to the protein, Tb^{3+} will be competed out of the binding pocket. As a result, one is expected to observe the gradual decrease of luminescence signal until it reaches saturation.

Acknowledgments

We appreciate the critical review by Natalie White. This work is supported in part by research from NIH GM081749 to Jenny J. Yang.

References

1. Berridge MJ, Bootman MD, Roderick HL (2003) Calcium signalling: dynamics, homeostasis and remodelling. Nat Rev Mol Cell Biol 4:517–529
2. Zhou Y, Frey TK, Yang JJ (2009) Viral calciomics: interplays between Ca^{2+} and virus. Cell Calcium 46:1–17
3. Chami M, Oules B, Paterlini-Brechot P (2006) Cytobiological consequences of calcium-signaling alterations induced by human viral proteins. Biochim Biophys Acta 1763:1344–1362
4. Kretsinger RH, Nockolds CE (1973) Carp muscle calcium-binding protein. II. Structure determination and general description. J Biol Chem 248:3313–3326
5. Yanyi C, Shenghui X, Yubin Z, Jie YJ (2010) Calciomics: prediction and analysis of EF-hand calcium binding proteins by protein engineering. Sci China Chem 53:52–60
6. Zhou Y, Yang W, Kirberger M, Lee HW, Ayalasomayajula G, Yang JJ (2006) Prediction of EF-hand calcium-binding proteins and analysis of bacterial EF-hand proteins. Proteins 65:643–655
7. Zhou Y, Tzeng WP, Yang W, Ye Y, Lee HW, Frey TK, Yang J (2007) Identification of a Ca2+-binding domain in the rubella virus nonstructural protease. J Virol 81:7517–7528
8. Ye Y, Lee HW, Yang W, Yang JJ (2005) Calcium and lanthanide affinity of the EF-loops from the C-terminal domain of calmodulin. J Inorg Biochem 99:1376–1383
9. Ye Y, Lee HW, Yang W, Shealy SJ, Wilkins AL, Liu ZR, Torshin I, Harrison R, Wohlhueter R, Yang JJ (2001) Metal binding affinity and structural properties of an isolated EF-loop in a scaffold protein. Protein Eng 14:1001–1013
10. Horrocks WD Jr (1993) Luminescence spectroscopy. Methods Enzymol 226:495–538
11. Brittain HG, Richardson FS, Martin RB (1976) Terbium (III) emission as a probe of calcium(II) binding sites in proteins. J Am Chem Soc 98:8255–8260
12. Ikura M, Clore GM, Gronenborn AM, Zhu G, Klee CB, Bax A (1992) Solution structure of a calmodulin-target peptide complex by multidimensional NMR. Science 256:632–638

Part II

Expression, Purification, Biochemical, and Structural Analyses and Cellular Localization

Chapter 4

New Aspects of Calmodulin–Calmodulin Binding Domains Recognition

Emilie Audran, Rania Dagher, Sophie Gioria, Marie-Claude Kilhoffer, and Jacques Haiech

Abstract

Understanding the role of calmodulin (CaM) in calcium signal transduction implies to describe the calcium-dependent molecular mechanism of interaction of CaM with the various CaM-binding domains (CBD). In order to fulfill this aim, we have developed a new strategy and the afferent techniques to quantify the interaction of CaM with any CBD as a function of calcium concentration. Excel software has been used to deconvolute the experimental data and to obtain the macroscopic constants characterizing the system. We are illustrating our approach on six different CaM/CBD. This strategy may be used to analyze the interaction between any calcium-binding protein and its targets.

Key words: Calcium, CaM, CaM-binding domain, Calcium-dependent interaction, Anisotropy, EF-hand

1. Introduction

Calcium is a universal messenger [1] found in all the reigns from archae to mammals. Calcium signal has been described and analyzed in numerous reports and by thousands of teams all around the world. Each "club of calcium aficionados" has built a specific jargon. Therefore, the same word may recover several meanings or a same concept may be described by many locutions. Following the pioneering work of Michael Berridge, we have tried to revisit the calcium semantics in order to lay a calcium Esperanto [2, 3].

Claus W. Heizmann (ed.), *Calcium-Binding Proteins and RAGE: From Structural Basics to Clinical Applications*, Methods in Molecular Biology, vol. 963, DOI 10.1007/978-1-62703-230-8_4, © Springer Science+Business Media New York 2013

Calcium signal may be divided into several steps:

1. A coding mechanism.
2. A decoding mechanism.
3. A transduction step leading to a cellular event.

The first step implies information coding which occurs through the shaping of a calcium signal, meaning a transitory and spatially localized calcium concentration modulation. Calcium concentration homeostasis is due to an exquisite equilibrium between calcium inflow and outflow. Calcium inflow is modulated through calcium channels whereas calcium outflows by means of calcium pumps and exchangers. The second step, the decoding mechanism, implies a set of calcium sensors (calcium-binding proteins) that are capable to detect the characteristics of the calcium signal, namely, the calcium signature. The third step occurs when a calcium-binding protein interacts with target proteins, the first step in the transduction of the signal which leads to the cellular event.

Calmodulin (CaM) is the stereotypic member of the calcium sensors family. The protein possesses four calcium-binding sites that are strongly coupled [4].

1.1. CaM Regulation of CaM Target Proteins and Domains

In 1975, Brostrom et al. [5] reported that this protein was an activator of brain adenylate cyclase. Results obtained thereafter, in the late seventies, indicated that CaM undergoes a very large conformational change after binding of two Ca^{2+} ions, under conditions where there is an apparent positive cooperativity between the first two sites occupied by calcium. Subsequently other targets have been identified such as the carrier Ca^{2+}-ATPase [6, 7], the kinase myosin light chain (MLCK) [8], phosphorylase kinase [9], and calcineurin [10]. Bovine brain CaM was the first to be sequenced by Watterson and colleagues [11]. Finally, the first crystallization of CaM was obtained in 1980 [12, 13].

In an effort to integrate the multiple regulatory activities of CaM [4] the concept of calcisome, a supramolecular complex carrying a cellular function regulated by calcium, was proposed in 1980 [3].

CaM can interact at the cellular level with more than 300 variable target sequences to regulate biological functions in response to calcium stimulation [14]. The interaction of CaM and its targets can be achieved in a Ca^{2+}-dependent or -independent way. Previous work [15–17] concerning the mechanism of target enzyme activation stressed the importance of the structural integrity of CaM for optimal activation. This observation can be correlated with the remarkable conservation of CaM during evolution. Today, more

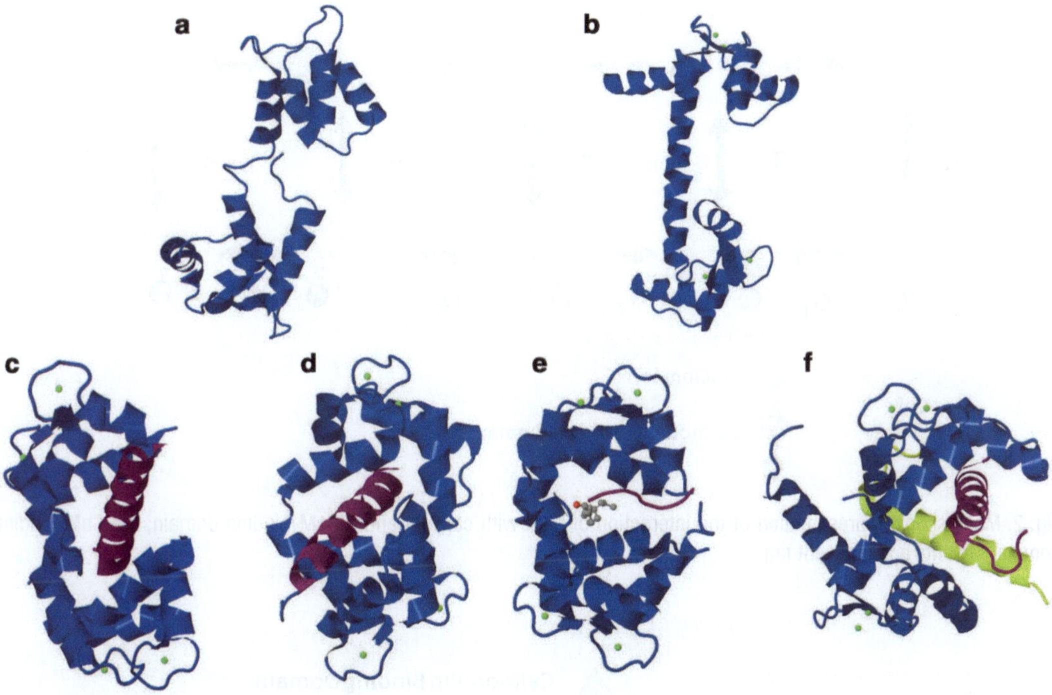

Fig. 1. NMR structure resolutions of CaM (*blue*) in complex with peptides representing the CaM-binding domain of CaM-binding proteins (*pink and green*). (**a**): CaM in its apo form (PDB entry 1CFD). (**b**): CaM loaded with calcium (PDB entry 3CLN). CaM in complex within: (**c**): alphaII-spectrin CaM-binding domain (PDB entry 2FOT), (**d**): CaM-dependent kinase I CaM-binding domain (PDB entry 2L7L), (**e**): CAP-23/NAP-22 CaM-binding domain (PDB entry 1L7Z), (**f**): a dimer of peptides from the Glutamate Decarboxylase CaM-binding domain (PDB entry 1NWD).

than 20 structures of CaM bound to different targets have been resolved. These include:

1. Two structures for the apo form.
2. Two structures with partial calcium occupancy.
3. 16 structures for the holo form.

These structures are used to compare the relationships between the different binding modes and target activation. Figure 1 presents the structures of the complexes between CaM and CaM-binding domains (CBD) that are studied in this chapter.

The interaction of CaM with a CBD cannot be described only on a structural basis and needs to be coupled with calcium binding to the protein (Fig. 2). We have therefore developed a strategy to decipher the calcium dependency of the interaction between a CBD and CaM.

The CBD is a peptide of about 20 amino acids tagged with a fluorescent molecule. To follow the binding of the fluorescent CBD, we are using fluorescence anisotropy (see next section for a primer on anisotropy).

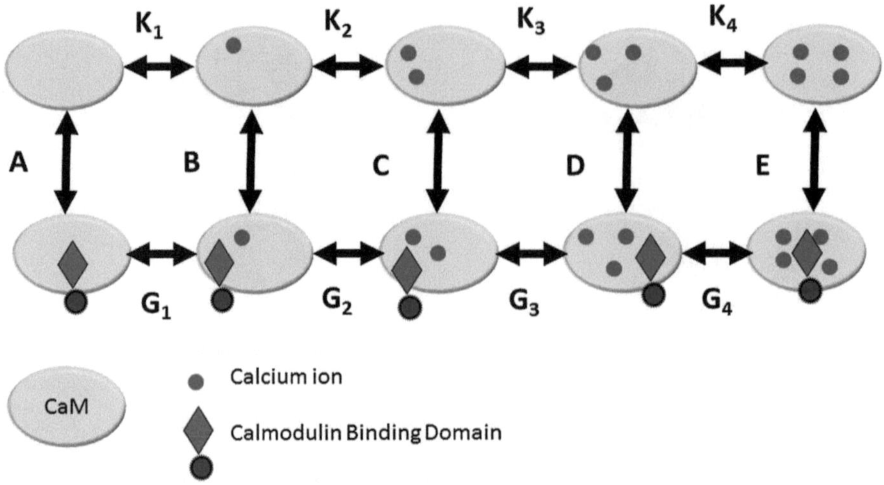

Fig. 2. Macroscopic representation of the interaction of CaM with calcium and a CaM-binding domain; the CaM-binding domain presents a fluorescent tag.

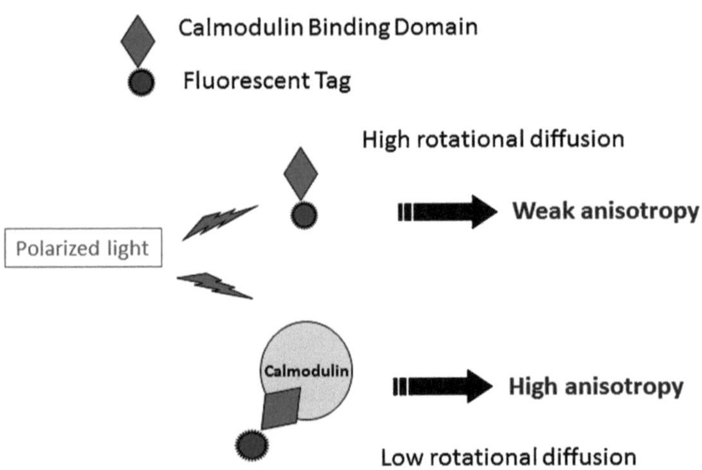

Fig. 3. Schematic representation of an experiment using anisotropy to follow the binding of a CaM-binding domain to CaM.

CBD binding to CaM is followed at different free calcium concentrations. Free calcium concentration is maintained in the experimental buffer by using EGTA calcium buffers (see Subheadings 2 and 3).

The macroscopic constants $\mathbf{K}_1$, $\mathbf{K}_2$, $\mathbf{K}_3$, and $\mathbf{K}_4$ describing the calcium binding to CaM are determined using microcalorimetry calcium titration in the experimental medium [18].

For a given free calcium concentration, we are titrating CaM with the fluorescent CBD using anisotropy. Such titration allows determining a macroscopic binding association constant (Kapp). This association constant is a function of free calcium. By following its variation, we are able to determine the other constants characterizing the system presented in Fig. 2 (namely, $\mathbf{G_1}$, $\mathbf{G_2}$, $\mathbf{G_3}$, $\mathbf{G_4}$, **A**, **B**, **C**, **D**, and **E**) and described by the following equation:

$$\mathrm{Kapp} = \frac{1 + K_1x + K_1K_2x^2 + K_1K_2K_3x^3 + K_1K_2K_3K_4x^4}{A(1 + G_1x + G_1G_2x^2 + G_1G_2G_3x^3 + G_1G_2G_3G_4x^4)},$$

where x is the free calcium concentration.

1.2. Fluorescence Anisotropy to Analyze the Interaction

The fluorescence polarization technique is a very useful tool for equilibrium binding studies when the free and bound species involved in the equilibrium have different rotational rates.

Fluorescent molecules absorb light in a given direction and emit it in another direction. When a population of fluorescent molecules is excited with polarized light (characterized by parallel and perpendicular components), molecules oriented in a direction close to the light orientation absorb in a preferential manner. If an excited molecule rotates during its excited state, the direction of the emitted light will be deviated: light is depolarized and the polarization degree, or anisotropy, is low. As a consequence, the rapid rotation of small fluorescent molecules in solution results in a loss of light polarization and the polarization degree depends on the extent to which the molecules rotate between excitation and emission. If such molecules bind to larger molecules, their rotation is slowed down and the polarization degree is thus increased, i.e., the polarization or anisotropy is high (Fig. 3) [18].

2. Materials

Prepare all aqueous solutions using ultrapure water (Milli-Q apparatus, Millipore) and analytical grade reagents in decalcified glassware or tubes; use water stored in decalcified glassware to prepare solutions. Prepare and store all reagents at room temperature (unless indicated otherwise).

2.1. Decalcification

For a 50 ml tube, add 5 ml EGTA 10 mM, pH 8, mix by inverting the tube five times, and wash the tube with water; repeat this step two times.

2.2. Chemicals

All chemicals were obtained from commercial suppliers and used without further purification. DMSO, KCl, Tris, MOPS, and EGTA were purchased from Sigma; Trichloroacetic acid was purchased from Fluka Biochemika. K_2EGTA and CaEGTA 10 mM solutions were purchased from Invitrogen (calcium calibration buffer kit).

Table 1
List of sequences of synthetic peptides. All peptides were labeled at their N-terminus with carboxytetramethylrhodamine. Peptides were purchased from Schafer-N (Copenhagen, Denmark)

Peptide	Sequence
alphaII-spectrin	SPWKSARLMVHTVATFNSI
CaM-dependent protein kinase I	IKKNFAKSKWKQAFNATAVVRHMRK
Neuromodulin	TKIQASFRGHITRKKLKGE
CAP-23/NAP-22 (neuronal axonal membrane protein, brain acid soluble protein)	GGKLSKKKKGYNVNDEKAKE
Glutamate decarboxylase (GAD)	HKKTDSEVQLEMITAWKKFVEEKKKK
Ca^{2+}-CaM-dependent kinase kinase alpha	VKLIPSWTTVILVKSMLRKRSFGNPF

2.3. Proteins and Peptides

Assays are carried out with human CaM purchased from ProtEra (Sesto Fiorentino, Italy); CaM is received at 1.4 mg/ml in 20 mM Tris–HCl, pH 8.0, 150 mM NaCl, 2 mM CaCl2. Synthetic peptides labeled at their N-terminus with carboxytetramethylrhodamine were purchased from Schafer-N (Copenhagen, Denmark) (sequences in Table 1); stock solutions are prepared at 10 mM in DMSO and stored at –20°C.

2.4. Analysis Buffers

MOPS buffer: 30 mM MOPS, 100 mM KCl, pH 7.2:

1. Weigh 3.14 g MOPS and 3.73 g KCl in a glass beaker, add 450 ml water, and mix.
2. Adjust pH to 7.2 with NaOH 1 N solution and make up to 500 ml with water.
3. Filter in a glass bottle and store at +4°C.

Tris solution, 1 M:

1. Weigh 6.06 g Tris in a 50 ml tube and add 50 ml water. Mix and store at room temperature.

Trichloroacetic acid solution, 1 g/ml:

1. Weigh 0.3 g in a 1 ml polypropylene tube and add 300 μl water.

EGTA 10 mM solution:

1. Weigh 1.9 g EGTA in a glass beaker, add 450 ml water, and mix.

2. Adjust pH to 8.0 with NaOH 1 N solution and make up to 500 ml with water.
3. Filter in a glass bottle and store at +4°C.

3. Methods

Carry out all procedures at room temperature unless otherwise specified. CaM dilutions are calculated to evaluate the binding of six peptides to CaM at different free Ca^{2+} concentrations.

3.1. Buffers (see Note 4)

1. Transfer 50 ml of MOPS buffer in a 50 ml tube and check pH (pH 7.2).
2. Transfer 19 ml of K_2EGTA 10 mM and 22 ml of CaEGTA 10 mM in a 50 ml tube. Equilibrate the solution to room temperature and check pH (pH 7.2).

Calcium calibration

To obtain the series of solutions at given free Ca^{2+} concentrations, calcium calibration solutions are mixed in varying proportions. The free Ca^{2+} concentrations are determined from

$$[Ca^{2+}] = K_d^{EGTA} \frac{CaEGTA}{K_2EGTA},$$ where K_d^{EGTA} is the dissociation constant of EGTA for Ca^{2+} in the experimental medium.

Prepare 4 ml of each calcium buffer:

3. Label eight 15 ml tubes with calcium concentrations: 0, 0.017, 0.038, 0.1, 0.6, 1.35, 5, 10 μM.
4. Perform serial dilutions by adding CaEGTA 10 mM solution to previous calcium buffer as described in Table 2 (see Note 7).

Table 2
Calcium buffer preparation by serial dilutions. Starting from tube 1, buffer in tube *n* is prepared by mixing the indicated volume of the 10 mM CaEGTA solution with the indicated volume of buffer taken from the tube *n*−1. Tube 0 corresponds to the K_2EGTA solution. In each tube, the final volume will be at least 4 ml

Tube number (*n*)	1	2	3	4	5	6	7	8
Calcium (μM)	0	0.017	0 .038	0.1	0.6	1.35	5	10
Buffer volume (μl) from tube *n*−1	14,584	10,584	7,760	4,725	2,300	2,900	1,800	2,000
10 mM CaEGTA volume (μl)	0	1,176	970	1,575	4,600	2,900	4,200	2,000

3.2. CaM Preparations (see Note 1)

In K_2EGTA and CaEGTA buffers:

1. Thaw one CaM solution tube (1 ml).
2. Add 30 μl of trichloroacetic acid solution, spin down for 2 min in a lab top centrifuge at 10,000 rpm (6,700 × *g*), and eliminate the supernatant with a pipette. Add 30 μl Tris 1 M and 970 μl water to the precipitate and dissolve by pipetting up and down. Repeat this step two times.
3. Separate the pellet in two parts: pipet 500 μl in one 1 ml tube and 500 μl in another 1 ml tube.
4. Add 15 μl of trichloroacetic acid solution in each tube, spin down for 2 min in a lab top centrifuge at 10,000 rpm, and eliminate the supernatant with a pipette.
5. Wash the two pellets by adding slowly 100 μl water, make it pass on the entire surface, and remove it. Repeat this step two times.
6. Resuspend one of the pellets in 100 μl K_2EGTA 10 mM solution and the other one in 100 μl CaEGTA 10 mM solution.

Dilution in K_2EGTA and CaEGTA buffers:

7. CaM spectroscopic measurement in K_2EGTA and CaEGTA solutions: Record steady-state absorption spectra on a Nanodrop spectrophotometer (Thermo Scientific) to determine the protein concentration. Carry out measurement at 280 nm using a molecular extinction coefficient of 2,560 M^{-1} cm^{-1} for CaM. Correct spectra for background before determining the concentration of the solution.
8. Transfer the appropriate volume of each CaM solution to 2 ml tubes to prepare 1,600 μl of CaM at 20 μM in CaEGTA and 1,350 μl of CaM at 20 μM in K_2EGTA.
9. Transfer and mix CaM exceeding solutions in a 1.5 ml tube and add 12 μl of 0.1 M $CaCl_2$ solution. Store the resulting saturated CaM solution at −20°C for further use (see Note 5).

CaM in calcium buffers:

Prepare 320 μl of each solution. CaM 20 μM in calcium buffers are prepared by cascade dilutions with the "CaM 20 μM/K_2EGTA 10 mM" and "CaM 20 μM/CaEGTA 10 mM" solutions.

10. Label eight 500 μl tubes with calcium concentrations (0, 0.017, 0.038, 0.1, 0.6, 1.35, 5, 10 μM).
11. Perform serial dilutions by adding the "CaM 20 μM/CaEGTA 10 mM" solution to the previous "CaM 20 μM/calcium buffer" solution as indicated in Table 3.

Table 3
CaM preparation in calcium buffers. Starting from tube 1, solution in tube *n* is prepared by mixing the indicated volume of "CaM 20 μM/CaEGTA 10 mM" solution with the indicated volume of "CaM 20 μM/calcium buffer" solution prepared taken from tube *n* – 1. Tube 0 corresponds to the CaM 20 μM/K_2EGTA 10 mM solution. In each tube, the final volume is at least 320 μl

Tube number (*n*)	1	2	3	4	5	6	7	8
Calcium (μM)	0	0.017	0.038	0.1	0.6	1.35	5	10
Solution from tube *n* – 1 (μl)	1,175	855	624	378	184	232	144	160
CaM 20 μM/10 mM CaEGTA (μl)	0	95	78	126	368	232	336	160

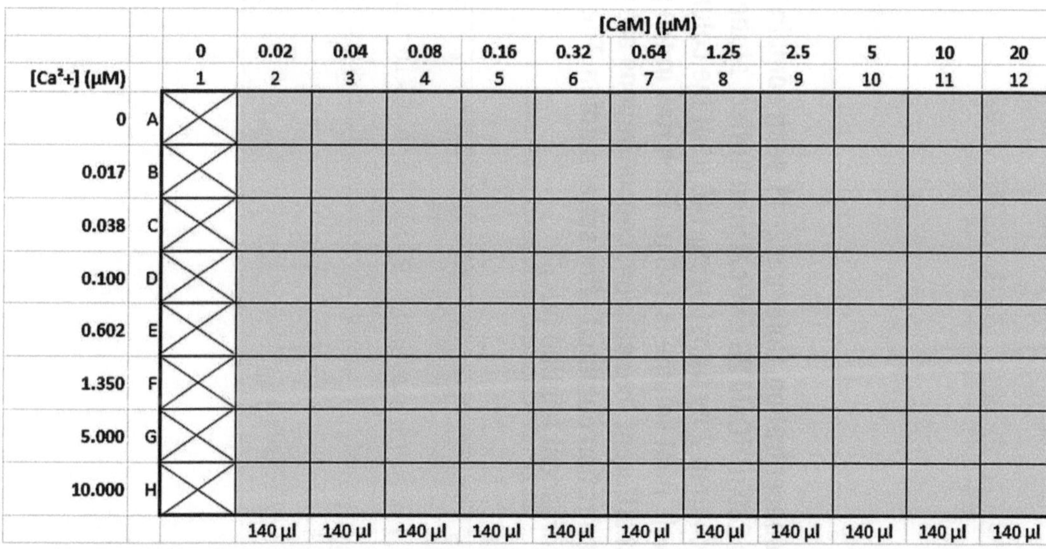

Fig. 4. "CaM/calcium buffers" plate plan. Column 1 is empty, columns 2–12 contain CaM at different concentrations in different calcium buffers; final volume is 140 μl per well.

CaM serial dilutions in calcium buffers:

Prepare a 96-well plate (Greiner 650201) as follows and according to the "CaM/calcium buffers" plate plan (Fig. 4):

12. Add 140 μl calcium buffers in columns 2–11, one calcium buffer in each row.
13. Add 280 μl CaM 20 μM in calcium buffer in column 12, one "CaM 20 μM/calcium buffer" solution per row.
14. Perform serial dilution from column 12 to column 2: Transfer 140 μl solution to the next one as described in Table 4.
15. Eliminate 140 μl solution from column 2 (final volume 140 μl).

In each well, the final volume is 140 μl (see Note 6).

Table 4
Serial CaM dilutions in one given free calcium concentration. Columns 2–11 each contain 140 μl calcium buffer at a given free calcium concentration and column 12 contains 280 μl "CaM 20 μM/calcium buffer" at the same free calcium concentration. The serial dilution is performed from column 12 to column 2 by transferring 140 μl of the column *n* to the column *n*−1; in the last column (*n*=2), 140 μl are eliminated. In each well, final volume is 140 μl. Each row in the plate gives a serial dilution of CaM at one given free calcium concentration. The eight different free calcium concentrations are present in the eight rows of the plate. The prepared [CaM] stands for the [CaM] concentration obtained after serial dilution. The final [CaM] corresponds to the working concentration obtained after addition of the fluorescent peptide (v/v)

Column (n)	12	11	10	9	8	7	6	5	4	3	2
Final [CaM] (M)	10×10^{-6}	5×10^{-6}	2.5×10^{-6}	1.25×10^{-6}	0.62×10^{-6}	0.31×10^{-6}	0.15×10^{-6}	0.08×10^{-6}	0.04×10^{-6}	0.02×10^{-6}	0.01×10^{-6}
Prepared [CaM] (M)	20×10^{-6}	10×10^{-6}	5×10^{-6}	2.5×10^{-6}	1.25×10^{-6}	0.62×10^{-6}	0.31×10^{-6}	0.15×10^{-6}	0.08×10^{-6}	0.04×10^{-6}	0.02×10^{-6}
Initial volume (μl)	280	140	140	140	140	140	140	140	140	140	140
Previous solution (μl)		140	140	140	140	140	140	140	140	140	140

3.3. Peptide Preparation (See Note 1)

1. Peptide spectroscopic measurements: Record steady-state absorption spectra on a Nanodrop spectrophotometer (Thermo Scientific) to determine each peptide concentration. Carry out measures at 565 nm using molecular extinction coefficients of 103,000 M^{-1} cm^{-1} and correct spectra for background before determining the solution concentrations.
2. Dilute each peptide in DMSO to 1×10^{-4} M using 1 μl peptide stock solution.
3. From the 1×10^{-4} M solution, dilute each peptide in MOPS buffer to 2×10^{-5} M by adding 5 μl of the previous dilution in 20 μl MOPS buffer (final volume 25 μl).
4. Dilute each peptide to 2×10^{-7} M solution in the eight previously prepared calibrated calcium buffers (K_2EGTA/CaEGTA solutions): Add 2.4 μl of the 2×10^{-5} M peptide solution to 237.6 μl of each of the calcium buffers (final volume: 240 μl per calcium buffer).
5. In a 96-well plate (Greiner 650201), add 240 μl MOPS buffer in column 1 and 205 μl peptide in calcium buffer solutions according to the plate plan (Fig. 5).

3.4. CaM/Peptide Binding in Calcium Buffers and Reading

For each peptide, two plates are prepared as duplicates.

1. In 12 costar ½ well plates (Corning Costar 3686, Acton, MA), named as assay plates, distribute robotically (BiomeK 2000, Beckman Coulter) the prepared peptide solutions as follows and according to the final plate plan (Fig. 6):
   (a) 15 μl from column 1 of the "peptide/calcium buffer" plate to column 1 of each assay plates

[Ca^{2+}] (μM)		1	2	3	4	5	6	7	8	9	10	11	12
0	A		Peptide 1	Peptide 2	Peptide 3	Peptide 4	Peptide 5	Peptide 6					
0.017	B		Peptide 1	Peptide 2	Peptide 3	Peptide 4	Peptide 5	Peptide 6					
0.038	C		Peptide 1	Peptide 2	Peptide 3	Peptide 4	Peptide 5	Peptide 6					
0.100	D		Peptide 1	Peptide 2	Peptide 3	Peptide 4	Peptide 5	Peptide 6					
0.602	E		Peptide 1	Peptide 2	Peptide 3	Peptide 4	Peptide 5	Peptide 6					
1.350	F		Peptide 1	Peptide 2	Peptide 3	Peptide 4	Peptide 5	Peptide 6					
5.000	G		Peptide 1	Peptide 2	Peptide 3	Peptide 4	Peptide 5	Peptide 6					
10.000	H		Peptide 1	Peptide 2	Peptide 3	Peptide 4	Peptide 5	Peptide 6					
		240 μl	205 μl	205 μl	205 μl	205 μl	205 μl	205 μl					

Fig. 5. "Peptides/calcium buffers" plate plan. Column 1 contains 240 μl MOPS buffer and columns 2–7 contain 205 μl of one peptide at 2×10^{-7} M in different calcium buffers.

		[CaM] (μM)											
		0	0.01	0.02	0.04	0.08	0.16	0.32	0.64	1.25	2.5	5	10
[Ca^{2+}] (μM)		1	2	3	4	5	6	7	8	9	10	11	12
0	A												
0.017	B												
0.038	C												
0.100	D												
0.602	E												
1.350	F												
5.000	G												
10.000	H												
		15 μl	7.5 μl "CaM / calcium buffer" + 7.5 μl "Peptide / calcium buffer"										

Fig. 6. "Assay plate" plan. Peptide fluorescence anisotropy will be evaluated as a function of CaM concentration in different calcium buffers; final volume is 15 μl per well. Plates are made in duplicate for each peptide.

(b) 7.5 μl from column 2 of the "peptide/calcium buffer" plate to columns 2 to 12 of 2 assay plates

(c) 7.5 μl from column 3 of the "peptide/calcium buffer" plate to columns 2 to 12 of 2 assay plates

(d) 7.5 μl from column 4 of the "peptide/calcium buffer" plate to columns 2 to 12 of 2 assay plates

(e) 7.5 μl from column 5 of the "peptide/calcium buffer" plate to columns 2 to 12 of 2 assay plates

2. Distribute robotically (Biomek FX, Beckman Coulter) the 7.5 μl of the prepared CaM solutions in each assay plate according to the same plate plan as for the "CaM/calcium buffers" plate.
3. Mix briefly the plate and incubate for 5 min at room temperature (see Note 3).
4. Place the plate in a Victor 3 apparatus (Perkin–Elmer Life and Analytical Sciences, Boston, MA) at 25°C and measure polarization degrees (FP) with an excitation wavelength set at 544 nm (bandwidth 15 nm) and an emission wavelength set at 600 nm (bandwidth 10 nm). For each plate, FP-measurements are performed with control wells containing the unbound fluorescent peptide (CaM, 0 μM) (see Notes 8 and 9).

3.5. Data Analysis

The polarization degree is defined by the equation $FP = (I_{//} - I_{\perp})/(I_{//} + I_{\perp})$, where FP is the fluorescence polarization degree and $I_{//}$ and $I_{\perp}$ are the fluorescence intensities of the vertically (//) and horizontally (⊥) polarized emissions, when the sample is excited with

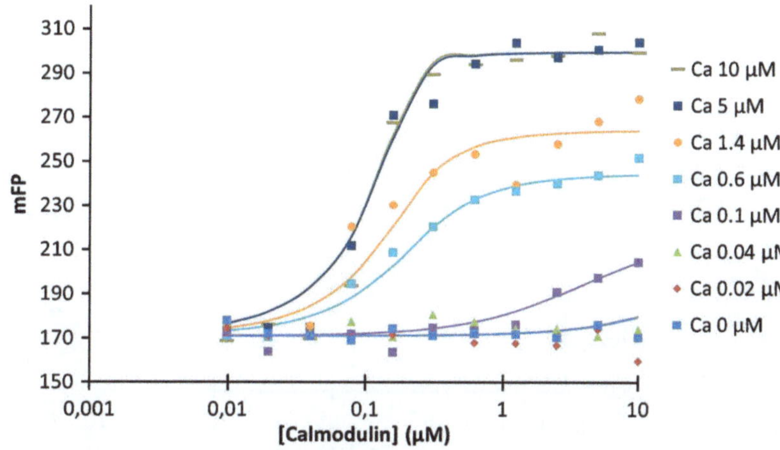

Fig. 7. Peptide Matrix for Alpha II spectrin CaM-binding domain. FP values are expressed as a function of [CaM] in each calcium buffer.

vertically polarized light. For each plate, background correction is performed by subtracting blank parallel and perpendicular components (means of eight wells) from the respective intensities. Total fluorescence of each well is calculated by $FT = I_{//} + 2\,I_{\perp}$ and mean of duplicate plates is calculated for both FP and FT.

Fluorescence polarization measurements are often expressed in anisotropy because these values are additive. Anisotropy values (A) are obtained from polarization (FP) as follows: $A = \frac{2FP}{3 - FP}$

The equilibrium dissociation constant (K_d) of the peptide binding to CaM, in each calcium buffer solution, is calculated by fitting the sigmoid dose-dependent curve as described in Dagher et al. [19]. This dose–response curve is corrected by the Q factor expressed as the median of FT value variation depending on CaM concentration in each calcium buffer: $Q = \text{median}\,((FT_{\text{bound peptide}}/FT_{\text{unbound peptide}}\ \text{in calcium buffer i})_{(i=1-8)})$ (Fig. 7). Data are adjusted with the Microsoft Excel Solver tool (excel folder could be obtained upon request from the authors).

CaM interaction with fluorescent peptides is studied as a function of calcium bound to the protein. Data are analyzed according to the Adair-Klotz scheme (Fig. 2) to allow fitting of all binding curves. This model allows the determination of A, B, C, D, and E values representing the peptide association constants to the different CaM complexes, respectively, CaM, CaCaM, Ca_2CaM, Ca_3CaM, and Ca_4CaM (deduced from Kapp expression, see calculation details in ref. 19): This experiment allows the distribution of CaM-binding peptides into four groups depending on the CaM-Ca_n ($n = 1, 2, 3,$ or 4) recognized by the peptides (Fig. 8 and Table 5).

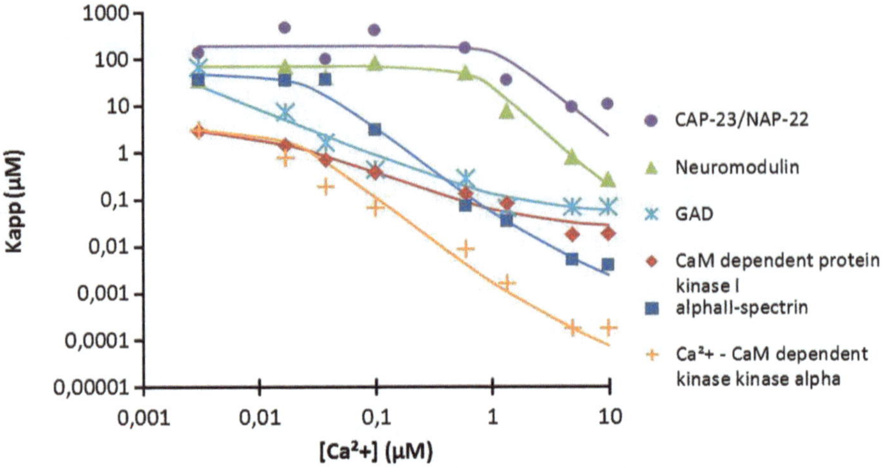

Fig. 8. Apparent dissociation constants of CAP-23/NAP-22 (neuronal axonal membrane protein, brain acid soluble protein), Neuromodulin, Glutamate Decarboxylase (GAD), CaM-dependent protein kinase I, Alpha II spectrin, and calcium CaM-dependent kinase kinase alpha titrated on CaM as a function of calcium concentration. Assays were conducted using calcium calibration buffers ranging from 0 to 10 mM. Fluorescent peptides were used at 10 µM final concentration and CaM final concentrations ranged from 0 to 10 µM. The equilibrium dissociation constants (Kapp) of peptides for CaM in each calcium buffer solution were determined and Kapp values were represented as a function of the free calcium concentrations.

Table 5
Dissociation constants relative to the interaction of CaM with CAP-23/NAP-22 (neuronal axonal membrane protein, brain acid soluble protein), Neuromodulin, Glutamate Decarboxylase (GAD), CaM-dependent protein kinase I, Alpha II spectrin, and calcium CaM-dependent kinase kinase alpha at different Ca^{2+} saturation degrees. The association constants (A, B, C, D, and E) were determined from peptide/CaM-binding curves at different calcium concentrations; A^{-1}, B^{-1}, C^{-1}, D^{-1}, and E^{-1} are the peptide dissociation constants for CaM, CaCaM, Ca_2CaM, Ca_3CaM, and Ca_4CaM, respectively. Fluorescent peptides were used at 10 µM final concentration and CaM final concentrations ranged from 0 to 10 µM; titrations were performed at fixed calcium concentrations. Data were fitted according to the Adair-Klotz scheme (Fig. 2)

CaM-binding protein	A^{-1} (µM)	B^{-1} (µM)	C^{-1} (µM)	D^{-1} (µM)	E^{-1} (µM)	Ca^{2+} number
CAP-23/NAP-22	1.92E+02	1.92E+02	1.92E+02	2.07E-01	2.07E-01	3
Neuromodulin	6.99E+01	6.99E+01	6.99E+01	1.78E-02	1.78E-02	3
GAD	6.86E+02	5.35E-02	5.35E-02	5.35E-02	5.35E-02	1
CaM-dependent protein kinase I	3.82E+00	2.50E-02	2.50E-02	2.50E-02	2.50E-02	1
Alpha II-spectrin	4.94E+01	4.94E+01	6.78E-04	6.78E-04	6.78E-04	2
Ca^{2+}-CaM-dependent kinase kinase alpha	3.20E+00	3.20E+00	2.13E-05	2.13E-05	2.13E-05	2

In our experimental conditions, different CBD bind to different calcium–CaM complexes and recognize CaM bound to 1, 2, or 3 calcium ions. Moreover, it is interesting to notice that the CaM-dependent kinases exhibit CBD able to bind the apoCaM in the μmolar range (the dissociation constant of the apoform of CaM for CaM-dependent protein kinase I and for CaM-dependent kinase kinase alpha is equal to 3.82 μM and 3.2 μM, respectively). The results suggest that CaM and those kinases may form a complex in the cell even at resting calcium concentration.

4. Notes

1. CaM solution may be stored at –20°C in the presence of calcium. When decalcified, CaM must be used readily (the same day).
2. The fluorescent peptide solutions have to be kept away from light (keep tubes wrapped in aluminum foil).
3. The plates have to be covered with aluminum foil to avoid light.
4. Calcium buffers are extremely sensitive to pH. It is important to accurately measure the pH of the solutions.
5. For CaM preparation, polyethylene low binding tubes are used (Trefflab réf 96.8180.9.02).
6. When working in the absence of calcium buffers, decalcified CaM solution needs to be handled with materials (tubes, tips) previously treated to remove calcium. This constraint is relieved when using calcium buffers.
7. The different buffers may be used within 1 month.
8. The plates may be kept for at least 1 day. The readings are stable for at least 24 h.
9. Check for air bubbles before plate readings. When present, air bubbles are removed by centrifuging the plate for 1 min at 1,000 rpm.

Acknowledgments

This work has been supported by grant from ANR CAPHE and by CNRS and University of Strasbourg (France). This work has been performed at the technological platform PCBIS (UMS CNRS 3286). S. G is member of this platform.

References

1. Haiech J, Moreau M (2011) The calcium signal: a universal carrier to code, decode and transduce information. Biochimie 93:v
2. Berridge MJ (2004) Calcium signal transduction and cellular control mechanisms. Biochim Biophys Acta 1742:3–7
3. Haiech J, Audran E, Feve M, Ranjeva R, Kilhoffer MC (2011) Revisiting intracellular calcium signaling semantics. Biochimie 93:2029–2037
4. Haiech J, Klee C, Demaille J (1981) Effects of cations on affinity of calmodulin for calcium-ordered binding of calcium ions allows the specific activation of calmodulin-stimulated enzymes. Biochemistry 20:3890–3894
5. Brostrom CO, Huang YC, Breckenridge BM, Wolff DJ (1975) Identification of a calcium-binding protein as a calcium-dependent regulator of brain adenylate cyclase. Proc Natl Acad Sci U S A 72:64–68
6. Gopinath RM, Vincenzi FF (1977) Phosphodiesterase protein activator mimics red blood cell cytoplasmic activator of (Ca^{2+}-Mg^{2+}) ATPase. Biochem Biophys Res Commun 77:1203–1209
7. Jarrett HW, Penniston JT (1977) Partial purification of the Ca^{2+}-Mg^{2+} ATPase activator from human erythrocytes: its similarity to the activator of 3′:5′-cyclic nucleotide phosphodiesterase. Biochem Biophys Res Commun 77:1210–1216
8. Yazawa M, Yagi K (1977) Calcium-binding subunit of myosin light chain kinase. J Biochem 82:287–289
9. Cohen P, Burchell A, Foulkes JG, Cohen PT (1978) Identification of the Ca2+-dependent modulator protein as the fourth subunit of rabbit skeletal muscle phosphorylase kinase. FEBS Lett 92:287–293
10. Klee CB, Haiech J (1980) Concerted role of calmodulin and calcineurin in calcium regulation. Ann N Y Acad Sci 356:43–54
11. Watterson DM, Sharief F, Vanaman TC (1980) The complete amino acid sequence of the Ca2+-dependent modulator protein (calmodulin) of bovine brain. J Biol Chem 255: 962–975
12. Cook WJ, Dedman JR, Means AR, Bugg CE (1980) Crystallization and preliminary X-ray investigation of calmodulin. J Biol Chem 255:8152–8153
13. Kretsinger RH (1980) Crystallographic studies of calmodulin and homologs. Ann N Y Acad Sci 356:14–19
14. Hoeflich KP, Ikura M (2002) Calmodulin in action: diversity in target recognition and activation mechanisms. Cell 108:739–742
15. Craig TA, Watterson DM, Prendergast FG, Haiech J, Roberts DM (1987) Site-specific mutagenesis of the alpha-helices of calmodulin – effects of altering a charge cluster in the helix that links the 2 halves of calmodulin. J Biol Chem 262:3278–3284
16. Crivici A, Ikura M (1995) Molecular and structural basis of target recognition by calmodulin. Annu Rev Biophys Biomol Struct 24:85–116
17. Chin D, Means AR (2000) Calmodulin: a prototypical calcium sensor. Trends Cell Biol 10: 322–328
18. Dagher R, Pigault C, Bonnet D, Boeglin D, Pourbaix C, Kilhoffer M, Villa P, Wermuth C, Hibert M, Haiech J (2006) Use of a fluorescent polarization based high throughput assay to identify new calmodulin ligands. Biochim Biophys Acta 1763:1250–1255
19. Dagher R, Peng S, Gioria S, Feve M, Zeniou M, Zimmermann M, Pigault C, Haiech J, Kilhoffer MC (2011) A general strategy to characterize calmodulin-calcium complexes involved in CaM-target recognition: DAPK and EGFR calmodulin binding domains interact with different calmodulin-calcium complexes. Biochim Biophys Acta 1813: 1059–1067

Chapter 5

Purification and Characterization of the Human Cysteine-Rich S100A3 Protein and Its Pseudo Citrullinated Forms Expressed in Insect Cells

Kenji Kizawa, Masaki Unno, Hidenari Takahara, and Claus W. Heizmann

Abstract

High quantity and quality of recombinant Ca^{2+}-binding proteins are required to study their molecular interactions, self-assembly, posttranslational modifications, and biological activities to elucidate Ca^{2+}-dependent cellular signaling pathways. S100A3 is a unique member of the S100 protein family with the highest cysteine content (10%). This protein, derived from human hair follicles and cuticles, is characterized by an N-terminal acetyl group and irreversible posttranslational citrullination by peptidylarginine deiminase causing its homotetramer assembly. Insect cells, capable of introducing eukaryotic N-terminus and disulfide bonds, are an appropriate host in which to express this cysteine-rich protein. Four out of ten cysteines in the recombinant S100A3 form two intramolecular disulfide bridges that modulate its Ca^{2+}-affinity. Three free thiol groups located at the C-terminus are predicted to form the high-affinity Zn^{2+}-binding site. Citrullination of specific arginine residues in native S100A3 can be mimicked by site-directed mutagenic substitution of Arg/Ala. This chapter details our procedures used for the purification and characterization of the human S100A3 protein and its pseudo citrullinated forms expressed in insect cells.

Key words: Baculovirus, Disulfide bridge, Insect cell, Peptidylarginine deiminase, S100A3, Calcium-binding, Zinc-binding, EF-hand

1. Introduction

The S100 protein family, with more than 20 members, constitutes the largest subgroup of the EF-hand-type Ca^{2+}-binding proteins (1, 2). S100 proteins are small acidic proteins with two distinct EF-hand-type motifs. S100 proteins are expressed at high levels in various human tissues, such as S100A1 in the heart (3, 4), S100A3 in hair follicles (5), and S100B in the brain (6, 7). The first isolated

Claus W. Heizmann (ed.), *Calcium-Binding Proteins and RAGE: From Structural Basics to Clinical Applications*, Methods in Molecular Biology, vol. 963, DOI 10.1007/978-1-62703-230-8_5, © Springer Science+Business Media New York 2013

S100 proteins were purified from bovine brain and named S100 on the basis of its solubility in 100% ammonium sulfate solution (6). Over the last few decades, several S100 genes belonging to this protein family (21 members in human) have been discovered and each official nomenclature has been updated (1, 8). To elucidate their metal-binding properties, tissue distribution, and biological functions, most human recombinant proteins have been produced in *Escherichia coli* and used for molecular, functional, and structural analyses and for production of specific antibodies.

S100A3 carries the highest cysteine content among all S100 protein family members. For preparation of recombinant S100A3 protein, we used a baculovirus-mediated, non-tagged expression strategy because of its many advantages over other expression systems. Insect cells are capable of introducing eukaryotic modifications and efficiently generating disulfide bonds in recombinant proteins (9). Four out of ten cysteines in S100A3 form two intramolecular disulfide bridges, which inversely modulate Ca^{2+}-affinity (10). Three more free thiol groups are located at the C-terminal region and form the high-affinity $(Cys)_3$His-type Zn^{2+}-binding site. As reported for other S100 proteins (11), the N-terminal methionine of S100A3 derived from human hair cuticles is replaced by an acetyl group (12). This type of N-terminal modification is conferred by the recombinant S100A3 protein in insect cells. A variety of purification strategies have been applied to recombinant S100 proteins. To facilitate protein purification, various types of purification tags (e.g., 6× His) and fused proteins (e.g., glutathione S-transferase, maltose-binding protein) are available (13). Critical studies on molecular structure and function have usually been performed after removal of these purification tags (14). Several surplus linker peptides remaining in the small S100 moiety may affect their assembly into larger structures, as well as their target binding and biological activities. However, using a purification procedure previously optimized with the endogenous products, it is relatively easy to isolate recombinant S100 proteins lacking purification tags (10, 15).

Endogenous S100 proteins, associated with specific cellular signaling and extracellular functions, have various posttranslational modifications, such as glutathionylation of S100A1 (16), citrullination of S100A3 (5), sumoylation of S100A4 (17), nitrosylation of S100A8 (18), carboxymethylation of S100A8/A9 (19), phosphorylation of S100A8/S100A9 (20) and S100A11 (21), and transamidation of S100A11 (22). These modifications are essential for target recognition (e.g., association of phosphorylated S100A11 with nucleolin) (21) or eliciting cytokine-like activities (e.g., chondrocyte activation by transamidated S100A11 via RAGE receptor) (22), but are difficult to reproduce in host cells. Therefore, mutagenic substitutions of Ser/Glu and Lys/Arg + Gln/Asn in recombinant S100 proteins are employed to simulate phosphorylation

(21) and to lose the capacity of transamidation (22), respectively. During hair cuticular differentiation, 1–3 out of 4 arginine residues in S100A3 are converted to citrullines by peptidylarginine deiminases. This naturally occurring and irreversible modification can be simulated by mutagenic amino acid substitutions. Substitution of Arg–51 in S100A3 with Ala can result in assembly of a homotetramer as well as the native form (5). In this chapter, we document protocols for the production of recombinant human S100A3 protein and its pseudo citrullinated forms in insect cells, as well as describe the purification procedures of non-tagged proteins preserving two disulfide bridges.

2. Materials

2.1. Molecular Biology

1. Full-length human S100A3 cDNA was previously cloned into the pMal–c2 expression vector ((14); see Note 1).
2. Advantage 2 PCR enzyme kit (Clontech).
3. QIAEX II (Qiagen).
4. DNA ligation kit ver. 2.1 (Takara).
5. QuickLyze Miniprep kit (Qiagen).
6. QuikChange II site-directed mutagenesis kit (Agilent Technologies).

2.2. Insect Cell Culture

1. *Spodoptera frugiperda*-derived cell line: IPLB–Sf21 cells (included in the BacPAK baculovirus expression system).
2. Complete Grace's medium: Combine 500 mL of Grace's medium (Invitrogen), 56 mL of fetal bovine serum (Cellculture Bioscience), 0.56 mL of gentamicin solution (50 mg/mL; Sigma), and 5.6 mL of amphotericin B solution (250 μg/mL; Sigma; see Note 2).
3. Cell culture flask (225 cm^2; Corning; Becton-Dickinson; see Note 3).

2.3. Recombinant Virus Production (See Note 4)

1. BacPAK baculovirus expression system (Clontech).
2. BacPAK baculovirus rapid titer kit (Clontech).
3. BacPAK Grace's Basic Medium (Clontech).
4. SeaPlaque agarose (Cambrex).

2.4. Recombinant Protein Purification

1. Q-Sepharose Fast Flow (GE-Healthcare).
2. Macro-Prep t-Butyl-HIC Support (Bio-Rad).
3. Specra/Por Dialysis Membrane MWCO: 3500 (Specrumlabs.com).

4. Centrifugal filter unit: Ultracel-3 K (Millipore).
5. HiLoad 16/60 Superdex 30 pg (16 mm×60 mm; GE-Healthcare).

2.5. Nu-PAGE and Western Blot Analyses

1. Nu-PAGE 4–12% Bis–Tris gel (1.0 mm×12 wells; Invitrogen).
2. Nu-PAGE LDS sample buffer (4×; Invitrogen).
3. MES SDS Running buffer (20×; Invitrogen).
4. PVDF membrane filter paper sandwich (Invitrogen).
5. Molecular makers, Mark 12 (Invitrogen), for protein staining.
6. Pre-stained marker, SeeBlue Plus2 (Invitrogen), for Western blot analyses.
7. Bovine serum albumin (BSA) Cohn fraction V, pH 7.0 (Wako Pure Chemical Industry).
8. A polyclonal rabbit anti-S100A3 antibody was previously produced by immunizing a multiple antigen peptide with the C-terminal sequence: LYCHEYF KDCPSEPPCSQ ((23); see Note 5).
9. AlexaFluor 488-labeled goat anti-rabbit IgG (Molecular Probes).

3. Methods

3.1. Construction of an S100A3 Recombinant Transfer Vector

3.1.1. S100A3 cDNA Amplification

1. Combine 40 μL distilled water, 5 μL 10× PCR buffer (Advantage 2 PCR enzyme system), 1 μL S100A3 cDNA template (10 ng/μL), 1 μL (10 μM) each of forward and reverse primers designed as described in Note 6, 1 μL dNTP mix (10 mM), and 1 μL polymerase mix.
2. Create a program for a thermal cycle with the following parameters: 95°C, 1 min; 95°C, 15 s×30 cycles; and 68°C, 3 min.
3. Combine 1 μL (100 ng) of the resultant PCR product or the pBacPAK9 transfer vector (included in BacPAK system) with 2 μL of 10× reaction buffer (final concentration 10 mM Tris–HCl buffer (pH 7.5) containing 7 mM each of $MgCl_2$ and 2-mercaptethanol), 50 mM NaCl, 0.01% BSA, 16 μL of distilled water, and 1 μL (10 U) of EcoR I. Incubate reaction mixture at 37°C for 1 h and then precipitate DNA with 70% ethanol.
4. Dissolve the precipitated DNA in 16 μL of distilled water and add 2 μL of 10× reaction buffer (final concentrations: 10 mM Tris–HCl (pH 7.5), 7 mM $MgCl_2$, 7 mM 2-mercaptethanol, 150 mM NaCl, 0.01% BSA, and 0.01% Triton X-100) and 1 μL (10 U) of Not I. Incubate at 37°C for 1 h.

5. Electrophorese the reaction mixtures using a 1.5% agarose gel for S100A3 cDNA and a 1% agarose gel for the pBacPAK9 vector.
6. Cut out from the gels the bands corresponding to S100A3 cDNA and lineated pBacPAK9 vector.

3.1.2. Purification of cDNA and Lineated Vector

1. Transfer gel slices containing S100A3 cDNA or lineated vector to a 1.5-mL tube.
2. Add QX1 buffer (300 μL/100 mg gel) and 10 μL QIAEX II suspension.
3. Melt the gel at 50°C for 10 min, and then spin at 10,000×*g* for 1 min.
4. Remove the supernatant and wash the pellet with 0.5 mL QX1 and twice with 0.5 mL buffer PE.
5. Elute S100A3 cDNA or lineated vector with 10 μL of 10 mM Tris–HCl buffer (pH 8.5) at 50°C for 5 min.
6. Centrifuge at 10,000×*g* for 30 s and place the supernatants on ice.

3.1.3. Construction of the Transfer Vector

1. Combine 15 ng of S100A3 cDNA, 50 ng of the lineated pBac-PAK9 vector, and an equivalent volume of T4 DNA ligase solution (Solution I from the DNA ligation kit ver. 2.1). Incubate at 16°C for 30 min.
2. Gently thaw competent cells (e.g., XL1-Blue cells) on ice, and transfer 50 μL to a prechilled 14-mL polypropylene round-bottom tube (BD Falcon).
3. Add 1 μL of the pBacPAK-S100A3 plasmid, swirl the reaction tube gently to mix, and incubate on ice for 30 min.
4. Heat shock for 45 s at 42°C and then place the reactions on ice for 2 min.
5. Add 0.5 mL SOC medium preheated to 42°C and incubate at 37°C for 1 h with shaking at approximately 180 rpm.
6. Plate 100 μL of the transformation culture onto agar plates containing ampicillin and incubate at 37°C for 16 h.
7. Select single colonies and grow overnight at 37°C in 1.5 mL LB medium containing ampicillin.
8. Isolate plasmid (pBacPAK-S100A3) using a DNA preparation kit (e.g., QuickLyze Miniprep kit).

3.2. Mutagenic Pseudo Citrullination Introduction into the Transfer Vector (See Note 7)

1. Site-directed mutagenesis using QuikChange II kit: Design the mutagenic primer sets using the QuikChange primer design program (available at https://www.genomics.agilent.com). Copy and paste the cDNA sequence conventionally starting from the initiating ATG codon (see Note 8).

2. Mix 39.5 μ/L of distilled water, 5 μL of 10× reaction buffer, 1 μL of the transfer vector (10 ng/μL), 1.25 μL each of forward and reverse primers (100 ng/μL), and 1 μL of *PfuUltra* HF DNA polymerase (2.5 U/μL).
3. Cycle with the following thermal parameters: 95°C, 30 s; 16 cycles: 95°C, 30 s; 55°C, 1 min; and 68°C, 6 min (see Note 9).
4. After cooling, add 1 μL of *Dpn* I restriction enzyme (10 U/μL) and incubate at 37°C for 1 h.
5. Perform XL1-Blue competent cell transformation, colony isolation, and plasmid purification according to steps 2–8 in Subheading 3.1.3.

3.3. Generation of S100A3 and Mutated Recombinant Baculovirus

3.3.1. Homologous Recombination in Insect Cells

1. Seed 1×10^6 Sf21 cells per well in a 6-well plate and incubate at 27°C for 1–2 h.
2. Wash the cells with 2 mL of Grace's basic medium twice.
3. Mix 43 μL of sterile H_2O, 2.5 μL of the constructed vector (100 ng/μL), and 2.5 μL of Bsu36 I digested BacPak6 virus DNA (Clontech) in a 1.5-mL sterile Eppendorf tube.
4. Add 2 μL of Bacfectin to the mixture and incubate at room temperature for 15 min to allow complexes to form with DNA.
5. Exchange the medium with 1.5 mL Grace's basic medium, and then add the Bacfection–DNA mixture dropwise while gently swirling the dish.
6. Incubate at 27°C for 5 h, and then add 1.5 mL of complete medium to each well.
7. After 72-h incubation, transfer the cotransfection supernatant to a sterile tube.

3.3.2. Recombinant Virus Isolation and Passage

1. Seed 1×10^6 Sf21 cells per well in a 6-well plate and incubate at 27°C for 4 h.
2. Inspect the dishes to ensure that the cells have attached to form an even monolayer of about 70–80% confluency.
3. Aspirate the medium and infect the cells with 100 μL of serial dilutions of the virus inoculum (1/10, 1/100, and 1/1,000 of the cotransfection supernatant) for 1 h at 27°C.
4. During the incubation of step 3, autoclave 12 mL of 2% SeaPlaque agarose in H_2O and cool to 37°C.
5. Combine the melted agarose with an equivalent volume of complete medium pre-warmed at 37°C (final concentration: 1% agarose).
6. Aspirate the virus inoculum, and gently overlay 1.5 mL of the 1% agarose in each well.

7. Once the agarose coagulates, overlay 1.5 mL of the complete medium in each well.
8. After 4–5 days of incubation at 27°C, add 1 mL of the 0.33% neutral red (5 mg/16.5 mL PBS, filtered).
9. After 2–3 h of incubation at 27°C, aspirate the stain, and leave the dish inverted in the dark overnight at room temperature.
10. Ensure that the cells are unstained by microscopy. Encircle well-isolated viral plaques with a pen on the outside bottom of the dish.
11. Pick the single plaques of recombinant virus by pushing on the blue tip and sucking a plug of agarose.
12. Transfer the agarose plug into the 0.5 mL of complete medium in a sterile tube, and leave the viruses to diffuse overnight at 4°C.
13. Seed 5×10^5 Sf21 cells per well of a 6-well dish, and incubate at 27°C for 1–6 h.
14. Remove the medium, gently add 100 μL of a plaque-pick, and incubate at 27°C for 1 h.
15. Add 2 mL of complete medium to each well and incubate for 3–4 days at 27°C.
16. Store the supernatant at 4°C until use for recombinant virus amplification in Subheading 3.3.4.

3.3.3. Evaluation of S100A3 Expression by Western Blot Analyses

1. Suspend the cells in 1 mL of PBS in the 6-well plate and transfer to a 1.5 mL Eppendorf tube.
2. Pellet the cells by centrifugation at $1,000 \times g$ for 1 min. After washing with PBS, resuspend the cells in 0.1 M Tris–HCl (pH 7.6) containing 1 mM EDTA and 0.1 M dithiothreitol (DTT).
3. Sonicate cells on ice using a hand-held sonicator (UR-20P, Tomy-Seiko), and centrifuge at $10,000 \times g$ for 15 min.
4. Mix 6.5 μL of the supernatant, 1 μL of 0.5 M DTT, and 2.5 μL of Nu-PAGE sample buffer and heat at 70°C for 10 min.
5. After flushing the tube, load the samples together with a prestained molecular marker (SeeBlue Plus2, 1.5 μL) onto a Nu-PAGE gel.
6. Electrophorese at 200 V for 35 min.
7. Transfer the separated proteins onto a PVDF membrane at 30 V for 1 h.
8. Immerse the membrane in 1 μg/mL rabbit anti-human S100A3 antibody (19) and 1% BSA in PBS at room temperature for 1 h.

9. After washing with PBST (5 min × 3 times), immerse the membrane in 2.5 μg/mL AlexaFlour 488-labeled goat anti-rabbit IgG and 1% BSA in PBS at room temperature for 1 h.
10. After washing with PBST (5 min × 3 times), detect fluorescent signals using the Typhoon system. Select cells expressing the target protein.
11. Split the passage 1 virus derived from the selected cells into aliquots (~75 μL), and store the vials at −70°C. Normally, virus titer should have risen to ~2×10^7 infectious unit (IFU)/mL at this stage (*see* Subheading 3.3.5).

3.3.4. Recombinant Virus Amplification

1. Culture Sf21 cells in a 225-cm^2 flask to 50–70% confluency (i.e., ~1.5×10^7 cells).
2. After removal of the medium, add ~7.5 mL of the fresh medium and ~75 μL of the passage 1 virus stock (~1.5×10^6 IFU, MOI = ~0.1).
3. After inoculation of the cells with the virus with gentle swirling of the flask and incubating the cells at 27°C for 1 h, add an additional 22.5 mL medium and continue further cultivation for 4–6 days at 27°C until the cells are well infected.
4. Centrifuge the culture medium, and store a part of the supernatant required for the current working stock for protein expression (in Subheading 3.4) at 4°C. Use this virus stock within 6 months.
5. Split the remaining passage 2 virus into aliquots (>1.5×10^6 IFU/vial) and store at −70°C. When the current working stock depletes, thaw an aliquot and use for generation of a new working stock (Steps 1–4 above).

3.3.5. Virus Titration (See Note 10)

1. Virus assay using the rapid titer kit: Seed 160 μL of Sf21 cell suspension (4×10^5 cells/mL) to one row of a 96-well microtiter plate (6.5×10^4 cells/well). Incubate the plate at 27°C for 1 h.
2. Dilute the virus stock with medium (Serial dilutions 10^{-3}, 10^{-4}, and 10^{-5}).
3. Aspirate the medium, and add 25 μL of the viral dilutions to the wells (one well with medium only as a negative control, three wells for 10^{-3}, and four wells each for 10^{-4} and 10^{-5}). Incubate the plate at 27°C for 1 h.
4. Aspirate the inoculums, and overlay 50 μL of methyl cellulose in each well. Incubate the plate at 27°C for 2 days.
5. Add 150 μL of 4% paraformaldehyde in PBS and incubate for 10 min at room temperature.
6. Shake out reagent and tap the plate lightly on paper towel and wash with 200 μL PBST (5 min × 3 times).

7. Add 50 μL goat serum and incubate for 5 min at room temperature (blocking).
8. Shake out reagent and tap the plate on a paper towel. Then, add 25 μL of diluted mouse gp64 antibody and incubate at 37°C for 25 min.
9. Repeat step 6.
10. Add 50 μL of diluted goat anti-mouse antibody–HRP conjugate and incubate at 37°C for 25 min.
11. Repeat step 6.
12. Add 50 μL of blue peroxidase substrate and incubate at room temperature. Foci of infection become visible within 10 min.
13. After 3 h, observe the cells using light microscopy and count the number of stained foci per well (at the dilution, this should yield 5–25 foci). Determine the virus titer using the following equation:

    Virus titer (IFU/mL) = average no. of foci per well × dilution factor × 40

    (e.g., 25 foci × 10^5 × 40 = 1 × 10^8 IFU/mL).

3.4. S100A3 Recombinant Protein Production

1. Culture Sf21 cells in a 225-cm^2 flask to 80% confluency (i.e., 2 × 10^7 cells).
2. Remove the medium and add the current working virus stock (3 × 10^8 IFU/6 mL medium, MOI = 10).
3. Swirl the flask gently and incubate at 27°C for 1 h.
4. Add an additional 19 mL of medium and continue further cultivation for 2 days at 27°C.
5. Harvest the infected cells and wash × 2 times with PBS in a 50-mL centrifuge tube. Then resuspend the cells in 5 mL of 0.1 M Tris–HCl (pH 7.6) containing 50 mM DTT, 1 mM EDTA, and 1 mM PMSF.
6. Sonicate cells using a Bioruptor (e.g., UCD 200TM; 200 W 30 s × 15 times) on ice.
7. Combine cell lysates collected from ten flasks. Collect the supernatant after centrifugation at 14,000 × *g* for 30 min.

3.5. Purification of the Recombinant Human S100A3 Protein

3.5.1. Anion Exchange Chromatography

1. Dilute the cell-lysate supernatant (60–90 mg protein) with an equal volume of distilled water to reduce ionic strength.
2. Apply the diluted sample onto a Q-Sepharose column (2 mL; GE-Healthcare, Uppsala, Sweden) pre-equilibrated with 20 mM Tris–HCl buffer (pH 7.6) containing 1 mM EDTA and 10 mM DTT (buffer A).
3. Wash the column with 10 mL of buffer A.

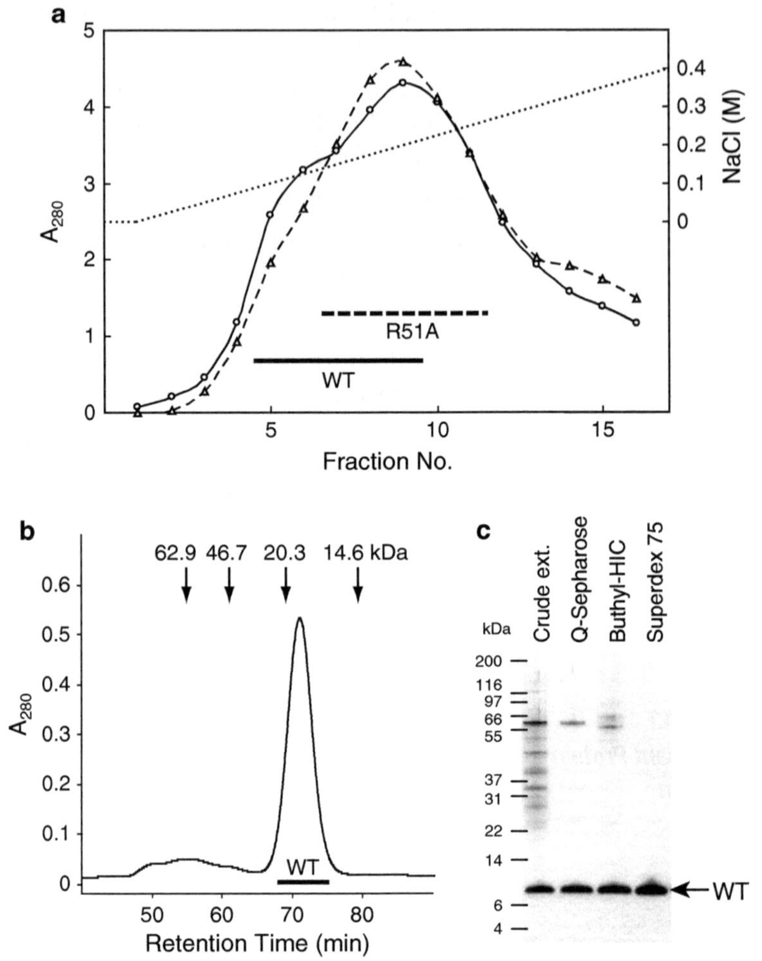

Fig. 1. Purification of recombinant S100A3 and R51A mutant proteins expressed in insect cells. (**a**) Anion exchange chromatograms. Shown are plotted elution profiles of wild-type S100A3 and R51A from a Q-Sepharose column (2 mL) and numbering the fractions aligned with a NaCl linear gradient only. Note that all mutant proteins used in this study eluted the same number of fractions depending on the charged residue substitution (i.e., R51A eluted with higher ionic strength than others). *Bars* show collected peaks. (**b**) A gel filtration chromatogram of WT-S100A3. Fractions after passage through a Macro-prep butyl resin column were loaded onto a Superdex-75 column (16 × 60 mm) with flow rate 1.0 mL/min. All S100A3 mutants eluted with comparable retention times. (**c**) Homogeneity of WT-S100A3 during purification. Shown are silver-stained Nu-PAGE gels loaded with aliquots of the crude and purified materials (50 ng protein was applied to each lane).

4. Elute the absorbed proteins containing S100A3 with a linear NaCl gradient (0–0.4 M) in buffer A (Fig. 1a).
5. Apply aliquots of fractions and a molecular maker (Mark 12) on a Nu-PAGE gel according to steps 4–6 in Subheading 3.3.3. Check the purity by protein staining (e.g., silver stain). Collect the fractions containing mainly S100A3.

3.5.2. Hydrophobic Column Chromatography

1. Add $(NH_4)_2SO_4$ to the pooled Q-Sepharose fractions to a final concentration of 1.5 M.
2. After centrifugation at 14,000 × *g* for 30 min, pass the resultant supernatant through a Butyl-IHC column (1 mL; Bio-Rad) pre-equilibrated with 1.5 M $(NH_4)_2SO_4$ containing buffer A. Elute the supernatant remaining in the column with 2 mL of the pre-equilibrated buffer.
3. Dialyze the thru pass fractions using Specra/Por (MW 3,500 cutoff) against 1 L of 20 mM Tris–HCl buffer (pH 7.6) containing 1 mM EDTA and 10 mM DTT at 4°C for 24 h. Exchange the buffer once.
4. Condense the dialyzed fraction to a volume <5 mL (maximum loading volume of the column used in the next step) by using a centrifugal filter unit (e.g., Ultracel-3K).

3.5.3. Size Exclusion Chromatography

1. Pre-equilibrate a HiLoad 16/60 Superdex-75 pg column (16 × 60 mm, GE-Healthcare) with 50 mM Tris–HCl (pH 7.6) containing 150 mM KCl, 10 mM DTT, and 1–10 μM EDTA at 4°C (see Note 11).
2. After filtration, load the sample (<5 mL) onto the column.
3. Flow the pre-equilibrated buffer at 1.0 mL/min and collect fractions corresponding to the main peak (Fig. 1b).
4. Condense the collected fractions using an Ultracel-3K.
5. Measure absorbance at 280 nm using the filtrated effluent as background ($A_{280,bg}$) (absorbance of DTT) and the condensed S100A3 (A_{280}) using a spectrophotometer (see Note 12). Determine the S100A3 concentration (C_{S100A3}) using the molecular weight ($MW_{S100A3} = 11{,}625\ gM^{-1}$) and molar coefficient ($\varepsilon = 14{,}500\ M^{-1}\ cm^{-1}$) according to the following equation (see Note 13):

   $$C_{S100A3}\ (mg/ml) = (A_{280} - A_{280,bg})/1.25.$$
6. Check the purity as described in step 5 of Subheading 3.5.1 (Fig. 1c).

4. Notes

1. Most full-length cDNAs of human and mouse S100 proteins are commercially available at http://www.genecopoeia.com/.
2. In insect cell cultivation at 27°C, we frequently encountered fungal contaminations rather than bacteria. Cellular proliferation and protein production in insect cells are not affected by Amphotericin B.

3. Herein, we describe monolayer cultures of the insect cells preferred for consistent recombinant protein expression. Adhesion of Sf21 cells changes slightly depending on the culture flask. Sf21 cells more tightly adhere to Becton-Dickinson flasks than to Corning products. Sf21 cells can be maintained for ~3 months (20–30 passages) by selecting an adequate flask for the type of cell under study.
4. Herein, we present an efficient method of producing recombinant virus by homologous recombination in insect cells (24). This method requires screening plaques to ensure the absence of contamination with wild-type virus (Subheading 3.3). The plaque isolation process can be omitted by using site-specific transposition (e.g., Bac-to-Bac, Invitrogen) or an engineered baculovirus containing a lethal mutation (e.g., flashBAC, Oxford expression system; BacMagic, Novagen). Investigators should select an expression system which can be adapted to their needs.
5. A number of antibodies immunized against synthetic peptide antigens and recombinant S100 proteins are commercially available, but verifying cross-reactivity with other S100 members is required.
6. Design the upstream primer sequence as follows:

   5'-CGGAATTCATGGCCAGGCCTCTGGAGCAGG-3'.

   The boxed sequence indicates the *EcoR* I site and the underlined sequence starts with the start codon of S100A3. Design the downstream primer as follows:

   ATAAGAATGCGGCCGCGGCAAGTCCAGATTG-3'. The boxed sequence indicates the *Not* I site and the underlined sequence is derived from the 3′-noncoding region of S100A3.
7. In vitro modification of S100A3 by peptidylarginine deiminase yields multiple forms of this protein. Mutagenic substitution of Arg/Ala is an advantageous recombinant way to study biological function and structural analysis of pseudo citrullinated proteins.
8. To replace amino acids located in the N-terminal region, input an upstream sequence with 3 × n nucleotides (to avoid a frameshift) of the transfer vector (pBacPAK9) followed by the inserted cDNA sequence until the primer design program. The upstream primer,

   5'-GGGAATTCATGGCCGCGCCTCTGGAGCAGG-3',

   and downstream primer,

   5'-CCTGCTCCAGAGGCGCGGCCATGAATTCCC-3',

   are computed for substitution of Arg-3 to Ala-3 in S100A3. The boxed sequences are derived from the transfer vector and the mutation sites are underlined.

9. Use a thermal parameter of 68°C per cycle, which is determined by computing 1 min/kb of plasmid length. In this case, a 5.5-kb pBacPAK9 plasmid plus a 0.4-kb S100A3 cDNA requires 6 min.
10. If routine repeats of virus amplification constantly give high titers of virus stock solution (>10^8 IFU/mL), the virus titration step may be omitted and, instead, proceed directly to protein production. Herein, we present an antibody-based assay (25). Original plaque assays or real-time PCR-based assays (BacPAK qPCR titration kit, Clontech) (26) are also available.
11. A minimum supplementation of EDTA that prevents binding of trace amounts of metal ions to S100A3 is encouraged. Note that an excess of the chelate added to the pre-equilibrated buffer affects the results of either metal-binding experiments or Ca^{2+}-dependent assays.
12. A cuvette of 1 cm width is not suitable for the absorbance measurement of protein solution at high concentrations. Use of a thinner cuvette or a non-cuvette-type photo-spectrometer (e.g., NanoDrop 1000; Thermo Fisher Scientific Inc.), which can measure small volumes without sample dilution, is encouraged.
13. In our experience, approximately 7 mg of wild-type S100A3 protein was obtained from 2×10^8 Sf21 cells. All mutant proteins examined, except for SS2-disrupted mutants, were purified in the same manner and yielded similar amounts (5–8 mg). C81A + C99A and R51A + C81A + C99A had lower yields (~3 mg) (10).

References

1. Kizawa K, Takahara H, Unno M, Heizmann CW (2011) S100 and S100 fused-type protein families in epidermal maturation with special focus on S100A3 in mammalian hair cuticles. Biochimie 93:2038–2047
2. Leclerc E, Heizmann CW (2011) The importance of Ca^{2+}/Zn^{2+} signaling S100 proteins and RAGE in translational medicine. Front Biosci S3:1232–1262
3. Haimoto H, Kato K (1988) S100a_0 ($\alpha\alpha$) protein in cardiac muscle. Isolation from human cardiac muscle and ultrastructural localization. Eur J Biochem 171:409–415
4. Schaub MC, Heizmann CW (2008) Calcium, troponin, calmodulin, S100 proteins: from myocardial basics to new therapeutic strategies. Biochem Biophys Res Commun 369:247–264
5. Kizawa K, Takahara H, Troxler H, Kleinert P, Mochida U, Heizmann CW (2008) Specific citrullination causes assembly of a globular S100A3 homotetramer: a putative Ca^{2+} modulator matures human hair cuticle. J Biol Chem 283:5004–5013
6. Moore BW (1965) A soluble protein characteristic of the nervous system. Biochem Biophys Res Commun 189:739–744
7. Donato R, Heizmann CW (eds) (2010) S100B protein in the nervous system and cardiovascular apparatus in normal and pathological conditions. Cardiovascular Psych Neurol 2010: 929712. doi:10.1155/2010/929712
8. Marenholz I, Lovering RC, Heizmann CW (2006) An update of the S100 nomenclature. Biochim Biophys Acta 1763:1282–1283

9. Brondyk WH (2009) Selecting an appropriate method for expressing a recombinant protein. Methods Enzymol 463:131–147
10. Unno M, Kawasaki T, Takahara H, Heizmann CW, Kizawa K (2011) Refined crystal structure of human Ca^{2+}/Zn^{2+}-binding S100A3 protein characterized by two disulphide bridges. J Mol Biol 408:477–490
11. Castagnola M, Inzitari R, Fanali C, Iavarone F, Vitali A, Desiderio C, Vento G, Tirone C, Romagnoli C, Cabras T, Manconi B, Sanna MT, Boi R, Pisano E, Olianas A, Pellegrini M, Nemolato S, Heizmann CW, Faa G, Messana I (2011) The surprising composition of the salivary proteome of preterm human newborn. Mol Cell Proteomics. doi:10.1074/mcp.M110.003467
12. Kizawa K, Troxler H, Kleinert P, Inoue T, Toyoda M, Morohashi M, Heizmann CW (2002) Characterization of the cysteine-rich calcium-binding S100A3 protein from human hair cuticles. Biochem Biophys Res Commun 299:857–862
13. Arnau J, Lauritzen C, Petersen GE, Pedersen J (2006) Current strategies for the use of affinity tags and tag removal for the purification of recombinant proteins. Protein Expr Purif 48:1–13
14. Föhr UG, Heizmann CW, Engelkamp D, Schäfer BW, Cox JA (1995) Purification and cation binding properties of the recombinant human S100 calcium-binding protein A3, an EF-hand motif protein with high affinity for zinc. J Biol Chem 270:21056–21061
15. Ostendorp T, Heizmann CW, Kroneck PM, Fritz G (2005) Purification, crystallization and preliminary X-ray diffraction studies on human Ca^{2+}-binding protein S100B. Acta Crystallogr F61:673–675
16. Goch G, Vdovenko S, Kozłowska H, Bierzyñski A (2005) Affinity of S100A1 protein for calcium increases dramatically upon glutathionylation. FEBS J 272:2557–2565
17. Miranda KJ, Loeser RF, Yammani RR (2010) Sumoylation and nuclear translocation of S100A4 regulates IL-1β mediated production of matrix metalloprotinase-13. J Biol Chem 285:31517–31524
18. Andrassy M, Igwe J, Autschbach F, Volz C, Remppis A, Neurath MF, Schleicher E, Humpert PM, Wendt T, Liliensiek B, Morcos M, Schiekofer S, Thiele K, Chen J, Kientsch-Engel R, Schmidt AM, Stremmel W, Stern DM, Katus HA, Nawroth PP, Bierhaus A (2006) Posttranslationally modified proteins as mediators of sustained intestinal inflammation. Am J Pathol 169:1223–1237
19. Lim SY, Raftery M, Cai H, Hsu K, Yan WX, Hseih HL, Watts RN, Richardson D, Thomas S, Perry M, Geczy CL (2008) S-nitrosylated S100A8: novel anti-inflammatory properties. J Immunol 181:5627–5636
20. Schenten V, Bréchard S, Plançon S, Melchior C, Frippiat JP, Tschirhart EJ (2010) iPLA2, a novel determinant in Ca^{2+}- and phosphorylation-dependent S100A8/A9 regulated NOX2 activity. Biochim Biophys Acta 1803:840–847
21. Sakaguchi M, Miyazaki M, Sonegawa H, Kashiwagi M, Ohba M, Kuroki T, Namba M, Huh NH (2004) PKCα mediates TGFβ-induced growth inhibition of human keratinocytes via phosphorylation of S100C/A11. J Cell Biol 164:979–984
22. Cecil DL, Terkeltaub R (2008) Transamidation by transglutaminase 2 transforms S100A11 calgranulin into a procatabolic cytokine for chondrocytes. J Immunol 180:8378–8385
23. Kizawa K, Ito M (2005) Characterization of epithelial cells in the hair follicle with S100 proteins. Methods Mol Biol 289:209–222
24. Kitts PA, Possee RD (1993) A method for producing recombinant baculovirus expression vectors at high frequency. Biotechniques 14:810–817
25. Kwon MS, Dojima T, Toriyama M, Park EY (2002) Development of an antibody-based assay for determination of baculovirus titers in 10 hours. Biotechnol Prog 18:647–651
26. Hitchman RB, Siaterli EA, Nixon CP, King LA (2007) Quantitative real-time PCR for rapid and accurate titration of recombinant baculovirus particles. Biotechnol Bioeng 96:810–814

Chapter 6

X-ray Structural Analysis of S100 Proteins

Günter Fritz

Abstract

X-ray crystallography is a potent and meanwhile fast technique to obtain detailed structural information of S100 proteins in their apo or metal ion-loaded state. S100 proteins crystallize in the absence or presence of Ca^{2+} and Zn^{2+} and the obtained crystals often diffract to high resolution yielding information on the ion-binding sites, conformation, and target interaction sites of the proteins. Here, I describe a general scheme to isolate and crystallize S100 proteins and the analysis of protein crystals using a modern synchrotron source.

Key words: S100 protein, Calcium, Zinc, Crystallization, X-ray, EF-hand

1. Introduction

The S100 protein family comprises more than 20 different members in human (1, 2). The S100 proteins are small 10–12 kDa Ca^{2+}- and Zn^{2+}-binding proteins sharing the EF-hand Ca^{2+}-binding motif (3). The S100 proteins display diverse expression patterns. Some S100 proteins are found almost ubiquitously in the organism whereas others show distinct expression in certain tissues, cell types, cell compartments, or at certain phases of the cell cycle (1). Most S100 proteins interact Ca^{2+} dependently with their target proteins controlling enzymatic activity, DNA binding, or other protein–protein interactions (4). Most S100 proteins constitute dimers, but a number of higher oligomers with particular different functions have been described in the recent years. Structural biology tries to elucidate the molecular requirements for target protein specificity of the S100 proteins and the conformational transitions triggered by Ca^{2+} or Zn^{2+} binding. Due to their small size, S100 protein dimers are well

Claus W. Heizmann (ed.), *Calcium-Binding Proteins and RAGE: From Structural Basics to Clinical Applications*, Methods in Molecular Biology, vol. 963, DOI 10.1007/978-1-62703-230-8_6,

suited for structural analysis by multidimensional nuclear magnetic resonance (NMR) spectroscopy. However, larger assemblies or protein–target complexes are still a major challenge in applying this technique; in contrast X-ray crystallography exhibits no limitations for the size of the molecule to be determined. Moreover, the technological advances in the area of X-ray crystallography are enormous and automated software pipelines make it easy, even for an inexperienced user, to determine a three-dimensional structure. In this article, a generalized procedure for heterologous expression and isolation of S100 proteins, crystallization in the apo or metal ion-bound state, and some general guidelines to determine the structure by X-ray crystallography are described.

2. Materials

All buffer and crystallization solutions should be prepared using double-distilled water or MilliQ-treated water with electric resistance of 18 MΩ.

2.1. Expression Medium

1. Phosphate-buffered DYT medium with glucose: 16 g tryptone (Applichem), 10 g yeast extract (Applichem), 4 g NaCl, 4 g Na_2HPO_4 is dissolved in 1 L deionized water and autoclaved for 20 min at 121 °C at 1 atm.
2. 20% (w/v) glucose stock solution is sterilized by filtration through a 0.2 μm filter (cellulose acetate, Millipore). Let the medium cool down after autoclaving and add glucose to a final concentration of 0.2% from the stock solution. Add appropriate antibiotics.
3. 100 mg/ml ampicillin stock solution (keep aliquots frozen at −20 °C).
4. DYT agar plates: Agar (Applichem) 16 g/L is added to the DYT medium prior to autoclaving. A magnetic stirrer is placed in the vessel allowing mixing of the solution after autoclaving. Let medium cool down to 60 °C and add glucose and antibiotics as described above (items 2 and 3). Mix gently with the magnetic stirrer and prepare plates.

2.2. Buffers for Purification of S100 Proteins Applying HIC and SEC Chromatography

1. Calcium buffer: 50 mM Tris–Cl, pH 7.6 containing 2 mM $CaCl_2$ and filtered through a 0.2 μm filter.
2. EDTA buffer: 50 mM Tris–Cl, pH 7.6 containing 4 mM EDTA and filtered through a 0.2 μm filter.
3. SEC buffer: 50 mM Tris–Cl, pH 7.6, 150 mM NaCl, filtered through a 0.2 μm filter.
4. Complete protease inhibitor (Roche).

2.3. Buffers for Crystallization

1. Pre-buffer: 10 mM Tris–Cl, pH 7.4.
2. Crystallization buffers ready to use from commercial suppliers are applied. Suppliers are Qiagen, Hampton Research, Jena Bioscience, and Molecular Dimensions.
3. Stock solutions for optimization of crystallization: 1.0 M solutions of the appropriate buffer are prepared at different pH values. Typically 50 ml of each buffer with a certain pH value are sterilized by filtration through a 0.2 μm filter and stored at 4 °C.
4. Stock solutions of PEG: Small PEGs like PEG400 to PEG800 are used directly without further dilution to prepare crystallization solutions. Stock solutions of larger PEGs are prepared as 50% (w/v) solutions in water and filtered through a 0.2 μm filter. PEG solutions used for crystallization are stored in the dark at 4 °C or –20 °C to prevent peroxide formation in the PEG by UV light.

2.4. Chromatography Columns

1. Hydrophobic interaction chromatography (HIC) matrix is Phenyl-Sepharose (GE-Healthcare).
2. Size exclusion chromatography (SEC) matrix is Superdex 75 (GE-Healthcare).
3. Columns for fast buffer exchange: NAP5 column (GE-Healthcare).

2.5. Crystallization Plates

1. Initial screening: 96-well crystallization plates. Supplier are Qiagen, Hampton Research, Jena Bioscience, Molecular Dimensions, and Greiner.
2. Fine screening: 24-well plates sitting drop devices for hanging Suppliers are Qiagen, Hampton Research, Jena Bioscience, and Molecular Dimensions.

2.6. Crystallization Screening Solutions

Ready-to-use screens are available from different suppliers like SIGMA, Qiagen, Hampton Research, Jena Bioscience, or Molecular Dimensions.

2.7. Equipment for Freezing Crystals and Analysis of Crystals

1. Nylon loops of different size mounted on a magnetic base ready to use at standard goniometer heads. Supplier is Hampton Research.
2. LithoLoops and micro meshes (MiTeGen). Suppliers are Molecular Dimensions and Jena Bioscience.
3. Glycerol (80%).
4. Ethylene glycol.
5. Glucose.

3. Methods

3.1. Expression of S100 Proteins

Most S100 proteins can be expressed heterologously in *E. coli* and purified at high yields without the help of any affinity tags. The cDNA of the S100 proteins is usually cloned into vectors (e.g., pGEMEX) under the control of a T7 promoter that allows high yield expression of the S100 protein upon induction of T7 polymerase in *E. coli* (DE3) strains by addition of IPTG to the culture medium. In the following expression based on a pGEMEX (Promega) vector carrying β-lactamase for ampicillin resistance is described.

1. Plate *E. coli* strain carrying a plasmid for S100 protein expression on DYT agar plate with 100 μg/ml ampicillin. Incubate overnight at 37 °C.
2. Inoculate 20 ml DYT supplemented with 0.2% glucose and 100 μg/ml ampicillin with a single colony from the agar plate and incubate culture on a shaker overnight at 37 °C.
3. Spin down *E .coli* cells from overnight culture at 4,000 × *g* for 5 min, discard supernatant to remove extracellular β-lactamases, and resuspend cells in 10 ml fresh medium.
4. Inoculate 1 L expression culture medium with the resuspended cells. Incubate under vigorous shaking at 37 °C. Take small samples of the culture every 40 min and measure $OD_{600\ nm}$.
5. When $OD_{600\ nm}$ of the culture reaches 0.6 add 1 mM IPTG and further incubate for 5 h.
6. Harvest cells by centrifugation at 5,000 × *g* for 20 min.
7. Freeze cell pellets and store until further use at −80 °C.

3.2. Purification of S100 Proteins

Upon Ca^{2+}-binding S100 proteins undergo a conformational change exposing a large hydrophobic surface to the solvent. This conformational change is fully reversible when Ca^{2+} is removed from the protein, e.g., by EDTA treatment. This Ca^{2+}-switch can be utilized for Ca^{2+}-dependent interaction of the S100 proteins with the hydrophobic matrix Phenyl-Sepharose. All chromatography steps should be performed at 4 °C.

1. Cell are thawed on ice and resuspended in 2 volumes per wet weight in ice-cold 50 mM Tris–Cl, pH 7.6.
2. Add 0.5 mM $MgCl_2$ and a small amount of DNAse I (SIGMA).
3. Cells are ruptured by one passage through a French pressure cell (Stansted) or a microfluidizer (Microfluidics).
4. Crude extract is centrifuged for 1 h at 100,000 × *g* at 4 °C.

5. Supernatant is diluted fivefold with ice-cold calcium buffer containing 1 mM PMSF, pH 7.6.
6. The diluted supernatant is applied to Phenyl-Sepharose column (26 mm × 50 mm) equilibrated in calcium buffer.
7. Wash column with calcium buffer until the absorption at 280 nm reaches baseline level again.
8. Elute S100 protein with EDTA buffer from Phenyl-Sepharose column.

   Prior to size exclusion chromatography the protein solution has to be concentrated to a volume corresponding to approx. 2% of the volume of the SEC column. Columns can be purchased ready to use at different sizes (GE-Healthcare) or custom made by an experienced user.
9. Concentrate S100 protein by ultrafiltration until the volume reaches ca. 25 of the volume of the SEC column.
10. Filter concentrated protein through a 0.2 μm filter and apply to SEC column equilibrated in SEC buffer.
11. Collect fractions containing S100 protein and combine fractions with pure protein.
12. Concentrate protein by ultrafiltration to ca 20 mg/ml and add 5% (v/v) glycerol. Prepare aliquots of 500 μl, shock freeze protein in liquid nitrogen, and store aliquots until further use at −80 °C.

3.3. Crystallization of S100 Proteins

Different ready-to-use crystallization screens are available commercially from different companies (SIGMA, QIAGEN, Hampton Research, Molecular Dimensions, Jena Bioscience). Most solutions from the different suppliers overlap largely in the content of different crystallization solutions, so a few crystallization screens from one supplier are usually sufficient to cover a broad range of potential crystallization conditions. Crystallization of proteins is based on a controlled salting out of the protein out of the solution. This is achieved by lowering slowly the solubility of the protein by addition of organic molecules like polyethylene glycol or kosmotropic salts (e.g., sulfate, tartrate, or phosphate) which are known to change the hydration sphere of the proteins.

3.4. Prior Crystallization Treatment of the S100 Protein

In order to avoid the interference of buffer components of the S100 protein stock with the components of the crystallization solutions a buffer exchange right before the crystallization experiment is recommended.

1. Thaw concentrated aliquot of S100 protein on ice. Optimal is a concentration of 20–30 mg/ml to reach a final concentration of 10–15 mg/ml in the next step.

2. Equilibrate NAP5 column (GE-Healthcare) with 4 column volumes of 10 mM Tris–Cl, pH 7.6.
3. Apply 500 μl of protein solution to the column and let the solution enter the column.
4. Elute S100 protein by addition of 1.0 ml of 10 mM Tris–Cl, pH 7.6. Thereby the protein is diluted twofold. Keep protein solution on ice.
5. Check concentration of protein prior to crystallization trials.
6. When trying to crystallize an S100 protein in the Ca^{2+}-loaded state typically 10–20 mM $CaCl_2$ is added to the protein solution after buffer change.

3.5. Setting Up Crystallization Trials

Initial screening for crystallization conditions is nowadays often performed using a nanodrop pipetting robot (e.g., Phoenix, Rigaku) that can handle drop sizes between 50 nl and 500 nl. An initial drop volume of 200 nl is a good compromise between reducing the protein amount needed and getting large enough crystals if the trial is successful. If no robot is available in the laboratory such a service is offered, e.g., at the EMBL (http://www.embl-hamburg.de/facilities/htpx/) and at other Synchrotron Centers like Paul-Scherer Institute Swiss Light Source (http://www.psi.ch/sls/pxiii/crystallization-facility). A 96-well crystallization plate offers a reservoir well for the crystallization buffer and a small well for the protein drop that is mixed with equal volume of the crystallization buffer. The volume of the reservoir is large (typically 50–100 μl) compared to the drop (0.2 μl). After placing the solutions the wells are sealed with a transparent tape. The plates are incubated at constant temperature, most often at 20 °C. Typically protein crystals show up after a minimum period of a few days or later after 6–8 weeks. In vapor diffusion crystallization, the protein drop is mixed with an equal volume of a crystallization buffer and the mixture is equilibrated against a reservoir of the crystallization buffer. The vapor pressure in the protein/crystallization buffer mixture is higher than in the reservoir which leads to a slow transfer of water from the drop to the reservoir. That way the protein concentration in the drop increases slowly until the maximum solubility is reached. At this point the protein forms crystals or precipitates.

The results of such initial screening are monitored these days by automatic documentation systems (e.g., CrystalTrak, Rigaku) providing high-resolution and -magnification images. The images are often accessible via a software toolkit (e.g., CrystalTrak, Rigaku) through the Internet.

Note on the selection of crystallization screens: When crystallization trials for Ca^{2+}-loaded S100 proteins are set up, one might avoid crystallization buffers containing high concentrations of sulfate, phosphate, or tartrate. The solubility of Ca^{2+} with each of these salts is very low leading to the formation of salt crystals which cannot be distinguished from protein crystals on the images.

Once a crystallization condition is recognized the experiment is usually repeated at a larger volume and the conditions a further varied by changing the concentration of the ions or organic molecules or the ph of the original crystallization buffer.

3.6. Crystal Mounting and Freezing

Crystals obtained are tested and analyzed in an X-ray beam. Typically protein crystals have sizes ranging between ca. 50 μm × 100 μm × 100 μm and 200 μm × 200 μm × 400 μm. For the manipulation of the small and very often fragile protein crystals a stereomicroscope is required. Crystals are mounted onto loops and are flash frozen in the cold nitrogen stream or by plunging into liquid nitrogen. Freezing of crystals is necessary if they are analyzed at high-energy X-ray beams of modern synchrotron sources. Upon exposure to the X-ray beam the protein crystals suffer from radiation damage, i.e., the crystal is destroyed by the exposure to X-rays. This process is largely slowed down if the crystal is frozen and kept at 100 K in a dry stream of cold nitrogen gas. When freezing the crystals in the crystallization solution one has to avoid the formation of water ice which leads to irreversible damage of the protein crystal and moreover water ice gives rise to an intensive diffraction pattern that overlays the diffraction pattern of the protein crystal. Therefore reagents like glycerol, ethylene glycol, or sugars which suppress water ice formation are added. The resulting frozen solutions form a glass that is translucent for the X-ray beam.

3.7. Crystal Analysis

The diffraction pattern of the crystals is recorded during rotation of the crystal relative to the X-ray beam in order to collect all possible diffracted X-rays. For reconstruction of the electron density of the protein, that gives rise to the diffraction, one needs the information on the amplitude and the phase of the scattered X-rays. During recording of the X-ray waves the phase information is lost and has to be assessed by further methods. The intensity of the single spots is integrated using semiautomatic software suites yielding a summary of all spot intensities in a so-called dataset. Software suites for spot integration are HKL2000 (http://www.hkl-xray.com/), XDS (5), mosflm (6), d*TREK (7), and automatic pipelines like xia2 or autoPROC (8).

3.8. Retrieving Phase Information

Phase information is obtained by two methods in X-ray crystallography: (1) analysis of heavy atom derivatives where an anomalous scatterer is introduced or (2) molecular replacement. For well diffracting crystals also the anomalous signal of the endogenous sulfur from methionine and cysteine residues can be exploited (9). In crystals of S100 proteins in complex with Ca^{2+} (9) or Zn^{2+} these atoms can serve as well as heavy atom to determine phases. For

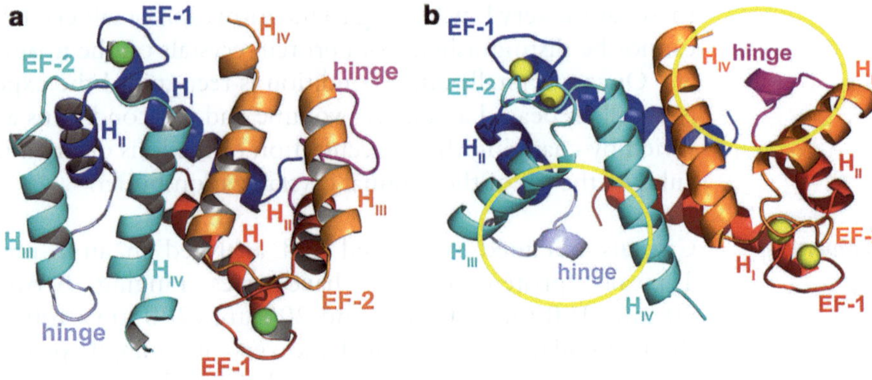

Fig. 1. Structure of S100 protein in the Ca2+-free state and Ca2+-loaded state. (**a**) Most S100 proteins consist of two subunits (*blue* and *red*, respectively) forming a homodimer. Each subunit comprises an S100-specific N-terminal EF-hand (EF1), a classical C-terminal EF-hand (EF2), and a linker region. The N-terminal EF-hand EF-1 (*dark colors*) is specific for the S100 proteins and can bind a Na^+ (*green spheres*) in the Ca^{2+}-free state. (**b**) Each EF-hand can bind one Ca^{2+} (*yellow spheres*) whereby Ca^{2+} binding induces a large conformational change: helix III is moving by about 90° opening the protein structure and exposing a large hydrophobic pocket in each subunit where target proteins are bound (*yellow circles*).

S100 protein, structure determination by molecular replacement is the method of first choice. In molecular replacement, a model of the real structure is used, to obtain first-phase information that is then further refined to obtain the real structure. Molecular replacement is very fast and successful if appropriate models are available. S100 proteins have a conserved basic architecture consisting of two EF-hands connected by a short flexible linker (Fig. 1). Structures of many S100 proteins in the apo state (10–13) or in the Ca^{2+}- and/or Zn^{2+}-loaded state (14–17) are available (Fig. 1) and can be used directly or after some modifications in molecular replacement trials. Molecular replacement calculations are easily performed in the CCP4 package with automatic pipelines like MrBUMP (18) or with the software suite PHASER (19). Refinement of structures is carried out with Refmac5 (20, 21), CNS (22, 23), or Phenix (24) and manual rebuilding is performed using the software Coot (25).

4. Notes

1. Expression yields for some S100 proteins can be increased using growing *E. coli* at lower temperature, e.g., at 30 °C and at the same time increasing the time for expression.
2. Some S100 proteins like S100A8 form inclusion bodies during expression. S100 proteins can be easily refolded by denaturation of the inclusion bodies in 6 M Gu-HCl and subsequent dialysis against 50 mM Tris–Cl, pH 7.6.

3. Several S100 proteins like S100A2, S100A3, or S100A4 have highly reactive cysteine residues at the surface which oxidize readily and form disulfide bonds. Addition of DTT to the purification buffers at a concentration of 1 mM usually prevents disulfide bond formation.
4. Protein concentration of pure S100 proteins can be assessed easily by their specific molecular extinction coefficients calculated from the amino acid sequence (http://web.expasy.org/protparam).
5. Oxidized DTT has an absorption band with a maximum at 280 nm that overlaps with the protein absorption. Therefore protein concentration determination as described in Note 4 should be performed before DTT is added.
6. Several S100 proteins like S100B form non-covalent multimers. For crystallization trials a monodisperse protein preparation is required. Different multimers can be separated in the final size exclusion purification step (26).
7. S100 proteins are highly soluble and can be concentrated to 50 mg/ml and more. Small aliquots can be easily frozen in PCR tubes.
8. Denaturation of protein during freezing is often prevented by addition of 10% glycerol to the solution. Always flash freeze protein stock aliquots in liquid nitrogen bath and store then at –80 °C.
9. Crystallization of S100 proteins with reactive cysteine residues might require the addition of 10 mM DTT for crystallization trials. Always prepare fresh DTT stock solutions and apply DTT directly prior to crystallization trials.
10. For crystallization of S100 proteins with Ca^{2+} bound to the EF-hands, $CaCl_2$ is added prior to crystallization to the protein after buffer exchange to a final concentration of 10–20 mM.
11. If Ca^{2+} is present avoid crystallization buffers with sulfate, phosphate, or tartrate. The solubility of Ca^{2+} with each of these salts is very low leading to the formation of salt crystals which cannot be distinguished at a first glance from protein crystals.
12. The dimeric architecture of many S100 proteins is often reflected in crystal packing. The symmetry axis is often a part of the crystal symmetry or gives rise to a pseudo symmetry.

Acknowledgments

This work was supported by grants FR14488/3-1 and FR1488/5-1 from the Deutsche Forschungsgemeinschaft (DFG).

References

1. Marenholz I, Heizmann CW, Fritz G (2004) S100 proteins in mouse and man: from evolution to function and pathology (including an update of the nomenclature). Biochem Biophys Res Commun 322:1111–1122
2. Marenholz I, Lovering RC, Heizmann CW (2006) An update of the S100 nomenclature. Biochim Biophys Acta 1763:1282–1283
3. Fritz G, Heizmann CW (2004) 3D structures of the calcium and zinc binding S100 proteins. In: Messerschmidt A, Bode W, Cygler M, Cygler M (eds) Handbook of metalloproteins, vol 3. Wiley, Chichester, pp 529–540
4. Santamaria-Kisiel L, Rintala-Dempsey AC, Shaw GS (2006) Calcium-dependent and -independent interactions of the S100 protein family. Biochem J 396:201–214
5. Kabsch W (2011) XDS. Acta Crystallogr D66:125–132
6. Battye TG, Kontogiannis L, Johnson O, Powell HR, Leslie AG (2011) iMOSFLM: a new graphical interface for diffraction-image processing with MOSFLM. Acta Crystallogr D67:271–281
7. Pflugrath JW (1999) The finer things in X-ray diffraction data collection. Acta Crystallogr D Biol Crystallogr 55:1718–1725
8. Vonrhein C, Flensburg C, Keller P, Sharff A, Smart O, Paciorek W, Womack T, Bricogne G (2011) Data processing and analysis with the autoPROC toolbox. Acta Crystallogr D Biol Crystallogr 67:293–302
9. Koch M, Diez J, Wagner A, Fritz G (2010) Crystallization and calcium/sulfur SAD phasing of the human EF-hand protein S100A2. Acta Crystallogr Sect F Struct Biol Cryst Commun 66:1032–1036
10. Koch M, Diez J, Fritz G (2008) Crystal structure of Ca^{2+} -free S100A2 at 1.6Å resolution. J Mol Biol 378:933–942
11. Otterbein LR, Kordowska J, Witte-Hoffmann C, Wang CL, Dominguez R (2002) Crystal structures of S100A6 in the Ca^{2+}-free and Ca^{2+}-bound states: the calcium sensor mechanism of S100 proteins revealed at atomic resolution. Structure 10:557–567
12. Drohat AC, Amburgey JC, Abildgaard F, Starich MR, Baldisseri D, Weber DJ (1996) Solution structure of rat apo-S100B(beta beta) as determined by NMR spectroscopy. Biochemistry 35:11577–11588
13. Potts BC, Smith J, Akke M, Macke TJ, Okazaki K, Hidaka H, Case DA, Chazin WJ (1995) The structure of calcyclin reveals a novel homodimeric fold for S100 Ca^{2+}-binding proteins. Nat Struct Biol 2:790–796
14. Ostendorp T, Diez J, Heizmann CW, Fritz G (2011) The crystal structures of human S100B in the zinc- and calcium-loaded state at three pH values reveal zinc ligand swapping. Biochim Biophys Acta 1813:1083–1091
15. Moroz OV, Blagova EV, Wilkinson AJ, Wilson KS, Bronstein IB (2009) The crystal structures of human S100A12 in apo form and in complex with zinc: new insights into S100A12 oligomerisation. J Mol Biol 391:536–551
16. Wilder PT, Varney KM, Weiss MB, Gitti RK, Weber DJ (2005) Solution structure of zinc- and calcium-bound rat S100B as determined by nuclear magnetic resonance spectroscopy. Biochemistry 44:5690–5702
17. Sastry M, Ketchem RR, Crescenzi O, Weber C, Lubienski MJ, Hidaka H, Chazin WJ (1998) The three-dimensional structure of Ca^{2+}-bound calcyclin: implications for Ca^{2+}signal transduction by S100 proteins. Structure 6:223–231
18. Keegan RM, Winn MD (2007) Automated search-model discovery and preparation for structure solution by molecular replacement. Acta Crystallogr D Biol Crystallogr 63:447–457
19. McCoy AJ (2007) Solving structures of protein complexes by molecular replacement with Phaser. Acta Crystallogr D Biol Crystallogr 63:32–41
20. Murshudov GN, Skubak P, Lebedev AA, Pannu NS, Steiner RA, Nicholls RA, Winn MD, Long F, Vagin AA (2011) REFMAC5 for the refinement of macromolecular crystal structures. Acta Crystallogr D Biol Crystallogr D67:355–367
21. Murshudov GN, Vagin AA, Dodson EJ (1997) Refinement of macromolecular structures by the maximum-likelihood method. Acta Crystallogr D Biol Crystallogr D53:240–255
22. Brunger AT, Adams PD, Clore GM, Gros P, Grosse-Kunstleve RW, Jiang J-S, Kuszewski J, Nilges N, Pannu NS, Read RJ, Rice LM, Simonson T, Warren GL (1998) Crystallography & NMR system: a new software suite for macromolecular structure determination. Acta Crystallogr D54:905–921
23. Brunger AT (2007) Version 1.2 of the crystallography and NMR system. Nat Protoc 2:2728–2733
24. Adams PD, Afonine PV, Bunkóczi G, Chen VB, Davis IW, Echols N, Headd JJ, Hung L-W, Kapral GJ, Grosse-Kunstleve RW, McCoy AJ, Moriarty NW, Oeffner R, Read RJ, Richardson DC, Richardson JS, Terwilliger TC, Zwart PH (2010) PHENIX: a comprehensive Python-based system for

macromolecular structure solution. Acta Crystallogr D Biol Crystallogr 66:213–221
25. Emsley P, Cowtan K (2004) Coot: model-building tools for molecular graphics. Acta Crystallogr D Biol Crystallogr D60:2126–2132
26. Ostendorp T, Heizmann CW, Kroneck PM, Fritz G (2005) Purification, crystallization and preliminary X-ray diffraction studies on human Ca2+-binding protein S100B. Acta Crystallogr F61:673–675

Chapter 7

Purification and Stable Isotope Labeling of the Calcium- and Integrin-Binding Protein 1 for Structural and Functional NMR Studies

Hao Huang and Hans J. Vogel

Abstract

The Calcium- and Integrin-Binding protein 1 (CIB1) has been identified as an important regulatory Ca^{2+}-binding protein that is involved in various cellular functions. Nuclear Magnetic Resonance (NMR) spectroscopy provides a powerful approach to study the structure, dynamics, and interactions of CIB1 and related proteins. Multidimensional NMR spectroscopy combined with various selective isotope labeling strategies has proven to be successful in the structure determination of CIB1. Moreover, the same approach allowed the detection of conformational changes when the protein binds different metal ions, and it facilitated the study of the interaction of CIB1 with the cytoplasmic domain of the human integrin αIIb subunit. In this protocol, we describe the purification and isotope labeling strategies for productive NMR studies of CIB1. The same isotope labeling strategies can be implemented to study numerous related regulatory calcium-binding proteins.

Key words: CIB1, Calcium- and integrin-binding protein, Methyl labeling, Deuteration, NMR spectroscopy, Isotope labeling

1. Introduction

The Calcium- and Integrin-Binding protein 1 (CIB1) is a member of the regulatory Ca^{2+}-binding helix-loop-helix or EF-hand protein family (1). CIB1 (sometimes also known as calmyrin or KIP) was originally discovered in an attempt to look for specific binding partners of the cytoplasmic domain of the human platelet integrin αIIb subunit by using yeast two-hybrid analysis (2). CIB1 is now known to be a ubiquitous protein with 191 residues (22 kDa) and it has several functions in cell signaling (3, 4). To date CIB1 has

Claus W. Heizmann (ed.), *Calcium-Binding Proteins and RAGE: From Structural Basics to Clinical Applications*, Methods in Molecular Biology, vol. 963, DOI 10.1007/978-1-62703-230-8_7, © Springer Science+Business Media New York 2013

been reported to interact with a number of protein targets, including the platelet integrin αIIb subunit (2), sphingosine kinase 1 (5), p21-activated kinase (6), apoptosis signal-regulating kinase (7), polo-like protein kinases (8), DNA-dependent protein kinase (9), oncoprotein LMO3 (10), calcineurin B subunit of the protein phosphatase calcineurin (11), and possibly presenilin-2 (12, 13).

CIB1 is a protein with a molecular weight (MW) of 22 kDa. In our work, the aim was to solve the structure of CIB1 and to study its interactions with its binding partners by solution-state Nuclear Magnetic Resonance (NMR) methods. The unique advantages of using solution-state NMR over other high-resolution structural methods are that it can provide structural as well as dynamics information, and that it can be easily used to study protein/protein or protein/ligand interactions. The latter is typically done by recording several Heteronuclear Single-Quantum Coherence (HSQC) spectra during titration experiments. When one wants to perform solution NMR studies of proteins, biosynthetic ^{15}N- and/or ^{13}C-stable isotope labeling is normally required to create multinuclear multidimensional NMR spectra with good resolution. Incorporation of isotope labels is normally achieved biosynthetically through expression of the protein in bacteria, which are grown in a defined medium containing the desired carbon and/or nitrogen sources. However, the typical ^{13}C-,^{15}N-labeling approach is limited to proteins that are smaller than 20 kDa. Fortunately, many regulatory calcium-binding proteins (CBPs), such as the ubiquitous calmodulin, fit within this size limit, and indeed many of the structures of EF-hand protein that are deposited in the Protein Data Bank (PDB) were determined by NMR spectroscopy rather than by X-ray crystallography methods. However, CIB1 and several related proteins are larger than 20 kDa. For these relatively large-sized CBPs, solution NMR studies will normally also require deuteration of the protein to improve the NMR relaxation properties and to obtain high-quality spectra (14). Upon deuteration, that is also achieved during biosynthesis, all the carbon-bound protons in the protein will be replaced by deuterons from the D_2O medium. The quality of the NMR spectra is usually significantly improved by deuteration and indeed the backbone sequential resonance assignment could be finished for nearly 90% of all the residues of Ca^{2+}-CIB1 (15). However, deuteration also removes the most abundant NMR probe, the protons, which prevents the subsequent structure determination from using the classical proton–proton Nuclear Overhauser Effect (NOE) approach. Fortunately, a strategy of using specific metabolic precursors to selectively label the methyl groups of Ile, Leu, and Val (I/L/V) residues in proteins has been developed by Kay and coworkers (16). This approach has been used in the structure determination of several high-molecular-weight proteins (17, 18) including CIB1 (19). Apart from structure determination, the I/L/V methyl group-labeled

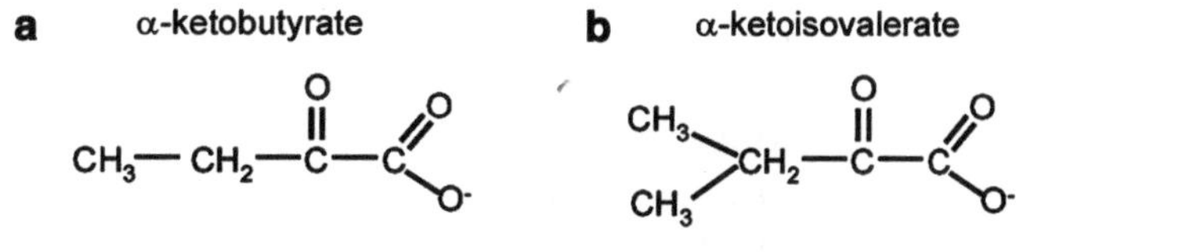

c
α-ketobutyrate ($^{13}C_4$;3,3-D_2)

*CH_3*CD_2*CO*COO-

d
α-ketoisovalerate (1,2,3,4-$^{13}C_4$;3,4,4,4-D_4)

*CH_3*CD(CD_3)*CO*COO-

e
α-ketobutyrate (methyl-^{13}C;3,3-D_2)

$^{13}CH_3CD_2$COCOO-

f
α-ketoisovalerate (3-methyl-^{13}C;3,4,4,4-D_4)

*CH_3CD(CD_3)COCOO-

The * indicates ^{13}C labelling

Fig. 1. Methyl group isotope labeling reagents used in this study.

and otherwise deuterated CIB1 protein has also been used in studies of its interactions with the cytoplasmic domain of the human integrin αIIb subunit. Moreover the solvent exposure of the methyl group bearing amino acid side chains in its hydrophobic pocket could be studied at residue-specific resolution using the nitroxide spin label TEMPOL (19, 20). Specifically, during biosynthesis of the protein in *Escherichia coli* α-ketobutyrate is required for Ile methyl group labeling and α-ketoisovalerate is a precursor that gives rise to Leu and Val methyl group labeling (16) (Fig. 1a, b). For medium- to large-sized CBPs and their protein complexes, this NMR/methyl group labeling approach can help us understand their roles in the calcium signaling pathways from both a structural and a dynamic perspective.

In this protocol, we describe the purification and stable isotope labeling strategies that have been used in the structural investigation of CIB1 using solution-state NMR. Moreover, we also describe the methodologies required to study the interactions of this protein with other proteins. The application of these experimental strategies also provides a path to study the structure/function relationships of the entire neuronal calcium sensor (NCS) family of CBPs, which are of a similar size (190–200 residues) to CIB1. These proteins are usually expressed in the central nervous system, where they have unique and important functions (21).

2. Materials

2.1. Bacterial Strain and Expression Vector

A synthetic CIB1 gene, with codons optimized for protein expression in *E. coli*, was cloned into the pET19b vector (Novagen). The entire construct contains an N-terminal His-tag followed by an enterokinase cleavage site and the complete CIB1 gene. For many proteins, the His-tag can be proteolytically removed after purification; however this is not the case for CIB1, because we found that the cleavage reduced the protein's solubility. The uncleaved His-tag increases the overall molecular weight of the expressed protein to ~24 kDa. The CIB1 plasmid was transformed into the *E. coli* strain ER2566 (New England Biolabs) for protein expression.

2.2. Protein Expression, Labeling, and Purification

1. M9/H_2O medium (1 L): 12 g Na_2HPO_4, 3 g KH_2PO_4, 0.5 g NaCl, 1 mM $MgSO_4$, 100 μM $CaCl_2$, 100 mg ampicillin, 10 mg $FeCl_2$, 6 mg $MnCl_2$, 1 mg thiamine, 3 g unlabeled glucose or $^{13}C_6$-glucose, 1 g NH_4Cl or ^{15}N-NH_4Cl, 10 mL of 100× vitamin mix (Sigma), and distilled H_2O. The phosphate buffer in 1 L H_2O (pH ~7.4–7.6) can be autoclaved (see **Note 1**). Additional ingredients can be sterilized by filtering through 0.22 μm pore size filters, and then added into 1 L of the autoclaved medium. Isopropyl β-d-1-thiogalactopyranoside (IPTG) is needed at a later stage for protein induction.
2. M9/D_2O medium (1 L): Use the same chemicals and filtration devices as used for the regular M9/H_2O medium, except use 99% D_2O to replace H_2O. Sterilization of the growth medium was achieved by filtering the final mixture through a 0.22 μm pore size filtration device.
3. Laboratory incubator for bacterial growth at 37°C.
4. Methyl group labeling reagents: 70 mg α-ketobutyrate ($^{13}C_4$;3,3-D_2) (Fig. 1c), 120 mg α-ketoisovalerate (1,2,3,4-$^{13}C_4$;3,4,4,4-D_4) (Fig. 1d), 70 mg α-ketobutyrate (methyl-^{13}C;3,3-D_2) (Fig. 1e), and 120 mg α-ketoisovalerate (3-methyl-^{13}C;3,4,4,4-D_4) (Fig. 1f) for 1 L medium; all of the above can be obtained from Cambridge Isotope Laboratories (CIL) or other suppliers of stable isotope-enriched material.
5. 100 mM phenylmethylsulfonyl fluoride (PMSF) in ethanol.
6. Laboratory large-capacity refrigerated centrifuge.
7. Ni-Sepharose 6 Fast Flow resin (Amersham Biosciences).
8. 6–8 kDa cut-off dialysis membrane (Spectrum Laboratories).
9. Resuspension buffer for purification: 50 mM HEPES buffer, 500 mM NaCl, pH 7.5.
10. Wash buffer for purification: 50 mM HEPES buffer, 500 mM NaCl, pH 7.5, 80 mM imidazole.
11. Elution buffer for purification: 50 mM HEPES buffer, 500 mM NaCl, pH 7.5, 300 mM imidazole.
12. UV/Visible-spectrophotometer, and both plastic and quartz cuvettes.
13. A commercially obtained synthetic peptide representing the cytoplasmic domain of the human integrin αIIb subunit (hereafter referred to as αIIb-p).

3. Methods

3.1. Expression of CIB1 in 1 L Regular M9/H_2O Medium

1. Prepare 10 mL overnight culture of CIB1 in LB broth.
2. Transfer the 10 mL overnight culture into a 3 L flask which contains 1 L M9 H_2O medium, and grow at 37°C with shaking until $OD_{600} = 0.8$, which typically takes 4–5 h. The OD_{600} absorbance can be measured using a UV/Visible spectrophotometer and disposable plastic cuvettes.
3. Add IPTG for a final concentration of 100 mg/L, and continue to grow at 37°C for 4 h (see Note 2).
4. Harvest the cells by centrifugation at $8{,}000 \times g$ and 4°C, and then freeze the cell pellet at −20°C for future purification.

3.2. Expression of CIB1 in 1 L M9/D_2O Medium (This Procedure Was Adapted from Tugarinov et al. 2006)

1. Prepare a 10 mL LB broth overnight culture of CIB1 and grow at 37°C.
2. Spin down the cells and resuspend a fraction of the cells in 50 mL M9/H_2O medium (containing unlabeled NH_4Cl and glucose) to achieve a starting OD_{600} of approximately 0.1, and continue to grow at 37°C until $OD_{600} = 0.6–0.8$.
3. Spin down the cells, resuspend all into 200 mL M9/D_2O medium (see Subheading 3.4 for choices of NH_4Cl and glucose with different isotopic labelings), and grow at 37°C until $OD_{600} = 0.6–0.8$.
4. Add 800 mL M9/D_2O medium into the 200 mL *E. coli* culture to have in total 1,000 mL medium, and continue to grow at 37°C.
5. If methyl group stable isotope labeling is not performed, continue to grow until OD = 0.5–0.6, add 100 mg IPTG for 1 L medium, and continue protein expression at 37°C until a final OD = 1.0–1.2 is reached, which might take 6–10 h (see Notes 3 and 4).
6. If methyl group stable isotope labeling is performed, continue growth until $OD_{600} = 0.3–0.4$, and then add the methyl labeling precursors; after growth for another hour, add IPTG to achieve a final concentration of 100 mg/L. Protein expression can proceed for 6–8 h at 37°C and the final OD600 will be around 1.0–1.2 (see Notes 3 and 4).

3.3. Purification of CIB1 Using a Ni-Column (See Note 5)

We used a one-step purification procedure and obtained high-purity CIB1 with a typical yield of 20–30 mg per 1 L of minimal medium, as assessed by Sodium Dodecyl Sulfate-PolyAcrylamide Gel Electrophoresis (SDS-PAGE) (Fig. 2).

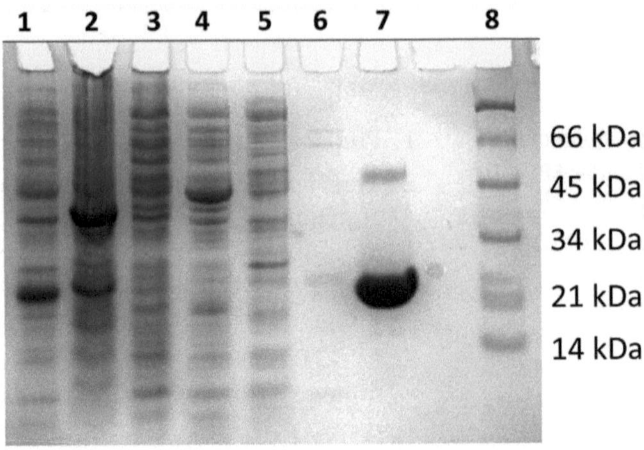

Fig. 2. SDS-PAGE for the purification of CIB1 using a Ni-column: 1. Supernatant after French press and centrifugation. 2. Pellet after French press and centrifugation. 3. Flow through during loading of the supernatant onto the Ni column. 4. Collection during washing with 0 mM imidazole. 5. Collection during washing with 40 mM imidazole. 6. Collection during washing with 80 mM imidazole. 7. Elution with 300 mM imidazole. 8. Protein molecular weight markers.

1. Resuspend cell pellets in 50 mL resuspension buffer. Add 50 μL of 100 mM PMSF to inhibit proteases (see Note 6).
2. Use a French press to break *E. coli* cells at 1,000 PSI for multiple times.
3. Centrifuge the cell lysate at 30,000 × *g* for 30–40 min.
4. During centrifugation, use the 100 mM $NiSO_4$ solution to charge the Ni-column which was filled with the Ni Sepharose Fast Flow resin.
5. Pre-equilibrate the Ni-column with 2–4 column volumes of resuspension buffer and wash off any free Ni^{2+} ions.
6. Load the supernatant of cell lysate after high-speed centrifugation onto the equilibrated Ni-column using a laboratory peristaltic pump.
7. Wash the column with 5 column volumes of wash buffer (see Note 7).
8. Elute CIB1 with the elution buffer; collect the fractions (fraction volume 1–1.5 mL) with positive A280 absorbance (using the elution buffer to blank the UV spectrophotometer). A UV/visible spectrophotometer and quartz cuvettes can be used to read the absorbance.
9. Dialyze CIB1 using a 6–8 kDa cutoff dialysis membrane in 4 L H_2O containing 4 g NH_4HCO_3 to maintain pH; change the dialysis buffer four to six times. Once the dialysis is finished, the protein sample can be lyophilized or concentrated using a 3 kDa

molecular cut-off filter (Millipore) and then flash frozen using liquid nitrogen followed by storage at –80°C (see Note 8).

10. Apart from the above protocol, Fast Protein Liquid Chromatography (FPLC) can also be successfully employed.

3.4. Labeling Strategies for Different NMR Experiments

1. Uniform ^{15}N-labeling or ^{15}N,^{13}C-labeling of CIB1: ^{15}N-labeling of a protein is usually the first step in protein NMR studies to check the overall folding of the protein and the quality of the NMR spectra by recording a ^{1}H,^{15}N-HSQC spectrum. Normally for proteins with an MW < 20 kDa, ^{15}N-,^{13}C-labeled protein can be used to complete the NMR backbone sequential assignment (this approach was not successful for CIB1, vide infra). For ^{15}N-labeling, $^{15}NH_4Cl$ and unlabeled glucose can be used to prepare the M9/H_2O medium and the previous procedure of using the M9/H_2O medium can be followed. For ^{15}N,^{13}C-labeling, $^{15}NH_4Cl$ and $^{13}C_6$-glucose should be used to prepare the M9/H_2O medium. If it turns out that the protein under study requires deuteration to achieve the proper NMR spectral quality, the protein deuteration procedures described below need to be used (as was the case for CIB1).
2. Backbone sequential resonance assignment: Uniformly ^{2}H-, ^{13}C-, and ^{15}N-labeled CIB1 was used for the backbone assignment of calcium-bound CIB1 (Ca^{2+}-CIB1). ^{13}C-glucose, ^{15}N-NH_4Cl, and 99% D_2O were used in the preparation of M9/D_2O medium (see Note 9). Overall ~87% deuteration level was achieved, which facilitated the backbone assignment of CIB1 using a suite of three-dimensional (3-D) NMR experiments (see Note 10). A typical ^{1}H,^{15}N-HSQC spectrum of CIB1 is shown in Fig. 3, with all the backbone assignments indicated.
3. Backbone dynamics: ^{2}H,^{15}N-CIB1 can be used for measurement of the backbone dynamics on picosecond–nanosecond timescale, e.g., backbone heteronuclear {^{1}H}^{15}N NOE dynamics, T1, T2, etc. The same sample can also be used to study microsecond motions through NMR relaxation dispersion experiments. This labeling usually gives improved signal/noise spectra over ^{15}N-labeled CIB1 which is not deuterated. For the preparation of the M9/D_2O medium, unlabeled glucose, ^{15}N-NH_4Cl, and 99% D_2O are needed (see Note 11).
4. Methyl side chain assignment: Uniformly ^{2}H,^{13}C- and Ile/Leu/Val methyl ^{1}H-labeled CIB1 was used to obtain the methyl group assignments. Unlabeled NH_4Cl (see **Note 12**), $^{13}C_6$-glucose, 99% D_2O, as well as the methyl labeling precursors α-ketobutyrate ($^{13}C_4$;3,3-D_2) (Fig. 1c) and α-ketoisovalerate (1,2,3,4-$^{13}C_4$;3,4,4,4-D_4) (Fig. 1d) were used to prepare this sample. The above-mentioned procedure of using M9/D_2O medium for methyl group labeling was employed (Fig. 4) (see Notes 13 and 14).

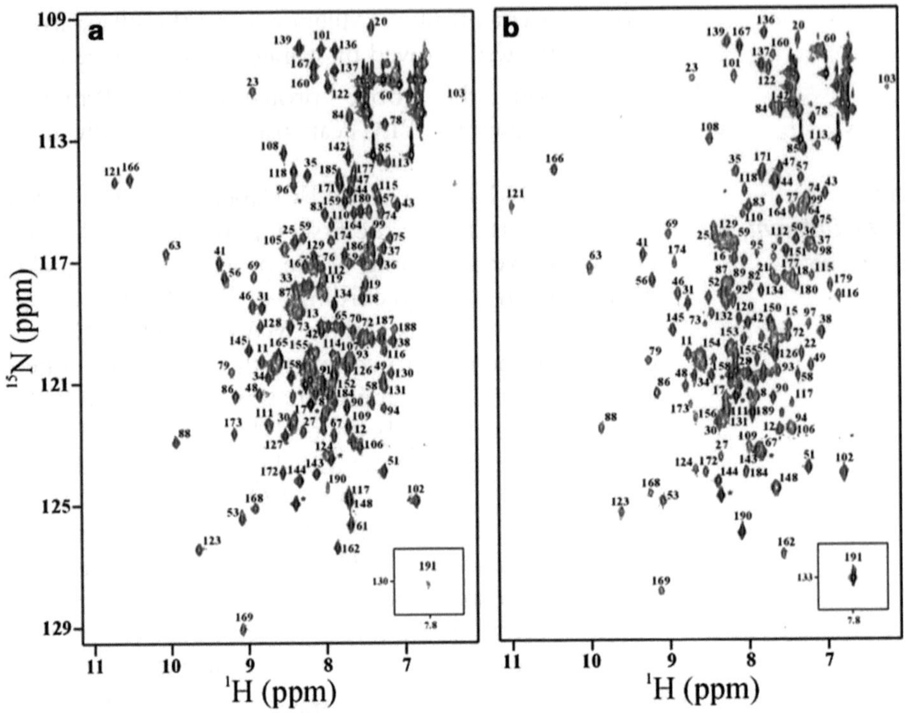

Fig. 3. $^{1}H,^{15}N$ HSQC spectra of Ca^{2+}-CIB1 (**a**) and Ca^{2+}-CIB1 in the complex with αIIb-p (**b**). The backbone assignment of the peaks is indicated by their residue number. The Chemical Shift Perturbation (CSP) can be extracted from these two sets of backbone assignment to characterize the interaction between Ca^{2+}-CIB1 and αIIb. (This research was originally published in the Journal of Biological Chemistry. Yamniuk AP, Ishida H, Vogel HJ. *J. Biol. Chem.* 2006; 281:26455–64 © the American Society for Biochemistry and Molecular Biology).

5. Stereospecific assignment of the methyl groups of Leu and Val: 10% ^{13}C-labeled CIB1 was used for the stereospecific assignment of the methyl groups of the Leu and Val residues in CIB1. Unlabeled NH_4Cl, 10% ^{13}C-glucose/90% unlabeled glucose, and M9/H_2O medium were used to prepare this sample. If in total 3 g glucose is used during expression, the ratio would be 0.3 g $^{13}C_6$-glucose and 2.7 g unlabeled glucose (see Note 15).
6. Acquisition of NOEs: Uniformly $^{2}H,^{15}N$- and I/L/V methyl ($^{1}H,^{13}C$)-labeled CIB1 were used to acquire NOEs using ^{13}C-edited and ^{15}N-edited NOESY spectra. D_2O, $^{15}NH_4Cl$, unlabeled glucose, as well as the methyl labeling precursors α-ketobutyrate (methyl-^{13}C;3,3-D_2) (Fig. 1e) and α-ketoisovalerate (3-methyl-^{13}C;3,4,4,4-D_4) (Fig. 1f) were used to prepare this sample. Since this sample is isotopically labeled on both the backbone NH and on the methyl groups of I/L/V, methyl–methyl NOEs, methyl–NH NOEs, and NH–NH NOEs can be acquired using both ^{15}N-edited and ^{13}C-edited NOESY experiments.
7. Methyl group cross-saturation experiments for studies of protein/protein interactions: Uniformly ^{2}H- and I/L/V methyl ($^{1}H,^{13}C$)-labeled CIB1 were used for this study. 99.9% D_2O

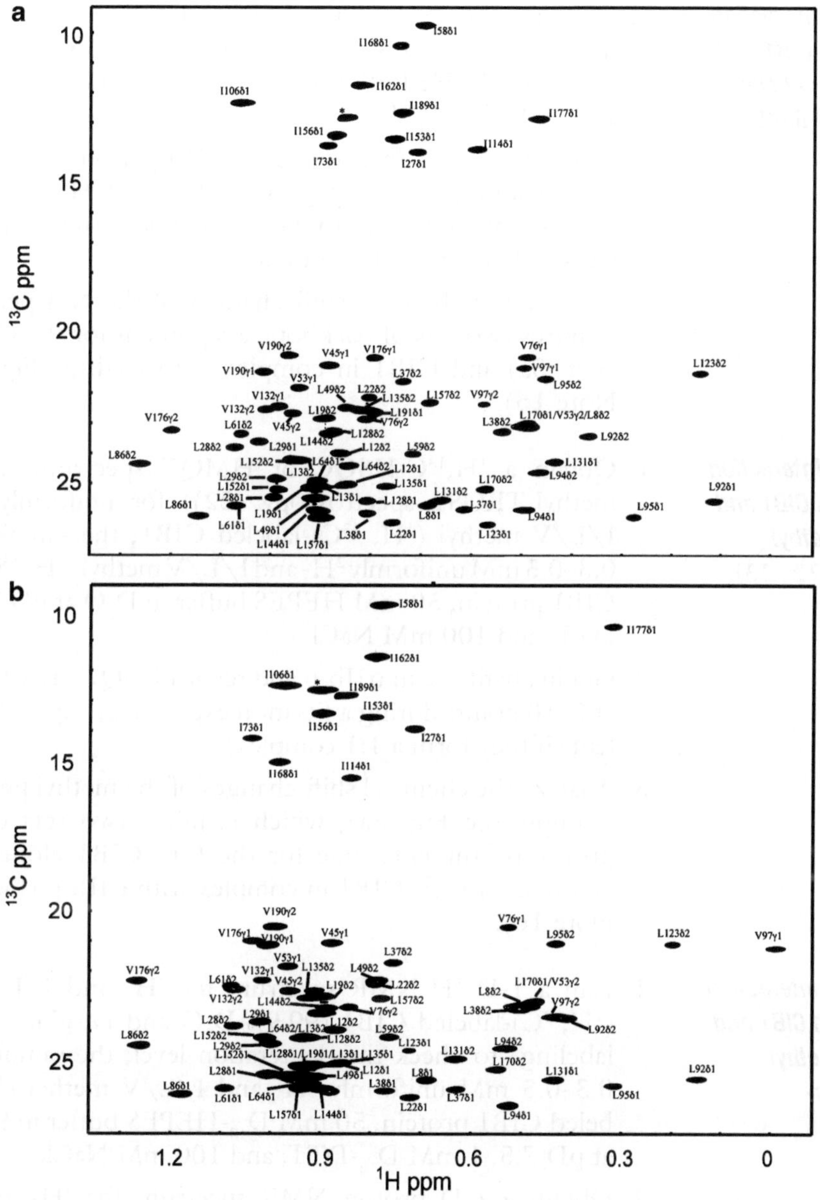

Fig. 4. ^1H,^{13}C-HSQC spectra of the methyl groups of Ca^{2+}-CIB1 (**a**) and Ca^{2+}-CIB1 in the complex with αIIb-p (**b**). The stereospecific assignments are indicated in the figure with the residue number. The methyl-group CSP can be extracted from these two sets of backbone assignments to characterize the interaction between Ca^{2+}-CIB1 and αIIb-p. (This research was originally published in the Journal of Biological Chemistry. Huang H, Ishida H, Yamniuk AP, Vogel HJ. *J. Biol. Chem.* 2011; 286:17181–92 © the American Society for Biochemistry and Molecular Biology).

and D_7-glucose as well as the methyl labeling precursors α-ketobutyrate (methyl-^{13}C;3,3-D_2) (Fig. 1e) and α-ketoisovalerate (3-methyl-^{13}C;3,4,4,4-D_4) (Fig. 1f) were used to prepare this sample. The use of 99.9% D_2O and D_7-glucose is required here to remove the maximum number of protons from the protein.

3.5. Protein Interaction Studies with CIB1 and αIIb-p by Backbone NH Chemical Shift Perturbation

1. Obtain a $^{1}H,^{15}N$-HSQC spectrum for $^{2}H,^{15}N$-labeled CIB1; the sample contains 0.3–0.5 mM $^{2}H,^{15}N$-labeled CIB1 protein, 50 mM HEPES buffer in H_2O at pH 7.5, 1 mM DTT, and 100 mM NaCl.
2. Gradually titrate in aliquots of αIIb-p from a concentrated stock solution of the peptide and record HSQC spectra at 7–10 points during a titration experiment (up to 1.2 equivalents if they form a 1:1 complex).
3. Analyze the chemical shift changes of the NH peaks, which requires two sets of backbone assignment for the CIB1 alone (Fig. 3a) and CIB1 in complex with αIIb-p (Fig. 3b) (see Note 16).

3.6. Protein Interaction Studies with CIB1 and αIIb-p by Methyl Group CSP (22, 23)

1. Obtain a $^{1}H,^{13}C$-HSQC or HMQC spectrum (also called methyl-TROSY spectroscopy (22)) for uniformly ^{2}H- and I/L/V methyl (^{1}H, ^{13}C)-labeled CIB1; the sample contains 0.3–0.5 mM uniformly ^{2}H- and I/L/V methyl ($^{1}H,^{13}C$)-labeled CIB1 protein, 50 mM HEPES buffer in D_2O at pD 7.5, 1 mM DTT, and 100 mM NaCl.
2. Gradually titrate in αIIb-p and record HSQC/HMQC spectra at 7–10 points during a titration experiment (up to 1.2 equivalents if they form a 1:1 complex).
3. Analyze the chemical shift changes of the methyl peaks (for an example see Fig. 5a), which requires two sets of methyl-group assignments: one for the Ca^{2+}-CIB1 alone (Fig. 4a) and one for Ca^{2+}-CIB1 in complex with αIIb-p (Fig. 4b) (see Note 16).

3.7. Protein Interaction Studies with CIB1 and αIIb-p by Methyl Group Cross-Saturation

1. Run a 1-D ^{1}H-NMR spectrum for ^{2}H- and I/L/V methyl ($^{1}H,^{13}C$)-labeled CIB1 (99.9% D_2O and D_7-glucose used for labeling) to check the deuteration level; the sample contains 0.3–0.5 mM uniformly ^{2}H- and I/L/V methyl ($^{1}H,^{13}C$)-labeled CIB1 protein, 50 mM D_{18}-HEPES buffer in 99.9% D_2O at pD 7.5, 1 mM D_{10}-DTT, and 100 mM NaCl.
2. Obtain a 1-D proton NMR spectrum for ^{2}H- and I/L/V methyl ($^{1}H,^{13}C$)-labeled CIB1 using the pulse sequence with selective pre-saturation on the aliphatic, Hα and Hβ spectral regions (3.5–8.5 ppm, centered at 6.0 ppm). A typical starting irradiation time will be between 1 and 2 s but researchers should adjust this carefully for a relatively strong irradiation while the signal of the methyl groups of CIB1 should be relatively unaffected.
3. Run a 2-D HSQC/HMQC spectrum on the ^{2}H- and I/L/V methyl ($^{1}H,^{13}C$)-labeled CIB1 sample with pre-saturation using the parameters obtained from step 2 above. This is a control

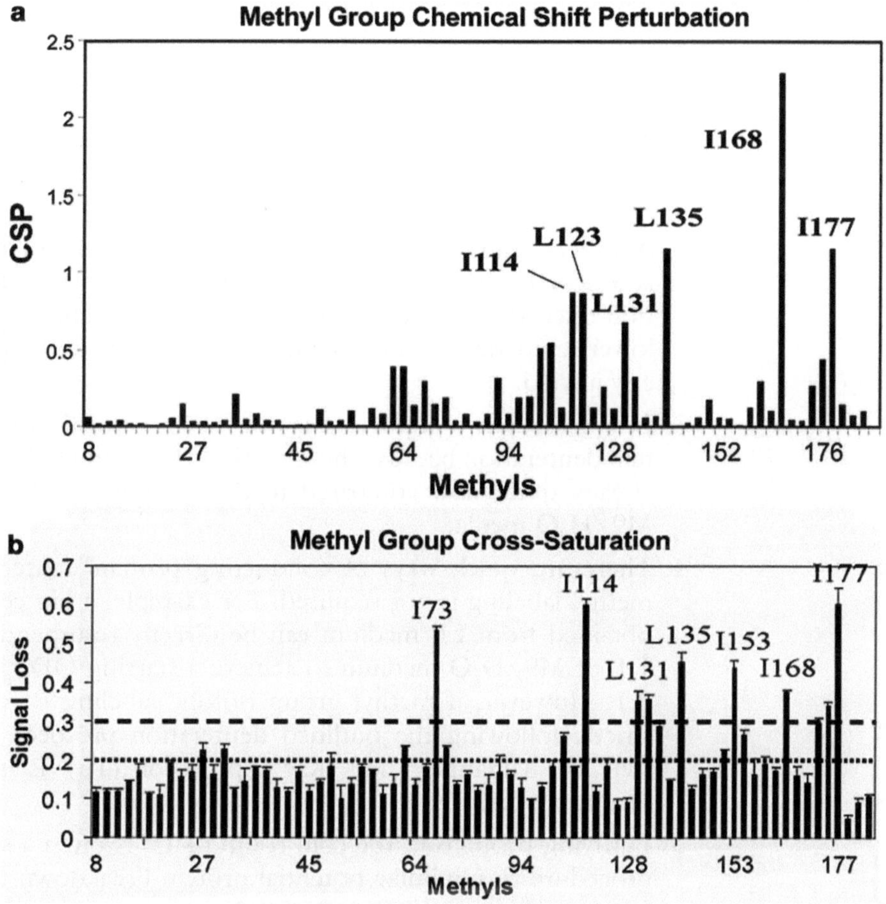

Fig. 5. The use of isotope-labeled methyl groups in the NMR studies of CIB1 with αIIb-p. (**a**) Chemical shift perturbation studies using methyl groups (this figure was originally published in the Journal of Biological Chemistry. Huang H, Ishida H, Yamniuk AP, Vogel HJ. *J. Biol. Chem.* 2011; 286:17181–92 © the American Society for Biochemistry and Molecular Biology). (**b**) Cross-saturation studies using methyl groups (this figure was originally published in the Journal of American Chemical Society. Huang H and Vogel HJ. *J. Am. Chem.* Soc. 2012;134: 3864–72 © American Chemical Society) (23). Consistent results were obtained for these two experiments and the data show that the amino acid side chains in CIB1 that interact with αIIb-p reside mainly in the C-lobe of the protein.

experiment to determine the effect of pre-saturation on CIB1 when the protonated binding partner is not present.

4. Add the synthetic αIIb-p (protonated protein partner) into the above NMR sample for saturation and obtain two sets of 2-D experiment as described above for this new complex sample of CIB1/αIIb-p for saturation time 0 s and 1–2 s (e.g., 1.5 s), respectively.
5. Analyze the intensity change of the ^{2}H- and I/L/V methyl (^{1}H,^{13}C)-labeled CIB1 in complex with αIIb-p both with and without the pre-saturation using the formula *Signal Loss* = 1 − $(I_{(Tsat=1.5)}/I_{(Tsat=0)})$. Typical results are shown in Fig. 5b.

4. Notes

1. The pH of the medium or buffer can be adjusted using small aliquots of 1 M NaOH or 1 M HCl stock solutions and a pH meter equipped with a calibrated glass electrode.
2. Another commonly used protein expression protocol in *E. coli* is to lower the temperature to 16–18°C for overnight induction once IPTG is added. The elongated induction period at lower temperature will sometimes help the protein fold properly in vivo.
3. Low-temperature induction is not recommended during protein deuteration because the growth of *E. coli* in D_2O is usually already quite slow compared to the growth in LB broth or M9/H_2O media.
4. There are quick ways of conducting protein deuteration if methyl labeling is not required. For example, a big cell pellet obtained from LB medium can be directly resuspended into 1 L of M9/D_2O medium to achieve a starting $OD_{600}=0.05$–0.1. However, if methyl group isotope labeling is required, strictly following the outlined deuteration protocol step by step or Tugarinov and Kay's protocol (16) is strongly recommended.
5. Purifying proteins in the cold room (4°C) is often a standard procedure to minimize potential protein breakdown through co-purifying proteolytic enzymes. However, in our studies, the CIB1 proteins with different isotope labelings were all purified at room temperature. We have not experienced any problems in purifying CIB1 at room temperature using the quick one-step Ni-column purification.
6. A protease inhibitor cocktail tablet (Sigma-Aldrich) can also be used to prevent nonspecific protein cleavage.
7. For wash buffer, typically 60–80 mM imidazole can be used.
8. For deuterated proteins, an unfolding/refolding process using denaturants sometimes needs to be performed to facilitate the D/H backward by hydrogen exchange of the nitrogen-bound deuterons. However, this step is not required for deuterated CIB1, for which 88% of the resonances in the backbone could be assigned without using an unfolding/refolding process (15).
9. If one uses $^{13}C_6,D_7$-glucose, a higher percentage (~95%) deuteration can be achieved, but the cost will be much higher than

using $^{13}C_6$-glucose. Therefore, $^{13}C_6$-glucose is recommended if expense is an issue.

10. The 3-D NMR experiments used for the assignment of CIB1 were Transverse relaxation optimized spectroscopy (TROSY) versions of HNCACB, HN(CA)CO, HN(CO)CACB, and HNCO (15).

11. Uniformly ^{15}N-labeled CIB1 can also be used for backbone heteronuclear $\{^1H\}^{15}N$ NOE dynamics. $^{15}NH_4Cl$, unlabeled glucose, and M9/H_2O medium could be used to prepare the M9/H_2O medium. It should be noted that the ^2H,^{13}C,^{15}N-labeled CIB1 sample cannot be used for the NMR dynamics measurements, because the ^{13}C-labeled backbone carbons would affect the relaxation of the backbone ^{15}N atoms.

12. When preparing samples for methyl group NMR studies, it is usually a good idea to label backbone amide nitrogens with $^{15}NH_4Cl$ as well even though ^{15}N-labeling is not required in every case. With ^{15}N-labeling of the backbone, the quality and correct folding of the protein sample can always be quickly assessed by running a ^1H,^{15}N-HSQC or TROSY-HSQC spectrum.

13. A recently published 3D multiple-quantum (MQ) (H) CCmHm-TOCSY experiment was employed to assign the methyl groups of CIB1 (24). This methyl assignment approach relies on the chemical shifts of Cα and Cβ, which have been obtained previously from the protein backbone assignment work (15, 25, 26).

14. In addition to the widely performed I/L/V methyl labeling, the methyl groups in the side chains of Met (27, 28) or Ala (29) can also be isotopically labeled and used for NMR studies. Apart from the specific methyl group labeling, a strategy of simply using the protein expressed with $^{13}C_6$-glucose in D_2O medium in combination with advanced NMR techniques can also be used to assign the methyl groups for most methyl-group containing residues (Thr, Ala, Ile, Leu, Val) in large proteins (30).

15. Constant time (CT)-HSQC can be used to stereospecifically assign the methyl groups of Leu and Val in CIB1.

16. When binding of the peptide to the protein is tight (Kd ≤ 1 μM), one usually obtains two separate sets of peaks (slow exchange) in NMR spectra for the two interconverting states of the protein. When the binding is weaker and fast exchange NMR conditions prevail, one can simply follow the shift of the assigned peaks in the spectra, and only the original assignment is needed.

Acknowledgments

This research is supported by an operating grant from the Canadian Institutes of Health Research. H.H. was the recipient of a Studentship award and H.J.V. is the holder of a Scientist award from the Alberta Heritage Foundation for Medical Research.

References

1. Gifford JL, Walsh MP, Vogel HJ (2007) Structures and metal-ion-binding properties of the Ca2+-binding helix-loop-helix EF-hand motifs. Biochem J 405:199–221
2. Naik UP, Patel PM, Parise LV (1997) Identification of a novel calcium-binding protein that interacts with the integrin alpha(IIb) cytoplasmic domain. J Biol Chem 272:4651–4654
3. Yamniuk AP, Vogel HJ (2006) Insights into the structure and function of calcium- and integrin-binding proteins. Calcium Bind Prot 1:150–155
4. Leisner TM, Yuan WP, DeNofrio JC, Liu J, Parise LV (2007) Tickling the tails: cytoplasmic domain proteins that regulate integrin alpha IIb beta 3 activation. Curr Opin Hematol 14:255–261
5. Jarman KE, Moretti PAB, Zebol JR, Pitson SM (2010) Translocation of sphingosine kinase 1 to the plasma membrane is mediated by calcium- and integrin-binding protein 1. J Biol Chem 285:483–492
6. Leisner TM, Liu MJ, Jaffer ZM, Chernoff J, Parise LV (2005) Essential role of CIB1 in regulating PAK1 activation and cell migration. J Cell Biol 170:465–476
7. Yoon KW, Cho JH, Lee JK, Kang YH, Chae JS, Kim YM, Kim J, Kim EK, Kim SE, Baik JH, Naik UP, Cho SG, Choi EJ (2009) CIB1 functions as a Ca2+-sensitive modulator of stress-induced signaling by targeting ASK1. Proc Natl Acad Sci USA 106:17389–17394
8. Kauselmann G, Weiler M, Wulff P, Jessberger S, Konietzko U, Scafidi J, Staubli U, Bereiter-Hahn J, Strebhardt K, Kuhl D (1999) The polo-like protein kinases Fnk and Snk associate with a Ca2+- and integrin-binding protein and are regulated dynamically with synaptic plasticity. EMBO J 18:5528–5539
9. Wu XT, Lieber MR (1997) Interaction between DNA-dependent protein kinase and a novel protein, KIP. Mutat Res 385:13–20
10. Hui L, Ji CN, Hui B, Lv TD, Ha XQ, Yang JS, Cai WQ (2009) The oncoprotein LMO3 interacts with calcium- and integrin-binding protein CIB. Brain Res 1265:24–29
11. Heineke J, Auger-Messier M, Correll RN, Xu JA, Benard MJ, Yuan WP, Drexler H, Parise LV, Molkentin JD (2010) CIB1 is a regulator of pathological cardiac hypertrophy. Nat Med 16:872–879
12. Blazejczyk M, Wojda U, Sobczak A, Spilker C, Bernstein HG, Gundelfinger ED, Kreutz MR, Kuznicki J (2006) Ca2+-independent binding and cellular expression profiles question a significant role of calmyrin in transduction of Ca2+-signals to Alzheimer's disease-related presenilin 2 in forebrain. Biochim Biophys Acta 1762:66–72
13. Stabler SM, Ostrowski LL, Janicki SM, Monteiro MJ (1999) A myristoylated calcium-binding protein that preferentially interacts with the Alzheimer's disease presenilin 2 protein. J Cell Biol 145:1277–1292
14. Gardner KH, Kay LE (1998) The use of H-2, C-13, N-15 multidimensional NMR to study the structure and dynamics of proteins. Annu Rev Biophys Biomol Struct 27:357–406
15. Yamniuk AP, Ishida H, Vogel HJ (2006) The interaction between calcium- and integrin-binding protein 1 and the alpha IIb integrin cytoplasmic domain involves a novel C-terminal displacement mechanism. J Biol Chem 281:26455–26464
16. Tugarinov V, Kanelis V, Kay LE (2006) Isotope labeling strategies for the study of high-molecular-weight proteins by solution NMR spectroscopy. Nat Protoc 1:749–754
17. Tugarinov V, Choy WY, Orekhov VY, Kay LE (2005) Solution NMR-derived global fold of a monomeric 82-kDa enzyme. Proc Natl Acad Sci USA 102:622–627
18. Wang X, Weldeghiorghis T, Zhang GF, Imperiali B, Prestegard JH (2008) Solution structure of Alg13: the sugar donor subunit of a yeast N-acetylglucosamine transferase. Structure 16:965–975
19. Huang H, Ishida H, Yamniuk AP, Vogel HJ (2011) Solution structures of Ca(2+)-CIB1

and Mg(2+)-CIB1 and their interactions with the platelet integrin alpha IIb cytoplasmic domain. J Biol Chem 286:17181–17192
20. Yuan T, Ouyang H, Vogel HJ (1999) Surface exposure of the methionine side chains of calmodulin in solution—a nitroxide spin label and two-dimensional NMR study. J Biol Chem 274:8411–8420
21. Ikura M, Ames JB (2006) Genetic polymorphism and protein conformational plasticity in the calmodulin superfamily: two ways to promote multifunctionality. Proc Natl Acad Sci USA 103:1159–1164
22. Takahashi H, Miyazawa M, Ina Y, Fukunishi Y, Mizukoshi Y, Nakamura H, Shimada I (2006) Utilization of methyl proton resonances in cross-saturation measurement for determining the interfaces of large protein-protein complexes. J Biomol NMR 34: 167–177
23. Huang H, Vogel HJ (2012) Structural basis for the activation of platelet integrin αIIbβ3 by calcium- and integrin-binding protein 1. J Am Chem Soc 134: 3864–3872
24. Yang DW, Zheng Y, Liu DJ, Wyss DF (2004) Sequence-specific assignments of methyl groups in high-molecular weight proteins. J Am Chem Soc 126:3710–3711
25. Yamniuk AP, Silver DM, Anderson KL, Martin SR, Vogel HJ (2007) Domain stability and metal-induced folding of calcium- and integrin-binding protein 1. Biochemistry 46:7088–7098
26. Yamniuk AP, Gifford JL, Linse S, Vogel HJ (2008) Effects of metal-binding loop mutations on ligand binding to calcium- and integrin-binding protein 1. Evolution of the EF-hand? Biochemistry 47:1696–1707
27. Gifford JL, Ishida H, Vogel HJ (2011) Fast methionine-based solution structure determination of calcium-calmodulin complexes. J Biomol NMR 50:71–81
28. Siivari K, Zhang MJ, Palmer AG, Vogel HJ (1995) NMR studies of the methionine methyl-groups in calmodulin. FEBS Lett 366:104–108
29. Isaacson RL, Simpson PJ, Liu M, Cota E, Zhang X, Freemont P, Matthews S (2007) A new labeling method for methyl transverse relaxation-optimized spectroscopy NMR spectra of alanine residues. J Am Chem Soc 129:15428–15429
30. Otten R, Chu B, Krewulak KD, Vogel HJ, Mulder FAA (2010) Comprehensive and cost-effective NMR spectroscopy of methyl groups in large proteins. J Am Chem Soc 132: 2952–2960

Chapter 8

Analysis of Calcium-Induced Conformational Changes in Calcium-Binding Allergens and Quantitative Determination of Their IgE Binding Properties

Nuria Parody, Miguel Ángel Fuertes, Carlos Alonso, and Yago Pico de Coaña

Abstract

The polcalcin family is one of the most epidemiologically relevant families of calcium-binding allergens. Polcalcins are potent plant allergens that contain one or several EF-hand motifs and their allergenicity is primarily associated with the Ca^{2+}-bound form of the protein. Conformation, stability, as well as IgE recognition of calcium-binding allergens greatly depend on the presence of protein-bound calcium ions.

We describe a protocol that uses three techniques (SDS-PAGE, circular dichroism spectroscopy, and ELISA) to describe the effects that calcium has on the structural changes in an allergen and its IgE binding properties.

Key words: Allergen, Polcalcin, Calcium, EF-hand, IgE binding

1. Introduction

Calcium-binding allergens include a wide variety of proteins that have been isolated from pollen, parasites, and fish (1). These allergens are highly cross-reactive and include proteins that have a wide variety of structural, functional, and calcium-binding properties.

One of the most epidemiologically relevant families of calcium-binding allergens is the polcalcin family, which includes pollen allergens from grasses, trees, and weeds. All these allergens bind calcium through interactions with EF-hand motifs. The polcalcin family contains at least 35 pollen allergens having two, three, or

Claus W. Heizmann (ed.), *Calcium-Binding Proteins and RAGE: From Structural Basics to Clinical Applications*, Methods in Molecular Biology, vol. 963, DOI 10.1007/978-1-62703-230-8_8, © Springer Science+Business Media New York 2013

four EF-hand motifs (1). These motifs consist of two perpendicularly placed α-helices and one interhelical loop forming a single calcium-binding site (2). Calcium binding to EF-hand motifs has shown to alter the secondary and tertiary structure of the proteins because there is a change in the orientation of the two amphipathic α-helices flanking the binding motif (3). These conformational changes might affect the allergen's stability as well as its IgE binding properties (4–7). A detailed knowledge of these molecular changes may be highly significant for the molecular design of hypoallergenic variants that can be used in allergen-specific immunotherapy.

This chapter contains a detailed protocol of three methods used to analyze the effect of calcium on the structural and IgE binding properties of a given allergen. Here, we focused on the analysis of the Cup a 4 protein, an allergen from the Arizona Cypress that was recently described in our laboratory (8). The polcalcin family Cup a 4 protein contains 4 EF-hand motifs and is highly cross-reactive with other calcium-binding allergens (unpublished results). The calcium binding to the Cup a 4 protein was proven by its different migration rate on SDS-PAGE in the presence and absence of calcium. A more detailed secondary structure analysis was carried out through circular dichroism spectroscopy. Finally, the IgE binding capability of the allergen in the presence or absence of calcium is quantitatively measured by an ELISA assay.

2. Materials

All solutions are prepared using deionized water and analytical grade reagents. Procedures are carried out at room temperature unless indicated otherwise. Basic chemicals were purchased from Sigma Aldrich unless indicated otherwise.

2.1. Dialysis

1. Normal dialysis buffer: 25 mM phosphate buffer pH 7.4.
2. High calcium dialysis buffer: 25 mM phosphate buffer pH 7.4 + 250 μM $CaCl_2$.
3. Calcium-free dialysis buffer: 25 mM phosphate buffer pH 7.4 + 250 μM EGTA (see Note 1).
4. Regenerated cellulose dialysis membranes with the appropriate molecular weight cut-off (Spectrum Europe B.V., Breda, The Netherlands).
5. Standard dialysis membrane closures (Spectrum Europe B.V.).
6. Dialysis reservoir, magnetic stirrer.

2.2. SDS-PAGE Electrophoresis

1. Stacking gel buffer: Tris–HCl 0.5 M, pH 6.8. Weigh 6.1 g Tris, dissolve in 75 ml water, and adjust pH with HCl. Add water up to 100 ml. Store at room temperature.
2. Resolving gel Buffer: Tris–HCl 1.5 M, pH 8.8. Weigh 18.1 g Tris, dissolve in 75 ml water, and adjust pH with HCl. Add water up to100 ml. Store at room temperature.
3. Acrylamide/bisacrylamide solution (30%) was purchased from Biorad Laboratories.
4. Ammonium persulfate was used as a 10% solution in water. Prepare a fresh 1 ml solution and store at 4°C for up to 1 month (see Note 2).
5. *N,N,N′,N′*-tetramethyl-ethylenediamine (TEMED).
6. 10% SDS solution in deionized water (see Note 3).
7. 5× Electrophoresis running buffer: 0.125 M Tris–HCl, 0.96 M glycine, 0.5% SDS. Weigh 15.12 g Tris, 72.06 g glycine and dissolve in 800 ml water. Add 5 ml SDS 10% and complete to 1 L with water. Store at room temperature.
8. 2× sample loading buffer: 0.125 M Tris–HCl pH 6.8, 4% SDS, 10% β-mercaptoethanol, 0.04% bromophenol blue, 20% glycerol. In a 15 ml Falcon tube add 2.5 ml Tris–HCl 0.5 M pH 6.8, 4 ml SDS 10%, 0.3 ml EDTA 0.5 M pH8, 1 ml β-mercaptoethanol (98%), 2.29 ml glycerol (87%), and 0.4 mg Bromophenol blue. Complete with water up to 10 ml. Freeze at −20°C in 1 ml aliquots (see Note 4).
9. SDS-PAGE was performed using the Mini-PROTEAN Tetra Electrophoresis System (Biorad).
10. Broad range molecular weight markers (Biorad).
11. Staining solution: 45% methanol, 7.5% acetic acid, 0.25% Coomassie brilliant blue (CBB) R250. Dissolve 2.5 g CBB in 450 ml methanol, add 75 ml glacial acetic acid, and complete with water up to 1 L.
12. Destaining solution I: 45% methanol, 7.5% acetic acid. In a graduated cylinder, add 450 ml methanol, 75 ml glacial acetic acid and complete with water up to 1 L.
13. Destaining solution II: 5% methanol, 7.5% acetic acid. In a graduated cylinder, add 50 ml methanol, 75 ml glacial acetic acid and complete with water up to 1 L.

2.3. Circular Dichroism Spectroscopy

1. Far-UV capable spectropolarimeter fitted with a thermostated cell holder and interfaced with a computer (see Note 5).
2. Spectrosil precision 0.1 cm cells.

2.4. ELISA Assay

1. 96 well flat bottom Nunc MaxiSorp™ plates (Nalge Nunc International, Rochester, NY, USA).

2. Phosphate-Buffered Saline (PBS) pH 7.4: 137 mM NaCl, 2.7 mM KCl, 10 mM $Na_2HPO_4 \times 2H_2O$, 2 mM KH_2PO_4.
3. Coating solutions: PBS + 250 μM calcium (add 1.25 ml of 100 mM $CaCl_2$ to 50 ml 10× PBS, complete with water to 500 ml), PBS + 250 μM EGTA (add 250 μl EGTA 0.5 M pH8 to 50 ml 10× PBS, complete with water to 500 ml).
4. Washing solutions: PBS + 0.5% Tween, PBS + 250 μM calcium + 0.5% Tween, PBS + 250 μM EGTA + 0.5% Tween.
5. Blocking solutions: PBS + 0.5% Tween + 5% Bovine Serum Albumin (BSA), PBS + 250 μM calcium + 0.5% Tween + 5% BSA, PBS + 250 μM EGTA + 0.5% Tween + 5% BSA.
6. Antibody diluting solutions: PBS + 1% Tween + 10% BSA, PBS + 500 μM calcium + 1% Tween + 10% BSA, PBS + 500 μM EGTA + 1% Tween + 10% BSA (see Note 6).
7. HRP conjugated anti-IgE antibody (Southern Biotech, AB, USA).
8. Substrate solution: Add four Dako OPD tablet (Dako Denmark A/S, Copenhagen) to 12 ml water, allow to dissolve in the dark, then add 12 μl H_2O_2 per added tablet (see Note 7).
9. 1 M H_2SO_4 solution: Prepare by adding 53.3 ml of 95–97% H_2SO_4 to 946.7 ml prechilled deionized water (see Note 8).

3. Methods

3.1. Sample Preparation

For all assays a purified allergen is needed. The allergen can originate from natural or recombinant sources and can be purified using several techniques (described in refs. 8–10). The protein of interest will be analyzed under three conditions: normal calcium, high calcium, and calcium-free. We have defined normal calcium as the status in which solutions are prepared for day-to-day use in the lab. High calcium conditions imply the presence of 250 μM $CaCl_2$ and in calcium-free buffers, EGTA is added to a final concentration of 250 μM.

1. Soak an appropriate length of dialysis tubing in deionized water for 20–30 min (see Note 9).
2. Clamp one end of the dialysis tube with a standard closure. Fill the tubing with water and check for leaks (see Note 10). Load the sample into the tube and leave space for a small bubble and clamp the other end. Check for leaks.
3. Immerse the clamped tube that contains the sample in a dialysis reservoir containing 2 L of the appropriate dialysis buffer (normal calcium, high calcium, or no calcium), check that the

clamped tube has slightly positive buoyancy, and add a magnetic stir bar. Place on a magnetic stirrer and set to low speed (see Note 11) in a cold room at 4°C.

4. Change dialysis buffer after 2 and 5 h and then leave overnight. Perform one final buffer change the next day and dialyze for an additional 2 h.

5. Carefully open one of the clamps, quantitate, and aliquot the dialyzed sample (see Note 12).

3.2. SDS-PAGE

1. Clean glass plates with a mild detergent and rinse with ethanol. Dry with a laboratory wipe. Set up glass plates in gel casting chamber.

2. 12% Resolving gel preparation: In a 15 ml polystyrene tube, mix 2.5 ml resolving buffer, 3.3 ml water, and 4 ml 30% acrylamide/bisacrylamide solution. Add 100 μl 10% SDS and mix gently by inversion. Add 50 μl APS 10% solution and 5 μl TEMED, and immediately add 4 ml to gel casting chamber. Overlay with 1 ml water. When gels have polymerized and immediately before stacking gel preparation, invert the casting chamber to pour out the overlaying water and gently dry with Whatman® paper.

3. Stacking gel preparation: Mix 1.24 ml stacking buffer with 3 ml water and 0.65 ml 30% acrylamide/bisacrylamide solution. Add 50 μl SDS 10% and mix gently by inversion. Add 25 μl APS 10% and 5 μl TEMED and pour immediately into gel casting chamber containing polymerized resolving gel. Insert a 10-well comb.

4. Sample preparation: To 1–2 μg dialyzed protein, add an equal volume of 2× sample loading buffer. Heat samples at 95°C for 5 min, transfer immediately to ice, and briefly spin in a microfuge to collect condensate.

5. Load samples in gel and add molecular weight marker. Run at 50 V for 5 min and then raise voltage to 200 V. The electrophoresis is finished when the dye front reaches the bottom of the gel.

6. Open the gel plates with the use of a plastic spatula and carefully transfer the gel into an appropriately sized container. Add staining solution, cover the container with aluminum foil, and stain for 20–30 min in an orbital shaker.

7. Discard staining solution, add destaining solution I, and incubate in an orbital shaker for 2–3 h, replacing the destaining solution after the first hour. Replace destaining solution I with destaining solution II and keep overnight at 4°C.

8. Analyze results: If the protein of interest binds calcium, different migration patterns may be observed (Fig. 1).

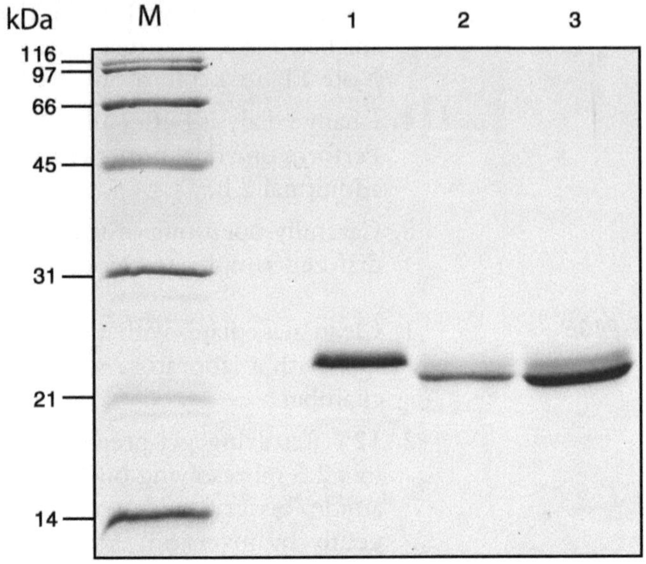

Fig. 1. SDS-PAGE analysis of Cup a 4 in the presence/absence of calcium. Lane 1. 250 μM EGTA. Lane 2. Normal calcium. Lane 3. 250 μM $CaCl_2$. 1 μg protein was loaded in each well. M: Broad range molecular weight markers (Biorad).

3.3. Circular Dichroism Spectroscopy

1. Prepare 25 μM dilutions of each of the protein samples: Normal, calcium-free (250 μM EGTA), and high calcium (250 μM $CaCl_2$) (see Note 13).
2. Switch on the CD spectrometer (see Note 5) and allow time for the electronics and lamp to stabilize (approximately 30 min).
3. Load the cell with protein sample and collect a CD spectrum using the following parameters: 25°C, 190–250 nm range at a rate of 50 nm/min, a bandwidth of 1.0 nm, and a response time of 2 s.
4. Collect three spectra per sample and calculate the mean spectrum.
5. Repeat steps 3 and 4 with the buffer used for each protein sample.
6. Subtract the buffer spectrum from the protein sample spectrum.
7. Normally, the spectrum ellipticity, θ_λ, is obtained in millidegrees; therefore it is necessary to convert the results to molecular ellipticity, $[\theta]_\lambda$, using the following formula:

$$[\theta]_\lambda = \frac{100.\theta_\lambda}{C.l}$$

where C is the molar concentration in mol/l and l is the cell length in cm.

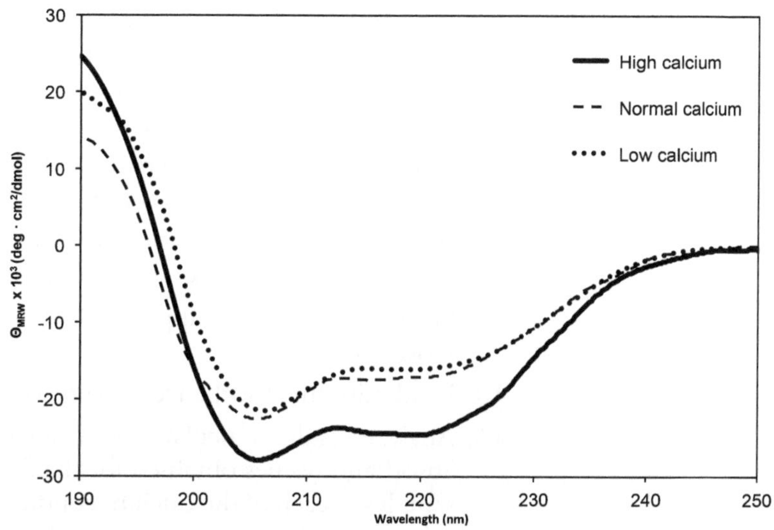

Fig. 2. Far UV CD spectra of 20 μM Cup a 4 in the presence/absence of calcium.

8. Plot-normal, no-calcium, and high-calcium spectra (Fig. 2 and Note 14).

3.4. Quantitative Analysis of IgE Binding Properties

All steps are carried out at room temperature unless stated otherwise.

1. Antigen coating: The allergen of interest is prepared at a final concentration of 7 μg/ml in one of the following solutions: PBS, PBS+250 μM calcium, or PBS+250 μM EGTA. Add 100 μl per well to a 96-well microplate (see Note 15). The number of wells depends on the number of sera to be analyzed, taking into account that each analysis will be performed in duplicate. Cover and incubate overnight at 4°C on a flat surface.
2. Empty each plate by inversion and dry by blotting on a paper towel. Wash three times by adding 200 μl of the selected washing solution and incubating for 5 min on a plate shaker at 200 rpm (see Note 16). Use the washing solution that corresponds to the calcium content selected for each assay (normal, no calcium, or high calcium).
3. Block for 2 h in 200 μl of the corresponding blocking solution (normal, no calcium, or high calcium). Remove blocking solutions by inversion and dry by blotting on a paper towel.
4. Prepare serum dilutions: Dilute each serum 1:1 in each of the antibody diluting solutions (normal, no calcium, or high calcium). Add 100 μl of each diluted serum and incubate for 2 h on a plate shaker at 200 rpm.
5. Empty each plate by inversion and dry by blotting on a paper towel. Wash three times as explained in step 2.

6. Add 100 μl of secondary antibody solution. This solution is prepared by diluting the anti-IgE antibody 1:800 in the corresponding blocking solution (normal, no calcium, or high calcium). Incubate for 1 h on a plate shaker at 200 rpm.
7. Empty each plate by inversion and dry by blotting on a paper towel. Wash three times as explained in step 2.
8. Add 100 μl substrate solution to each well and incubate for 20–30 min in the dark (see Note 17).
9. Stop reaction by adding 50 μl of 1 M H_2SO_4 per well (see Note 18).
10. Read 450 nm absorbance in a 96-well micro plate reader.
11. Analyze results: Calculate the mean and standard deviation of absorbance values obtained for the five negative control serum samples in each of the calcium conditions. The final absorbance value for each sample is calculated by the following formula:

$$\mathrm{Abs_n} = \mathrm{Abs} - (\mathrm{Abs_{neg}} + 2\mathrm{SD}),$$

where $\mathrm{Abs_n}$ is the normalized absorbance value, Abs is the O.D. value for each sample, $\mathrm{Abs_{neg}}$ is the mean absorbance calculated for the five negative control sera, and SD is the standard deviation calculated for the five negative control sera. The $\mathrm{Abs_n}$ calculated by this method are considered reactive against the tested allergen when they are positive. Plot the $\mathrm{Abs_n}$ values (Fig. 3).

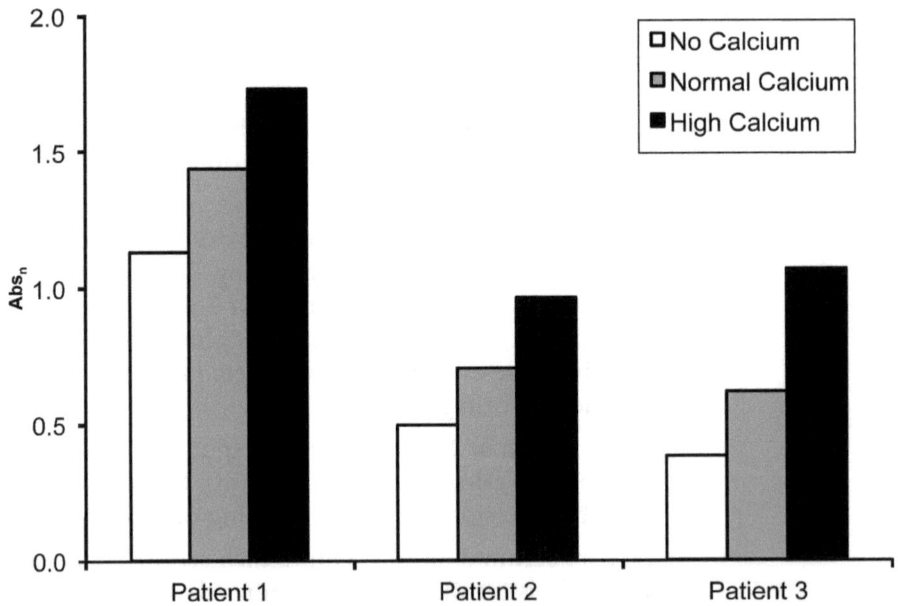

Fig. 3. IgE binding activity of sera from three allergic patients to Cup a 4 in the three calcium conditions: high, normal, and no calcium. Patients were selected in the context of the EU CRAFT Cyprall project QLK-CT-2002-71661, serum was obtained after written consent. All patients were positive by skin prick test to *C. arizonica* extract prepared in native conditions.

4. Notes

1. The best way to prepare these buffers is using 0.1 M phosphate buffer pH 7.4 (11): Add 250 ml to three graduated cylinders. Complete the first one to 1 L with water to obtain normal dialysis buffer. Add 2.5 ml 100 mM $CaCl_2$ to the second one and complete to 1 L with water (high-calcium dialysis buffer). Add 500 µl EGTA 0.5 M, pH8 and complete to 1 L with water to obtain calcium-free dialysis buffer.
2. Prepare 10 ml of a 10% solution by adding 1 g ammonium persulfate to 10 ml water, and store aliquots at −20° C. Once thawed, APS solution can be stored at +4°C for 1 month.
3. An easy way to prepare 10% SDS without frothing is by weighing 50 g of SDS and adding it on top of a beaker containing 350 ml water. The next day, the SDS will be dissolved and can be completed to 500 ml with water. Always wear a mask while weighing SDS.
4. This buffer can be prepared at 4× when the protein concentration of the sample is low. This allows loading of larger sample volumes.
5. We use a JASCO J-600 spectropolarimeter (Jasco Europe SLR, Cremella, Italy) and a NESLAB RTE-100 water bath (ThermoElectron Measurement Systems, Karlsruhe, Germany).
6. Since the sera will be diluted 1:1, the detergent and blocking agent concentrations in these solutions are doubled.
7. 12 ml is enough for one 96-well plate, and scale up or down, accordingly. This solution must be used within 60 min and should be kept in the dark. The original Dako instructions suggest adding a much lower quantity of H_2O_2 (5 µl). We find, however, that it is better to use 12 µl in order to avoid the reduction of activity that takes place after long-term storage of H_2O_2. This solution is toxic to aquatic organisms, and should be disposed according to local regulations.
8. Wear gloves and protect your eyes with safety goggles while preparing this solution. Dilution of concentrated acid should always be done in a fume hood. To avoid strong exothermic reactions, add concentrated acid to water slowly; never add water to a concentrated acid.
9. We use the Dialysis Tubing Calculator at the Spectrum Laboratories Web site: http://www.spectrumlabs.com/dialysis/dtCalc.html. Always use gloves to handle the dialysis membrane because the membrane is susceptible to cellulolytic microorganisms.

10. Always perform all dialysis sample handling procedures over a beaker in case the membrane slips. Use a paper towel to check for leaks: gently set the clamp with the tube on a paper towel and observe if moisture appears.
11. It is very important to check the dialysis frequently during the first hour to make sure that the membrane is not moving too quickly or touching the magnetic stirrer in the bottom of the dialysis reservoir.
12. If the protein is going to be used only for ELISA or SDS-PAGE analysis, quantitation by colorimetric methods (i.e., Bradford, BCA, or Lowry) is precise enough. However, if CD analysis is going to be performed, a more precise quantitation method based on A280 with an empirically determined extinction coefficient (12) is recommended.
13. Take into account that the protein samples were dialyzed against buffers that contained 500 μM EGTA and $CaCl_2$. Dilute with the appropriate buffer in each case. We recommend a first dilution step with 25 mM pH 7.4 phosphate buffer to correctly adjust calcium and EGTA concentrations. Samples can then be diluted to the final concentration with the appropriate buffer.
14. Further analysis of the secondary structure data can be performed using the dichroweb server (13) and analyzed as has been previously described (14).
15. The number of wells used depends on the number of sera to be analyzed. For each serum, three conditions will be tested (250 μM EGTA, residual calcium, and 250 μM calcium) and duplicates should be performed; count at least six wells per sample. For practical reasons each calcium condition is performed on a separate plate, include at least five nonreactive negative control sera as well as two no serum control wells in each plate.
16. We have found that washing in a 23 × 16 × 8 cm plastic food container (wash bowl) yields the lowest background and highest reproducibility results: Fill a wash bowl with 200 ml washing buffer and place the ELISA plate inside. Push it down so that all the wells are covered by the washing buffer, move the wash bowl sideways three or four times, then invert the plate in a sink, and blot dry with a paper towel. Perform this step three times and use a different wash bowl for each calcium condition.
17. The developing reaction can be carried out on the bench top, while the plates are covered with Styrofoam boxes. Check every 5 min until sufficient color is observed.
18. When working with a large number of samples, it is recommended to pipet the sulfuric acid in the same order as the developing solution was piped.

References

1. Wopfner N, Dissertori O, Ferreira F et al (2007) Calcium-binding proteins and their role in allergic diseases. Immunol Allergy Clin North Am 27:29–44
2. Grabarek Z (2006) Structural basis for diversity of the EF-hand calcium-binding proteins. J Mol Biol 359:509–525
3. Fuertes MA, Perez JM, Soto M et al (2001) Calcium-induced conformational changes in Leishmania infantum kinetoplastid membrane protein-11. J Biol Inorg Chem 6:107–117
4. Ledesma A, Gonzalez E, Pascual CY et al (2002) Are Ca2+-binding motifs involved in the immunoglobin E-binding of allergens? Olive pollen allergens as model of study. Clin Exp Allergy 32:1476–1483
5. Neudecker P, Nerkamp J, Eisenmann A et al (2004) Solution structure, dynamics, and hydrodynamics of the calcium-bound cross-reactive birch pollen allergen Bet v 4 reveal a canonical monomeric two EF-hand assembly with a regulatory function. J Mol Biol 336: 1141–1157
6. Niederberger V, Hayek B, Vrtala S et al (1999) Calcium-dependent immunoglobulin E recognition of the apo- and calcium-bound form of a cross-reactive two EF-hand timothy grass pollen allergen, Phl p 7. FASEB J 13:843–856
7. Verdino P, Westritschnig K, Valenta R et al (2002) The cross-reactive calcium-binding pollen allergen, Phl p 7, reveals a novel dimer assembly. EMBO J 21:5007–5016
8. Pico de Coana Y, Parody N, Fuertes MA et al (2010) Molecular cloning and characterization of Cup a 4, a new allergen from Cupressus arizonica. Biochem Biophys Res Commun 401:451–457
9. Ariano R, Spadolini I, Panzani RC (2001) Efficacy of sublingual specific immunotherapy in Cupressaceae allergy using an extract of Cupressus arizonica. A double blind study. Allergol Immunopathol (Madr) 29:238–244
10. Iacovacci P, Afferni C, Butteroni C et al (2002) Comparison between the native glycosylated and the recombinant Cup a1 allergen: role of carbohydrates in the histamine release from basophils. Clin Exp Allergy 32: 1620–1627
11. Maniatis T, Fritsch EF, Sambrook J (1982) Molecular cloning: a laboratory manual. Cold Spring Harbor Laboratory, Cold Spring Harbor, NY
12. Gill SC, von Hippel PH (1989) Calculation of protein extinction coefficients from amino acid sequence data. Anal Biochem 182: 319–326
13. Whitmore L, Wallace BA (2004) DICHROWEB, an online server for protein secondary structure analyses from circular dichroism spectroscopic data. Nucleic Acids Res 32:W668–W673
14. Clarke D (2011) Circular dichroism and its use in protein-folding studies. Methods Mol Biol 752:59–72

Chapter 9

Longistatin, an EF-Hand Ca^{2+}-Binding Protein from Vector Tick: Identification, Purification, and Characterization

Anisuzzaman, M. Khyrul Islam, M. Abdul Alim, and Naotoshi Tsuji

Abstract

EF-hand Ca^{2+}-binding motif, a structural component of the EF-hand protein, functions as a calcium sensor and/or buffer in the cytosol of the cell. However, in a few exceptional cases, the EF-hand proteins are secreted from cells and play crucial roles extracellularly. We have identified longistatin, an EF-hand Ca^{2+}-binding protein, from the salivary glands of the tick, *Haemaphysalis longicornis*. Longistatin possesses an N-terminal sequence of unknown structure and two EF-hand motifs in the C-terminus, which conserve a calmodulin-like canonical structure. Longistatin shows distinct changes in its migration during electrophoresis through SDS-PAGE gel containing calcium or ethylenediaminetetraacetic acid (EDTA). Both recombinant and endogenous forms of longistatin can be stained with rutheninum red, demonstrating that longistatin is a Ca^{2+}-binding protein.

Key words: Ticks, *Haemaphysalis longicornis*, Longistatin, EF-hand motif, Ca^{2+}-binding protein

1. Introduction

The EF-hand motif was first discovered by R.H. Kretsinger over three decades ago by the structural analysis of parvalbumin, a Ca^{2+}-binding protein first isolated from carp muscle (1). EF-hand motif is also known as helix-loop-helix motif. The calcium-incorporating loop region of a canonical EF-hand consists of 12 residues that usually starts with an aspartic acid and essentially ends with a glutamic acid (1–3). However, in unconventional EF-hands, the number of amino acids in the loop region may vary from 11 to 14 residues. For example, the N-terminal EF-hand of S100 proteins has a 14-residue loop (4–7) and osteonectin has an EF-hand with a 13-residue loop (2, 8). Some EF-hand proteins such as calpain,

Claus W. Heizmann (ed.), *Calcium-Binding Proteins and RAGE: From Structural Basics to Clinical Applications*, Methods in Molecular Biology, vol. 963, DOI 10.1007/978-1-62703-230-8_9, © Springer Science+Business Media New York 2013

grancalcin, and ALG-2 contain a Ca^{2+}-binding loop consisting of 11 residues (9–12). Despite the enormous diversity in the composition of the Ca^{2+}-binding loop region, the affinity for Ca^{2+}, although differing in degree, is still preserved, and they bind Ca^{2+}. As an exception, the ten-residue loop of EF-hand of KChIP1, the Kv4K^{+}channel interacting protein, and the 20-residue loop of EF-hand1 of CIB1, a protein related to the neuronal Ca^{2+} sensor, cannot bind calcium (2). In the canonical EF-hand motifs, residues 1, 3, 5, 7, 9, and 12 of the Ca^{2+}-binding loop chelate the Ca^{2+} and are denoted as X, Y, Z, –Y, –X, and –Z, respectively (1, 3). Although EF-hand motif usually occurs in pairs, alpha and beta parvalbumins have three EF-hands; grancalcin, calpain, and ALG-2 have five EF-hands (2); and, more exceptionally, a few Ca^{2+}-binding proteins have one EF-hand (13, 14). The unpaired EF-hand motif of parvalbumins cannot bind Ca^{2+} but plays critical roles in stabilizing the functional pair (2, 4). The penta-EF-hand proteins usually exist as homodimers through interaction between the fifth EF-hands; thus, the Ca^{2+}-binding capability of the fifth EF-hand motif is preserved (9, 11, 12). A group of EF-hand proteins contain 4 EF-hand motifs (calmodulin and troponin C) (2, 15, 16) but some Ca^{2+}-binding proteins contain six, eight, or even 12 EF-hand motifs (2). On the other hand, smaller EF-hand proteins, such as S100 proteins, osteonectin and multiple coagulation factor deficiency 2 like protein (MCFD2) possess two EF-hand motifs (2, 7, 17).

EF-hand Ca^{2+}-binding proteins and calcium (Ca^{2+}) are essential for life since they control substantial parts of cell functions, starting from the cell's birth, through mitosis to its end with apoptosis (18, 19). Almost every function of our body, such as heartbeat, breathing, vision, locomotion, thinking, and memory processing, is controlled by Ca^{2+} (4, 18–27). Cells respond to extracellular stimuli through the transient change in Ca^{2+} concentration. This signal transduction is sensed by Ca^{2+}-binding proteins. Generally, EF-hand Ca^{2+}-binding motifs act as calcium signaling molecules and/or buffers in the cytosol of the cell (15, 28–30). However, some Ca^{2+}-binding proteins such as S100A2, S100A4, S100A6, and S100B are present in the nucleus, where they control gene expression (7, 31–33). On the other hand, in a few exceptional cases, EF-hand Ca^{2+}-binding proteins are secreted from the cells and exert critical roles extracellularly (7, 29, 34, 35). EF-hand proteins have been reported from a wide spectrum of organisms from unicellular, simple prokaryotes to multicellular, complex mammals (4, 30, 36–38), and more than 600 Ca^{2+}-binding proteins have been discovered (31). Recently, we have identified an EF-hand Ca^{2+}-binding protein from the vector tick, *Haemaphysalis longicornis*, a notorious blood-sucking ectoparasite affecting humans and animals (38–43), and we have named this protein longistatin. Longistatin is a 17.8-kDa soluble protein with an N-terminal sequence of unknown structure and two calmodulin-like functional EF-hand motifs at the C-terminus (3), and has been shown to be linked with the blood-feeding success of ticks from host animals. Longistatin, a secreted EF-hand protein, degrades fibrinogen and activates plasminogen; thus, exerts pivotal roles in the feeding mechanisms of ticks by maintaining the fluidity of blood during the acquisition of blood-meals from hosts (44).

To identify EF-hand Ca^{2+}-binding proteins, several methods are employed such as radiological methods, using radioisotopes; gel mobility shift assays (GMSA), utilizing the changes in the migration pattern of apo- and holoforms of proteins; and direct staining of Ca^{2+}-binding protein on SDS-PAGE/PAGE gel or on nitrocellulose membrane using ruthenium red stain and stains-all. All of these methods have relative advantages and disadvantages. For example, although Ca^{45} autoradiography of SDS-PAGE gel/nitrocellulose membrane is used to detect Ca^{2+}-binding proteins, the low affinity of some proteins for Ca^{2+} (e.g., calsequestrin), the high background radioactivity, and health hazards for the handlers make this method unsuitable for routine detection of Ca^{2+}-binding proteins. On the other hand, cationic carbocyanine dye, stains-all, binds with various Ca^{2+}-binding proteins and gives a blue color, but it cannot detect several Ca^{2+}-binding proteins (e.g., sarcoplasmic reticulum Ca^{2+}-ATPase). The specificity of the stain is also questionable, as it shows reactivity to acidic sialoglycoproteins (e.g., glycophorin) (45–47). However, in GMSA, Ca^{2+}-binding is clearly evident by the changes in the movement of the respective proteins in Ca^{2+}-bound or -unbound state (3, 48). On the other hand, ruthenium red specifically and directly binds to the Ca^{2+}-binding sites and stains them (3, 45). Furthermore, these two methods are easier, require less equipment, and are more reliable for the routine detection of Ca^{2+}-binding proteins from diverse sources. We have used GMSA and ruthenium red staining techniques for the detection of Ca^{2+}-binding to longistatin.

2. Materials

Prepare all buffers using ultrapure water (prepare by purifying deionized water to attain a sensitivity of 18 MΩ cm at 25°C) and analytical grade reagents. Unless otherwise indicated, water indicates ultrapure water. Dispose off waste materials strictly following waste disposal regulations. Handle all biological materials as potential sources of infection. Use goggles and protective clothing whenever needed.

2.1. Longistatin Preparation

1. Phosphate-buffered saline (PBS, stock solution): Take 10.9 g of *di*-sodium hydrogenphosphate (Na_2HPO_4), 85.0 g of

sodium chloride (NaCl), and 3.9 g of sodium dihydrogen-phosphate dihydrate ($NaH_2PO_4{\cdot}2H_2O$) in a 1 L beaker. Dissolve the components in ~900 ml of water by stirring. Adjust the pH to 7.2 and finally, adjust the volume to 1 L by adding water. Autoclave the buffer and keep at room temperature (RT).

2. Phosphate-buffered saline (PBS, working solution): Pour 900 ml of water in a 1 L cylinder and add 100 ml of stock solution, then mix thoroughly. Autoclave the working solution and keep at RT.
3. DNA loading buffer: Take 2.25 ml of sterile glycerol in a 14 ml Falcon tube and add 0.2 ml of 0.5 M EDTA (pH 8.0), 1.25 ml of 1% xylene cyanol, and 1.25 ml of 1% bromophenol blue, then mix thoroughly in water to a total volume of 5 ml. Store the buffer at 4°C making 1 ml aliquot.
4. DNA ladder: Both 100 bp and 1 kb DNA ladders are commercially available (Invitrogen, CA, USA).
5. QIAGEN DNA extraction kit (QIAGEN Sciences, MD, USA): Used for DNA extraction from agarose gel.
6. Ampicillin: Dissolve 1 g ampicillin in 20 ml of water (50 mg/ml) and filter under a flame using a filter of 0.2 μm size. Store the antibiotic at −20°C making 500 μl aliquot.
7. Agar plate: Pour 100 ml of water in a 500 ml conical flask and add 4 LB-Agar Medium capsules (Q-Bio gene, OH, USA) and autoclave. Wait until the temperature returns to 50°C. Add ampicillin to a final concentration of 50 μg/ml and pour the melted agar on a plastic Petri dish and allow to solidify. Each liter of medium contains 10 g of tryptone-B, 5 g of yeast extract-B, 10 g of NaCl, and 15 g of agar-B. Store the agar plate at 4°C.
8. Tris-acetate (TAE) buffer (stock solution): Take 242 g of Tris base, 57.1 ml of absolute acetic acid, and 100 ml of 0.5 M EDTA (pH 8.0) in a 1 L beaker. Dissolve the salts in a total volume of 1 L of water by stirring and keep at RT.
9. TAE buffer (working solution): Measure 980 ml of water in a 1 L cylinder and add 20 ml of stock solution, then mix thoroughly. Keep the buffer at RT.
10. Ethidium bromide: Take 313 μl of stock solution of ethidium bromide (Nacalai Tesque, Kyoto, Japan) and add 4.687 ml of TAE buffer (working solution) and mix thoroughly. Keep the solution at RT.
11. 1% Agarose gel: Pour 100 ml of TAE buffer in a 500 ml conical flask and add 1 g of SeaKem Agarose® (Lonza, ME, USA). Apply heat until the agarose is dissolved, and then add two drops of ethidium bromide and mix properly. Set the agarose

plate and pour the melted agarose. Set the comb and allow to solidify. Store the gel at 4°C submerging in TAE buffer.

12. Super Optimal Broth (SOB medium): Pour 500 ml of water in a 2 L conical flask and add 14 g of Difco™ SOB medium powder (Becton, Dickinson and Company, NJ, USA), then mix thoroughly and autoclave. Each liter of SOB medium contains 20 g of tryptone, 5 g of yeast extract, 0.5 g of NaCl, 2.4 g of magnesium sulfate anhydrous, and 0.186 g of potassium chloride.
13. Isopropyl-β-d-thiogalactopyranoside (IPTG): Take 2.38 g of IPTG and dissolve in a total volume of 100 ml of water by stirring. Filter the solution under a flame using a filter of 0.2 μm size and store at –20°C making 10 ml aliquot.
14. Egg white lysozyme: Dissolve 1 g of egg white lysozyme in a total volume of 20 ml of water (50 mg/ml) and filter under a flame using a filter of 0.2 μm size. Store the enzyme at –20°C making 500 μl aliquot.
15. TALON™ resin (Clontech, CA, USA): Used to purify the recombinant protein.
16. Equilibration buffer (stock solution): Take 35.5 g of *di*-sodium hydrogenphosphate anhydrus (Na_2HPO_4) and 87.6 g of sodium chloride (NaCl) in a 1 L beaker. Dissolve the salts in ~900 ml of water by stirring. Adjust the pH to 7.0 by adding HCl. Finally, adjust the volume to 1 L by adding water and keep at RT.
17. Equilibration buffer (working solution): Measure 400 ml of water and add 100 ml of stock solution in a 500 ml cylinder, then mix thoroughly and store at 4°C.
18. Elution buffer (working solution): Take 7.1 g of Na_2HPO_4, 17.52 g of NaCl, and 10.2 g of imidazole in a 1 L beaker and dissolve in ~900 ml of water by stirring. Adjust the pH to 7.0 by adding HCl. Finally, adjust the volume to 1 L by adding water and store at 4°C.
19. 20 mM MES with 0.1 M NaCl buffer: Take 3.91 g of 2-(*N*-morpholino)ethanesulfonic acid (MES) and 5.58 g of NaCl in a 1 L beaker and dissolve in ~900 ml of water. Adjust the pH to 5.0 by adding HCl. Finally, adjust the volume to 1 L by adding water and store at 4°C.
20. 20% Ethanol with 0.1% azide: Measure 80 ml of water, and add 20 ml of absolute ethanol and 0.1 g of sodium nitrite (NaN_3), then mix thoroughly and store at 4°C.
21. Centrisart® (Sartorius, Goettingen, Germany): Used to concentrate protein.
22. Slide-A-Lyser Dialysis Cassette (Pierce, IL, USA): Used to dialyze protein.

23. Micro-BCA reagents (Thermo Scientific, IL, USA): Used to determine the concentration of protein.
24. Silver staining kit (DAIICHI Pure Chemical Co. Ltd., Tokyo, Japan): Used to stain proteins on SDS-PAGE gel.
25. Tris-buffered saline (TBS, stock solution): Take 60.0 g of Tris base and 87.7 g of NaCl in a 1 L beaker and dissolve in ~800 ml of water by stirring. Adjust the pH to 8.0 with HCl and finally, adjust the volume to 1 L by adding water. Keep the buffer at RT.
26. TBS (working solution): Pour 900 ml of water in a 1 L cylinder. Add 100 ml of TBS (stock solution) and mix thoroughly. Keep the buffer at RT.
27. Tris-buffered saline with Tween 20 (TBS-T, stock solution): Take 60.0 g of Tris base and 87.7 g of NaCl in a 1 L beaker and dissolve in ~800 ml of water by stirring. Adjust the pH to 8.0 with HCl. Add 10 ml of Tween 20 and mix thoroughly. Finally, adjust the volume to 1 L by adding water. Keep the buffer at RT.
28. TBS-T (working solution): Measure 900 ml of water in a 1 L cylinder. Add 100 ml of TBS-T (stock solution) and mix thoroughly. Keep the buffer at RT.
29. 1 M Magnesium chloride ($MgCl_2$): Take 9.52 g of $MgCl_2$ in a 200 ml bottle and mix thoroughly in water to a total volume of 100 ml by stirring. Keep the solution at RT.
30. Mouse anti-6 X His monoclonal antibody (Nacalai Tesque): Used to detect polyhistidin-tagged recombinant protein.
31. Alkaline phosphatase-conjugated goat anti-mouse IgG (Jackson ImmunoResearch, PA, USA): Used as secondary antibody.
32. AP buffer: Pour 100 ml of 1 M Tris–HCl (pH 9.0) in a 1 L bottle; add 30 ml of 5 M NaCl and 1 ml of 1 M $MgCl_2$, then mix thoroughly in water to a total volume of 1 L. Keep the buffer at RT.
33. 0.2% Ponceau-S: Take 495 ml of water in a 500 ml bottle and add 5 ml of absolute acetic acid and 1 g of Ponceau-S, then mix thoroughly by stirring. Keep the stain at RT.
34. Nitroblue tetrazolium/5-bromo-4-chloro-3-indolyl phosphate (BCIP/NBT, Promega, WI, USA): Used in immunoblotting (Fig. 1).

2.2. Gel Mobility Shift Assays

1. Solution A: Commercially available (Bio-Rad Laboratories, Inc., CA, USA) and contains 30% acrylamide.
2. Solution B: Take 181.71 g of Tris base in a 1 L beaker, add 800 ml of water, and dissolve by vigorous stirring with a magnetic stirrer. Adjust the pH to 8.8 by adding HCl

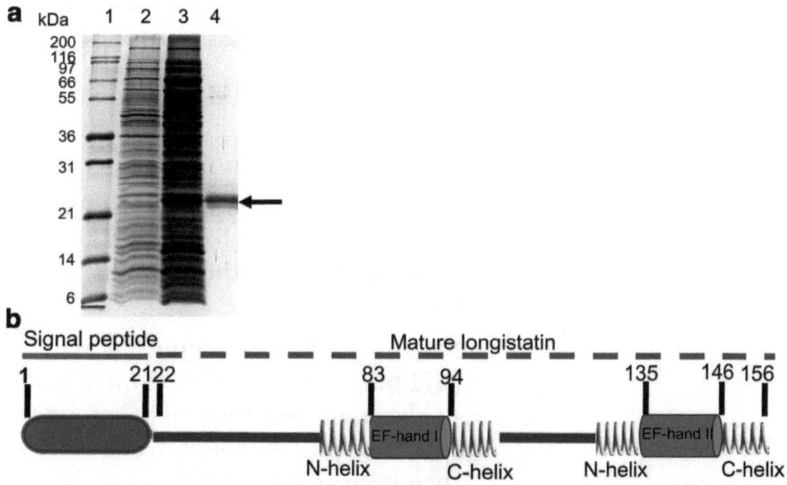

Fig. 1. Purity and structure of recombinant longistatin. (**a**) Monitoring the purification of longistatin by SDS-PAGE analysis. Purified longistatin and *Escherichia coli* lysate were electrophoresed on 12.5% SDS-PAGE under reducing conditions. The gel was developed with silver stain. *Lane 1*, molecular weight marker; *lane 2*, *E. coli* lysate before induction; *lane 3*, *E. coli* lysate 3 h after induction with isopropyl-β-d thiogalactopyranoside (IPTG); and *lane 4*, purified longistatin. (**b**) Schematic diagram of the protein structure of longistatin. Longistatin is composed of 156 amino acids and has a signal peptide, which is predicted to be cleaved at Ala[21]–Gln[22]. Longistatin contains two canonical EF-hand motifs at the C-terminus.

carefully and finally, adjust the volume to 1 L by adding water. Filter the solution using a filter of 0.2 μm size and store at 4°C until use.

3. Solution C: Place 60.57 g of Tris base in a 1 L beaker, add 800 ml of water, and dissolve by vigorous stirring with a magnetic stirrer. Adjust the pH to 6.8 by adding HCl carefully and finally, adjust the volume to 1 L by adding water. Filter the solution using a filter of 0.2 μm size and store at 4°C until use.
4. 10% Sodium dodecyl sulfate (SDS): Dissolve 10 g of SDS in water to a total volume of 100 ml and keep at RT until further use.
5. Ammonium peroxodisulfate (APS): Dissolve 10 g of APS (Wako, Tokyo, Japan) in water to a total volume of 100 ml and store at −20°C making 1 ml aliquots until further use.
6. 100 mM calcium chloride ($CaCl_2$): Dissolve 11.1 g of $CaCl_2$ in water to a total volume of 1 L and keep at RT.
7. 0.5 M Ethylenediaminetetraacetic acid (EDTA): Take 93.05 g of EDTA in a 500 ml beaker and add 300 ml of water and stir vigorously. Adjust the pH to 8.0 with 10 N NaOH. Finally, adjust the volume to 500 ml by adding water and autoclave. The solution can be kept at RT up to 6 months.

8. *N,N,N′,N′*-tetramethylethylenediamine (TEMED, Nacalai Tesque): Used to prepare SDS-PAGE gel.
9. Running buffer (stock solution): Take 29 g of Tris base, 144 g of glycine, and 10 g of SDS in a 1 L beaker and dissolve in water in a total volume of 1 L, then keep at RT.
10. Running buffer (working solution): Pour 900 ml of water in a 1 L cylinder and add 100 ml of stock solution and mix thoroughly. Keep the solution at RT. Running buffer can be reused.
11. Sample buffer (reducing): Take 2.1 ml of 1 M Tris–HCl (pH 6.8), 0.86 g of SDS, 2 ml of beta-mercaptoethanol, 2.8 ml of glycerine, and 1 mg of bromophenol blue in a 14 ml Falcon tube, and mix thoroughly in a total volume of 10 ml of water, then keep at RT.
12. Coomassie brilliant blue (CBB) stain: Pour 300 ml of methanol, 100 ml of absolute acetic acid, 100 μl of glycerol, and 600 ml of water in a 1 L glass bottle. Add 1 g of CBB and dissolve properly by stirring overnight with a magnetic stirrer, and keep at RT.
13. CBB destain: Take 300 ml of methanol, 100 ml of absolute acetic acid, 100 μl of glycerol, and 600 ml of water in a 1 L cylinder and mix thoroughly. Keep the destain at RT (Fig. 2).

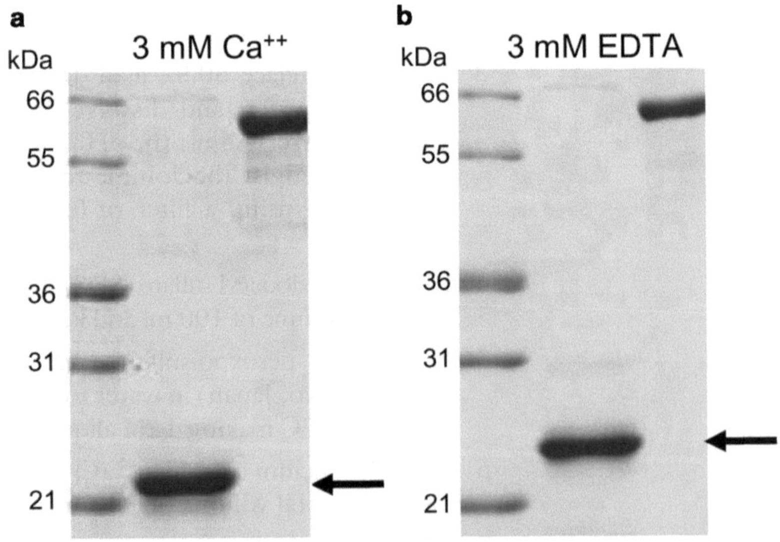

Fig. 2. Ca^{2+}-binding specificity of recombinant longistatin by gel mobility shift assay. Longistatin (2 μg) was electrophoresed in the presence of 3 mM Ca^{2+} (**a**) or 3 mM EDTA (**b**) through a 10% SDS-PAGE gel. BSA (2 μg) was used as negative control. *Lane 1*, molecular weight marker; *lane 2*, longistatin (*arrow*); *lane 3*, BSA (*control*). Longistatin migrated faster in the presence of Ca^{2+}.

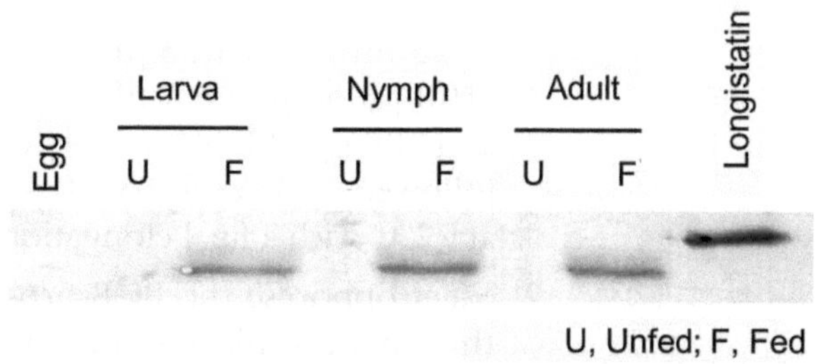

Fig. 3. Ca^{2+}-binding of longistatin visualized by ruthenium red staining. Purified recombinant longistatin (2 μg; containing a his-tag) and ticks' salivary gland lysate (20 μg) at different life-cycle stages were subjected to SDS-PAGE analysis followed by blotting onto nitrocellulose membranes. Nitrocellulose membranes were then stained with ruthenium red for 15 min at room temperature. *U* unfed, *F* fed.

2.3. Ruthenium Staining

1. Fiber pads: Commercially available (Bio-Rad Laboratories, Inc.) and size of the fiber pads, used in Mini Trans-Blot Cell® (Bio-Rad Laboratories, Inc.), is 10.8 cm × 8.0 cm × 0.6 cm.
2. Nitrocellulose membrane: Cut the commercially available nitrocellulose membrane (Whatman, Dassel, Germany) into pieces of suitable size.
3. Transfer buffer (stock solution): Take 30.28 g of Tris base and 144.2 g of glycine in a 1 L beaker, and dissolve in a total volume of 1 L of water, then keep at RT.
4. Transfer buffer (working solution): Pour 100 ml of stock solution, 700 ml of water, and 200 ml of methanol in a 1 L cylinder, mix properly and keep at 4°C for 12 h before use. Transfer buffer can be reused three to four times.
5. Ruthenium red buffer: Take 3.83 g of potassium chloride (KCl) and 0.48 g of magnesium chloride ($MgCl_2$) in a 1 L beaker and add 900 ml of water, then dissolve properly. Add 10 ml of 1 M Tris–HCl (pH 7.5) and adjust the volume to 1 L by adding water, then keep at RT.
6. Ruthenium red stain: Measure 1 L of ruthenium red buffer in a 1 L bottle, add 25 mg of ruthenium red (Calbiochem, CA, USA) and dissolve by stirring. Keep the stain at RT (Fig. 3).

3. Methods

3.1. Longistatin Production

3.1.1. Recombinant Longistatin Expression

1. Amplify the open reading frame (ORF) of longistatin by PCR from pBS/longistatin using a set of primers, 5′CCGCTC GAGCGGGCAGGCCGGCGACCAGCAG 3′ and 5′GGAAT TCCCTAA ATTTGGTTGGTCAGGTC 3′, which contain *Xho*I and *Eco*RI restriction sites, respectively. Each 100 μl of

PCR mixture contains 20 μl of Go Tag flexi buffer (5×), 4 μl of 25 mM $MgCl_2$, 8 μl of 2.5 mM dNTP, 1 μl of each primer (100 pmol/μl), 0.5 μl of Go Tag DNA polymerase enzyme, and 500 ng of DNA template. Conduct PCR for 4 min at 95°C followed by 35 cycles of 30 s at 95°C, 30 s at 57°C, and 1.5 min at 72°C with a final elongation at 72°C for 5 min (see Note 1).

2. Electrophorese the PCR product using 1% agarose gel. To do this, mix a sufficient amount of loading buffer with the PCR product (usually 1 μl of 10× loading buffer for 9 μl of PCR product) and load into the wells of the gel. Load 3–5 μl of 100 bp ladder at one side and the same amount of 1 kb ladder at another side of the gel. Run the machine at 100 V for ~30 min. Visualize the bands with an UV illuminator and cut the target band.
3. Extract DNA using QIAGEN DNA extraction kit following the manufacturer's instructions. Briefly, transfer the target DNA band into an Eppendorf tube, add QG buffer (supplied in Kit Box) at three times the volume of the gel, incubate at 50°C for 10 min, and vortex. Add 2-propanol at a volume to that equal of the gel and vortex.
4. Transfer the mixture into the column (supplied in Kit Box) and centrifuge at 10,000 ×*g* for 2 min then discard the flow-through. Add 750 μl of PE buffer (supplied in Kit Box) and centrifuge in the same way as described above and discard the flow-through. Centrifuge the whole assembly at 10,000 ×*g* for 2 min.
5. Transfer the column into a new Eppendorf tube and elute the column with 20–30 μl of water by centrifuging at 10,000 ×*g* for 2 min and immediately transfer the eluted DNA into ice. The purified DNA can be stored at −20°C.
6. Digest both, the purified PCR product and the plasmid vector (pTrcHisB), using *Xho*I and *Eco*RI restriction enzymes by incubating overnight at 37°C. Electrophorese the digested products, excise the target band, and extract the DNA as described above. Insert the purified PCR product into the *Xho*I and *Eco*RI sites of the vector pTrcHisB by incubating at 16°C for 30 min. It is better to use insert (digested and purified PCR product) and plasmid at a 3:1 molar ratio.
7. Transform the resultant plasmid into competent cells of *Escherichia coli* Top10Fα′strain following the conventional method. Briefly, thaw the *E. coli* for 15 min in ice. Mix the construct by pipetting on ice, add 5 μl of constructed plasmid into the competent cells, then mix gently, and incubate for 30 min on ice.
8. Apply heat shock at 42°C in a water bath for 30 s fitting the tube to a floater and transfer immediately into ice and keep for

1 min. Add 200 μl of SOC medium under a flame and shake vigorously at 250–270 rpm and 37°C for 30–45 min fixing the tube horizontally. Spread the transfected cells onto an agar plate with a sterilized spreader and incubate overnight at 37°C. Check the colony in the following morning.

9. Identify positive colony doing PCR using 10–20 colonies. To do this, pick up colonies one by one using sterile 200 ml pipette-tips and mix thoroughly in a reaction mixture of 20 μl separately and do PCR following the same procedure as described above. Electrophorese the PCR product in 1% agarose gel and check the insert under UV-illuminator. In addition, check the insert by restriction enzyme digestion and sequencing of the construct.
10. Take one positive colony and check the recombinant protein production by pilot expression. To do this, culture the transfected *E. coli* in 100 ml of SOB medium containing ampicillin (50 μg/ml, final concentration) at 37°C under vigorous shaking (200 rpm) until an OD of 1 at 600 nm (OD_{600}) is achieved. To induce recombinant protein expression, add IPTG to 1 mM concentration and culture for an additional 3–4 h at 37°C on a shaking incubator. Collect 1 ml of culture before adding IPTG and every hour after the induction of protein expression.
11. Centrifuge the culture at 10,000 × *g* and 4°C for 30 min and discard the supernatant. Sonicate the pellet properly for 2 min in 100 μl of PBS with an ultrasonic processor (VP-5 S) and centrifuge the suspension at 10,000 × *g* and 4°C for 30 min. Collect the supernatant and electrophorese in 12.5% SDS-PAGE under reducing conditions. Stain one gel with CBB and destain with CBB-destain following the same procedures as described in Subheading 3.2. Wash the gel with water and check the pattern of recombinant expression.
12. Besides, check the recombinant protein by immunoblotting using mouse anti-6 X His monoclonal antibody. Briefly, after electrophoresis, transfer the proteins onto the nitrocellulose membrane following the same procedures as described in Subheading 3.3.
13. Cut the nitrocellulose membrane with a knife to separate the marker lanet from the antigenic lane. Stain the marker using 5 ml of Ponceau-S for 15 min on a platform shaker at RT. After 15 min staining, wash the marker lane with water and allow to dry.
14. In parallel, take the antigenic part on a plastic tray and block with 5% non-fat skim milk for 45 min on a platform shaker. Discard the blocking agent and incubate the membrane with mouse anti-6 X His monoclonal antibody (1:1,000, in TBS) overnight on a platform shaker at 4°C.

15. In the following morning, wash the membrane three times with TBS-T each for 5 min at RT on a platform shaker. Incubate the membrane with alkaline phosphatase-conjugated goat anti-mouse IgG (1:1,000) for 1 h at RT on a platform shaker. Wash the membrane three times with TBS-T in the same way as mentioned above. Then, wash the membrane with AP buffer at RT for 5 min on a shaker.
16. Visualize the protein bound to the secondary antibody with nitroblue tetrazolium/5-bromo-4-chloro-3-indolyl phosphate (BCIP/NBT) in the dark. As soon as the expected protein band is visible discard the buffer and wash the membrane with water. Air-dry the membrane and check the protein expression.
17. After checking the recombinant production through pilot expression, express the polyhistidine-tagged longistatin in a large scale in *E. coli* culture. Briefly, grow the transfected cells in 1 L of SOB medium containing ampicillin (50 µg/ml, final concentration) at 37°C under vigorous shaking (200 rpm) until an OD of 1 at 600 nm (OD_{600}) is achieved. To induce recombinant protein expression, add IPTG to 1 mM concentration and culture for an additional ~3 h at 37°C on a shaking incubator (see Note 2).
18. Centrifuge the culture at 10,000 × *g* and 4°C for 30 min and discard the supernatant. Resuspend the pellet in 10 ml of equilibration buffer. Add egg white lysozyme (100 µg/ml, final concentration) to the cell suspension (see Note 3) and incubate in ice for 15 min.
19. Sonicate the suspension for 2 min on ice with an ultrasonic processor (VP-15S) followed by immediate freezing at −80°C and then thawing at 37°C for 15 min in each case. After three cycles of sonication, freezing and thawing, centrifuge the *E. coli* lysate at 24,000 × *g* and 4°C for 30 min and collect the supernatant (see Note 4).

3.1.2. Purification of Recombinant Longistatin

1. Purify recombinant protein using TALON™ resin under native conditions following the manufacturer's instructions. Briefly, immediately transfer the required amount of thoroughly resuspended TALON resin (2 ml for ~3 mg of recombinant protein) into a 14 ml Falcon tube and centrifuge at 700 × *g* for 2 min. Discard the supernatant, wash the pellet with 10 bed-volumes of equilibration/wash buffer, and centrifuge at 700 × *g* for 2 min. Discard the supernatant and repeat the washing procedures once again.
2. Add the previously harvested *E. coli* lysate to the resin and gently agitate at RT for 20 min on a platform shaker to allow the polyhistidine-tagged protein to bind to the resin. Centrifuge

the suspension at 700 × *g* for 5 min and carefully remove the supernatant as much as possible without disturbing the pellet.

3. Add 10–20 bed-volumes of equilibration buffer and gently agitate at RT for 10 min on a platform shaker to wash thoroughly. Centrifuge at 700 × *g* for 5 min and discard the supernatant carefully and repeat the washing once again.
4. Add one bed-volume of equilibration buffer to the resin, mix thoroughly by vortexing, transfer the resin to a 2 ml gravity-flow column, and allow the resin to settle down. Remove the end cap, allow the buffer to drain until it reaches the top of the resin bed, and wash the column once with 5 bed-volumes of equilibration buffer.
5. Elute the protein with 5 ml of elution buffer and collect in 500 μl fractions in separate Eppendorf tubes (see Notes 5 and 6).

3.1.3. Concentration of the Recombinant Protein

1. Concentrate the eluted recombinant protein using Centrisart® having a mol. wt cut-off of 10 kDa. To do this, remove the filter of the Centrisart® and load ~2 ml of eluate carefully so that no air bubbles are formed.
2. Insert the filter gently and centrifuge at ~2,400 × *g* and 4°C. Standardize the time required to obtain the desired concentration.

3.1.4. Dialysis of Longistatin

1. Extensively dialyze the concentrated protein at 4°C with successive changes of 20 mM Tris–HCl (pH 7) and a decreasing concentration of NaCl (500–250 mM) using a Slide-A-Lyser Dialysis Cassette with a mol. wt cut-off of 10 kDa.
2. Briefly, take a Slide-A-Lyser Dialysis Cassette, inject the protein into the cassette using an 18G needle and aspirate the air with a syringe. Float the cassette with a buoy on the dialysis fluid, rotate using a magnetic stirrer and complete dialysis with several (~4 times) changes of dialysis buffer (see Note 7).

3.1.5. Determination of Protein Concentrations (micro-BCA Protein Assay)

1. Determine the concentration of the protein using micro-BCA reagent following the manufacturer's instructions. To do this, take eight Eppendorf tubes and mark them by numbering (I–VIII).
2. In the first 5 tubes (I–V), take 1, 0.5, 0.25, 0.125, and 0.025 mg/ml of bovine serum albumin (BSA), respectively, in a total volume of 25 μl of PBS.
3. In the next two tubes (VI–VI) take 24 μl and 49 μl of PBS and add 1 μl of sample protein in each tube to make 1: 25 and 1: 50 dilution, respectively. As negative control, add only 25 μl of PBS in the last tube (VIII).
4. In a 14 ml Falcon tube add 2 ml of solution A and 40 μl of solution B (supplied in Kit Box) and mix thoroughly by vortexing. Transfer 200 μl of the mixture in each tube, including

the control tube, and mix by pipetting. Incubate at 37°C for 30 min and measure the concentration using a spectrophotometer.

3.1.6. Analysis of Protein Purity

Purity of the collected protein can be judged by SDS-PAGE followed by CBB or silver staining. It is better to use the silver staining method as it is most sensitive. The procedure is as follows.

1. Electrophorese the protein in 12.5% SDS-PAGE under reducing conditions, carefully remove the gel and place onto a plastic tray.
2. Treatment with fixation solution-I: Measure 50 ml of absolute methanol in a 100 ml cylinder, add 40 ml of water and 10 ml of absolute acetic acid, then mix thoroughly. Pour fixation solution-I on the gel and shake gently on a platform shaker for 10 min.
3. Treatment with fixation solution-II: Take 30 ml of absolute methanol in a 100 ml cylinder, add 55 ml of water, 10 ml of absolute acetic acid, and 5 ml of reagent-1 (supplied in Kit Box), then mix thoroughly. Discard fixation solution-I, pour fixation solution-II on the gel, and shake gently on a platform shaker for 15 min.
4. Pretreatment: Pour 50 ml of absolute methanol in a 100 ml cylinder, add 55 ml of water and 5 ml of reagent-2 (supplied in Kit Box), then mix thoroughly. Discard fixation solution-II, pour the pretreatment buffer on the gel, and shake gently on a platform shaker for 10 min. After 10 min, discard the pretreatment buffer, wash the gel with water for 5 min on a platform shaker, and repeat washing twice.
5. Silver staining: Measure 90 ml of water, 5 ml of reagent-3 (solution A, supplied in Kit Box), and 5 ml of reagent-4 (solution B, supplied in Kit Box) and mix thoroughly. Pour the stain on the gel and shake gently on a platform shaker for 15 min. After 15 min, discard the stain, wash the gel with water for 5 min on a platform shaker and repeat washing twice.
6. Developer: Take 95 ml of water and add 5 ml of reagent-5 (supplied in Kit Box), and mix thoroughly. Pour the developer on the gel and shake gently on a platform shaker until the desired protein band appears.
7. Stopping reagent: Add 5 ml of stopping solution (reagent-6, supplied in Kit Box) to the developer and shake gently on a platform shaker for further 10 min, then discard the buffer and wash the gel with water.

3.1.7. Preparation of Ticks' Salivary Gland Lysate

1. Collect salivary glands from 10 to 20 unfed or partially fed (72 h) adult ticks. To do this, fix the ventral surface of the ticks

(keeping the dorsal shell upward) on a watch glass (~9 cm in diameter) with glue and allow proper adherence. Cut the lateral margin of the tick with a blade under a dissecting microscope and remove the dorsum carefully. Carefully wash the ticks with ice-cool PBS, collect salivary glands, and keep them in ice-cool PBS.

2. Transfer the collected salivary glands into a sterilized pestle, pour liquid nitrogen, and then let the liquid nitrogen to evaporate. Grind up the salivary glands into a fine powder and transfer it into an Eppendorf tube. Add ice-cool, sterile PBS and mix properly, and sonicate tissues with an ultrasonic processor (VP-5S) for ~2 min (see Note 8).
3. Centrifuge the suspension at 10,000 × *g* and 4°C for 15 min. Collect the supernatant in an Eppendorf tube and store at −80°C until further use.
4. Collection of salivary glands from larva and nymph is not possible because they are too tiny to dissect. Therefore, whole-tick lysate is collected to detect native protein following the same procedures as described above.

3.2. Gel Mobility Shift Assays

1. Prepare 10% SDS-PAGE gel with 3 mM Ca^{2+} or 3 mM EDTA. To do this, prepare 10% resolving (running) gel containing 3 mM Ca^{2+} (see Note 9). Briefly, measure 6 ml of solution A (30% acrylamide) and add 6.66 ml of water, 4.5 ml of solution B (1.5 M Tris–HCl, pH 8.8), 180 μl of 10% SDS, 200 μl of 10% APS, 543 μl of 100 mM $CaCl_2$, and 10 μl of TEMED in a 50 ml Falcon tube, then mix thoroughly. Pour the mixture between the prearranged glass slabs. Immediately add 200 μl of butanol to make the surface smooth and even, and keep standing until the gel solidifies. The size of the resolving gel is 6.5 cm × 9 cm.
2. Prepare 4% stacking gel containing 3 mM Ca^{2+}. Take 670 μl of solution A, and add 3.45 ml of water, 630 μl of solution B, 50 μl of 10% SDS, 50 μl of 10% APS, 150 μl of 100 mM $CaCl_2$, and 5 μl of TEMED in a 14 ml Falcon tube, then mix thoroughly. Remove the butanol from the resolving gel and wash the surface with water. Pour the mixture, set the comb, and wait until the stacking gel solidifies (see Note 10).
3. Prepare 10% resolving gel with 3 mM EDTA (see Note 11). Measure 6 ml solution A and add 7.09 ml of water, 4.5 ml of solution B, 180 μl of 10% SDS, 200 μl of 10% APS, 108.6 μl of 0.5 M EDTA, and 10 μl of TEMED in a 50 ml Falcon tube, then mix thoroughly. Pour the mixture between the prearranged glass slabs. Immediately add 200 μl of butanol to make the surface smooth and even and keep standing until the gel solidifies.

4. Prepare 4% stacking gel with 3 mM EDTA. Take 670 μl of solution A, and add 3.57 ml of water, 630 μl of solution B, 50 μl of 10% SDS, 50 μl of 10% APS, 30 μl of 0.5 M EDTA, and 5 μl of TEMED in a 14 ml Falcon tube, then mix thoroughly. Remove the butanol from the resolving gel and wash the surface of the gel with water. Pour the mixture and set the comb, and keep standing until the stacking gel solidifies.
5. Take 2 μg of purified and dialyzed longistatin in an Eppendorf tube. Add an appropriate amount of 4× reducing sample buffer (one-fourth of total volume) and mix thoroughly. Boil the mixture for 5 min in boiling water (see Note 12).
6. Use bovine serum albumin (BSA) or any known protein(s) that does not bind with calcium as a negative control.
7. Set the previously prepared Ca^{2+} or EDTA containing 10% SDS-PAGE gel into a vertical electrophoresis apparatus.
8. Pour a sufficient amount of running buffer into the electrophoresis machine and carefully remove air bubbles, if present.
9. Load the protein mixture and 10 μl of marker proteins (e.g., Mark12™) in separate wells.
10. Electrophorese the protein(s) at 60–80 V until the protein crosses the stacking gel and then at 120–180 V until the run is completed.
11. Remove the gel carefully and place on a plastic tray. Stain the gel with 0.1% CBB for 1 h on a platform shaker (see Note 13).
12. Wash the gel thoroughly with CBB-destain on a platform shaker until the gel becomes transparent.
13. Take photograph(s) using a suitable digital camera.

3.3. Staining with Ruthenium Red

1. Prepare 12.5% SDS-PAGE gel. To do this, prepare 12.5% resolving gel. Briefly, take 7.5 ml of solution A and add 5.72 ml of water, 4.5 ml of solution B, 180 μl of 10% SDS, 200 μl of 10% APS, and 10 μl of TEMED in a 50 ml Falcon tube and mix thoroughly. Pour the mixture between the prearranged glass slabs and keep standing until solidification. Immediately add 200 μl of butanol to make the surface smooth and even. The size of the resolving gel is 6.5 cm × 9 cm.
2. Prepare 4% stacking gel. Briefly, take 670 μl of solution A and add 3.6 ml of water, 630 μl of solution B, 50 μl of 10% SDS, 50 μl of 10% APS, and 5 μl of TEMED in a 14 ml Falcon tube, then mix thoroughly. Remove the butanol from the gel and wash the surface of the resolving gel with water. Pour the mixture, set the comb, and keep standing until the stacking gel solidifies.
3. Electrophorese salivary glands lysate and/or recombinant protein(s) using 12.5% SDS-PAGE gel under reducing conditions following the same procedures as described above.

4. Transfer the protein(s) onto nitrocellulose membrane using ice-cool transfer buffer at 100 V for 1 h. Briefly, take a cassette on a plastic tray and pour a sufficient amount of transfer buffer onto the tray. Place a piece of fiber pad on the tray and soak it properly with transfer buffer. Cover the foam with a piece of filter paper of appropriate size. Place the nitrocellulose membrane on the filter paper. Carefully remove the gel from the slab and place the gel on the membrane. Cover the gel with another piece of filter paper and place a piece of fiber pad over the filter paper.
5. Gently press the whole arrangement to remove air bubbles. Carefully transfer the whole setup on a cassette keeping the membrane on the white side of the cassette, and close the clamp.
6. Take an immunoblot apparatus (Mini Trans-Blot Cell®) and pour sufficient amount of transfer buffer into the immunoblot device. Insert the previously arranged whole setup of the gel and nitrocellulose membrane into the immunoblot cell and turn on the magnetic stirrer.
7. Insert an icebox into the immunoblot cell on the black side of the cassette and place the immunoblot cell onto a magnetic stirrer. Set the voltage and time and finally, run the trans-blot cell.
8. After 1 h, remove the membrane and wash gently with ruthenium red buffer. Stain the nitrocellulose membrane with ruthenium red stain for 15 min or more at RT on a platform shaker.
9. Take photograph(s) immediately using a digital camera (see Note 14).

4. Notes

1. Annealing temperature and number of PCR cycles are not fixed. Those are variable for different primer sets. Usually annealing temperature is ~5°C lower than the lower melting temperature of the primers used. However, the protocol of PCR must be standardized for each molecule.
2. Avoid over-culturing of the transfected *E. coli* since over-culture increases the amount of bacterial proteins, which will hamper the purification processes. However, the duration and temperature of the culture must be standardized for each protein by pilot expression to ensure the desired concentration and purity.

3. If the recombinant protein is inactivated by lysozyme then addition of the enzyme is not recommended.
4. Sonication of *E. coli* suspension must be carried out on ice with intermittent pauses to ensure better activity of the harvested protein. Usually three bursts of sonication each of 12 s on ice, with 25 s intervals between each burst are performed. In addition, carefully adjust the temperature during thawing. If the protein is more thermolabile then thaw bacterial suspension at a lower temperature.
5. After elution of the recombinant protein, check the purity of each fraction by SDS-PAGE analysis. The protein with the highest purity is harvested and further characterized.
6. TALON resin can be reused three to four times, if the same protein is eluted. After use, the resin must be washed with 20 mM MES buffer (pH 5.0) containing 0.1 M NaCl until color returns to the original color of the resin (normally light rose in color). Then, rinse the resin with five bed-volumes of distilled water and store at 4°C in 20% non-buffered ethanol containing 0.1% azide. Do not store TALON resin with bound imidazole. During reuse, discard the alcohol and wash the resin extensively with wash buffer three to four times.
7. During dialysis, pH of the dialysis fluid and the suitable temperature must be standardized for each protein.
8. Time required for proper sonication is variable depending on the consistency of the tissues. Hard tissues require more time than the soft tissues. Therefore, time must be standardized.
9. In Gel mobility shift assays (GMSA), the concentrations of calcium and EDTA are not fixed. They can vary from micromolar to milimolar range (up to 5 mM). Initially, lower concentration should be used.
10. If calcium is used in higher concentrations (>2 mM) and the gel is kept for a few hours (~2 h) then calcium salts may precipitate. This will, however, not influence further processing and still the gel can be stored up to 3 days at RT. We recommend, however, to use the minimum effective concentration of calcium. Concentration of calcium/EDTA needs to be standardized for each protein.
11. In some cases, EDTA can be replaced by ethylene glycol tetraacetic acid (EGTA) which is a more specific chelator for calcium.
12. In GMSA, SDS-PAGE can also be performed under nonreducing conditions.
13. Instead of CBB, silver stain can be used in GMSA.
14. In ruthenium red staining, photographs of the gels should be taken as quickly as possible to avoid bleaching due to drying.

References

1. Kretsinger RH, Nockolds CE (1973) Carp muscle calcium-binding protein. II. Structure determination and general description. J Biol Chem 248:3313–3326
2. Grabarek Z (2006) Structural basis for diversity of the EF-hand calcium-binding proteins. J Mol Biol 359:509–525
3. Anisuzzaman, Islam MK, Miyoshi T, Alim MA, Hatta T, Yamaji K, Matsumoto Y, Fujisaki K, Tsuji N (2010) Longistatin, a novel EF-hand protein from the ixodid tick Haemaphysalis longicornis, is required for acquisition of host blood-meals. Int J Parasitol 40:721–729
4. Chazin WJ (2011) Relating form and function of EF-hand calcium binding proteins. Acc Chem Res 44:171–179
5. Choi KC, Jeung EB (2008) Molecular mechanism of regulation of the calcium-binding protein calbindin-D9k, and its physiological role(s) in mammals: a review of current research. J Cell Mol Med 2:409–420
6. Marenholz I, Heizmann CW, Fritz G (2004) S100 proteins in mouse and man: from evolution to function and pathology (including an update of the nomenclature). Biochem Biophys Res Commun 322:1111–1122
7. Leclerc E, Heizmann CW (2011) The importance of Ca^{2+}/Zn^{2+} signaling S100 proteins and RAGE in translational medicine. Front Biosci S3(3):1232–1262
8. Hohenester E, Maurer P, Hohenadl C, Timpl R, Jansonius JN, Engel J (1996) Structure of a novel extracellular Ca^{2+}-binding module in BM-40. Nat Struct Biol 3:67–73
9. Blanchard H, Grochulski P, Li Y, Arthur JS, Davies PL, Elce JS, Cygler M (1997) Structure of a calpain Ca^{2+}-binding domain reveals a novel EF-hand and Ca^{2+}-induced conformational changes. Nat Struct Biol 4:532–538
10. Lin GD, Chattopadhyay D, Maki M, Wang KK, Carson M, Jin L, Yuen PW, Takano E, Hatanaka M, DeLucas LJ, Narayana SV (1997) Crystal structure of calcium bound domain VI of calpain at 1.9 A resolution and its role in enzyme assembly, regulation, and inhibitor binding. Nat Struct Biol 4:539–547
11. Jia J, Han Q, Borregaard N, Lollike K, Cygler M (2000) Crystal structure of human grancalcin, a member of the penta-EF-hand protein family. J Mol Biol 300:1271–1281
12. Jia J, Tarabykina S, Hansen C, Berchtold M, Cygler M (2001) Structure of apoptosis-linked protein ALG-2: insights into Ca^{2+}-induced changes in penta-EF-hand proteins. Structure 9:267–275
13. Reddy VS, Day IS, Thomas T, Reddy AS (2004) KIC, a novel Ca2+ binding protein with one EF-hand motif, interacts with a microtubule motor protein and regulates trichome morphogenesis. Plant Cell 16:185–200
14. Evenäs J, Malmendal A, Forsén S (1998) Calcium. Curr Opin Chem Biol 2:293–302
15. Schaub MC, Heizmann CW (2008) Calcium, troponin, calmodulin, S100 proteins: from myocardial basics to new therapeutic strategies. Biochem Biophys Res Commun 369:247–264
16. Grabarek Z (2011) Insights into modulation of calcium signaling by magnesium in calmodulin, troponin C and related EF-hand proteins. Biochim Biophys Acta 1813:913–921
17. Zheng C, Liu HH, Zhou J, Zhang B (2010) EF-hand domains of MCFD2 mediate interactions with both LMAN1 and coagulation factor V or VIII. Blood 115:1081–1087
18. Berridge MJ, Bootman MD, Lipp P (1998) Calcium – a life and death signal. Nature 395:645–648
19. Kahl CR, Means AR (2003) Regulation of cell cycle progression by calcium/calmodulin-dependent pathways. Endocr Rev 24:719–736
20. Barger-Lux MJ, Heaney RP (1994) The role of calcium intake in preventing bone fragility, hypertension, and certain cancers. J Nutr 124: 1406S–1411S
21. Belizán JM, Villar J, Repke J (1988) The relationship between calcium intake and pregnancy-induced hypertension: up-to-date evidence. Am J Obstet Gynecol 158:898–902
22. Völkers M, Rohde D, Goodman C, Most P (2010) S100A1: a regulator of striated muscle sarcoplasmic reticulum Ca^{2+}handling, sarcomeric, and mitochondrial function. J Biomed Biotechnol 2010:178614. doi:10.1155/2010/178614
23. Desjardins JF, Teichert-Kuliszewska K, Parker T (2010) S100A1: a pluripotent regulator of cardiac and vascular function. Can J Cardiol Suppl A:9A–12A
24. Weiss JL, Hui H, Burgoyne RD (2010) Neuronal calcium sensor-1 regulation of calcium channels, secretion, and neuronal outgrowth. Cell Mol Neurobiol 30:1283–1292
25. Fritz G, Botelho HM, Morozova-Roche LA, Gomes CM (2010) Natural and amyloid self-assembly of S100 proteins: structural basis of functional diversity. FEBS J 277:4578–4590
26. Rohde D, Brinks H, Ritterhoff J, Qui G, Ren S, Most P (2011) S100A1 gene therapy for heart failure: a novel strategy on the verge of clinical trials. J Mol Cell Cardiol 50:777–784

27. Stengel A, Goebel M, Taché Y (2011) Nesfatin-1: a novel inhibitory regulator of food intake and body weight. Obes Rev 12: 261–271
28. Lewit-Bentley A, Réty S (2000) EF-hand calcium-binding proteins. Curr Opin Struct Biol 10:637–643
29. Tsigelny I, Shindyalov IN, Bourne PE, Südhof TC, Taylor P (2000) Common EF-hand motifs in cholinesterases and neuroligins suggest a role for Ca^{2+} binding in cell surface associations. Protein Sci 9:180–185
30. Krebs J, Heizmann CW (2007) Calcium-binding proteins and the EF-hand principle. In: Krebs J, Michalak M (eds) Calcium: a matter of life or death, vol 41, 1st edn. Elsevier, UK, pp 51–93
31. Ikura M, Ames JB (2006) Genetic polymorphism and protein conformational plasticity in the calmodulin superfamily: two ways to promote multifunctionality. Proc Natl Acad Sci USA 103:1159–1164
32. Boye K, Maelandsmo GM (2010) S100A4 and metastasis: a small actor playing many roles. Am J Pathol 176:528–535
33. Wolf S, Haase-Kohn C, Pietzsch J (2010) S100A2 in cancerogenesis: a friend or a foe? Amino Acids 41:849–861. doi:10.1007/s00726-010-0623-2
34. Pottgiesser J, Maurer P, Mayer U, Nitsch R, Mann K, Timpl R, Krieg T, Engel J (1994) Changes in calcium and collagen IV binding caused by mutations in the EF hand and other domains of extracellular matrix protein BM-40 (SPARC, osteonectin). J Mol Biol 238:563–574
35. Bianchi R, Kastrisianaki E, Giambanco I, Donato R (2011) S100B protein stimulates microglia migration via RAGE-dependent up-regulation of chemokine expression and release. J Biol Chem 286:7214–7226
36. Nagamune K, Moreno SN, Chini EN, Sibley LD (2008) Calcium regulation and signaling in apicomplexan parasites. Subcell Biochem 47:70–81
37. DeFalco TA, Bender KW, Snedden WA (2009) Breaking the code: Ca^{2+} sensors in plant signalling. Biochem J 425:27–40
38. Orans J, Johnson MD, Coggan KA, Sperlazza JR, Heiniger RW, Wolfgang MC, Redinbo MR (2010) Crystal structure analysis reveals Pseudomonas PilY1 as an essential calcium-dependent regulator of bacterial surface motility. Proc Natl Acad Sci USA 107: 1065–1070
39. Sonenshine DE (1991) Biology of ticks, vol 1. Oxford University Press, New York
40. Fujisaki K, Kawazu S, Kamio T (1994) The taxonomy of the bovine *Theileria* spp. Parasitol Today 10:31–33
41. Hoogstraal H, Roberts FH, Kohls GM, Tipton VJ (1968) Review of *Haemaphysalis (Kaiseriana) longicornis* Neumann (resurrected) of Australia, New Zealand, New Caledonia, Fiji, Japan, Korea, and northeastern China and USSR, and its parthogenetic and bisexual populations (Ixodoidea, Ixodidae). J Parasitol 54:1197–1213
42. Tsuji N, Miyoshi T, Battsetseg B, Matsuo T, Xuan X, Fujisaki K (2008) A cysteine protease is critical for *Babesia* spp. transmission in *Haemaphysalis* ticks. PLoS Pathog 4:e1000062
43. Ho T, Htwe KK, Yamasaki N, Zhang GQ, Ogawa M, Yamaguchi T, Fukushi H, Hirai K (1995) Isolation of *Coxiella burnetii* from dairy cattle and tick, and some characteristics of isolates in Japan. Microbiol Immunol 39: 663–671
44. Anisuzzaman, Islam MK, Miyoshi T, Alim MA, Hatta T, Yamaji K, Matsumoto Y, Fujisaki K, Tsuji N (2010) Longistatin, a plasminogen activator, is key to the availability of blood-meals for ixodid ticks. PLoS Pathog 7: e1001312
45. Charuk JH, Pirraglia CA, Reithmeier RA (1990) Interaction of ruthenium red with Ca^{2+}-binding proteins. Anal Biochem 188: 123–131
46. Maruyama K, Nonomura Y (1984) High molecular weight calcium binding protein in the microsome of scallop striated muscle. J Biochem 96:859–870
47. Schibeci A, Martonosi A (1980) Detection of Ca^{2+}-binding proteins on polyacrylamide gels by 45Ca autoradiography. Anal Biochem 104:335–342
48. Klee CB, Crouch TH, Krinks MH (1979) Calcineurin: a calcium- and calmodulin-binding protein of the nervous system. Proc Natl Acad Sci USA 76:6270–6273

Chapter 10

Super-Resolution Microscopy of the Neuronal Calcium-Binding Proteins Calneuron-1 and Caldendrin

Johannes Hradsky, Marina Mikhaylova, Anna Karpova, Michael R. Kreutz, and Werner Zuschratter

Abstract

Calcium (Ca^{2+}) signaling in neurons is mediated by plethora of calcium binding proteins with many of them belonging to the Calmodulin family of calcium sensors. Many studies have shown that the subcellular localization of neuronal EF-hand Ca^{2+}-sensors is crucial for their cellular function. To overcome the resolution limit of classical fluorescence and confocal microscopy various imaging techniques have been developed recently that improve the resolution by an order of magnitude in all dimensions. This new microscope techniques make co-localization studies of Ca^{2+}-binding proteins more reliable and help to get insights into the macromolecular organization of intracellular structures and signaling pathways beyond the diffraction limit of visible light.

Key words: STED microscopy, Synaptic proteins, Cytoskeleton, Golgi apparatus, Calneuron, Caldendrin, Calcium, EF-hand

1. Introduction

1.1. Compartmentalization of Ca^{2+} Signaling in Neurons and Role of Calmodulin-Like Neuronal Calcium-Binding Proteins

Ca^{2+} signaling in neurons exhibits steep spatiotemporal gradients with a spatial resolution of micro- and nanodomains (1, 2). Occurrence of Ca^{2+} gradients is strongly related to Ca^{2+} influx from extracellular space or induced cytosolic release from intracellular stores (endoplasmic reticulum/ER, Golgi complex, and other). Gradients derived from sequestration of Ca^{2+} ions by ER, mitochondria, and locally enriched Ca^{2+} buffer proteins can be generated (2, 3). Multiple Ca^{2+} sensor proteins play a crucial role

Johannes Hradsky and Marina Mikhaylova have contributed equally to the work.

Claus W. Heizmann (ed.), *Calcium-Binding Proteins and RAGE: From Structural Basics to Clinical Applications*, Methods in Molecular Biology, vol. 963, DOI 10.1007/978-1-62703-230-8_10, © Springer Science+Business Media New York 2013

in transducing Ca^{2+} signals into alterations of cellular processes. Especially a large variety of Calmodulin-like Ca^{2+}-binding proteins is expressed in the central nervous system. The groups of neuronal calcium sensors (NCS) and neuronal calcium-binding proteins (nCaBPs) derived from their ancestor Calmodulin gave rise a variety of proteins that exhibit different Ca^{2+}-binding properties, tissue, cell type, and subcellular localization profiles (see ref. 4 for review).

Calmodulin (CaM) is a very abundant small cytosolic protein that can be recruited to the different neuronal compartments via interaction with its protein targets (5). Differently from CaM, NCS proteins are associated with intracellular membranes via co-translational N-terminal lipid modification by myristoyltransferase (NCS-1, Hippocalcin, Recoverin, KChIPs, VILIPs, and others) (4, 6). A Ca^{2+}-induced myristoyl switch triggers the conformational changes and membrane recruitment of some NCS proteins (Recoverin, VILIPs (4, 7)). nCaBP proteins also can be *N*-myristoylated (like s- and l-CaBP1) but additionally, adopt different strategies for targeting to a particular cell compartment. For example, Calneuron-1 and Calneuron-2 (also called CaBP8 and CaBP7, respectively) are unique members of nCaBP group because their targeting to Golgi membranes and post-Golgi vesicular compartments is provided by the C-terminal transmembrane domain (TMD) (8–11). Interestingly, they are the only members of EF-hand protein superfamily that can be classified as tail-anchored proteins (9, 11). Calneurons are initially post-translationary inserted into the endoplasmic reticulum (ER) with assistance of TRC40/ASNA1 chaperon and then, due to the length of the TMD and associations with particular phospholipids, are transported to the Golgi complex (11). Calneurons are enriched at the Golgi but they also can be found along the constitutive secretory pathway and at the plasma membrane (9–11). At the Golgi network, Calneurons decode local calcium concentrations via regulating the activity of phosphatidylinositol 4-kinase IIIβ (PI-4KIIIβ), thus providing a Ca^{2+}-dependent control for the regulated local synthesis of phosphatidylinositol 4-phosphate (PI(4)P) and for the exit of vesicles from the trans-Golgi network (4, 10, 12). At the plasma membrane of bovine chromaffin cells, Calneuron-1 is involved in transduction of Ca^{2+} signaling and regulating the activity of voltage gated Ca^{2+} channels (VGCC) (13). A mutant Calneuron-1 lacking the C-terminal transmembrane domain did not inhibit currents generated through N-, L-, and P/Q-type channels indicating that proper targeting of Calneuron-1 is required for its function (13). Caldendrin is a founding member of nCaBP group of Ca^{2+}-sensors. Caldendrin is abundant in brain where it shows prominent immunostaining of the neurons in the cortex, hippocampus, and hypothalamus (14–16). Caldendrin is enriched in somatodendritic compartments and is enriched in the postsynaptic density fraction after induction of kainate-induced seizures (17). In terms of its EF-hand organization, Caldendrin/CaBP1 is the closest homologue of CaM in brain (14, 18, 19). Caldendrin can sense Ca^{2+} in the same dynamic range as CaM and it has, like CaM, two independent EF-hand domains connected via a linker. Differently from CaM, Caldendrin has an extended amino-terminus enriched in basic amino acids. Another important difference between Caldendrin and CaM relates to amino acid substitutions within the second EF-hand loop that render this motif incapable of Ca^{2+}-binding (14). Interestingly, CaM cannot substitute Caldendrin in terms of its interactions. Binding of Caldendrin to IP3 receptors and L-type Ca^{2+} channels ($Ca_v1.2$) modulate the properties of these channels in Ca^{2+}-dependent manner. Binding of Caldendrin, but not of CaM, has been demonstrated for the light chain 3 of microtubule-associated protein MAP1A/B (20) and for the synapto-nuclear messenger protein Jacob (21). Additionally, the amino-terminus of Caldendrin contains proline-rich regions that could serve as an interface for binding of SH3-domain proteins (4). Visualization of the precise subcellular localization of NCS and nCaBPs is important to appreciate their function. Additionally, dynamic aspects of this localization, like transport along the secretory trafficking pathway for Calneurons or activity-induced synaptic redistribution of Caldendrin can be addressed by using high-resolution light microscopy.

1.2. High-Resolution Light-Microscopy Beyond the Diffraction Limit

During the past two decades the traditional fluorescence microscopy has developed from a discipline interested in the description of static microstructures to a tool of molecular cell biology aiming to analyze physiologically relevant molecular processes on the cellular and systems network level. Exploring the function of macromolecular networks and the Ca^{2+}-mediated effects on the cellular machinery requires high-resolution microscope techniques that allow image acquisition below the diffraction limit of visible light (22). In the past, subcellular localizations of immunostained structures were only possible by time-consuming electron microscopy.

With the introduction of new optical techniques like 4pi (23), Structured Illumination (SIM) (24, 25), Spatial Illumination Microscopy (SMI) (26), Spectral-Precision-Distance Microscopy (SPDM) (27, 28), Ground State Depletion Microscopy (GSD) (29), Stimulated Emission Depletion (STED) (30, 31) or Photoactivated Localization Microscopy (PALM/STORM) (32, 33) a significant improvement of spa-

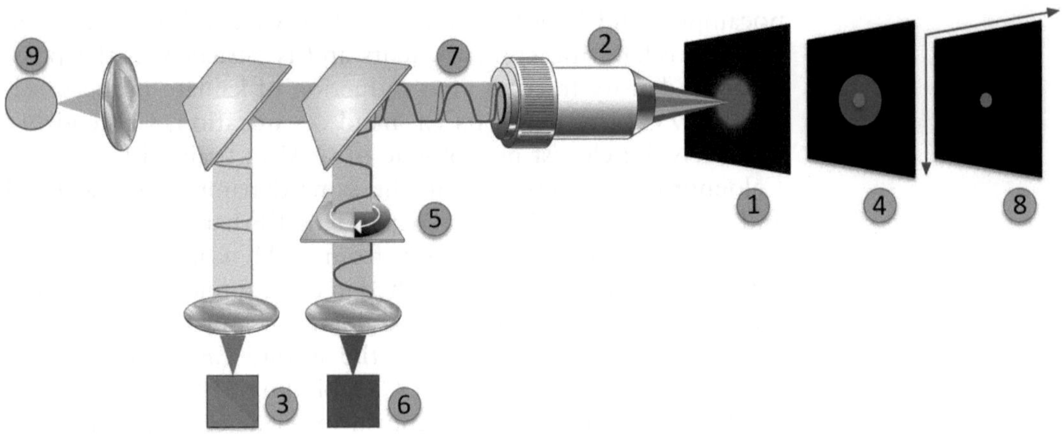

Fig. 1. Principle of STED resolution enhancement. In a conventional confocal laser scanning microscope, the sample is sequentially illuminated by a diffraction limited spot (1) via a high numerical aperture objective (2). In a Ti-Sapphire-based STED microscope, this illumination spot (produced by a pulsed excitation laser (3) of 532 or 640 nm) is overlayed by a perfectly aligned doughnut-shaped second laser pulse of 750 nm (4) produced by a Vortex phase filter (5) in the beam-path of the Ti-Sapphire depletion laser (6). The STED laser pulse immediately follows the excitation pulse with a short delay (7). As result, the excited molecules, in the doughnut-shaped area, return to the ground state without emitting spontaneous fluorescence. Because the excited molecules in the center remain unaffected from the depletion process the effective illumination spot becomes smaller (8). The power of the depletion laser (6) determines the size of the central area from where fluorescence can still be detected (9). In other words, the resolution increases with the power of the depletion laser. Diagram courtesy of Leica Microsystems, Germany.

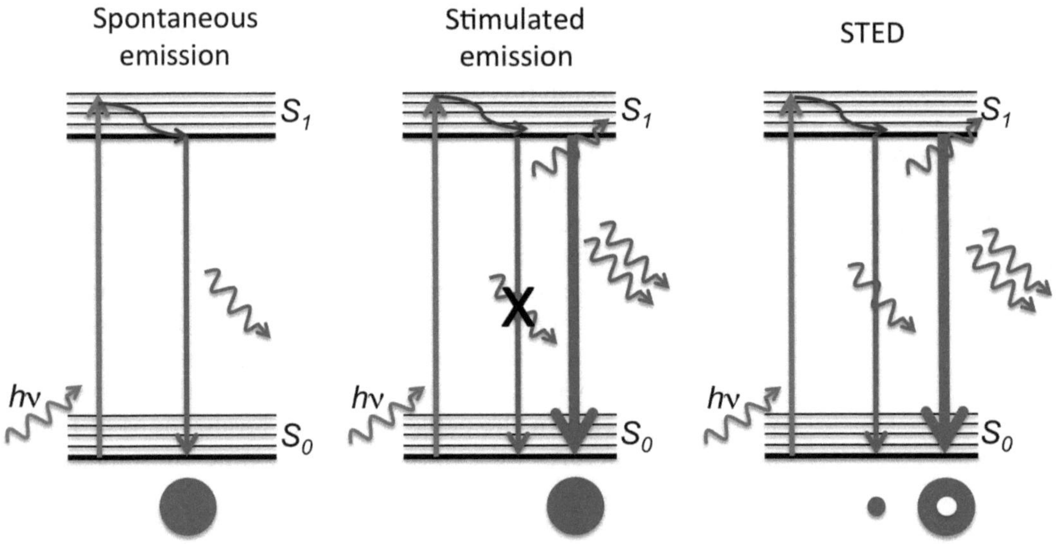

Fig. 2. Principle of spontaneous fluorescence and stimulated (induced) fluorescence emission. Spontaneous emission (*left panel*): When a fluorescent molecule absorbs light quanta, it will go to a higher energy level. From here, it will relax back spontaneously to the ground state after a short period of time (i.e., the fluorescence lifetime) by emitting a photon of less energy (i.e., red-shifted wavelength). The emission results in a diffraction limited spot. Stimulated emission (*center panel*): From the excited state a fluorescent molecule can be forced to return back to the ground state, when the electron in the excited state is stimulated by another photon, whose energy matches the difference between the electron's energy and the ground state. In this case the electron will not relax spontaneously, but will immediately return to the ground state by emitting an additional photon of the same wavelength as the incident photon. *Right panel*: In a STED microscope spontaneous fluorescence emission is restricted to the inner area of an excitation spot, whereas stimulated emission depopulates the excited molecules in an outer ring.

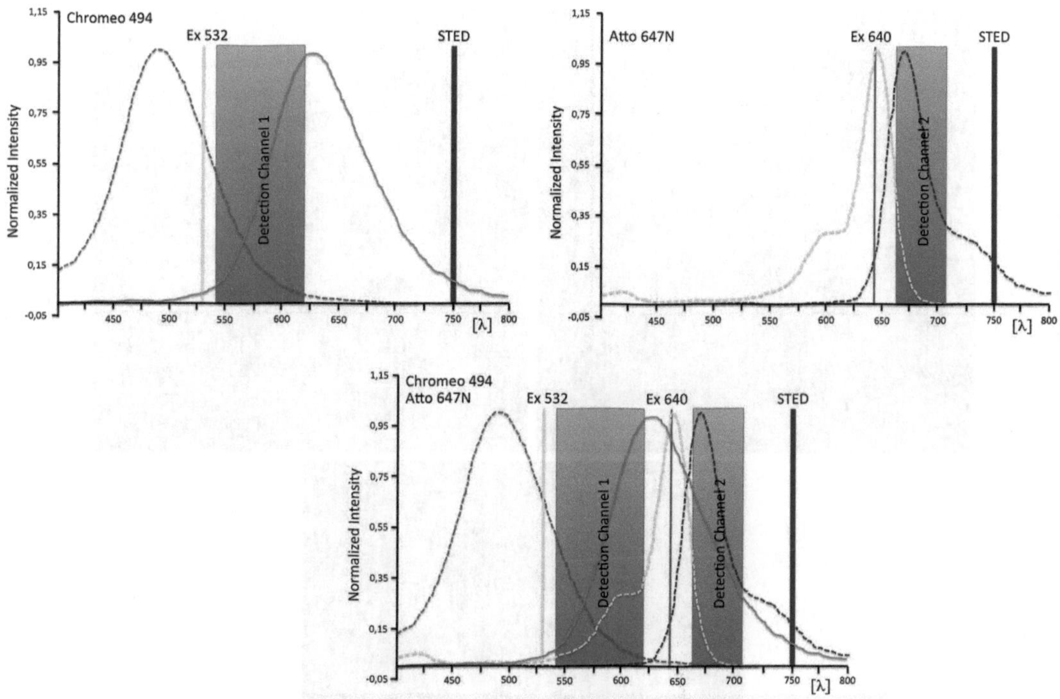

Fig. 3. Absorption and Emission spectra of Chromeo 494 (*left*) and Atto 647N (*right*) in relation to the excitation wavelength, the detection bandwidth and the depletion (STED) laser pulse. The *bottom graph* summarizes the situation in a Ti-Sapphire 2-channel STED microscope.

tial, i.e., structural resolution beyond the diffraction limit of classical microscopy (defined by Ernst Abbe in 1873 (34)) was recently achieved. Most of these concepts take advantage of the fact that fluorescence can be switched between a dark state and a bright state, either in a stochastic (PALM/STORM/GSD) or in a deterministic way (STED).

Here, we show by using two channel STED microscopy (see Notes 1 and 3, Figs. 1–3) that the subcellular distribution of the calcium-binding proteins, Calneuron and Caldendrin, can be determined with much higher precision than before (Figs. 4–8). In immunocytochemical co-localization studies, we demonstrate distinct labeling patterns of these Ca^{2+}-binding proteins along Golgi-membranes (Figs. 4 and 5), cytoskeleton-markers (Figs. 4, 5, 8) and in relation to pre- and postsynaptic scaffolding proteins (Figs. 6–8), which previously could not be resolved.

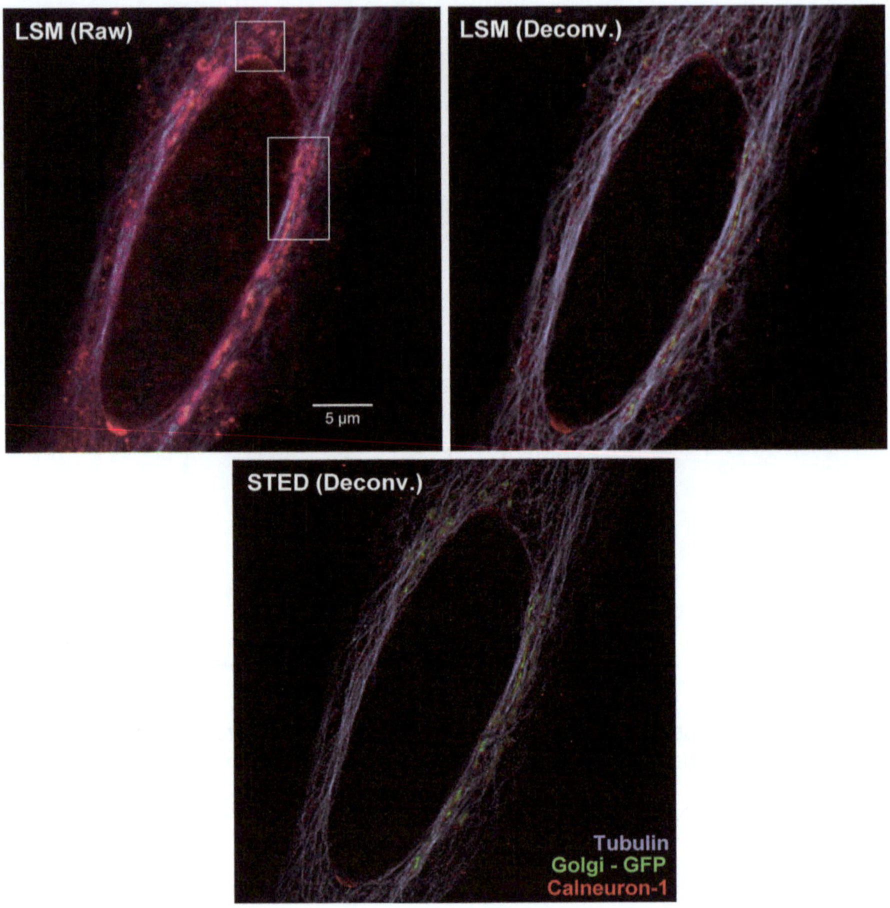

Fig. 4. HeLa and Cos7 cells were successfully transfected with the untagged Calneuron-1 construct (in pcDNA3.1) and GFP-Sec61β or GFP-Monomeric-Golgi. Subsequently, cell cultures were incubated with antibodies against β-Tubulin III and Calneuron-1 and stained using Chromeo 494 and Atto 647N as dyes suitable for Ti-Sapphire-based STED imaging. Series of STED and confocal images were taken from triple stained cells according to above protocol. Figure shows the difference between the confocal (*left panel*) and the STED (*bottom panel*) image of a HeLa cell stained with Chromeo 494 (*blue*, indicating β-Tubulin III), Atto 647N (*red*, indicating Calneuron-1) and GFP-monomeric-Golgi (*green*). Whereas the deconvolved confocal image (*right panel*) already demonstrates some optimization of the blurred confocal raw image (*left*), a significant improvement of the tubular cytoskeleton and the Calneuron-1 immunostainings becomes visible only in the STED image (*bottom*). Scale bar indicates 5 µm.

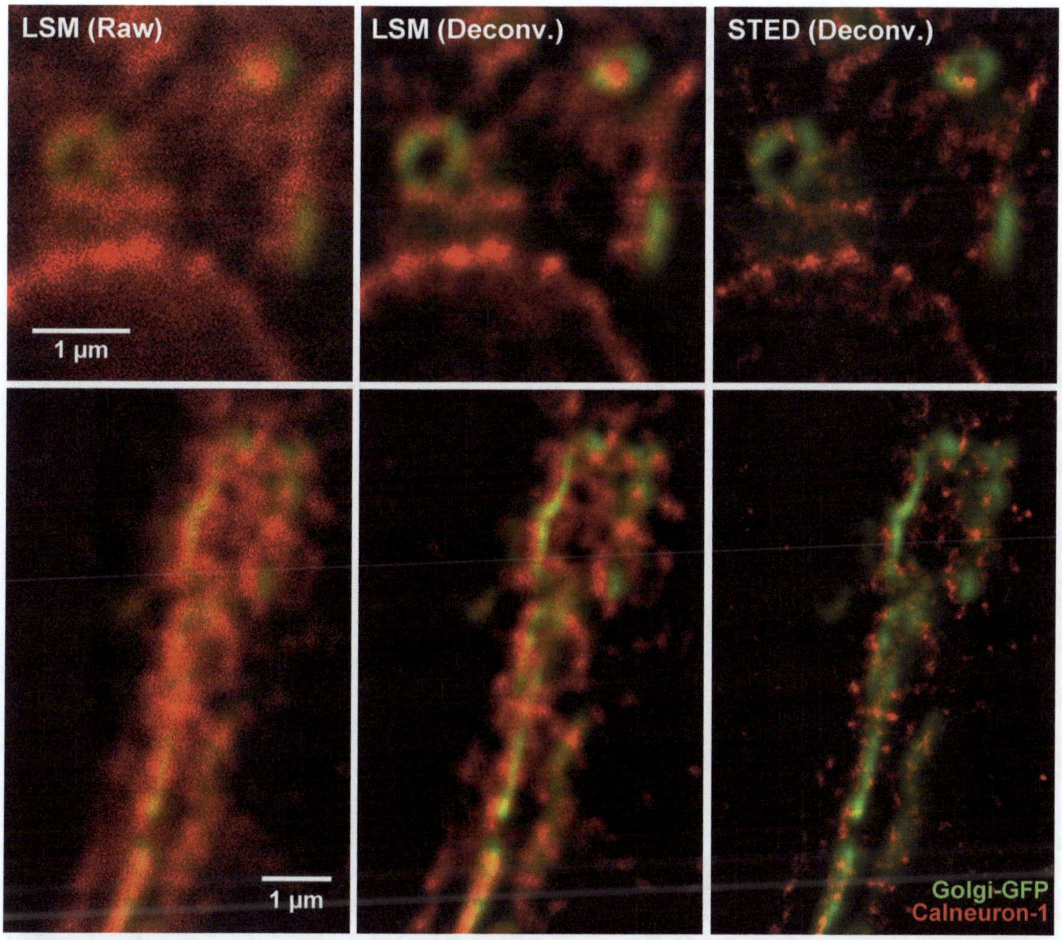

Fig. 5. Detail magnifications of the GFP-labeled Golgi apparatus (*green*) and Calneuron-1 (*red*) depicted in Fig. 10.4. The details revealed distinct foci of Calneuron-1 association along the Golgi-network (*green*) in the picture acquired in STED mode (*right panel*). In contrast to the STED-image scans recorded in the classical confocal mode (*left and center panel*) did not show such details but appeared mainly blurred due to their lower resolution. Deconvolution of the confocal raw image (*left panel*) uncovered some more details (*center panel*) but could not compete with the resolution of the STED image (*right panel*). For instance, as it would be expected, Calneuron-1 staining was mostly associated with the Golgi membrane from cytosolic leaflet but not inside of the Golgi lumen (STED images, *right panel*). Scale bar indicates 1 μm.

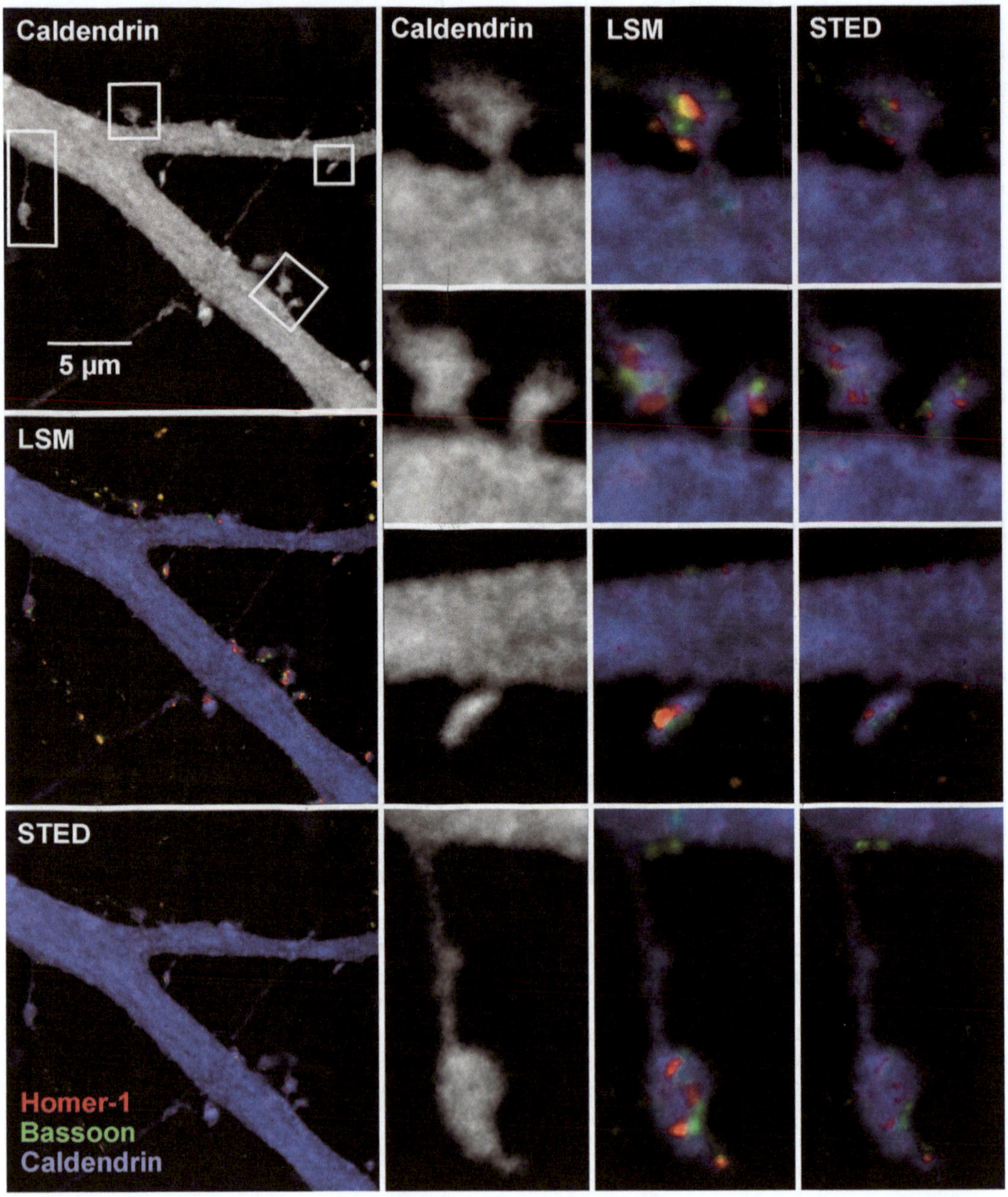

Fig. 6. To demonstrate the potential of STED imaging for characterizing the dynamics of calcium-binding proteins in neurons, we performed immunostainings against the pre- and postsynaptic scaffolding proteins, Bassoon and Homer-1, in hippocampal cell cultures, previously transfected with a GFP-fusion construct of Caldendrin. The dendritic branch of a neuron transfected with Caldendrin-GFP (*blue*) shows specific localizations of Bassoon (*green*) and Homer (*red*) mainly at spiny synapses of different size and morphology. Caldendrin is present within spines where it co-localizes with Homer. Comparison between confocal (*center panels*) and STED (*right panel*) images show that blurred structures of confocal images can be resolved as distinct contact zones within the STED images. Scale bar indicates 5 μm.

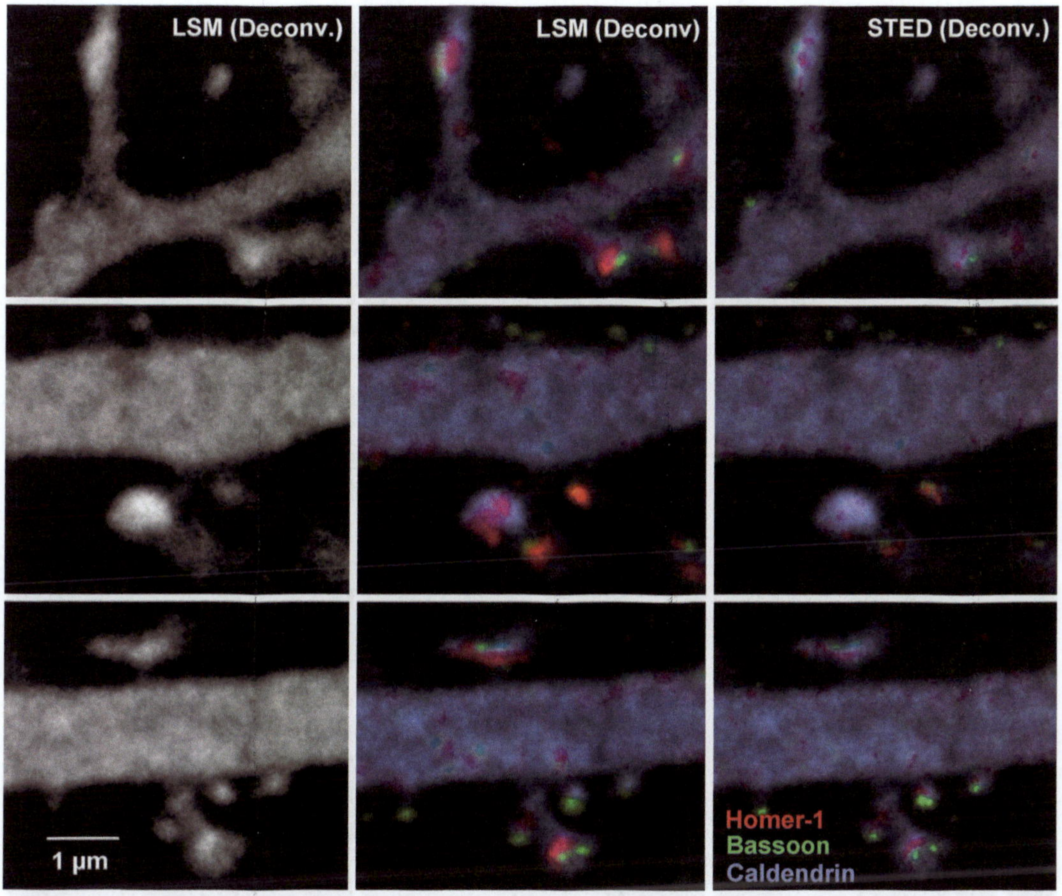

Fig. 7. Another example of Bassoon (*green*) and Homer-1 (*red*) immune-positive contacts at spines of a Caldendrin-GFP transfected cell (*blue*). Whereas some of the spiny contacts displayed an irregular or complementary distribution of Caldendrin, others exhibited a strong enhancement of the calcium-binding protein, which then correlated with the distribution of Homer. This might indicate a correlation in the activity levels of single synapses. Comparison between confocal (*left, center panels*) and STED (*right panel*) images revealed a more precise distribution and higher resolution of both synaptic markers in the STED images. Frequently, blurred structures of confocal images could be resolved as distinct contact zones within the STED images. Scale bar indicates 1 μm.

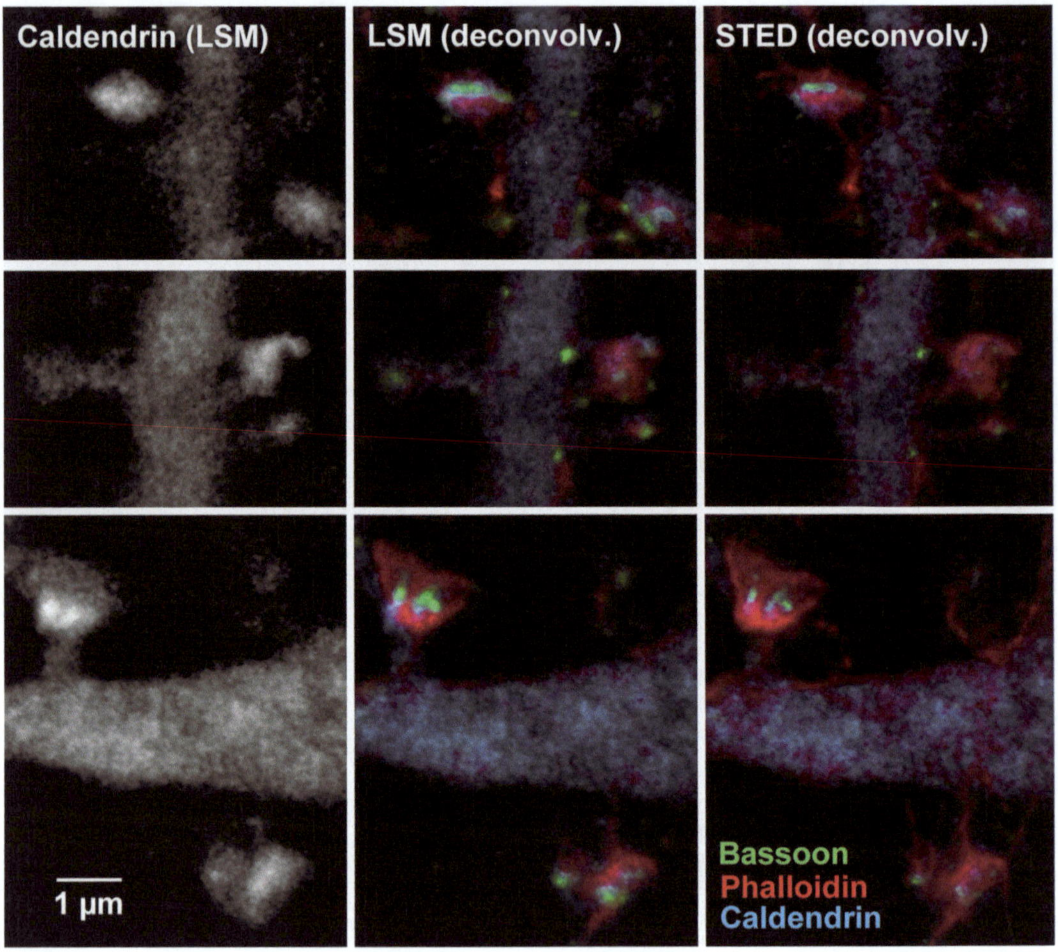

Fig. 8. To analyze the interaction of Caldendrin (*blue*) with the actin-cytoskeleton we exchanged Homer antibody-staining by Phalloidin-Atto 647N (*red*) incubation. STED imaging (*right*) showed that Caldendrin (*blue*) is rather complementary or associated than fully overlapping with Phalloidin (*red*) staining. Again the STED images (*right*) showed a considerable higher resolution in which individual contact zones of the presynaptic marker Bassoon (*green*) could be distinguished. Scale bar indicates 1 µm.

2. Materials

2.1. Cell Culture and Transfection Reagents

1. NB medium→Neurobasal® Medium (Gibco).
2. NB(−)→NB, 1× B27 (Gibco), 0.5 mMl-Glutamine.
3. NB(+)→NB(−), 100 U/mL Penicillin.
4. Opti-MEM® *Reduced Serum Medium* (Gibco).
5. DMEM→Dulbecco's Modified Eagle Medium (Gibco).
6. DMEM(+)→DMEM, 10% fetal calf serum (FCS), 2 mM l-glutamine, 100 U/mL penicillin, 100 µg/mL streptomycin.

7. Lipofectamine™ 2000 *Transfection Reagent* (Invitrogen).
8. PolyFect Transfection Reagent (Qiagen).

2.2. Cell Culture Material

1. 12-well Multidish (e.g., Nunclon™Δ, NUNC).
2. 18 mm coverslips (e.g., Thermo CB00180RA1).

2.3. Staining Reagents

1. PBS→10 mM $NaPO_4$ pH 7.4.
2. 4% PFA→4% PFA in PBS.
3. Permeabilization buffer pH 7.4→PBS, 0.25% Triton X-100.
4. Blocking buffer→PBS, 2% Glycine, 2% BSA, 0,2% Gelatine, 50 mM NH_3Cl, pH 7.4.
5. ProLong® Gold Antifade Reagent (Molecular Probes).

2.4. Staining Material

1. Forceps with angled tip (e.g., Dumont #5/45).
2. Microscope slides 3×1 in. (e.g., Diagonal, 360 535 00).
3. Chipboard slide folder (e.g., Ceesem, 51.850043).
4. Parafilm "M" (Pechiney Plastic Packaging).
5. Lightproofed, humidified chamber (Tables 1 and 2).

Table 1
Primary antibodies

Antigen	Host	Dilution	Company
Bassoon	ms	1:800	Stressgene
Homer 1	rb	1:500	SynapticSystems
Calneuron-1	rb	1:400	Homemade, affinity purified
β-Tubulin III	ms	1:800	Sigma

See Notes 8 and 9

Table 2
Secondary antibodies

Antigen	Host	Dilution	Company
ATTO 647N anti-Rabbit IgG	Goat	1:200	Active Motif
Chromeo™ 494 anti-Mouse IgG	Goat	1:50	Active Motif

See Note 5

3. Methods

3.1. Transfection of Hippocampal Primary Culture with Lipofectamine2000

1. Exchange of media: Rat hippocampal primary neurons are prepared at embryonic day 18 and plated on 18 mm coverslips coated with poly-d-lysine at density 30,000 cells per 1 mL of media as described previously (21). For transfection the conditioned NB+-media, where the neurons were growing until day in vitro 8 (DIV8), has to be replaced by 1 mL of freshly prepared NB– (without the antibiotics). The old media is collected and stored in a Falcon tube in the incubator.
2. Preparation of transfection mix: Transfection is done using 2 μl Lipofectamine2000-reagent and 2 μg of a Caldendrin-EGFP construct (in pEGFP-N1 vector) for each well. Both components are diluted and incubated separately for 5 min in 50 μl Optimem at room temperature (RT). The Lipofectamine2000 containing Optimem is then mixed with the DNA followed by another 20 min of incubation (RT).
3. Transfection of neurons: After the addition of the transfection mix the neurons are transferred back to the incubator where they are kept at 5% CO_2 and 37 °C for another 2 h. Finally the NB–/Optimem transfection mix is replaced by the conditioned media collected prior to transfection and the cells are left in the incubator till DIV16.

3.2. Transfection of HeLa- and COS-7 Cell Cultures with Polyfect

In contrast to the transfection of neurons, an exchange of media prior to transfection is not required. About 24 h after splitting, the cells are ready for transfection. In this approach a double transfection with untagged Calneuron-1 construct (in pcDNA3.1) and GFP-Monomeric-Golgi (subcloned from pDsRed-Monomer-Golgi in frame with EGFP-N1/Clontech) is performed. Therefore, 1.5 μg of each construct (in total 3 μg/well) and 6 μl of Polyfect-reagent are added to 100 μl of RT DMEM– media and kept for 10 min incubation at RT. To the preincubated mix, 100 μl of 37 °C DMEM+ media is added. The transfection mix is then applied to the DMEM+ cultivation media of the cells, which are then kept for another 48 h in the incubator followed by the fixation.

3.3. Staining of Cell Cultures on 18 mm Coverslips for STED-Microscopy

1. Fixation of the cells: Hippocampal primary neurons, HeLa and Cos-7 cells are fixed using 4% paraformaldehyde (PFA). After removal of the cultivation media, cells are kept for 10 min at RT with 4% PFA, followed by three times washing with PBS. The fixation should be carried out in the cultivation plates, whereas for the following steps, the coverslips have to be transferred to a piece of Parafilm placed on microscope slides in a humidified, lightproofed chamber.

2. Permeabilization of the cells: To permeabilize the cells, 150 μl permeabilization buffer is applied to each coverslip. The permeabilization is stopped after 8 min RT incubation, by another three times PBS wash.
3. Blocking and primary antibody labeling: To avoid unspecific binding of the primary antibody the cells are blocked for 1 h at RT in blocking buffer. During the blocking step, dilutions of the primary antibody are prepared in blocking buffer (50 μl per coverslip). To reduce the amount of antibody used for the staining, incubation is carried out putting the coverslips in invers orientation to the antibody solution. The incubation with the primary antibody is performed overnight at 4 °C (see Notes 8 and 9).
4. Secondary antibody labeling: After removal of the primary antibody the coverslips are flipped back and washed five times with PBS. In the meanwhile, dilutions of the fluorescence-labeled secondary antibodies are prepared in blocking buffer. For Ti-Saphir-STED-imaging secondary antibodies tagged with Chromeo 494 and Atto 647N dyes are strongly recommended at dilutions of 1:50 and 1:200, respectively. Keep in mind: To maintain the stability of the fluorophores, it is required to reduce the exposure of them to light to a minimum. After washing, secondary antibody dilutions are applied analogous to the primary antibodies, whereas the incubation is carried out only for 2 h at RT, followed by another five times of washing (see Notes 5 and 6).
5. Embedding: Finally the stained coverslips have to be mounted on microscope slides. To avoid any steric problems while examining the cells on a microscope, it is advisable to put only one coverslip on each slide. Therefore, a drop of RT Prolong-reagent has to be put in the center of the slide in which the coverslip, after a short rinse in distilled water (to remove any salt precipitations), gets finally flipped. After embedding and labeling of the slides, it is best to store them at RT in a light sealed chipboard slide folder. 24 h After embedding, the cells are ready for microscopy (see Note 7).

3.4. Image Acquisition of Stained Cell Cultures by STED-Microscopy

For STED-imaging of Chromeo 494 and Atto 647N dyes (see Fig. 3, Note 5) with emission maxima around 600 nm (channel 1) and 680 nm (channel 2), respectively, we used a commercial 2-channel STED microscope from Leica Microsystems, Germany (see Notes 1–4). Since the Leica STED microscope is fully software controlled, acquiring STED images is nearly as easy as confocal imaging. The following workflow protocol gives an overview about the critical steps and different approaches to acquire high-resolution images.

1. After switching on microscope, computer, scanner, lasers, and starting the LAS software open the configuration menu/laser and activate the multi-photon (MP) laser, the pulsed 532 and 640 nm laser, the 594 nm laser, and the Argon laser (if required for triple labeling in confocal mode). Open the shutter of the MP laser (red light is visible) (see Note 4).
2. Let the system and the MP laser warm up for at least 1 h to ensure mode lock and to stabilize focus and the optical components (from thermal drift).
3. Make sure that the 100× Oil Plan Apo NA 1.4 STED objective is inserted into the beam path.
4. Open configuration menu/STED and let the auto alignment routine run, which takes only few minutes. This routine optimizes the adjustment between the excitation lasers and the STED laser and should be repeated every 2 h in order to minimize thermal drifts within the beam paths. Because the sample is not illuminated during the beam alignment, there is no need to remove it from the microscope stage.
5. Since Chromeo 494 and Atto 647 N emit their fluorescence in the red range of the spectrum (Fig. 3), it is difficult to find an area of interest by eye with wide field epifluorescence illumination. Therefore, it is advisable to have a third label, like GFP or Alexa 488, within the sample that helps to find the focus and to navigate on the sample until one has found the region of interest for STED imaging (see Note 5).
6. Before scanning the best position one should perform a test run with a dispensable structure to find the optimized settings without bleaching the significant areas from which STED images should be acquired. In contrast to Chromeo 494 with low quantum yield Atto 647N is a rather stable fluorophore that allows repeated scanning. Hence, it is recommendable to use the Atto 647N channel for fine adjustments at higher zoom factors (see Note 5).
7. Confocal as well as STED images are recorded sequentially by scanning the focused beam with a galvo-mirror across the specimen.

   A typical image series consists of four channels, which have to be acquired in sequential mode. Channel 1: Atto 647N (LSM mode), channel 2: chromeo 494 (LSM mode), channel 3: Atto 647N (STED mode), channel 4: Chromeo 494 (STED mode). Eventually in a 5th channel GFP or Alexa 488 can be recorded as counterstaining in LSM mode (see Notes 2, 3 and Figs. 4–8).
8. Atto 647N and Chromeo 494 have to be captured by Avalanche photodiodes (APDs), whereas the counterstaining

(i.e., GFP or Alexa 488) should be imaged by one of the internal photomultiplier tubes (PMTs).

9. Adjusting the scan parameters according to the signal to noise ratio within the sample requires fine adjustment of each channel while scanning. To avoid bleaching fast scanning (1,000 Hz) is recommended at low resolution (256 × 256 pixel) with a look up table (glow over/under) that indicates saturated pixel in blue and background values in green.
10. To find the optimal intensity of the excitation laser start scanning in LSM mode and then toggle between LSM and STED mode by activating the button at the STED slider. Increasing the power of the STED laser should result in higher depletion, visible as smaller structures indicating improvement of resolution. Care must be taken to avoid bleaching from strong laser excitation as well as re-excitation from the depletion laser (see Notes 1–3).
11. If no clear improvement between the STED and the LSM image is visible the beam alignment has to be controlled. If excitation and depletion lasers are well aligned a missing STED effect might also be caused by a mismatch in the temporal synchronization between excitation and STED laser pulse. In this case the default value of the delay time between both pulses has to be tested manually in the configuration/STED menu.
12. Once optimal settings for the individual channels have been found, they can be taken over with slight adaptation for a whole series of images. For high-quality STED recording, two opposite strategies can be applied: Either the scan speed can be reduced to 10, 100 or 200 Hz with no or little averaging (4× is a good value) or the scan speed is set to 1,000 or 1,400 Hz and the image is averaged up to 64 times. With both settings we have obtained reasonable results, if the staining is homogeneous without any hot spots in the sample. Because APDs are very sensitive but lack high dynamic range as known from PMTs, they will shut off immediately after attaining saturation. Since the risk of entering saturation is much higher with lower scanning speed, a repetition rate of 1,000 Hz and 64 times line averaging is recommend as optimal scanning parameters.
13. In contrast to the 532 nm excitation laser, which is controlled by an AOTF, the intensity of the pulsed 640 nm laser has to be adjusted manually before channels toggle between LSM (intensity value at the power supply: 5.25) and STED (Intensity value: 5.3) mode. For this reason it is advised to scan alternating with 532 and 640 nm excitation.
14. To avoid undersampling and thus loss of information, STED imaging requires a pixel size below 30 nm. This is important, because due to the Nyquist theorem, the sampling frequency

should be at least two times higher than the highest spatial frequencies in the sample. Practically it means that an adequate data representation is fulfilled if the pixel size is two to three times below the theoretical resolution. In STED mode, a pixel size of 25 nm is achieved with an 100× Oil NA 1.4 objective at an image format of 1,024 pixel with a zoom factor of 6.

15. Acquisition of 3D stacks is principally possible with the STED system, but one has to take into account that the axial resolution of the commercial setup is still the same as in confocal mode (500–700 nm). Moreover, repeated scanning boosts bleaching, particularly of the Chromeo 494 dye. To minimize bleaching during stack acquisition fast scanning (1,400 or 8,000 Hz using the resonant scanner) in combination with enhanced averaging (i.e., 64× line average) is strongly recommended.

3.5. Image Processing

To improve image quality raw data of confocal and STED images are deconvolved using Autoquant deconvolution software (Media Cybernetics, Inc., Bethesda, USA) with a theoretical PSF. Subsequently, images are processed using ImageJ (National Institutes of Health, USA) for merging channels and conversion into 8 Bit RGB images. Contrast and brightness levels of individual channels are finally adapted by Photoshop CS 5 (Adobe. System, Inc., San Jose, USA).

4. Notes

1. Spontaneous versus stimulated emission. In fluorescence microscopy the process of absorption and emission can be described as cycle between lower (ground) or higher (excited) electronic energy states of an atom. If the atom absorbs a photon of a certain frequency ν (that equals the energy difference between the two states) an electron makes a transition to the excited electronic state (Fig. 2, left panel). From here it will relax back spontaneously to the ground state after a short period of time (i.e., the fluorescence lifetime) by emitting a photon of less energy (i.e., red-shifted wavelength). Mathematically, the energy difference between the excited (E2) and the ground (E1) state is given by the equation:

$$E2 - E1 = h\nu ,$$

where ν is the frequency and h is Planck's constant. Typical excited state lifetimes of fluorescent dyes are in the range of picoseconds to nanoseconds.

In contrast to spontaneous relaxation stimulated (or induced) emission occurs when an electron in the excited state encounters a photon whose energy matches the difference between the electron's energy and the ground state (Fig. 2 center panel). In this case the photon will stimulate the electron to return immediately to the lower energy state by emitting an additional photon of the same phase, frequency, direction, and polarization, as the incident photon. Stimulated emission was theoretically predicted first by Albert Einstein in 1917 (35). It can be used to augment the external field (as in lasers) or to deplete spontaneous fluorescence specifically by switching fluorescent molecules off in a certain area as in STED microscopy (Figs. 1 and 2 right panel).

2. Resolution limit in confocal microscopy. In confocal laser scanning microscopy, the excitation beam is focused by an objective into a diffraction limited spot, which is sequentially moved across the specimen by a scanning mirror. The emitted fluorescence is then collected through the objective and guided to a detector (usually a photomultiplier tube (PMT)). The light distribution of the focal spot is determined by the point-spread function (PSF) which, in an ideal case, shows a typical pattern called "Airy disk." Its smallest size is defined by the diffraction limitation of light. The diffraction limitation was first described by Ernst Abbe in 1873 (34) in his formula:

$$d = \frac{\lambda}{2n \sin\alpha}$$

where d is the resolvable characteristic value, λ is the wavelength, n is the refractive index of the medium, and $\sin\alpha$ the semi-aperture angle of the objective. Because of this resolution barrier the excitation spot for illuminating fluorescent structures in a scanning microscope cannot be smaller than approximately half the wavelength of visible light, i.e., for blue excitation: »200 nm.

3. Resolution in STED microscopy. In a scanning STED (STimulated Emission Depletion) microscope Abbe's resolution limit is circumvented by annihilating excited dye molecules in the periphery of the excitation spot, thus confining the fluorescence emission to a smaller area in the center. The STED concept of deactivating dye molecules was first described in 1994 by Stephan Hell (30). In a scanning microscope, it is realized by modulating the wavefront of a second red-shifted laser with a patterned phase plate (vortex pattern), so that a doughnut shaped ring around the excitation spot is illuminated. As effect the fluorescence emission of excited molecules is increasingly quenched in the outer rim by the second (STED)

laser, whereas the emission in the center of the excitation spot is widely unaffected (Figs. 1 and 2 right panel). In principal, excitation and quenching can be done by cw- or pulsed lasers, respectively. Whereas cw-lasers are cheaper than pulsed lasers and require only a precise spatial alignment between excitation and quenching beams, pulsed lasers are more complex and necessitate, in addition to the spatial alignment, a precise temporal cyclic succession of both excitation and de-excitation laser pulses to assure that the arrival of both pulse trains are precisely synchronized. Using powerful depletion lasers forces almost all of the excited molecules in the outer ring to revert to the ground state, leaving only a tiny fraction of excited molecules in the center region of the excitation spot the chance to fluoresce. In other words, the size of the scanning spot from where fluorescent emission can be detected is defined by the precise arrival and the intensity of the de-excitation beam, i.e., the higher the depletion power, the smaller the fraction of fluorescent molecules in the remaining central spot and the better the resolution. Mathematically, the resolution improvement beyond the diffraction limit is described by the following extension of Abbe's formula:

$$d = \frac{\lambda}{2n\sin\alpha\sqrt{(1+\frac{I}{I_s})}}$$

where I is the intensity of the STED laser and I_s the saturation intensity of the selected fluorophore, i.e., the laser intensity at which the fluorescence depleted by 50% (22).

4. Configuration of the 2 channel STED setup. Figure 1 shows a scheme of a Titanium Sapphire laser-based STED microscope used for imaging dyes with emission maxima between 600 and 700 nm. For 2 channel excitation, diode lasers (PicoQuant, Berlin) emitting pulses of ≈70 ps at 532 nm (channel 1) and 640 nm (channel 2) with a repetition rate of 80 MHz were used. The STED pulses of 750–760 nm (repetition rate of 80 MHz) were generated by the Ti-Sapphire laser (Cameleon Ultra II, Coherent Inc., Santa Clara, USA) that provided also the triggering signal for the precise synchronization of the STED laser with the excitation lasers. Temporal adjustment of excitation and quenching pulses was managed via a delay unit to guarantee that every excitation pulse is immediately followed by a depletion pulse. The STED laser pulses of ≈100 fs were first prestretched to ≈1 ps by guiding the beam through a glass block and then stretched to ≈300 ps via a 100 m long polarization-maintaining glass fiber. Subsequently, the collimated STED beam was guided through a vortex

Table 3
Photophysical properties of STED dyes

Dye	Abs (nm)	Em (nm)	ε (M^{-1} cm^{-1})	QY	Stoke's shift (nm)
ATTO 647N	644	669	150,000	65	25
Chromeo 494	494	628	55,000	7[a]/25[b]	134

Abs absorption maximum in water, *Em* emission maximum in water, ε molar extinction coefficient, *QY* quantum yield, *Stoke's shift* Em-Abs

[a]Goat–anti-mouse conjugate or Goat–anti-rabbit conjugate

[b]BSA-conjugate

patterned phase plate and then overlaid with the excitation beams via a dichroic mirror. Calibration between excitation and STED pulse is done automatically by an alignment tool provided by the software. Both beams were focused into the sample by a 100× Pl APO objective (Leica-Microsystems, Wetzlar, Germany), and the emitted fluorescence signal was registered by two avalanche photon diodes (APDs, Perkin Elmer, Inc., San Jose, USA). To enhance focus stability and minimize thermal drift, the microscope was equipped with light tight customized heatable incubation chamber (Pecon, Erbach, Germany).

5. Dyes used for STED microscopy. Although principally almost all fluorophores can be depleted by stimulated emission, practically the photophysical properties, particularly the photostability, and the available laser lines for pulsed excitation shrink dramatically the number of fluorophores to be considered suitable. A list of dyes recommended for cw-STED and/or Ti-Sapphire STED microscopy can be found online at http://nanobiophotonics.mpibpc.mpg.de/old/dyes/. In our hands the best results, so far, were obtained with Atto 647N (Atto-Tec, Siegen, Germany) excited at 640 nm (Fig. 3 center panel). For co-localization studies, Chromeo 494 (Active Motif Chromeon GmbH, Tegernheim, Germany), excited at 532 nm, was used as second STED-fluorophore (Fig. 3 left) in association with Atto 647 N (Fig. 3 right panel). Due to the large Stoke's shift of Chromeo 494 this dye has a low quantum efficacy and care must be taken to avoid fast bleaching (see Table 3). Both immunostainings could be combined with previous transfection of genetically encoded fluorophores, like eGFP, without considerable crosstalk. The latter can be used as third label to define cell borders or areas of interest and recorded in confocal mode only.
6. Direct versus indirect labeling. Since the size and the number of the fluorophores linked to an antibody influence the

dimensions of the structure to be recorded, direct labeling of primary-antibodies is advised. Compared to indirect immune-fluorescence with directly labeled antibodies, the fluorophore is considerably closer to the structure of interest, thereby contributing to an improvement of the resolution. However, a drawback of direct immune-fluorescence is an attenuation of signal due to less fluorophores.

7. Embedding and mounting. STED images are acquired by selected high numerical aperture objectives (100× Plan Apo 1.4 NA oil), which are designed to be used with coverslips of 170 μm thickness and on the assumption that also the refractive indices of embedding and immersion media equal $n = 1.52$. Since any mismatch in the refractive index of the beam path will dramatically reduce the resolution, care must be taken that the embedding medium of the sample, the coverslip, and the immersion medium do not alter the refractive index perceptibly. Hence, for super-resolution microscopy we recommend the usage of immersion oil from Zeiss or Leica (F-518) with a refractive index of 5.18 and coverslips with a homogeneous thickness (170 μm ± 5 μm), like from Zeiss # 1.5 (Carl Zeiss, Jena, Germany) or Marienfeld No 1.5H (http://www.marienfeld-superior.com).

   As embedding medium in combination with an antifade reagent, we have some good experience with Mowiol + Dabco (1,4-Diazabicyclo-octane) in Glcyerol/Tris-buffer, pH 8.5. Mowiol (Mowiol 4.88, Calbiochem) is a solution of polyvinyl alcohol, which hardens overnight after slide preparation, and does not require the coverslips to be sealed with nail varnish. The addition of Dabco or *p*-Phenylenediamine (PPD) is recommended to reduce bleaching of fluorescent probes. A protocol for Mowiol preparation is presented at http://www.niaid.nih.gov/LabsAndResources/labs/aboutlabs/rtb/biologicalImaging/protocols/Pages/mowiolPreparation.aspx. An alternative Glycerol-based mounting medium used for STED-imaging is ProLong Gold (Invitrogen). Prolong Gold is suitable for Atto 647N and Chromeo 494 but not so good for eGFP and other fluorescent proteins. With time it polymerizes so that a sealing is not necessary. A good overview about mountants and antifades can be found at http://www.biocenter.helsinki.fi/bi/lmu/images/Mountants.pdf.

   As non-hardening mounting medium with antioxidant features we have successfully applied TDE (2,2 -Thiodiethanol, Sigma Aldrich). TDE allows a fine adjustment of the refractive index between water (1.33) and oil (1.52) by changing the ratio between water and TDE in the mixture of an ascending series (36).

8. Primary antibodies used for presynaptic staining. To demonstrate the potential of STED imaging for the subcellular

localization of calcium-binding proteins in neurons, we performed immunostainings against pre- and postsynaptic scaffolding proteins in hippocampal cell cultures previously transfected with a GFP-fusion construct of Caldendrin (Figs. 6–8). As presynaptic marker, we chose Bassoon, a giant protein of 420 kDa, which is supposed to be a major constituent of the so-called cytomatrix at the active zone found at both excitatory and inhibitory synapses (37–40). Ultrastructural and immunocytochemical studies have shown that Bassoon is always present in nascent synapses, indicating that it plays an essential role during assembly of active zones. Here, we recognized Bassoon by a monoclonal antibody followed by indirect immune-fluorescence with Chromeo 494 as dye.

9. Primary antibodies used for postsynaptic staining. As postsynaptic marker we chose Homer, which consists of several isoforms (Homer-1, 2, 3) and is localized at the postsynapse of excitatory synapses (41). It acts as an adaptor protein to link multiple targets, such as type 1 metabotropic glutamate receptors, IP3 receptors, ProSAP/Shanks, and others (42–44). By inducing the clustering of these proteins, Homer was suggested to organize distinct signaling pathways and to modulate calcium signaling and dendritic spine morphology (45). In our study, Homer was identified by an affinity-purified rabbit antibody against Homer-1 that was visualized by anti-rabbit Atto 647N. Caldendrin-GFP expressed in adult hippocampal primary neurons was unevenly filling up complete neurons with most of GFP fluorescence in the soma (but not the nucleus), dendrites and dendritic spines (Figs. 6–8).

Acknowledgments

This work was supported by the following grants: BMBF "Novel Optics" VDI 13N10077 and DFG SFB 854 TPZ (WZ) and DFG Kr1879/3-1 and SFB 854 TP7 (MK).

References

1. Berridge MJ, Bootman MD, Lipp P (1998) Calcium—a life and death signal. Nature 395:645–648
2. Berridge MJ, Bootman MD, Roderick HL (2003) Calcium signalling: dynamics, homeostasis and remodelling. Nat Rev Mol Cell Biol 4:517–529
3. Pozzo-Miller LD, Pivovarova NB, Leapman RD et al (1997) Activity-dependent calcium sequestration in dendrites of hippocampal neurons in brain slices. J Neurosci 17: 8729–8738
4. Mikhaylova M, Hradsky J, Kreutz MR (2011) Between promiscuity and specificity: novel roles of EF-hand calcium sensors in neuronal Ca2+ signalling. J Neurochem 118:695–713
5. Ikura M, Ames JB (2006) Genetic polymorphism and protein conformational plasticity in

the calmodulin superfamily: two ways to promote multifunctionality. Proc Natl Acad Sci U S A 103:1159–1164

6. Burgoyne RD (2007) Neuronal calcium sensor proteins: generating diversity in neuronal Ca^{2+} signalling. Nat Rev Neurosci 8:182–193
7. Fries R, Reddy PP, Mikhaylova M et al (2010) Dynamic cellular translocation of caldendrin is facilitated by the Ca2+-myristoyl switch of recoverin. J Neurochem 113:1150–1162
8. McCue HV, Burgoyne RD, Haynes LP (2009) Membrane targeting of the EF-hand containing calcium-sensing proteins CaBP7 and CaBP8. Biochem Biophys Res Commun 380: 825–831
9. McCue HV, Burgoyne RD, Haynes LP (2011) Determination of the membrane topology of the small EF-Hand Ca^{2+}-sensing proteins CaBP7 and CaBP8. PLoS One 6:e17853
10. Mikhaylova M, Reddy PP, Munsch T et al (2009) Calneurons provide a calcium threshold for trans-Golgi network to plasma membrane trafficking. Proc Natl Acad Sci USA 106:9093–9098
11. Hradsky J, Raghuram V, Reddy PP et al (2011) Post-translational membrane insertion of tail-anchored transmembrane EF-hand Ca2+ sensor Calneurons requires the TRC40/Asna1 protein chaperone. J Biol Chem 286:36762–36776
12. Mikhaylova M, Reddy PP, Kreutz MR (2010) Role of neuronal Ca2+-sensor proteins in Golgi-cell-surface membrane traffic. Biochem Soc Trans 38:177–180
13. Shih PY, Lin CL, Cheng PW et al (2009) Calneuron I inhibits Ca^{2+} channel activity in bovine chromaffin cells. Biochem Biophys Res Commun 388:549–553
14. Seidenbecher CI, Langnaese K, Sanmartí-Vila L et al (1998) Caldendrin, a novel neuronal calcium-binding protein confined to the somato-dendritic compartment. J Biol Chem 273:21324–21331
15. Laube G, Seidenbecher CI, Richter K et al (2002) The neuron-specific Ca^{2+}-binding protein caldendrin: gene structure, splice isoforms and expression in the rat central nervous system. Mol Cell Neurosci 19:459–475
16. Bernstein HG, Seidenbecher CI, Smalla KH et al (2003) Distribution and cellular localization of caldendrin immunoreactivity in adult human forebrain. J Histochem Cytochem 51:1109–1112
17. Smalla KH, Seidenbecher CI, Tischmeyer W et al (2003) Kainate-induced epileptic seizures induce a recruitment of caldendrin to the postsynaptic density in rat brain. Brain Res Mol Brain Res 19:159–162
18. Mikhaylova M, Sharma Y, Reissner C et al (2006) Neuronal Ca^{2+} signaling via caldendrin and calneurons. Biochim Biophys Acta 1763:1229–1237
19. McCue HV, Haynes LP, Burgoyne RD (2010) The diversity of calcium sensor proteins in the regulation of neuronal function. Cold Spring Harb Perspect Biol 2. doi:10.1101/cshperspect.a004085
20. Seidenbecher CI, Landwehr M, Smalla KH et al (2004) Caldendrin but not calmodulin binds to light chain 3 of MAP1A/B: an association with the microtubule cytoskeleton highlighting exclusive binding partners for neuronal Ca^{2+}-sensor proteins. J Mol Biol 336:957–970
21. Dieterich DC, Karpova A, Mikhaylova M et al (2008) Caldendrin-Jacob: a protein liaison that couples NMDA receptor signalling to the nucleus. PLoS Biol 6:e34, Erratum in: PLoS Biol. 7 (2009):e1000022
22. Hell SW (2007) Far-field optical nanoscopy. Science 316:1153–1158
23. Hell SW, Stelzer EHK, Lindek S et al (1994) Confocal microscopy with an increased detection aperture: type-B 4Pi confocal microscopy. Opt Lett 19:222–224
24. Gustafsson MG (2000) Surpassing the lateral resolution limit by a factor of two using structured illumination microscopy. J Microsc 198: 82–87
25. Gustafsson MG (2005) Nonlinear structured-illumination microscopy: wide-field fluorescence imaging with theoretically unlimited resolution. Proc Natl Acad Sci USA 102: 13081–13086
26. Heintzmann R, Cremer C (1999) Lateral modulated excitation microscopy: improvement of resolution by using a diffraction grating. Proc SPIE 3568:185–196
27. Esa A, Edelmann P, Kreth G et al (2000) Three-dimensional spectral precision distance microscopy of chromatin nanostructures after triple-colour DNA labelling: a study of the BCR region on chromosome 22 and the Philadelphia chromosome. J Microsc 199: 96–105
28. Lemmer P, Gunkel M, Baddeley D et al (2008) SPDM: light microscopy with single-molecule resolution at the nanoscale. Appl Phys B 93: 1–12
29. Hell SW, Kroug M (1995) Ground-state-depletion: a concept for breaking the diffraction resolution limit. Appl Phys B 60: 495–497
30. Hell SW, Wichmann J (1994) Breaking the diffraction resolution limit by stimulated emission. Opt Lett 19:780–782

31. Klar TA, Hell SW (1999) Subdiffraction resolution in far-field fluorescence microscopy. Opt Lett 24:954–956
32. Betzig E, Patterson GH, Sougrat R et al (2006) Imaging intracellular fluorescent proteins at nanometer resolution. Science 313:1642–1645
33. Rust MJ, Bates M, Zhuang X (2006) Sub-diffraction-limit imaging by stochastic optical reconstruction microscopy (STORM). Nat Methods 3:793–795
34. Abbe E (1873) Beiträge zur Theorie des Mikroskops und ihrer mikroskopischen Wahrnehmung. Arch Mikrosk Anat 9:411–468
35. Einstein A (1917) Zur Quantentheorie der Strahlung. Physik Zeitschr 18:121–128
36. Staudt T, Lang MC, Medda R et al (2007) 2,2 -thiodiethanol: a new water soluble mounting medium for high resolution optical microscopy. Microsc Res Tech 70:1–9
37. Cases-Langhoff C, Voss B, Garner AM et al (1996) Piccolo, a novel 420 kDa protein associated with the presynaptic cytomatrix. Eur J Cell Biol 69:214–223
38. Tom Dieck S, Sanmartí-Vila L, Langnaese K et al (1998) Bassoon, a novel zinc-finger CAG/glutamine-repeat protein selectively localized at the active zone of presynaptic nerve terminals. J Cell Biol 142:499–509
39. Dresbach T, Hempelmann A, Spilker C et al (2003) Functional regions of the presynaptic cytomatrix protein bassoon: significance for synaptic targeting and cytomatrix anchoring. Mol Cell Neurosci 23:279–291
40. Richter K, Langnaese K, Kreutz MR et al (1999) Presynaptic cytomatrix protein bassoon is localized at both excitatory and inhibitory synapses of rat brain. J Comp Neurol 408:437–448
41. Dani A, Huang B, Bergan J et al (2010) Superresolution imaging of chemical synapses in the brain. Neuron 68:843–856
42. Brakeman PR, Lanahan AA, O'Brien R et al (1997) Homer: a protein that selectively binds metabotropic glutamate receptors. Nature 386:284–288
43. Yuan JP, Kiselyov K, Shin DM et al (2003) Homer binds TRPC family channels and is required for gating of TRPC1 by IP3 receptors. Cell 114:777–789
44. Tu JC, Xiao B, Naisbitt S et al (1999) Coupling of mGluR/Homer and PSD-95 complexes by the Shank family of postsynaptic density proteins. Neuron 23:583–592
45. Sala C, Piech V, Wilson NR et al (2001) Regulation of dendritic spine morphology and synaptic function by Shank and Homer. Neuron 31:115–130

Part III

Interaction of Ca^{2+}-Binding Proteins with Targets and Receptors/RAGE

Chapter 11

NMR Studies of the Interaction of Calmodulin with IQ Motif Peptides

Steven M. Damo, Michael D. Feldkamp, Benjamin Chagot, and Walter J. Chazin

Abstract

Calmodulin (CaM) is a ubiquitous EF-hand calcium sensor protein that transduces calcium signals in a wide range of signaling pathways. Structural analysis of complexes with peptides has provided valuable insights into the remarkable variety in the way in which CaM interacts with and activates its targets. Among these various targets, CaM has been shown to be an essential component of a calcium-sensing regulatory apparatus for a number of voltage-gated ion channels. NMR spectroscopy has proven to be a powerful tool for the structural characterization of CaM–peptide complexes, in particular for the study of IQ motifs, which bind CaM at the basal level of calcium in cells and thereby serve to localize CaM to its sites of action. We describe here methods for the robust expression and purification of CaM isotopically enriched for NMR analysis, as well as for the complex of CaM with a peptide derived from the IQ motif sequence of the human cardiac sodium channel $Na_V1.5$. We also describe methods for NMR analysis of titrations of CaM with IQ motif peptides to determine the stoichiometry of the complex and to identify the residues at the binding interface.

Key words: Calmodulin, $Na_V1.5$, IQ motif, ^{15}N-^{1}H HSQC, NMR spectroscopy, Protein complex, Co-expression, Calcium binding protein, Calcium signaling, EF-hand

1. Introduction

EF-hand calcium binding proteins are used to regulate, shape, and transduce intracellular calcium signals (1, 2). The ubiquitous calcium sensor calmodulin (CaM) integrates calcium signal transduction, modulating the activity of more than 50 different targets (3). It performs this function through coupled binding of calcium and distinct CaM-binding domains in these target proteins. CaM is

Claus W. Heizmann (ed.), *Calcium-Binding Proteins and RAGE: From Structural Basics to Clinical Applications*, Methods in Molecular Biology, vol. 963, DOI 10.1007/978-1-62703-230-8_11, © Springer Science+Business Media New York 2013

composed of two classical EF-hand domains connected by a flexible linker. Binding of calcium causes a shift within each domain from a closed to an open conformation, which leads to exposure of a substantial hydrophobic surface that is integral to its interaction with targets. These conformation changes combined with a high degree of flexibility in the linker between the domains provide CaM with a remarkable ability to interact with targets in a wide range of different modes. Given the significant molecular size and complexity of CaM targets, peptides derived from the CaM-binding domain/s of targets have been used extensively to mimic the interaction of CaM with full length targets (4–8). NMR spectroscopy has been used to directly validate this reductionist approach for studying CaM (9).

Calcium signals are tightly regulated in time and duration because otherwise they would activate so many signaling pathways and overwhelm the cell. This tight regulation means that the calcium sensor proteins need to be able to respond rapidly to signals. One way that CaM ensures its rapid response is by pre-localizing to its sites of action. The IQ motif is a conserved sequence (IQxxxBGxxxB, B = Lys or Arg) in CaM targets that is widely used for this purpose (10). The key characteristics of IQ motifs providing this function is that they bind relatively tightly to CaM at basal levels of intracellular calcium and the interaction changes when calcium levels are raised (11). Voltage-gated ion channels are one important class of targets whose activity is modulated by intracellular calcium signals through the action of CaM. The channels have IQ motifs in their C-terminal intracellular domains, which are known to recruit CaM.

One of the most well-studied voltage-gated ion channels modulated by calcium is the human cardiac voltage-gated sodium channel ($Na_V1.5$). $Na_V1.5$ is essential in the initial phase of action potential and a core regulatory element of the beating of the heart (12). CaM plays a critical role in the calcium regulation of $Na_V1.5$ and is pre-localized to the channel via a C-terminal IQ motif. Physical interaction of CaM with the IQ motif has been directly demonstrated both in the presence and in the absence of calcium (13). In the absence of calcium, the C-terminal domain of CaM binds the IQ motif with the N-lobe remaining free in solution (4). Although no structure has been determined at saturating calcium, NMR studies have shown that both the N- and C-terminal domains are engaged (13).

NMR spectroscopy has played a critical role in characterizing the interaction of CaM with its many cellular targets. NMR is well suited for the study of protein interactions because the resonance frequency observed is exquisitely sensitive to changes in chemical environment, for example, when a ligand is titrated into a solution. In the past 10–15 years, ^{15}N-^{1}H HSQC NMR spectroscopy has become a widely used technique to observe residue specific changes in the chemical environment of backbone amide groups within proteins, and this approach has been widely used in studies of target peptide fragments binding to CaM.

The use of peptide fragments for the reductionist approach has certain advantages. Titrations of peptides into solutions of ^{15}N-enriched proteins monitored by ^{15}N-^{1}H HSQC spectra require relatively modest quantities of the peptides, which can be synthesized in a cost effective manner. However, for more in-depth NMR studies, it is advantageous to monitor changes not only in the protein but also in the peptide, which means producing the peptide with isotopic enrichment. Due to the much higher cost of producing such samples, the labeled peptide is usually produced recombinantly in a bacterial expression system. This approach faces certain technical challenges as in general it is difficult to express small peptides alone due to their susceptibility to proteolytic cleavage and low expression levels. These obstacles can be overcome through fusion to a carrier protein (14). Furthermore, co-expression of protein complexes offers the advantages of pre-assembly of the complex, generally accurate stoichiometry and more facile purification.

Here we describe methods for the production of ^{15}N-enriched CaM and titration of CaM with a synthetic $Na_V1.5$ IQ motif peptide ($Na_V1.5_{IQp}$) monitored by ^{15}N-^{1}H HSQC NMR spectroscopy. We show how to quantify changes in amide chemical shifts observed in the ^{15}N-^{1}H HSQC spectrum of ^{15}N-labeled apo CaM to dissect the stoichiometry of peptide binding, as well as to identify residues perturbed upon peptide binding. In addition, we describe co-expression and purification of a uniformly ^{15}N enriched CaM-IQ peptide complex. The methods described here are generally applicable to CaM complexes. Given the large and still increasing number of CaM targets, the use of these approaches to characterize of CaM–peptide complexes will continue to be valuable for understanding the molecular basis for transduction of calcium signals by CaM and the activation of intracellular signaling pathways.

2. Materials

2.1. Production of ^{15}N Apo CaM for NMR Studies

Expression of ^{15}N CaM

1. CaM pET15 plasmid (see Note 1).
2. BL21(DE3) competent cells (Invitrogen): 100 μL aliquots stored at −80°C.
3. Ampicillin (AMP) 1,000× stock solution 100 mg/mL in Milli-Q (MQ) water (filtered to a resistance of 18.3 MΩ cm). Sterilize by 0.2 μM filtration.

4. LB AGAR plates: 10 g/L tryptone, 10 g/L NaCl, 5 g/L yeast extract, 15 g/L agar dissolved in MQ water and autoclaved at 121°C for 20 min. Add AMP stock at a 1:1,000 dilution when the medium has cooled to 50–60°C.
5. SOC Recovery medium: 20 g/L tryptone, 0.5 g/L NaCl, 5 g/L yeast extract, 2.5 mM KCl, 5 mM $MgCl_2$, 5 mM $MgSO_4$, 20 mM glucose dissolved in MQ H_2O, and autoclaved at 121°C for 20 min.
6. LB medium: 10 g/L tryptone, 10 g/L NaCl, 5 g/L yeast extract autoclaved at 121°C for 20 min.
7. ^{15}N labeled M9 medium: Per L of medium. Add 0.5 g NaCl, 3.0 g KH_2PO_4, 6.0 g Na_2HPO_4, and 1.0 g $^{15}NH_4Cl$ autoclaved at 121°C for 20 min. Add 1.0 mL of 2 M $MgSO_4$ (sterile filtered 0.2 μM), 100 μL of 1 M $CaCl_2$ (sterile filtered, 0.2 μM), 20 mL of 20% glucose (sterile filtered, 0.2 μM).
8. IPTG (Gold Biotechnology): 1 M solution in MQ H_2O. Sterilize by filtration at 0.2 μM.

Purification of Apo CaM

1. Refrigerated ÄKTA FPLC (GE Life Sciences) purification system and accessories (see Note 2).
2. Two Phenyl sepharose 6 Fast Flow columns (prepacked 25 mL, Amersham).
3. Phenyl sepharose (PS) Buffer A: 50 mM Tris–HCl pH 7.5, 1 mM $CaCl_2$.
4. Phenyl sepharose (PS) Buffer B: 50 mM Tris–HCl pH 7.5, 500 mM NaCl, and 1 mM $CaCl_2$.
5. Phenyl sepharose (PS) Buffer C: 50 mM Tris–HCl pH 7.5, 1 mM EDTA.
6. Dialysis tubing: 3 kDa molecular weight cutoff (MWCO).
7. NMR buffer N: 20 mM HEPES pH 6.8, 100 mM KCl, 1 mM EDTA, and 0.01% NaN_3 (see Note 3).

2.2. Titration of Apo CaM with Peptide and Analysis of NMR Data

Sample preparation and NMR data acquisition

1. $Na_v1.5_{IQp}$ (Genscript): corresponding to residues E1901-L1927 of $Na_v1.5$.
2. NMR buffer N: 20 mM HEPES pH 6.8, 100 mM KCl, 1 mM EDTA, and 0.01% NaN_3.
3. D_2O.
4. 9 in. glass pipette (Sigma-Aldrich).
5. 5 mm OD Precision borosilicate glass tubes (for use with a 5 mm probe) (Wilmad glass company).
6. 500 MHz or higher field spectrometer equipped with 5 mm ^1H,^{13}C,^{15}N room temperature or cryo probe (e.g., Bruker Avance III).

Analysis of NMR Data

1. Computer (running Mac OS X or linux operating system).
2. NMR data processing software (e.g., NMRPipe) (15).
3. NMR spectral analysis software (e.g., Sparky or NMRView) (16, 17).

2.3. Production of Apo CaM–$Na_V1.5_{IQp}$ Complex for NMR Studies

Co-expression of CaM–His_6-SUMO–$Na_V1.5_{IQp}$ fusion complex

1. CaM pET15 plasmid.
2. IQ pBG102 plasmid (see Notes 4 and 5).
3. BL21(DE3) competent cells (Invitrogen): 100 μL aliquots stored at −80°C.
4. Ampicillin (AMP) 1,000× stock solution 100 mg/mL in MQ H_2O. Sterilize by 0.2 μM filtration.
5. Kanamyacin (KAN) 1,000× stock solution 35 mg/mL in MQ H_2O. Sterilize by filtration at 0.2 μM.
6. LB AGAR plates: 10 g/L tryptone, 10 g/L NaCl, 5 g/L yeast extract, 15 g/L agar dissolved in MQ water, and autoclaved at 121°C for 20 min. Add AMP and KAN stocks at a 1:1,000 dilution when the medium has cooled to 50–60°C.
7. SOC Recovery medium: 20 g/L tryptone, 0.5 g/L NaCl, 5 g/L yeast extract, 2.5 mM KCl, 5 mM $MgCl_2$, 5 mM $MgSO_4$, 20 mM glucose dissolved in MQ H_2O, and autoclaved at 121°C for 20 min.
8. LB medium: 10 g/L tryptone, 10 g/L NaCl, 5 g/L yeast extract autoclaved at 121°C for 20 min.
9. ^{15}N labeled M9 medium: Per L of medium. Add 0.5 g NaCl, 3.0 g KH_2PO_4, 6.0 g Na_2HPO_4, and 1.0 g $^{15}NH_4Cl$ autoclaved at 121°C for 20 min. Add 1.0 mL of 2 M $MgSO_4$ (sterile filtered 0.2 μM), 100 μL of 1 M $CaCl_2$ (sterile filtered, 0.2 μM), 20 mL of 20% glucose (sterile filtered, 0.2 μM).
10. IPTG (Gold Biotechnology): 1 M solution in MQ H_2O. Sterilize by filtration at 0.2 μM.

Co-purification of Apo CaM–$Na_V1.5_{IQp}$ complex

1. Refrigerated ÄKTA FPLC purification system and accessories (GE Life Sciences).
2. Ni-NTA (prepacked 25 mL column Amersham).
3. Ni-NTA Buffer A: 50 mM Tris–HCl pH 7.5, 300 mM NaCl, and 10 mM Imidazole (see Note 6).
4. Ni-NTA Buffer B: 50 mM Tris–HCl pH 7.5, 300 mM NaCl, and 500 mM Imidazole.
5. His_6-Human rhinovirus 3C (H3C) protease. 10 Units per mg of substrate (see Note 7).
6. Dialysis membrane (3 kDa MWCO).

7. Centrifugal concentrator (3 kDa MWCO).
8. HiLoad Superdex 75 16/600 column (GE Life Sciences).
9. S75 buffer: 50 mM Tris–HCl pH 7.5 and 300 mM NaCl.

3. Methods

3.1. Production of ^{15}N Apo CaM for NMR Studies

Here we describe our protocol for the robust expression and purification of ^{15}N labeled CaM for NMR studies. Purification of CaM involves two hydrophobic interaction chromatography steps, exploiting the significant changes in hydrophobic surface accessibility of apo CaM and calcium-loaded CaM. Typical yields are 30–40 mg/L for M9 media.

Expression of ^{15}N CaM

1. Thaw 100 μL of BL21(DE3) competent cells on ice, add 50 ng of CaM pET15b. Incubate on ice for 20 min.
2. Heat shock the cells at 42°C for 30 s and incubate on ice for 2 min.
3. Add 900 μL of SOC recovery medium. Incubate for 1 h at 37°C with shaking (220 rpm).
4. Plate 150 μL of cells onto LB-agar ampicillin plates and incubate at 37°C overnight.
5. Inoculate a single colony into 100 mL of LB media supplemented with AMP. Incubate overnight (12–16 h) at 37°C.
6. Add 20 mL of LB starter culture to 1 L of sterile ^{15}N M9 media. Incubate at 37°C with shaking (220 rpm) until OD_{600} reaches 0.6.
7. Induce cells by adding IPTG to a final concentration of 1 mM (1 mL of 1 M IPTG solution added to 1 L of cell culture) and incubate at 37°C for 4 h with shaking (220 rpm).

Purification of Apo CaM

1. Harvest cells by centrifugation at 5,000 × *g* for 10 min.
2. Resuspend cell paste in PS buffer C (10–15 mL of buffer per gram of cell paste) and transfer to a stainless steel beaker. Chill this beaker in a secondary container with melting ice.
3. Sonicate the resuspended cell paste with a macrotip set to 50 W output power for total processing time of 5 min (5 s pulse bursts with 5 s recovery/off time).
4. Clear cell lysate by centrifugation at 48,000 × *g* for 20 min.
5. Decant the lysate and filter through a 0.45-μM membrane.

6. Equilibrate one prepacked 25 mL phenyl-sepharose 6 Fast Flow column with 1 column volume (CV) of MQ H_2O and 2 CVs of PS buffer C.
7. Load the filtered lysate onto the phenyl sepharose column at a flow rate of 1.0 mL/min. Collect the flow through in one fraction.
8. Wash the column with 1 CV of PS buffer C.
9. Combine the flow through and wash fractions, and adjust $CaCl_2$ concentration to 5 mM.
10. Equilibrate the second prepacked 25 mL phenyl-sepharose 6 Fast Flow column with 1 CV of MQ H_2O and 2 CVs of PS buffer A.
11. Load the combined flow through and wash fractions onto the second phenyl sepharose column.
12. Wash the column with 4 CVs of PS buffer A, followed by 4 CVs of PS Buffer B, and 4 CVs of PS Buffer A.
13. Elute the protein with 4 CVs of PS Buffer C with 5 mL fractionation (see Note 8).
14. Determine fractions containing CaM by SDS-PAGE.
15. Pool the fractions and dialyze against buffer N. Concentrate sample to 0.1–1 mM for NMR studies (see Note 9).

3.2. Titration of Apo CaM with Peptide and Analysis of NMR Data

Sample preparation and NMR data acquisition

1. Solubilize the lyophilized $Na_V1.5_{IQp}$ with buffer N containing 10% D_2O to a concentration of 1–5 mM or as high a concentration permitted by the peptide of study to minimize the dilution of apo CaM during the course of the titration series (see Note 10). It may be necessary to adjust the pH of the peptide solution to ensure that a pH effect is not encountered upon peptide titration. It is likely that the volume of peptide containing solution will be 1 mL or less, we recommend a micro-pH probe or spotting of the peptide solution onto pH paper to verify the solution pH.
2. Add 50 μL of D_2O to 450 μL of 100 μM ^{15}N-labeled apo CaM (>95% purity) in buffer N.
3. Using a 9 in. glass pipette, carefully transfer the CaM solution to a clean NMR tube (see Note 11).
4. Insert the apo CaM containing NMR tube into the sample spinner, verify tube depth, insert and lower tube into NMR spectrometer.
5. Set and equilibrate the temperature of the sample, then lock, tune, and shim the spectrometer and use simple 1-dimensional ^{1}H and ^{15}N experiments to determine 90° pulse lengths for these nuclei.

6. Load a suitable ^{15}N-^{1}H HSQC pulse program, input 90° ^{1}H and ^{15}N pulse lengths and set the receiver gain. Collect an initial ^{15}N-^{1}H HSQC spectrum to access the quality and signal strength of the apo CaM sample. Typically increment values of 1,024 and 128 points are used in the direct and indirect dimensions, respectively, with a scan setting of 16 to obtain a reasonable spectrum. Due to the expected chemical shifts to be observed upon peptide addition, it is recommended to set the spectral width in ^{1}H and ^{15}N dimensions to be extended beyond the apo CaM peaks observed in this initial spectrum. Although the spectral widths required are peptide dependent, we suggest extending the spectral width 1 and 5 ppm beyond that of the observed apo CaM crosspeaks in the ^{1}H and ^{15}N dimensions as a starting point.
7. Collect an ^{15}N-^{1}H HSQC spectrum of apo CaM using the optimized settings determined in **step 5** to serve as a reference point.
8. Eject the sample from the NMR spectrometer and make the first peptide addition. We have found removal of the sample with a 9 in. glass pipette into a 1.5 mL Eppendorf tube allows for confident addition and mixing of the peptide. Anticipating a 1:1 peptide:apo CaM-binding stoichiometry, we suggest making additions in an incremental manner, where a minimum of four additions (0.25 molar equivalents of peptide) are used to reach a 1:1 ratio.
9. After transfer of the sample back to the NMR tube and insertion into the NMR spectrometer, allow for the temperature to equilibrate, re-shim the sample, and acquire the first titration point. Repeat this process until apo CaM amide crosspeaks cease to shift (fast exchange) or that peaks corresponding to apo CaM are no longer present (slow exchange) in the ^{15}N-^{1}H HSQC spectrum upon addition of peptide. We suggest at least one additional point where an excess amount of peptide is added to ensure saturation.

Analysis of NMR Data

1. Process the ^{15}N-^{1}H HSQC titration data using NMRpipe (15) or other software.
2. Assign amide crosspeaks of apo CaM in titration spectra using spectral viewing and analysis software such as Sparky or NMRView (16, 17).
3. An incremental change in apo CaM amide cross peak position in response to peptide addition is indicative of fast chemical exchange (typically associated with Kd ~10 μM or weaker binding affinity), assignment of the peptide bound state can be made by simply tracking the incremental movement of perturbed peaks (Fig. 1). If slow chemical exchange is observed

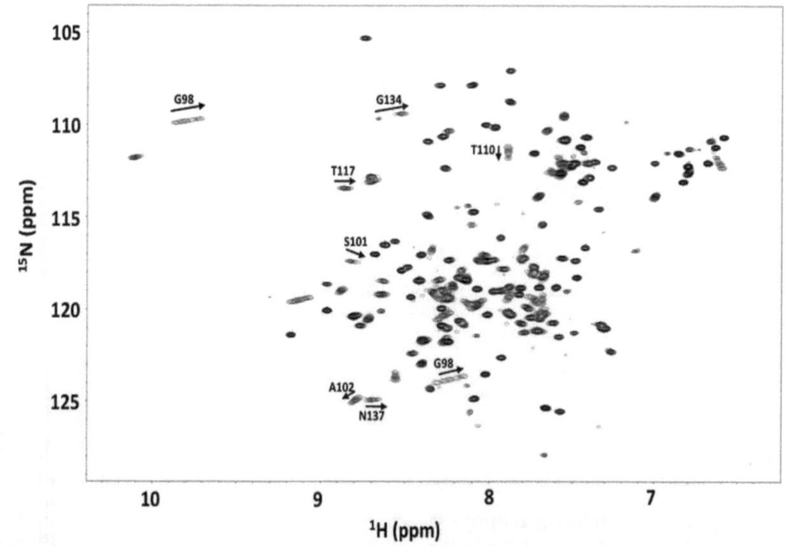

Fig. 1. Overlaid ^{15}N-^{1}H HSQC Spectra of apo CaM Titrated with $Na_V1.5_{IQp}$. $Na_V1.5_{IQp}$ is a known binding partner of apo CaM, which upon addition to apo CaM shifts a select number of apo CaM amide cross peaks. The direction and magnitude of the chemical shift perturbation of a select number of apo CaM amide cross peaks are indicated with *arrows*.

(typically associated with Kd ~1 μM or stronger binding affinity), the titration spectra will display population-weighted sets of amide cross peaks whose intensity correspond to the relative amount of free and peptide bound apo CaM present in the sample. Intermediate exchange is observed in between the fast and slow exchange regimes and is characterized by broadening and disappearance of signals over the course of the titration, followed by reappearance of signals at a new location, as in the case of slow exchange. In the case of intermediate and slow exchange, three-dimensional NMR experiments to assign the new cross peaks from the complex are required to determine the identity of the apo CaM signals in the complex. It is common to increase temperature in an attempt to shift peaks in intermediate chemical exchange to fast chemical exchange in an effort to characterize binding in detail and to simplify the resonance assignment in the peptide bound state.

4. Quantify the peptide binding induced perturbation of apo CaM amide cross peaks with Eq. 1 and plot the Δppm value determined (*y*-axis) versus residue number (*x*-axis) as shown in Fig. 2.

$$\Delta\text{ppm} = \sqrt{(\Delta\delta_H)^2 + (\Delta\delta_N \square .2)^2}. \quad (1)$$

In this equation, Δppm refers to the linear change of an amide cross peak from its initial starting position in the reference spectrum, while $\Delta\delta_H$ and $\Delta\delta_N$ refer to the individual changes in the ^{1}H and ^{15}N dimensions. A scaling factor of

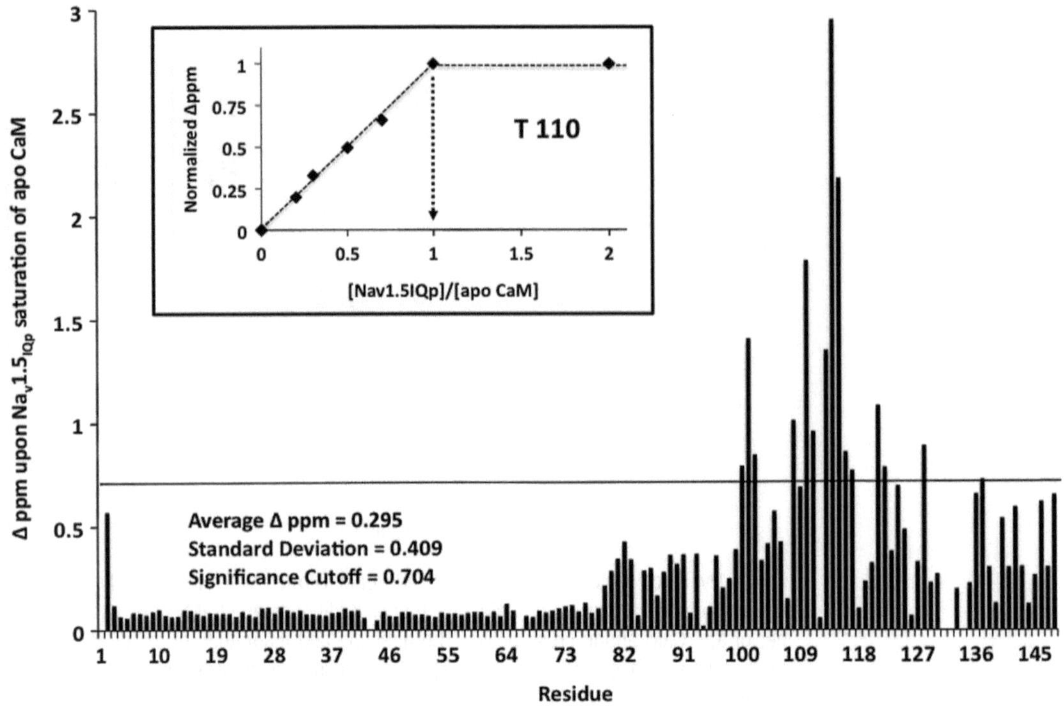

Fig. 2. Quantified apo CaM Chemical Shift Perturbations Induced upon $Na_v1.5_{IQp}$ Binding. Using Eq. 1, the change in chemical shift of apo CaM amide cross peaks were calculated and plotted. The horizontal bar at 0.704 represents the significance cutoff value used to identify apo CaM residues whose chemical shift were deemed significant. Shown in the inset in the upper left is a stoichiometry plot calculated for residue T110 over the course of the titration series. To help guide the eye, a line and arrow are placed in the data to highlight the 1:1 stoichiometry achieved once the normalized change in chemical shift has reached a plateau, signifying an $Na_v1.5_{IQp}$ saturation of apo CaM

0.2 is used to account for differences in the span of ^1H and ^{15}N chemical shifts.

5. Determine the Δppm value to use as a cutoff for what is classified as significant chemical shift perturbation to identify apo CaM residues that are either directly or allosterically involved in peptide binding. We suggest using a value of one standard deviation over the average of all chemical shifts as a cutoff (Fig. 1).
6. Different treatments are required to examine the stoichiometry of peptide binding to apo CaM when rate of chemical exchange is in the fast or slow exchange binding regime. In the case of fast exchange, the position of the amide cross peak is the weighted average of the chemical shifts of the peptide free and peptide bound apo CaM states. During slow exchange, separate apo CaM amide cross peaks whose intensity is directly proportional to the ratio of peptide free to peptide bound apo CaM are observed. To determine the binding stoichiometry for peptide binding in fast chemical exchange,

plot the chemical shift for each titration point (y-axis) against the ratio of [peptide]/[apo CaM] (x-axis). When slow chemical exchange is observed, plot the normalized ratio of the intensities of the peptide free to peptide bound apo CaM amide crosspeaks for a given resonance (y-axis) versus the ratio of [peptide]/[apo CaM] (x-axis).

7. The stoichiometry of the peptide:apo CaM complex can then be determined as illustrated in Fig. 2.

3.3. Production of Apo CaM–$Na_V1.5_{IQp}$ Complex for NMR Studies

This section describes the co-expression and co-purification of uniformly ^{15}N labeled CaM–$Na_V1.5_{IQp}$ complex for NMR studies. The $Na_V1.5_{IQp}$ is expressed as an N-terminally His_6-tagged SUMO fusion protein (His_6-SUMO–$Na_V1.5_{IQp}$). The complex is formed in vivo and isolated via a three step purification procedure. The CaM–His_6-SUMO–$Na_V1.5_{IQp}$ fusion complex is purified by Ni-NTA chromatography. This is followed by cleavage and removal of the His_6-SUMO fusion tag by another Ni-NTA column followed by size exclusion chromatography. Typical yields are 5–10 mg of complex per L of M9 medium.

Co-expression of CaM–His_6-SUMO–$Na_V1.5_{IQp}$ fusion complex

1. Thaw 100 μL of BL21(DE3) competent cells on ice, add 100 ng each of CaM pET15b and IQ pBG102 vectors. Incubate on ice for 20 min (see Note 12).
2. Heat shock the cells at 42°C for 30 s and incubate on ice for 2 min.
3. Add 900 μL of SOC recovery medium. Incubate for 1 h at 37°C with shaking (220 rpm).
4. Plate 150 μL of diluted cells onto LB-agar AMP/KAN plates and incubate at 37°C overnight.
5. Inoculate a single colony into 100 mL of LB media supplemented with AMP/KAN. Incubate overnight (12–16 h) at 37°C.
6. Add 20 mL of LB starter culture to 1 L of sterile ^{15}N M9 media. Incubate at 37°C with shaking (220 rpm) until OD_{600} reaches 0.6.
7. Incubate cells at 20°C for 1 h with shaking at 220 rpm (see Note 13).
8. Induce cells by adding IPTG to a final concentration of 1 mM (1 mL of 1 M IPTG solution added to 1 L of cell culture) and incubate at 20°C overnight with shaking (220 rpm).

Co-purification of Apo CaM–$Na_V1.5_{IQp}$ complex

1. Harvest cells by centrifugation at 5,000 × g for 10 min.
2. Resuspend cell paste in Ni-NTA buffer A (10–15 mL of buffer per gram of cell paste) and transfer to a stainless steel beaker.

Chill this beaker in a secondary container with melting ice (see Note 14).

3. Sonicate the resuspended cell paste with a macrotip set to 50 W output power for total processing time of 5 min (5 s pulse bursts with 5 s recovery/off time).
4. Clear cell lysate by centrifugation at 48,000 × *g* for 20 min.
5. Decant the lysate and filter through a 0.45-μM membrane.
6. Equilibrate the prepacked 25 mL Ni-NTA column with 1 CV of MQ H_2O and 2 CVs of Ni-NTA buffer A.
7. Load the filtered lysate onto the Ni-NTA column at a flow rate of 1.0 mL/min.
8. Wash unbound lysate from the column with 4 CVs of Ni-NTA buffer at a flow rate of 2.5 mL/min.
9. Elute protein complex with a 4 CV gradient (0–100% Ni-NTA Buffer B, 10–500 mM imidazole), with 5 mL fractionation.
10. Determine fractions containing CaM–His_6-SUMO–$NaV1.5_{IQp}$ complex by SDS-PAGE.
11. Prepare 4 L of S75 buffer in a glass beaker equilibrated at 4°C.
12. Pool relevant fractions, combine with H3C protease, and seal in dialysis tubing (3 kDa MWCO). Transfer dialysis tubing to glass beaker containing 4 L of S75 buffer. Incubate at 4°C overnight.
13. Filter the dialyzed sample through a 0.45-μM membrane.
14. Equilibrate the prepacked 25 mL Ni-NTA column with 1 CV of MQ H_2O and 2 CVs of Ni-NTA buffer A.
15. Load the cleaved dialyzed solution onto the Ni-NTA column at a flow rate of 1.0 mL/min. Collect the flow through.
16. Wash unbound protein from the column with 4 CVs of Ni-NTA buffer at a flow rate of 2.5 mL/min with 5 mL fractionation.
17. Determine fractions containing CaM-IQ complex by SDS-PAGE.
18. Equilibrate S75 with 1 CV of MQ H_2O and 1 CV of S75 Buffer.
19. Pre-rinse two centrifugal concentrators (15 mL, 3 kDa MWCO) with MQ H_2O) by centrifuging at 3,700 × *g* for 15 min at 4°C. Concentrate the pooled Cam-IQ complex fractions to ~3–5 mL.
20. Load the concentrate onto the Superdex 75 column at 1 mL minute. Elute the protein isocratically for 1 CV with 5 mL fractionation.
21. Determine fractions containing CaM-IQ complex by SDS-PAGE.

22. Pool the fractions and dialyze against 20 mM HEPES pH 6.8, 100 mM KCl, 1 mM EDTA, and 0.01% NaN_3. Concentrate sample to 0.1–1 mM for NMR studies.

4. Notes

1. cDNA encoding for human CaM was sublconed between the *Nco*I and *Bam*HI restriction sites of a pET15b vector (ampicillin resistance Novagen).
2. To preserve columns and chromatography equipment, all buffers should be 0.2 μM filtered immediately prior to use.
3. Due to the high Ca^{2+}-binding affinity of CaM, all buffer solutions should be prepared with MQ water to reduce the risk of Ca^{2+} contamination of apo CaM.
4. cDNA encoding for Human $Na_V1.5$ IQ motif E1901-L1927 subcloned between *Bam*HI and *Xho*I restriction sites of a pET27 derivative in house vector (L. Mizoue CSB Vanderbilt University). This derivative codes for an N-terminal His_6-tagged SUMO followed by a flexible linker containing a H3C protease cleavage site. The IQ motif is expressed as a fusion C-terminal fusion to SUMO (His_6-SUMO–$NaV1.5_{IQp}$). The final product is isolated as a 31-residue polypeptide that contains Gly-Pro-Gly-Ser fused to its N-terminus after cleavage of the His_6-SUMO.
5. The choice of His_6-SUMO as a fusion partner for $Na_V1.5_{IQp}$ simplifies the purification by taking advantage of robust Ni-NTA chromatography methods and allows for the separation of excess CaM.
6. The use of 10 mM imidazole in the loading buffer greatly reduces nonspecific binding of endogenous *E. coli* proteins to the Ni-NTA resin.
7. We produce N-terminal His_6-tagged H3C protease in house (L. Mizoue CSB Vanderbilt University). As a commercially available alternative (GST-tagged) PreScission Protease (GE Lifesciences) can be used.
8. Typically, CaM will elute completely within the first 40 mL.
9. CaM can be quantified by measuring the absorbance at 276 nm and using the extinction coefficient 3,006 M^{-1} cm^{-1}.
10. In cases of extremely hydrophobic peptides, an organic solvent such as dimethyl sulfoxide (DMSO) can be used as an additive to increase solubility.
11. Clean and rinse the NMR tube with a 1% solution of nitric acid to remove any Ca^{2+} contamination present within the NMR

tube followed by extensive rinsing with buffer N to remove any residual 1% nitric acid.

12. The use of higher concentrations of plasmid DNA increases the efficiency of the double transformation.
13. Induction at lower temperature greatly enhances the solubility of the His_6-SUMO–$NaV1.5_{IQp}$ fusion.
14. It is critical to ensure the resuspended cell paste remains at cold temperatures during lysis or protein precipitation will occur leading to reduced yields.

References

1. Chazin WJ (2011) Relating form and function of EF-hand calcium binding proteins. Acc Chem Res 44:171–179
2. Bhattacharya S, Bunick CG, Chazin WJ (2004) Target selectivity in EF-hand calcium binding proteins. Biochim Biophys Acta 1742:69–79
3. Klee CB, Newton DL, Ni WC, Haiech J (1986) Regulation of the calcium signal by calmodulin. Ciba Found Symp 122:162–182
4. Chagot B, Chazin WJ (2011) Solution NMR structure of apo-calmodulin in complex with the IQ motif of human cardiac sodium channel $Na_V1.5$. J Mol Biol 406:106–119
5. Fallon JL, Baker MR, Xiong L, Loy RE, Yang G, Dirksen RT, Hamilton SL, Quiocho FA (2009) Crystal structure of dimeric cardiac L-type calcium channel regulatory domains bridged by Ca^{2+}* calmodulins. Proc Natl Acad Sci U S A 106:5135–5140
6. Feldkamp MD, Yu L, Shea MA (2011) Structural and energetic determinants of apo calmodulin binding to the IQ motif of the $Na_V1.2$ voltage-dependent sodium channel. Structure 19:733–747
7. Kim EY, Rumpf CH, Fujiwara Y, Cooley ES, Van Petegem F, Minor DL Jr (2008) Structures of Ca_V2 Ca^{2+}/CaM-IQ domain complexes reveal binding modes that underlie calcium-dependent inactivation and facilitation. Structure 16:1455–1467
8. Mori MX, Vander Kooi CW, Leahy DJ, Yue DT (2008) Crystal structure of the Ca_V2 IQ domain in complex with Ca^{2+}/calmodulin: high-resolution mechanistic implications for channel regulation by Ca^{2+}. Structure 16:607–620
9. Kranz JK, Lee EK, Nairn AC, Wand AJ (2002) A direct test of the reductionist approach to structural studies of calmodulin activity: relevance of peptide models of target proteins. J Biol Chem 277:16351–16354
10. Bahler M, Rhoads A (2002) Calmodulin signaling via the IQ motif. FEBS Lett 513:107–113
11. Black DJ, Leonard J, Persechini A (2006) Biphasic Ca^{2+}-dependent switching in a calmodulin-IQ domain complex. Biochemistry 45:6987–6995
12. Yu FH, Catterall WA (2003) Overview of the voltage-gated sodium channel family. Genome Biol 4:207.1–207.7
13. Shah VN, Wingo TL, Weiss KL, Williams CK, Balser JR, Chazin WJ (2006) Calcium-dependent regulation of the voltage-gated sodium channel hH1: intrinsic and extrinsic sensors use a common molecular switch. Proc Natl Acad Sci U S A 103:3592–3597
14. Huth JR, Bewley CA, Jackson BM, Hinnebusch AG, Clore GM, Gronenborn AM (1997) Design of an expression system for detecting folded protein domains and mapping macromolecular interactions by NMR. Protein Sci 6:2359–2364
15. Delaglio F, Grzesiek S, Vuister GW, Zhu G, Pfeifer J, Bax A (1995) NMRPipe: a multidimensional spectral processing system based on UNIX pipes. J Biomol NMR 6:277–293
16. Goddard TD, Kneller DG (2008) SPARKY, 3rd edn. University of California, San Francisco
17. Johnson BA, Blevins RA (1994) NMR View: a computer program for the visualization and analysis of NMR data. J Biomol NMR 4: 603–614

Chapter 12

Biochemical and Immunological Detection of Physical Interactions Between Penta-EF-Hand Protein ALG-2 and Its Binding Partners

Kanae Osugi, Hideki Shibata, and Masatoshi Maki

Abstract

Many nonenzymatic cellular proteins exert their functions by interacting with other proteins or macromolecules. Analysis of the physical interactions of proteins is an important step to understand their functions, and the information obtained is helpful for predicting the roles of the proteins in cells. Here we describe three biochemical and immunological methods for the detection of interactions between ALG-2 (a penta-EF-hand Ca^{2+}-binding protein, also known as PDCD6) and its target proteins: (1) glutathione-*S*-transferase (GST) pulldown assay, (2) co-immunoprecipitation assay, and (3) Far Western blot analysis using biotinylated ALG-2. Dependency of Ca^{2+} for interaction is examined by inclusion of $CaCl_2$ or EGTA in buffers used for binding assays.

Key words: Protein–protein interaction, Calcium-binding protein, GST-pulldown, Co-immunoprecipitation, Far Western, Biotinylation, EF-hand

1. Introduction

There are several methods for large-scale screening of novel interacting proteins, such as yeast two-hybrid, phage display and mass spectrometry analysis of pulldown or co-immunoprecipitation products. However, when these methods are used, it is not clear which proteins are true binding partners and which proteins are bona fide binding factors due to nonspecific interactions. We need to confirm the authenticity of physical interactions using different methods.

ALG-2 (apoptosis-linked gene 2, also known as PDCD6) is a 22-kDa Ca^{2+}-binding protein containing five serially repetitive EF-hand motifs called the penta-EF-hand domain (1, 2). Upon bind-

Claus W. Heizmann (ed.), *Calcium-Binding Proteins and RAGE: From Structural Basics to Clinical Applications*, Methods in Molecular Biology, vol. 963, DOI 10.1007/978-1-62703-230-8_12, © Springer Science+Business Media New York 2013

ing to Ca^{2+}, ALG-2 changes its conformation and Ca^{2+}-dependently interacts with various proteins (3, 4). ALIX (ALG-2-interacting protein X, also named AIP1 and PDCD6IP) is the first protein reported as an ALG-2-interacting factor (5, 6). ALG-2 directly binds to the C-terminal proline-rich region (PRR) of ALIX at a specific site (7). Generally, the PRR is structurally flexible and often found in other known ALG-2-interacting proteins that have a motif similar to that found in ALIX, PLSCR3, and Sec31A (8, 9). MISS (MAPK-interacting and spindle-stabilizing protein-like) also contains a similar sequence in the PRR and has been predicted to be a novel ALG-2-binding protein in our laboratory by scoring the similarity of the motif (unpublished data).

In this chapter we describe three different methods using lysates of cultured cells expressing candidate proteins and demonstrate the detection of interaction between ALG-2 and MISS that is fused to green fluorescence protein (GFP). GFP-fused ALIX and unfused GFP protein are used as positive and negative controls. First, we introduce a pulldown assay using recombinant GST-fused human ALG-2 and the lysates of HEK293T cells expressing GFP-fusion proteins. Next, we describe a co-immunoprecipitation assay using anti-GFP antibody, in which we can detect binding between GFP-fused proteins and endogenous ALG-2. Finally, we show a Far Western blot analysis using biotinylated ALG-2 (bio-ALG-2) as a probe that interacts with proteins blotted on a solid membrane filter, being a simple method to detect a direct physical interaction (7–10).

2. Materials

2.1. GST-Pulldown

1. Buffer TNE: Dilute stock solutions of 1 M Tris–HCl (pH 7.5), 5 M NaCl, 0.5 M EDTA (pH 8) with chilled water to 20 mM Tris–HCl, 150 mM NaCl, and 1 mM EDTA.
2. Glutathione Sepharose 4B: Take an aliquot of the beads 50% suspension in 20% ethanol, stored in a refrigerator; GE Healthcare and equilibrate with buffer TNE.
3. GST and GST-ALG-2 proteins: Purify recombinant proteins expressed in *Escherichia coli* BL21 using glutathione Sepharose 4B beads essentially according to the manufacturer's instruction. Dialyze the purified proteins three times against buffer TNE at 4°C for 6 h each time (see Note 1).
4. GST-ALG-2-immobilized beads (0.5 mg protein per mL bed volume): Add 200 μg of purified GST-ALG-2 (800 μL TNE) to buffer-equilibrated glutathione Sepharose 4B beads (bed volume, 400 μL) (see Subheading 2.1, item 2) in a 1.5-mL microfuge tube. Gently mix by rotation for 2.5 h. Spin down

the beads at 1,000 rpm using a high-speed refrigerated microcentrifuge (e.g., TOMY MX-150, rotor TMP-11) for 1 min. Discard 400 μL of the supernatant to make 50% suspension of beads. Store in the presence of protease inhibitors (e.g., 0.1 mM concentration of pefabloc, a commercial name for AEBSF {4-(2-Aminoethyl) benzenesulfonyl fluoride hydrochloride}) at 4°C until use. Similarly, prepare GST-immobilized beads for control experiments (see Note 2).

5. Human embryonic kidney (HEK) 293T cells (RIKEN BioResource Center Cell Bank; also available from American Type Culture Collection) expressing GFP fusion proteins: Grow HEK293T cells in Dulbecco's modified Eagle's medium (DMEM) supplemented with 0.5 mg/mL glutamine, 5% fetal bovine serum (FBS), 100 units/mL penicillin, and 100 μg/mL streptomycin at 37°C under humidified air containing 5% CO_2. One day after seeding in 9-cm dishes, transfect cells with plasmid DNAs for expression of GFP-fusion proteins by the conventional calcium phosphate precipitation method or liposome method (see ref. 11).
6. Phosphate-buffered saline (PBS): 137 mM NaCl, 2.7 mM KCl, 8 mM Na_2HPO_4, and 1.5 mM KH_2PO_4 (pH 7.4), diluted from 10× PBS stock solution (stored at room temperature) with water. Prepare at the time of use or store at 4°C for less than 1 month.
7. Tween-20: 10% (w/v) stock solution. Store at 4°C.
8. PBST: PBS containing 0.1% Tween-20. Store at 4°C.
9. Lysis buffer H: 20 mM HEPES–NaOH (pH 7.4), 142.5 mM KCl, 1.5 mM $MgCl_2$, 0.2% (w/v) Nonidet P (NP)-40 (stored at 4°C). Just before use, add protease inhibitors to the following final concentrations using stock solutions stored at −20°C: 0.1 mM pefabloc (100 mM stock), 3 μg/mL leupeptin (3 mg/mL stock), 1 μM E-64 (1 mM stock), 1 μM pepstatin (1 mM stock in ethanol), and 0.2 mM phenylmethylsulfonylfluoride (PMSF, 200 mM stock in methanol).
10. 200 mM ethyleneglycoltetraacetic acid (EGTA): Dissolve 3.8 g of EGTA with approximately 20 mL of 1 M sodium hydroxide solution and then adjust pH to 8.0 and fill up with water to 50 mL.
11. Pulldown wash buffer: Add 4 mM $CaCl_2$ or 200 mM EGTA (pH 8.0) to lysis buffer H and make the final concentration to 0.1 or 5 mM for each experimental condition in accordance with the pulldown condition in the presence of $CaCl_2$ or EGTA.

2.2. Western Blot Analysis

1. 5×SDS polyacrylamide gel electrophoresis (PAGE) sample buffer: 0.31 M Tris–HCl (pH 6.8), 10% (w/v) SDS, 50% (w/v) glycerol, 25% (v/v) 2-mercaptoethanol (2-ME),

0.0125% (w/v) Bromophenol blue (BPB). Dispense aliquots into screw-capped vials while carefully avoiding contamination of keratins from finger tips. Dilute the solution with water and prepare 2 × SDS and 1 × SDS sample buffers for convenience. Store at −20°C.

2. Gel staining solution: 0.125% (w/v) Coomassie brilliant blue R-250 (CBB), 50% (v/v) methanol, 10% (v/v) acetic acid. Dissolve 0.625 g of CBB in 250 mL of methanol by stirring overnight, and then add 200 mL of water and 50 mL of acetic acid. Store in a chemical fume hood.
3. Gel destaining solution: 25% (v/v) methanol and 7% (v/v) acetic acid. Store in a chemical fume hood.
4. Transfer buffer: 25 mM Tris, 192 mM glycine, 15% (v/v) methanol, 0.01% (w/v) SDS. Store at room temperature.
5. Transfer membrane: a polyvinylidene difluoride (PVDF) membrane (Immobilon-P from Millipore or equivalent).
6. Sodium azide (NaN_3) solution: 2% (w/v) stock solution, stored at room temperature (see Note 3).
7. Blocking buffer: PBST containing 5% (w/v) skim milk. Prepare at the time of use.
8. Primary and secondary antibodies: affinity-purified rabbit anti-human ALG-2 polyclonal antibody (home-made, *see* ref. 12), mouse anti-GFP monoclonal antibody (clone B2, Santa Cruz Biotechnology, Santa Cruz, CA, USA), horseradish peroxidase (HRP)-conjugated goat antibodies from Jackson Immunoresearch Laboratories (West Groove, PA, USA) against rabbit immunoglobulin G (IgG) and against mouse IgG + IgM. Use of secondary antibody specific for mouse IgG with minimum cross-reactivity with IgG of other species is recommended in the case of detecting proteins of interest with molecular masses similar to those of heavy and light chains of IgG (~50 and ~25 kDa, respectively).

2.3. Co-immunoprecipitation

1. Lysis buffer T: 50 mM Tris–HCl (pH 7.5), 150 mM NaCl, 1.5 mM $MgCl_2$, 0.2% (w/v) Triton X-100 (stored at 4°C). Just before use, add protease inhibitors as described for preparation of lysis buffer H (see Subheading 2.1, item 9).
2. Dynabeads Protein G: Equilibrate magnetic beads coated with protein G (Invitrogen) in lysis buffer T before use.
3. Antibodies: rabbit anti-GFP antiserum suitable for immunoprecipitation (Invitrogen/Molecular Probes, Carlsbad, CA, USA); rabbit γ-globulins (Sigma-Aldrich, St Louis, MO, USA) as control IgG.
4. Special equipment: a magnetic separation stand from Invitrogen or equivalent.

2.4. Far Western

1. Recombinant human ALG-2: Expressed in *Escherichia coli* BL21 (DE3) pLysS and purified (see ref. 8).
2. Buffer $HTME_1$: 10 mM HEPES–NaOH (pH 8.0), 0.1% (w/v) Tween 20, 10 mM 2-ME, and 1 mM EDTA.
3. Buffer $HTME_{0.1}$: 10 mM HEPES–NaOH (pH 8.0), 0.1% (w/v) Tween 20, 10 mM 2-ME, and 0.1 mM EDTA.
4. Coupling buffer: 200 mM $NaHCO_3$, 500 mM NaCl, 0.1% (w/v) Tween 20, and 10 mM 2-ME
5. Biotinylation reagent: Dissolve 1 mg of Biotin-$(AC_5)_2$Sulfo-Osu (Dojindo Laboratories, Kumamoto, Japan) in 100 μL of cold water. Prepare at the time of use.
6. Tris-buffered saline (TBS): 10 mM Tris–HCl (pH 7.5) and 150 mM NaCl. Prepare by diluting 10× TBS stock solution stored at room temperature.
7. TBST: TBS containing 0.1% (w/v) Tween-20.
8. Blocking buffer: TBST containing 1% (w/v) gelatin and 0.02% NaN_3. Prepare TBS containing 1% (w/v) gelatin, dissolve by autoclaving, and then add Tween-20 and NaN_3 to 0.1% and 0.02%, respectively. Store at 4°C.
9. TBSTC: TBST containing 0.1 mM $CaCl_2$.
10. Binding buffer: TBSTC containing 0.1% gelatin. Prepare TBS containing 0.1% (w/v) gelatin, dissolve by autoclaving, and then add Tween-20 and $CaCl_2$ to 0.1% and 0.1 mM, respectively. Store at 4°C.
11. TBSC: TBS containing 0.1 mM $CaCl_2$.
12. Streptavidin-POD: Dilute 1 mg/mL of streptavidin-conjugated horseradish peroxidase (Rockland, Inc., Gilbertsville, PA, USA) stored at 4°C to 500 ng/mL with the binding buffer at the time of use.

3. Methods

Use a chilled microfuge for centrifugation and ice-cold solutions unless otherwise indicated.

3.1. GST-Pulldown

1. Before preparation of cell lysates, equilibrate GST-beads with lysis buffer H. Take 20 μL of 50% suspension of GST-ALG-2-immobilized beads for each sample.
2. Harvest cultured HEK293T cells expressing proteins that are to be analyzed. Remove the medium with an aspirator and wash cells in 3 mL of PBS per one 9-cm dish. Add 1 mL of PBS to the dish, scrape cells with a rubber policeman, and transfer them to a 1.5-mL microfuge tube.

3. Spin down the cells at 5,000 rpm for 1 min using a high-speed refrigerated microcentrifuge (e.g., TOMY MX-150, rotor TMP-11). Discard the supernatant (see Note 4).
4. Add 1.2 mL of lysis buffer H to the pelleted cells, mix, and leave for 30 min on ice (mix every 10 min).
5. Centrifuge the cell lysate at 14,000 rpm for 10 min. Transfer the supernatants to an empty tube (see Note 5). At least 1 mL of the supernatant (cleared lysate) should be obtained from cells of one dish for each protein. When plural experimental conditions are examined, combine the obtained cleared lysates from a few dishes in a 15-mL tube and mix well before dispensing.
6. Take 50 μL from the cleared lysate, mix with 50 μL of 2 × SDS-PAGE sample buffer, and heat at 95°C for 2 min. Use this sample as "Input" for Western blot analysis.
7. Add 500 μL of the cleared lysates to the buffer-equilibrated glutathione Sepharose beads (20 μL of 50% suspension) immobilizing 5 μg of GST (control) or GST-ALG-2 (see Subheading 2.1, item 4). Add 12.5 μL of 4 mM $CaCl_2$ (for Ca^{2+} condition) or 200 mM EGTA (for Ca^{2+}-free condition) to the final concentrations of 0.1 and 5 mM, respectively. Gently mix by rotating the tubes for 2 h at 4°C (see Note 6).
8. Spin down the beads at 3,000 rpm for 1 min. After the centrifugation, pay attention not to disturb the bed of beads and carefully remove the supernatants. Add 500 μL of ice-cold washing buffer (same as the binding buffer) containing either 0.1 mM $CaCl_2$ or 5 mM EGTA to the beads in accordance with the experimental conditions, mix gently for a few seconds, spin down, and remove the supernatants. Repeat this washing step three times.
9. Finally, add 50 μL of 1 × SDS-PAGE sample buffer to the recovered beads. Mix and heat the tubes at 95°C for 2 min. Spin down and use supernatants as "Pulldown" products for Western blot analysis (see Note 7).
10. Resolve the proteins in input samples and pulldown products by SDS-PAGE (see ref. 13) in duplicate gels, and use one gel for staining with Coomassie brilliant blue R-250 (CBB) and the other gel for Western blot analysis.
11. After electrophoresis, soak the gel in CBB-staining solution for ~30 min with gentle shaking and then transfer to a destaining solution in a chemical fume hood.
12. For Western blotting, first, electrophoretically transfer the proteins from the gel to a sheet of PVDF membrane with a semi-dry blotter using Transfer buffer at 100 mA for 90 min. Next, block the blotted membrane with PBST containing 5% (w/v) skim-milk by incubation at 37°C for 1 h. After rinsing twice with PBST, incubate the membrane with mouse monoclonal

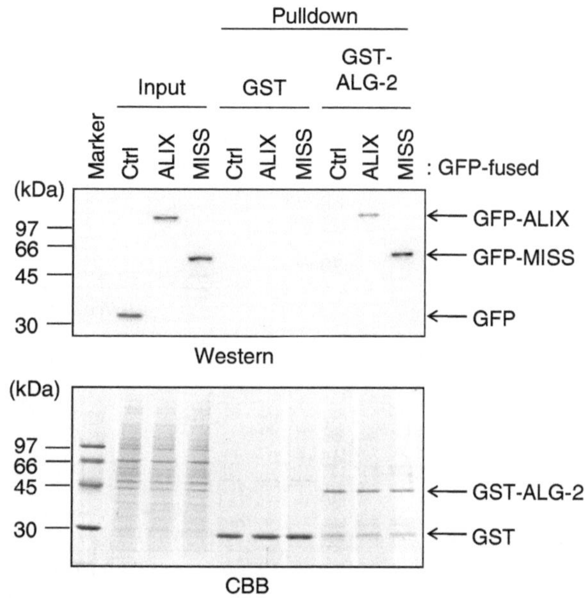

Fig. 1. GST-pulldown assay. HEK293T cells expressing unfused GFP, GFP-ALIX, and GFP-MISS were lysed, and supernatants (cleared lysates) were incubated with glutathione Sepharose beads carrying GST or GST-ALG-2 in the presence of 0.1 mM $CaCl_2$. The beads were then pelleted by low-speed centrifugation, followed by washing with lysis buffer, containing 0.1 mM $CaCl_2$. The pellets (pulldown products) and cleared lysates were subjected to SDS-PAGE and analyzed by Western blotting using an anti-GFP antibody (*Western*) (*upper panel*). Although similar intensities of immunoreactive signals were detected for GFP (*Ctrl*), GFP-ALIX (*ALIX*), and GFP-MISS (*MISS*) in the lanes of cleared lysate (*Input*), only GFP-ALIX and GFP-MISS gave positive signals in the lanes for pulldown products (*Pulldown*). A parallel gel was stained with Coomassie brilliant blue R-250 (*CBB*) (*lower panel*). Bands of approx. 30 kDa, observed in the lanes of GST-ALG-2, correspond to degraded products, probably GST moiety of GST-ALG-2 (see Note 1). In these analyses, protein samples corresponding to 1% of total cleared lysates and 10% of pulldown products were loaded to each lane on 12.5% gels of SDS-PAGE. The relative amounts of cleared lysate proteins analyzed for pulldown (Figs. 1 and 2) or immunoprecipitation assays (Figs. 3 and 4) are expressed as Input 5%.

antibody against GFP (0.2 μg/mL IgG) as a probe in the presence of PBST containing 0.1% (w/v) bovine serum albumin at 4°C overnight, and after rinsing the membrane, incubate it with a secondary antibody of goat anti-mouse IgG + IgM conjugated with horseradish peroxidase (HRP) (0.4 μg/mL) at 37°C for 1 h (see Note 3).

13. After successively washing with PBST twice at 37°C for 5 min and with PBS twice at 37°C for 5 min, develop immunoreactive signals by the chemiluminescent method using a commercially available peroxidase substrate such as Super Signal West Pico Chemiluminescent Substrate (Pierce/Thermoscientific) and detect chemiluminescence by a LAS-3000mini lumino-image analyzer (Fujifilm) (see Figs. 1 and 2).

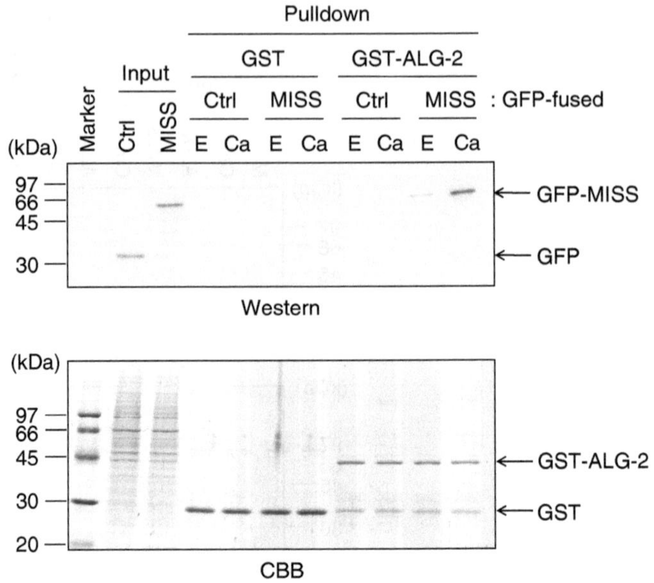

Fig. 2. Ca^{2+}-dependent GST-pulldown of ALG-2-interacting protein. Pulldown assay of GFP-MISS using GST-ALG-2 was performed in the presence of either 5 mM EGTA (*E*) or 0.1 mM $CaCl_2$ (*Ca*) and pulldown products were subjected to Western blot analysis (*upper panel*). Staining was performed with CBB and is described in the legend to Fig. 1.

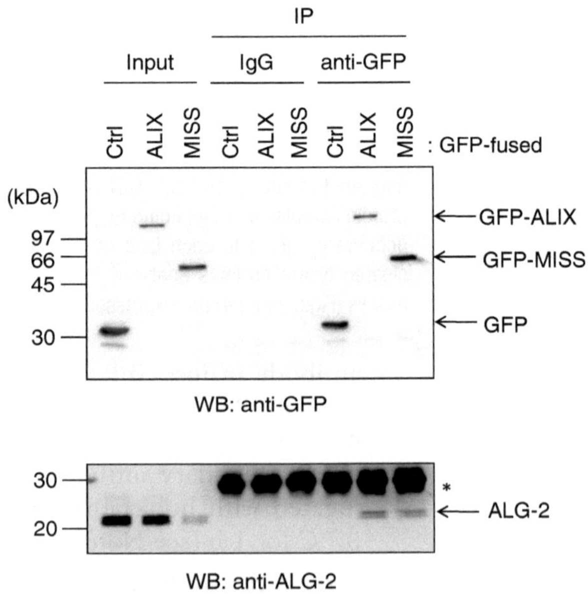

Fig. 3. Co-immunoprecipitation assay. HEK293T cells expressing unfused GFP, GFP-ALIX, and GFP-MISS were lysed, and cleared lysates were incubated with either control IgG or anti-GFP antibody (rabbit antiserum being used in this experiment) in the presence of 0.1 mM $CaCl_2$. Precipitates of the immuno-complex were collected with Dyna beads Protein G. Cleared lysates (*Input*) and immunoprecipitates (*IP*) were analyzed by Western blotting (*WB*) using mouse anti-GFP monoclonal antibody (*upper panel*) and rabbit anti-ALG-2 polyclonal antibody (*lower panel*). *Asterisk*, immunoglobulin G (IgG) light chain. Input 5%.

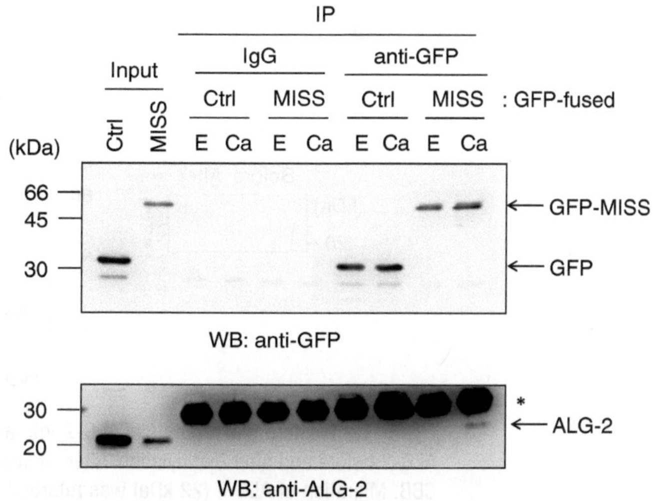

Fig. 4. Ca^{2+}-dependent co-immunoprecipitation of ALG-2. Immunoprecipitation was performed in the presence of 5 mM EGTA (*E*) and 0.1 mM $CaCl_2$ (*Ca*) as described in the legend to Fig. 3 using the cleared lysates of cells expressing GFP (*Ctrl*) or GFP-MISS (*MISS*). *Asterisk*, IgG light chain. Input 5%.

3.2. Co-immunoprecipitation

1. Prepare the cleared lysate of DNA-transfected cells as described in Subheading 3.1, steps 2–5 except for the use of lysis buffer T instead of lysis buffer H (see Note 5).
2. Take 50 μL from the cleared lysate for Western blot analysis to use it as an input sample as described in Subheading 3.1, step 6. Dispense 500 μL each of the cleared lysates into four tubes. Add 12.5 μL of 4 mM $CaCl_2$ or 200 mM EGTA. Add 2.5 μL of rabbit control IgG (1 mg/mL) or antibody against GFP. Gently mix by rotating tubes at 4°C for 1 h.
3. Add the antibody-treated cleared lysates to tubes containing 10 μL of slurry of buffer-equilibrated Dyna beads Protein G (see Note 7). Continue rotation for 1 h or longer (see Note 6).
4. Set the tubes on a magnetic separation stand for 30 s and collect magnetic beads at the side walls of the tubes. Remove supernatants (see Note 8). Wash the beads twice with 500 μL of buffer T containing $CaCl_2$ or EGTA in accordance with the experimental conditions.
5. Add 50 μL of 1×SDS-PAGE sample buffer to the collected beads and heat at 95°C for 2 min. Spin down and use supernatants as samples of immunoprecipitation (IP) products (see Note 9).
6. Analyze the IP products by Western blotting using mouse anti-GFP monoclonal antibody and rabbit anti-ALG-2 polyclonal antibody as probes (see Figs. 3 and 4).

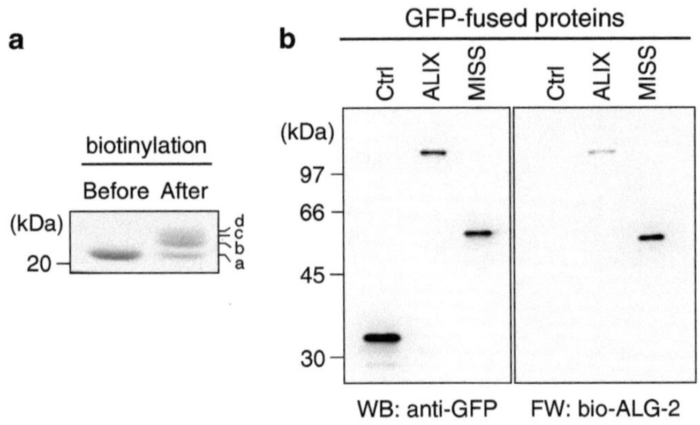

Fig. 5. Far Western blot analysis of ALG-2-interacting proteins. (**a**) SDS-PAGE analysis of ALG-2 before and after biotinylation. After electrophoresis, a 12.5% gel was stained with CBB. Migration of ALG-2 (22 kDa) was retarded and the band was shifted upward after biotinylation: band *a*, unmodified protein; bands *b-d*, biotinylation at a single site and multiple sites. (**b**), Far Western blot analysis of immunoprecipitation (IP) products. IP products prepared as described in the legend to Fig. 3 were subjected to Far Western blot analysis using biotinylated (bio)-ALG-2 as a probe. *Left*, Western blot analysis (*WB*) using anti-GFP monoclonal antibody. *Right*, Far Western blot analysis (*FW*) using bio-ALG-2.

3.3. Biotinylation of ALG-2

1. Dialyze recombinant human ALG-2 (~2 mg/mL) against 500 mL each of buffer $HTME_1$, buffer $HTME_{0.1}$ and coupling buffer at 4°C for 6 h three times (see Note 10).
2. Add 60 μL of 10 mg/mL Biotin-$(AC_5)_2$Sulfo-Osu to 180 μg of recombinant human ALG-2 in 440 μL of coupling buffer (see Note 11). Mix and incubate at 25°C for 2 h.
3. Remove the biotinylation reagent by dialysis against 10 mM Tris–HCl (pH 7.5), 150 mM NaCl, 0.1% Tween-20, 10 mM 2-ME, and 0.1 mM EDTA at 4°C for 6 h. Repeat dialysis three times.
4. Collect the dialysate and store the biotinylated ALG-2 (bio-ALG-2) at 4°C in the presence of 0.02% NaN_3. Check the efficiency of biotinylation by SDS-PAGE followed by CBB staining (Fig. 5a; Note 12).

3.4. Far Western Blot Analysis

1. Resolve the immunoprecipitation products by SDS-PAGE and transfer the proteins from the gels to PVDF membranes (see Subheading 3.2, step 5 and Note 13). Process one membrane for Western blot analysis using anti-GFP antibody as described in Subheading 3.1, steps 12–13 (see Fig. 5b). For Far Western analysis, block the other membrane by immersing the blots in Blocking buffer (TBST containing 1% gelatin and 0.02% NaN_3) at 37°C for 1 h. Rinse briefly twice with TBSTC at room temperature.

2. Just before use, dilute the stock solution of bio-ALG-2 to 20 ng/mL with 3 mL of the binding buffer (TBSTC containing 0.1% gelatin). Incubate the membrane with bio-ALG-2 at 4°C overnight.
3. Wash the membrane twice by incubation with TBSTC at 37°C for 5 min.
4. Incubate the membrane with 500 ng/mL streptavidin-POD in the binding buffer at 37°C for 1 h (see Note 14).
5. Wash the membrane twice by incubation with TBSTC at 37°C for 5 min.
6. Wash the membrane twice by incubation with TBSC at 37°C for 5 min.
7. Determine the amount of bound bio-ALG-2 by the chemiluminescent method as described for Western blot analysis except for inclusion of 0.1 mM $CaCl_2$ during chemiluminescence development (Fig. 5c and Note 15).

4. Notes

1. When solutions of purified recombinant proteins of ALG-2 and GST-ALG-2 are stored for a long period of time (~6 months) at 4°C, proteins may be precipitated. Immobilize the purified proteins as soon as possible. Since GST-fusion proteins are prone to be digested by proteases at the linker region during cell culture and purification; unfused GST proteins are also co-purified to some extent (see Fig. 1).
2. Omission of protease inhibitors will cause degradation of GST-fusion proteins during storage with time even at 4°C. Generally, affinity purification, using glutathione Sepharose beads, is sufficient for GST-pulldown assays. For storage of the immobilized beads, add protease inhibitors (e.g., 0.1 mM pefabloc) and store at 4°C. When EDTA is omitted from the buffer, sodium azide (final concentration, 0.02%) should be added as a protecting reagent.
3. Do not add sodium azide as a preservative to buffers that are used for incubation with horseradish peroxidase. The reagent inhibits the peroxidase reaction.
4. The cell pellets obtained can be temporally stored at −80°C.
5. Check the efficiency of recovery of expressed proteins in the cleared lysate by comparing the amounts of those proteins in the total cell lysate by Western blot analysis. The recovery efficiency depends on localization of the proteins (cytosol,

membrane-bound, organelles, etc.), lysis procedure (gentle mixing or sonication), and detergents contained in the lysis buffer. For poorly soluble proteins, use higher concentrations of nonionic detergents (such as 1% NP-40 or 1% Triton-X100). Alternatively, use a radioimmunoprecipitation assay (RIPA) buffer: 50 mM Tris–HCl (pH 7.5), 150 mM NaCl, 0.1% SDS, 0.5% sodium deoxycholate, 1% NP-40 (or Triton X-100) supplemented with protease inhibitors. However, it is necessary to dilute the obtained cleared lysate at least five times with RIPA buffer without detergents to reduce the concentration of SDS for the GST-ALG-2 pulldown assay. In the case of immunoprecipitation, RIPA buffer may affect immunoreaction efficiencies of some antibodies.

6. To detect weakly binding proteins, extend the incubation time from 1 to 2 h to overnight.
7. Protein G Sepharose can be used instead of Dyna beads Protein G. However, the former gives a higher background for ALG-2 than the latter, probably due to either weak adsorption of ALG-2 to Sepharose beads or penetration of the protein into the bead matrix in the presence of Ca^{2+}.
8. Save the supernatants for estimation of immunoprecipitation efficiency of the antibody used in the experiments by checking how much protein of interest remains in the supernatants. Add more antibodies if the efficiency is poor.
9. Products of pulldown and immunoprecipitation can be stored at −20°C. When using lysis buffer H, during storage even on ice for a short period of time, white precipitates appear in the SDS-PAGE samples due to potassium dodecyl sulfate, potassium ions of which are derived from lysis buffer H. Dissolve the precipitates by warming up before loading to the wells of the polyacrylamide gels. Lysis buffer containing sodium chloride instead of potassium chloride can be used to prevent precipitation.
10. Since amino groups of proteins (ε-NH_2 of Lys, α-NH_2) are used for coupling, avoid the use of buffers containing Tris and amines.
11. Nonionic detergent (0.1% Tween-20) is included to prevent aggregation of modified ALG-2 during coupling. However, the detergent is not effective to prevent its aggregation when a higher concentration of biotin-$(AC_5)_2$Sulfo-Osu is used.
12. There are six lysine residues in the ALG-2 molecule (Lys-37, 77, 90, 112, 116, 137). Biotinylation of ALG-2 causes retardation of migration in SDS-PAGE. Appearance of three additional bands (Fig. 5a) suggests incorporation of biotin to two to three sites at different degrees. Effects of biotinylation at these sites on the binding capacity are not known. Since there

is only one cysteine residue (Cys-155) in ALG-2, a site-specific biotinylation at this site is possible by using an SH-reactive maleimide derivative of biotin. Based on the 3D structure of the complex between ALG-2 and ALIX peptide, Cys-155 is located near a hydrophobic pocket that is important for accepting the ALIX peptide. Thus, biotinylation of Cys-155 may interfere with binding of ALG-2 to target proteins.

13. Since a variety of endogenous biotinylated proteins are present in the cells, they bind to streptavidin-POD and are detected as nonspecific bands for bio-ALG-2 interaction in the Far Western analysis. To eliminate nonspecific signals, it is recommended to use immunoprecipitated samples instead of total or cleared cell lysates. However, to detect high-affinity ALG-2-binding proteins expressed in large amounts, use of total or cleared lysates will give enough reliable data (see ref. 14).
14. When recombinant proteins, with hexahistine (6xHis)-tagged near the ALG-2-binding site, are used as binding partners, addition of 1 mM $NiCl_2$ and 10 mM 2-ME to the binding buffer will enhance the subsequent signals (see ref. 7).
15. Since Ca^{2+} is required for ALG-2 to bind its target proteins, removal of Ca^{2+} in the reaction mixture will cause a rapid dissociation of the complex. Addition of $CaCl_2$ (final concentration, 0.1 mM) to the chemiluminescent substrate solution stabilizes the complex and signal.

Acknowledgments

This work was supported by a Grant-in-Aid for Scientific Research (B) (to MM). We thank members of the Laboratory of Molecular and Cellular Regulation for valuable suggestions.

References

1. Maki M, Narayana S, Hitomi K (1997) A growing family of the Ca^{2+}-binding proteins with five EF-hand motifs. Biochem J 328:718–720
2. Maki M, Kitaura Y, Satoh H, Ohkouchi S, Shibata H (2002) Structures, functions and molecular evolution of the penta-EF-hand Ca^{2+}-binding proteins. Biochim Biophys Acta 1600:51–60
3. Maki M, Yamaguchi K, Kitaura Y, Satoh H, Hitomi K (1998) Calcium-induced exposure of a hydrophobic surface of mouse ALG-2, which is a member of the penta-EF-hand protein family. J Biochem 124:1170–1177
4. Maki M, Suzuki H, Shibata H (2011) Structure and function of ALG-2, a penta-EF-hand calcium-dependent adaptor protein. Sci China Life Sci 54:770–779
5. Missotten M, Nichols A, Rieger K, Sadoul R (1999) Alix, a novel mouse protein undergoing calcium-dependent interaction with the apoptosis-linked-gene 2 (ALG-2) protein. Cell Death Differ 6:124–129
6. Vito P, Pellegrini L, Guiet C, D'Adamio L (1999) Cloning of AIP1, a novel protein that associates with the apoptosis-linked gene ALG-2 in a Ca^{2+}-dependent reaction. J Biol Chem 274:1533–1540

7. Suzuki H, Kawasaki M, Inuzuka T, Okumura M, Kakiuchi T, Shibata H, Wakatsuki S, Maki M (2008) Structural basis for Ca^{2+} -dependent formation of ALG-2/Alix peptide complex: Ca^{2+}/EF3-driven arginine switch mechanism. Structure 16:1562–1573
8. Shibata H, Suzuki H, Kakiuchi T, Inuzuka T, Yoshida H, Mizuno T, Maki M (2008) Identification of Alix-type and non-Alix-type ALG-2-binding sites in human phospholipid scramblase 3. J Biol Chem 283:9623–9632
9. Shibata H, Inuzuka T, Yoshida H, Sugiura H, Wada I, Maki M (2010) The ALG-2 binding Site in Sec31A influences the retention kinetics of Sec31A at the endoplasmic reticulum exit sites as revealed by live-cell time-lapse imaging. Biosci Biotechnol Biochem 74:1819–1826
10. Katoh K, Suzuki H, Terasawa Y, Mizuno T, Yasuda J, Shibata H, Maki M (2005) The penta-EF-hand protein ALG-2 interacts directly with the ESCRT-I component TSG101, and Ca^{2+}-dependently co-localizes to aberrant endosomes with dominant-negative AAA ATPase SKD1/Vps4B. Biochem J 391:677–685
11. Sambrook J, Russell DW (2001) Molecular cloning: a laboratory manual, vol 3, 3rd edn. Cold Spring Harbor Laboratory Press, New York. Chapter 16
12. Shibata H, Suzuki H, Yoshida H, Maki M (2007) ALG-2 directly binds Sec31A and localizes at endoplasmic reticulum exit sites in a Ca^{2+}-dependent manner. Biochem Biophys Res Commun 353:756–763
13. Sambrook J, Russell DW (2001) Molecular cloning: a laboratory manual, vol 3, 3rd edn. Cold Spring Harbor Laboratory Press, New York. Appendix 8.40
14. Shibata H, Yamada K, Mizuno T, Yorikawa C, Takahashi H, Satoh H, Kitaura Y, Maki M (2004) The penta-EF-hand protein ALG-2 interacts with a region containing PxY repeats in Alix/AIP1, which is required for the subcellular punctate distribution of the amino-terminal truncation form of Alix/AIP1. J Biochem 135:117–128

Chapter 13

Measuring Binding of S100 Proteins to RAGE by Surface Plasmon Resonance

Estelle Leclerc

Abstract

Surface plasmon resonance (SPR) is a label-free biophysical method that allows to measure the binding parameters (ka, kd, K_D) of the interaction between two molecules. In this method, one protein/molecule (ligand) is immobilized on the surface of a sensor chip, while the other molecule (analyte) is in solution.

We describe here the use of SPR to measure the binding parameters of the interaction between S100 proteins and isolated domains of the receptor for advanced glycation endproducts (RAGE). In particular, we present the protocols that allow to measure the binding of S100B to RAGE V domain, and the binding of S100A1 and S100A6 to RAGE V-C1-C2 domain fused to a Glutathione S-Transferase (GST) moiety (GST-RAGE).

Key words: Surface plasmon resonance, SPR, S100B, S100A1, S100B, Receptor for advanced glycation endproducts, RAGE, Calcium-binding proteins, EF-hand

1. Introduction

Calcium is an important secondary messenger and calcium-binding proteins react to changes in intracellular calcium concentration by changing their conformation, for most of them, resulting in the interaction with specific target proteins (1–3). The EF-hand calcium-binding proteins (CaBP) form a large family of calcium-binding proteins, with more than 200 members (4, 5). EF-hand CaBPs play important roles in many human diseases including Alzheimer's disease and cancer (6, 7). In order to understand the mechanisms of signal transduction triggered by the CaBPs, it is essential to understand how they interact with their target proteins. Several label-free biophysical methods are used to study the interaction of the CaBPs with their targets. In this chapter, we

Claus W. Heizmann (ed.), *Calcium-Binding Proteins and RAGE: From Structural Basics to Clinical Applications*, Methods in Molecular Biology, vol. 963, DOI 10.1007/978-1-62703-230-8_13,

focus on Surface Plasmon Resonance (SPR) methods and describe protocols that have been used to study the interaction of several S100 proteins (S100B, S100A1, and S100A6) with several domains of the receptor for advanced glycation endproducts (RAGE) on a BIACORE/GE 3000 instrument (GE Healthcare).

The SPR technology utilizes the properties of certain metals (e.g., gold) that can generate plasmons. In defined conditions, the photons of an incident light will generate resonance phenomenon in the plasmons of the metal, resulting in surface plasmon resonance and the generation of evanescent waves by the metal (Fig. 1). The angle of the reflected light will vary upon the environment at

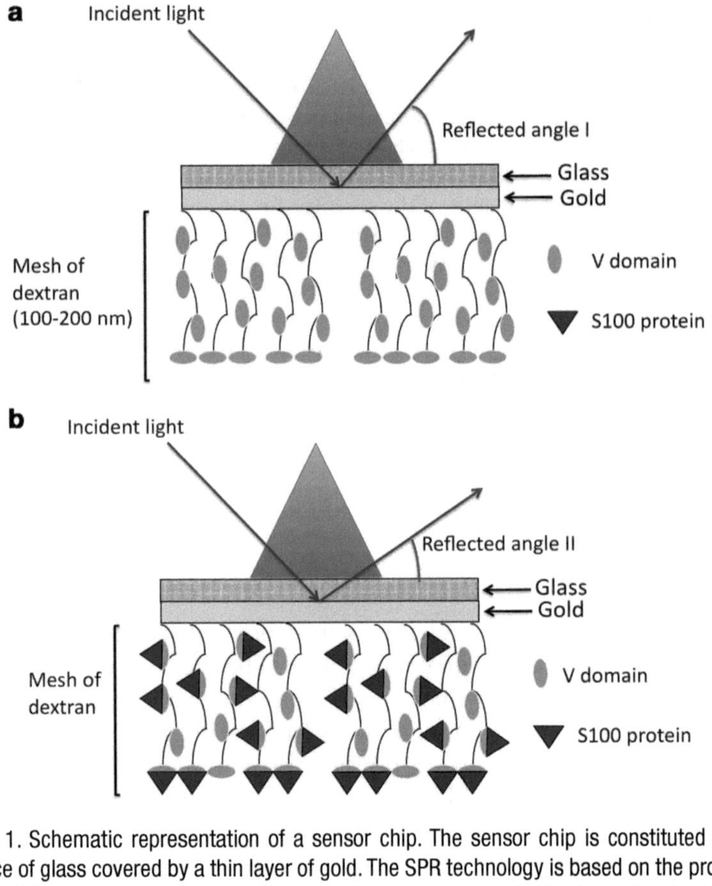

Fig. 1. Schematic representation of a sensor chip. The sensor chip is constituted by a piece of glass covered by a thin layer of gold. The SPR technology is based on the properties of the gold surface that can generate plasmons after excitation by an incident light. The angle of the reflected light depends of the environment of the gold layer (**a**). For instance, in the presence of the bound ligand, the V domain in our case, the light is reflected at an angle I. When the analyte S100B binds to the ligand, the refractive index at the gold interface changes and the light is now reflected by an angle II (**b**). The changes of angle of the reflected light are converted in resonance units (RU). The most commonly used sensor chips possess a dextran layer that forms a tridimensional mesh (e.g., CM5 from BIACORE/GE) allowing formation of small size oligomers of ligands. Modified from the BIACORE/GE Technology Handbook (1998). Note that the relative dimensions of the glass layer, the gold and the dextran layer have not been conserved in the figure.

Table 1
Non-exhaustive list of SPR instruments currently commercially available

Manufacturer	Platform
Biacore/GE Healthcare	1000, 2,000, 3000 X100, A100, T100, S51 X, Flexchip
Bio-Rad	ProteOn XPR36
Reichert	SR7000DC, SR7500DC
GWC Technologies	SPRimager
Neosensors	IAsys
Nomadics	SensiQ Pioneer, Pioneer ST
Sierrasensors	SPR-2
Sensata Technologies/Nomadics	Spreeta Evaluation

the interface of the metal. For instance, the presence of bound proteins or other molecules at the surface of the metal will result in a change of the refractive index at the interface of the metal and will lead to the deviation of the reflected light by a certain angle (Fig. 1). This change in angle is then converted into resonance units (RU). The SPR data are presented in the forms of sensorgrams that describe changes of RU over time. Both association (ka) and dissociation (kd) rates can be extracted from the sensorgrams.

In most SPR instruments (see Table 1), the sensor chip is constituted by a piece of glass covered by a thin layer of gold. The surface of the gold is modified with diverse chemical groups (carboxylic groups, thiols) to allow the covalent attachment of proteins (ligands) onto the surface. The most commonly used sensor chips possesses a dextran layer on top of the surface that forms a tridimensional mesh allowing immobilized proteins to form dimers or small oligomers (Fig. 1).

We describe here the protocols used to study the interaction of three S100 proteins (S100B, S100A1, and S100A6) with isolated domains of the RAGE.

2. Materials

All the solutions described in this protocol are prepared with deonized water. After preparation, the solutions should be filtered and degassed before use (see Note 1).

2.1. Immobilization Reagents

1. Immobilization solution: 10 mM Na Acetate pH 5. For 100 ml, resuspend 0.14 g Na acetate (trihydrate form) into 80 ml water. After complete resuspension, adjust pH with acetic acid. Complete to 100 ml with water. Dilute RAGE domains into the Na acetate buffer in order to obtain a final concentration of 50 μg/ml.
2. Running buffer: 50 mM Tris–HCl, 150 mM NaCl, 5 mM $CaCl_2$, 0.005% Tween 20 pH 7.5 (see Note 2). For 1 L, weight 6.05 g Tris-Base, 8.77 g NaCl, 0.73 g $CaCl_2$ and dissolve all components in about 800 ml water. Adjust the pH with HCl to 7.5. Add 50 μl Tween 20. Adjust the volume to 1 L.
   Note that this running buffer should not be used for the immobilization of the ligands since it contains primary amino groups that will compete with the amino groups of the proteins which are used for the amine coupling. During the immobilization use a phosphate or HEPES buffer instead, without any calcium. For 1 L of 50 mM HEPES, 150 mM NaCl, pH 7.5, weight 11.9 g HEPES (MW: 238.3), 8.77 g NaCl and dissolve all components in 800 ml water. Adjust the pH with KOH and then the volume to 1 L.
3. Coupling reagents: EDC and NHS. EDC and NHS can be purchased from BIACORE/GE Healthcare in powder forms (Amine coupling kit, Cat # BR-1000-50). 750 mg EDC (1-ethyl-3-(3-dimethylpropyl)-carboiimide) is provided and should be resuspended in 10 ml water. 115 mg *N*-hydroxysuccinimide (NHS) is provided and should be resuspended in 10 ml water as well (see Note 3).
4. Blocking solution: Ethanolamine. The ethanolamine solution pH 8.5 is also provided in a liquid form. All the solutions can be aliquoted and stored at −20°C. Frozen aliquots can be used for up to 1 year.

2.2. Regeneration Solutions

1. Solution 1: 100 mM EDTA pH 8.5. For 100 ml, weigh 3.72 EDTA-disodium (MW: 372.25) and dissolve it in 70 ml water. Adjust the pH to 8.5 with NaOH (see Note 4). EDTA is a very convenient regeneration buffer when studying calcium-dependent interactions.
2. Solution 2: 20 mM Glycine pH 2.5. Dissolve 0.15 g Glycine into 80 ml water. Adjust the pH with HCl to 2.5. Complete to 100 ml.
3. Solution 3: 100 mM HCl. Add 0.83 ml commercial HCl (37% = 12 N = 12 M) into 100 ml water.
4. Solution 4: 1 M NaCl. Weigh 2.84 g NaCl and dissolve in water. Adjust the volume to 100 ml.
5. Anti-GST-antibody. The antibody used for capturing GST-RAGE can be purchased from BIACORE/GE (Cat # 27-4577-01) (see Note 5).

2.3. Sensor Chips

1. Direct immobilization. CM5 sensor chips (BIACORE/GE) or their equivalents from other providers are covered with a dextran layer that forms a tridimensional mesh (BIACORE/GE). The selected RAGE domain is covalently attached to the pre-activated surface using amine chemistry (see Subheading 3).
2. Immobilization by capture. CM5 sensor chips or their equivalents are also used when capturing the RAGE domain with an antibody.

3. Methods

3.1. Pre-concentration

Prior to performing a covalent immobilization, perform a pre-concentration procedure that will estimate the amount of protein to use in order to obtain a desired level of immobilization. Target for 1,000 RU (see Note 6). Start a new run with a flow rate of 5 μl/min. Prepare 100 μl of domain (50 μg/ml). Inject 20 μl and observe the sensorgram. If the sensorgram reaches over 1,000 RU by injecting 20 μl, regenerate the surface with regeneration solution 3 (HCl) and repeat the injection with a lower volume. If 20 μl are not sufficient to reach 1,000 RU, multiply the injections until you reach 1,000 RU. Following this pre-concentration step, inject the regeneration solution 3 over the flow cells. The RAGE domains are not covalently bound and will be eliminated from the surface easily.

3.2. Covalent Immobilization

After estimating the amount of RAGE domain that needs to be injected over the flow cells in order to obtain 1,000 RU, prepare the coupling reagents (see Note 7). Add 30 μl EDC to 30 μl NHS. Mix properly (see Note 8). Start a new run and set the flow rate at 5 μl/min. Inject 35 μl of the EDC:NHS mixture. Inject then the appropriate volume of RAGE V domain (14,400 d) as determined from the pre-concentration experiment. Inject 35 μl of the blocking solution (ethanolamine). Following the covalent immobilization of the protein onto the surface, regenerate the surface with regeneration solution 4 (NaCl). The high salt concentration solution will eliminate all non-covalent nonspecific interactions that could have formed during the immobilization (see Note 9).

3.3. Capture of RAGE-GST with Anti-GST Antibody

The anti-GST antibody is covalently immobilized onto a CM5 type surface using amine chemistry. For the immobilization, set the flow rate to 5 μl/min. Inject 35 μl of the EDC:NHS mixture. Dilute the anti-GST antibody by adding 3 μl antibody to 77 μl Na Acetate pH 5. Inject 20 μl of the antibody solution over the surface. Inject then 35 μl of the ethanolamine solution. The surface is then regenerated for 1 min contact (5 μl) with regeneration solution 2 (Glycine).

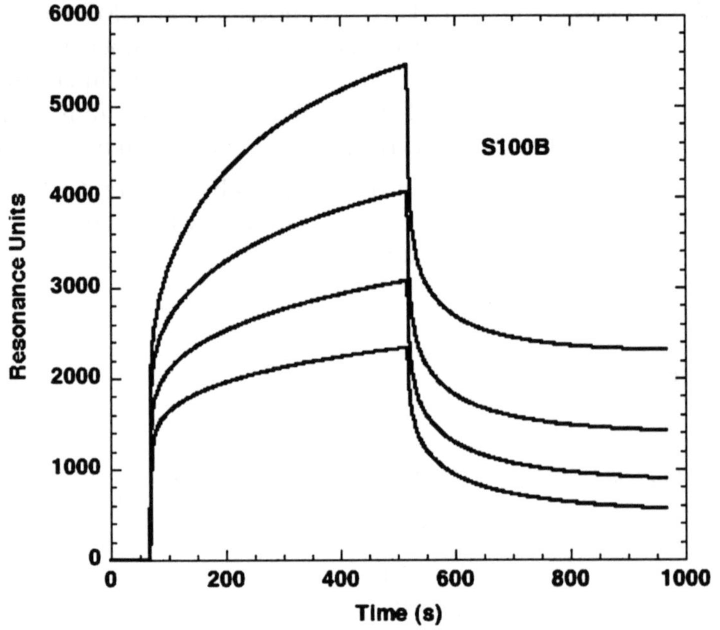

Fig. 2. Sensorgrams corresponding to the binding of S100B to directly immobilized RAGE V domain.

3.4. Samples Run with Directly Immobilized RAGE Domains

Prepare the dilution series of the analyte. The following procedure was used to obtain the sensorgrams presented in Fig. 2. Label five 1.7 ml tubes. In tube 1, prepare 360 μl of 10 μM S100B in the running buffer (see Note 10). In tubes 2–4, add 180 μl running buffer. Take 180 μl from tube 1 and add to tube 2. Mix and take 180 μl from tube 2 and add to tube 3, mix again. Repeat this procedure until tube 4 (0.5 μM). Set the tubes in the instrument. Set-up the run according to the instrument's guidelines.

For BIACORE/GE 3000, inject the sample for 3 min at 50 μl/min, followed by 5 min dissociation time. Add the regeneration step, 2 times 1 min contact with the regeneration solution 2 (EDTA), repeat this cycle with the next sample. The analyte samples should be injected from the lowest to the highest concentration to avoid carrying over between samples. Injection of running buffer only can be inserted between samples to serve as negative controls. Samples can be duplicated and serve as reproducibility controls. An example of a program for BIACORE/GE 3000 is presented in Table 2a. One can estimate the signal intensity of the binding using the following formula (Eq. 1):

$$R_{max} = \text{immobilized RU} \times S \times MW_{Analyte} / MW_{ligand}, \quad (1)$$

where R_{max} is the maximal RU signal obtained upon binding of the analyte to the ligand. Immobilized RU is the amount of ligand on the surface. S is the stoichiometry of the binding between the analyte and the ligand. If one analyte binds two ligands, S = 0.5. MW is the molecular weight of either the ligand or the analyte

Table 2
Examples of programs for a BIACORE/GE 3000 instrument. (A) Program for measuring the binding of S100 to directly immobilized RAGE domain. (B) Program for measuring the binding of S100 to GST-RAGE captured onto the surface using the anti-GST antibody

Program A: Binding of S100 to directly immobilized V domain						
The S100 samples at the indicated concentration (u = micromolar) are in positions r1a1 to r1a5. The flow rate (FLOW) is set-up at 50 µl/min. The regeneration solution 1 (EDTA) is at position r2f3. At the end of the run, the instrument is automatically set on stand-by mode with a flow rate of 5 µl/min. This program is designed to run five samples. More samples can be easily added by adding a new line in the program. The program that was used to measure the binding of S100B to RAGE V domain had only four samples.						
DEFINE LOOP cycle						
LPARAM	%sample	%position	%volume	%name	%conc	
TIMES 1						
		sample1	r1a1	150	S100	0.62u
		sample2	r1a2	150	S100	1.25u
		sample3	r1a3	150	S100	2.5u
		sample4	r1a5	150	S100	5u
		sample5	r1a5	150	S100	10u
END						
DEFINE APROG analyze						
PARAM	%sample	%position	%volume	%name	%conc	
KEYWORD	sample	%sample				
Flow	50					
* INJECT		%position	%volume			
	-0:15	RPOINT	baseline	–b		
	2:50	RPOINT	plateau			
	6:00	INJECT	r2f3	50		
MAIN						
	FLOW	50				
	DETECTION	1,2,3,4				
	LOOP	cycle	STEP			
	APROG	%sample	%position	%volume	%name	%conc
	ENDLOOP					
	APPEND standby					
END						

(continued)

Table 2 (continued)

Program B: Binding of S100 to GST-RAGE captured by anti-GST antibody					
The S100 samples at the indicated concentration (u = micromolar) are in positions r1a1 to r1a5. The flow rate (FLOW) is set-up at 50 μl/min. The GST-RAGE solution is in position r1d1. The regeneration solution 2 (Glycine) is at position r2f4. At the end of the run, the instrument is automatically set on stand-by mode with a flow rate of 5 μl/min. The program that was used to run the binding of S100A1 and S100A6 to GST-RAGE had only four samples. The protein concentrations used are slightly different than those presented here (*See* Subheading 3.5).					
DEFINE LOOP cycle					
LPARAM	%sample	%position	%volume	%name	%conc
TIMES 1					
	sample1	r1a1	150	S100	0.62u
	sample2	r1a2	150	S100	1.25u
	sample3	r1a3	150	S100	2.5u
	sample4	r1a5	150	S100	5u
	sample5	r1a5	150	S100	10u
END					
DEFINE APROG analyze					
PARAM	%sample	%position	%volume	%name	%conc
KEYWORD	sample	%sample			
FLOW	50				
	* INJECT	%position	%volume		
	-0:15	RPOINT	baseline	-b	
	2:50	RPOINT	plateau		
	8:00	INJECT	r2f3	50	
	INJECT	r2f4	50		
END					
DEFINE APROG CAPTURE					
	FLOW	50			
	FLOW PATH	1,2,3,4			
	INJECT	r1d1	150		
MAIN					
	FLOW	50			
	DETECTION	1,2,3,4			

(continued)

Table 2 (continued)

	LOOP cycle	STEP				
	APROG CAPTURE					
	APROG	%sample	%position	%volume	%name	%conc
	ENDLOOP					
	APPEND standby					
END						

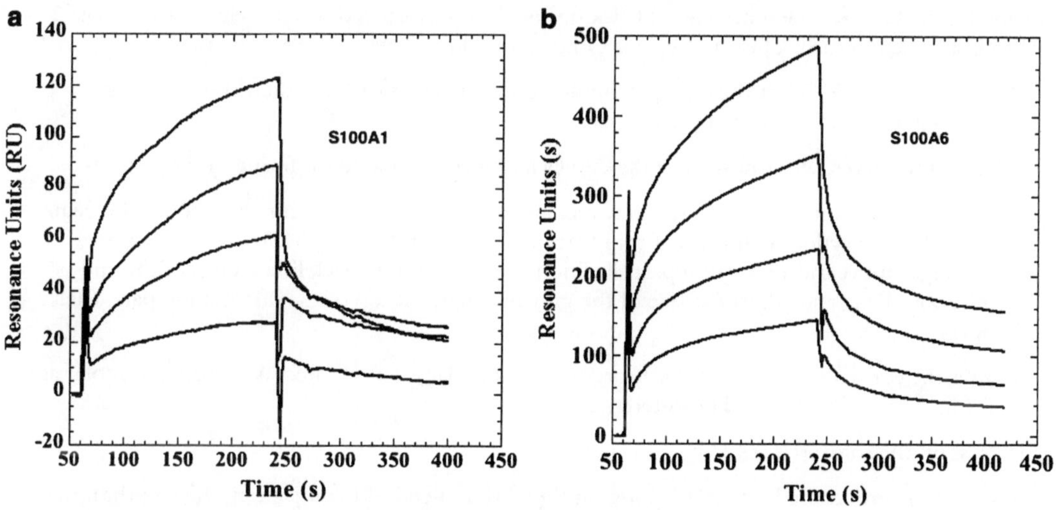

Fig. 3. Sensorgrams corresponding to the binding of S100A1 (**a**) and S100A6 (**b**) to GST-RAGE captured by anti-GST antibody. Only the binding of S100A1 and S100A6 to bound GST-RAGE is shown. About 3,000 RU of the anti-GST antibody were immobilized onto the surface. Injection of GST-RAGE onto the surface, not shown here, resulted in a further increase of about 500 RU onto the surface. For comparison, binding of GST-RAGE onto the reference surface resulted in an increase of only 20 RU.

3.5. Sample Run with Captured GST-RAGE

The following procedures were used to obtain the sensorgrams presented in Fig. 3a, b. Label five 1.7 ml tubes. In tube 1, prepare 360 μl of 11 μM S100A1 in the running buffer. In tubes 2–4, add 180 μl running buffer. Take 180 μl from tube 1 and add to tube 2. Mix and take 180 μl from tube 2 and add to tube 3, mix again. Repeat this procedure until tube 4 (1.47 μM). For S100A6, prepare 360 μl of 23 μM solution in the running buffer. Add 180 μl from tube 1 to tube 2 that contains 180 μl running buffer. Mix and take 180 μl from tube 2 and add to tube 3, mix again. Repeat this procedure until tube 4 (2.87 μM). After each analyte injection, the analyte and the capture GST-RAGE is detached from the surface using regeneration solution 2 (Glycine). Therefore GST-RAGE must be reinjected prior each analyte solution. Prepare 0.8 ml solution of

Table 3
Analysis of the sensorgrams using BIAevaluation software on BIACORE/GE 3000 instrument

1. Under File menu, open the sensorgrams to be fitted (the file has .blr extension). Select curve only. Click ok to select the highlighted curves
2. Click on the "plot overlay" icon: the curves appear in the window
3. With the right button of the mouse, click on the beginning of the regeneration part and drag the black box that appears until the end of the sensorgram, select the cut option in the edit menu
4. Repeat this procedure to define a small part of the baseline, before the association part of the sensorgram, click on the "Y-transform" icon and select "Zero at average of selection." Click on "Replace original" or "add as new." The curves are normalized
5. Subtract the blank or reference curve. Click on the "Y-transform icon", and select curve-curve 2 (blank subtraction run). Click on Replace Original or add as new
6. Click on the fit icon. Select the option "simultaneously kinetics ka/kd", then choose the kinetic model of choice and follow the instructions:
(a) Adjust the curves, select next since the curves have been prepared for fitting in the steps 3–5
(b) Two thin black lines and two large gray rectangles appear on top the curves. Choose the beginning of the injection with the first thin black line. Drag and adjust the size of the first gray rectangle above the association part. Position the second thin black line on top of the end of injection. Drag and adjust the size of the gray rectangle on top of the dissociation part. Click next
(c) Enter the concentration for each sensorgram: nanomolar is abbreviated as n and micromolar as u, for instance, 25 nM would be entered 25n
(d) Press fit: the software tries to give the best fit
(e) Repeat the procedure by slightly changing the start and end of the injection, the length of the association and dissociation part
(f) All the fitting parameters are given in the window below the fit. Click on the parameters menu (under the windows) to see how good the fit is. Small residues (<1% of the signal) are indicative of a good fit

50 μg/ml GST-RAGE (61,000 d) in running buffer. Prepare the analyte solution as described in Subheading 3.4. An example of a program for BIACORE/GE 3000 is presented in Table 2b.

3.6. Data Analysis

The analysis of the sensorgram is performed using the data analysis software provided by BIACORE/GE. Several fitting models are proposed where the association and dissociation phases are analyzed simultaneously. It is also possible to analyze the association and dissociation phases separately to estimate the dissociation and association rate when the curves do not fit any model. We will not discuss the different models available to the research in this protocol, we will only briefly describe the models used to analyze the data presented. A quick procedure to analyze the data is shown in Table 3.

1. S100B to directly immobilized V domain:

   S100B binding to V domain could not be fitted with the 1:1 Langmuir model. We used an alternative model described as parallel reaction model for heterogeneous ligand. This model follows Eq. 2 described below:

$$A + B1 \leftrightarrow AB1 \text{ and } A + B2 \leftrightarrow AB2 \quad (2)$$

   In this Eq. 2, one analyte molecule (A = S100B) can bind independently two ligand molecules B1 or B2 (B = V domain) with K_d1 the dissociation constant of the first equilibrium, and K_d2 the corresponding parameter for the second event. This analysis gave two binding constants of different value: $K_d1 = 550$ nM and $K_d2 = 470$ nM (see more details in refs. 8, 9).

2. S100A1 to capture RAGE:

   S100A1 binding to GST-RAGE could be best fitted with a Langmuir 1:1 model and showed an affinity of $K_d = 2.5$ μM.

3. S100A6 to capture RAGE:

   Binding of S100A6 to GST-RAGE resulted in sensorgrams that could also be best fitted with the parallel reaction model for heterogeneous ligand. The fit of the data gave $K_{D1} = 4.5$ μM for about 72% of the population and $K_{D2} = 3.5$ μM for about 28% of the population.

3.7. Conclusion

We present here data of binding of S100 proteins to two isolated domains of RAGE. The smaller V domain (14,400 d) and the larger domain fused with GST GST-RAGE (61,000 d). We show that different strategies can be employed to study the interaction of S100 proteins and RAGE: direct immobilization of the domain or capture through an anti-GST antibody. Other strategies are also available. For instance, it is possible to first biotinylate the V domain and then immobilize it onto a streptavidin covered surface. These strategies can be used for studying the binding of the S100 proteins to other target proteins. Finally, it is important to note that these studies are all facilitated by the calcium-dependent nature of the interaction, allowing the use of mild conditions (EDTA) for the regeneration of the surface.

4. Notes

1. A quick and efficient way of degassing a solution is to filter the solution with the help of a vacuum pump. This procedure filters the solution as well.
2. Do not use a phosphate buffer with calcium ions. Calcium will form insoluble calcium phosphate complexes.

3. Although GE Healthcare provides EDC, NHC, and ethanolamine in a convenient format, it is less expensive to purchase larger amount of the compounds and to weigh the corresponding amounts.
4. EDTA has a poor solubility in water and is very acidic. Adding NaOH will help the dissolution.
5. Although the anti-GST antibody provided by BIACORE/GE is expensive, we recommend to purchase it from them. Indeed, we found that the bound GST-RAGE dissociated very slowly from the anti-GST antibody immobilized on the surface. Other "cheaper" antibodies might dissociate too quickly from the GST moiety.
6. Although the guidelines provided by BIACORE/GE recommend to have a low level (≈100 RU) of immobilized protein on the surface of the sensor chip in order to perform accurate kinetic studies, we did not observed significant binding of the analyte (S100B) to the ligand (RAGE V domain) unless we reached an immobilization threshold of 1,000 RU. We hypothesize that small size oligomers such as dimers must be formed at the surface of the sensor chips for allowing binding of S100B.
7. When using the BIACORE/GE 3000 instrument that allows four flow cells, the protein of interest should be immobilized on flow cells 2, 3, and 4. Flow cell 1 should be kept as reference cell. The surface of flow cell 1 should be activated with EDC:NHS in the same conditions as the other flow cells. Following the activation, two options are considered: (1) the surface is immediately blocked with ethanolamine. (2) An irrelevant protein is injected over the surface, such as bovine serum albumin. We chose the first option in the experiments shown here.
8. When preparing small volumes of solutions, make sure that there is no bubble trapped at the bottom of the tube. Indeed, in this case, the needle that is used to aspirate the solutions needs to dip into some liquid. Also, prepare the EDC:NHS mixture just prior the immobilization. The mixture looses efficiency after 30 min.
9. This step is very important. If not performed, there will be a large change in RU following the first injection of the analyte.
10. The samples should be prepared from a high concentration stock (>100 μM). If the analyte is not available at high concentration, the sample should be dialyzed against the running buffer before use to avoid large "jumps" in the sensorgram dues to changes in the refractive index of the buffer (bulk effect).

References

1. Berridge MJ, Lipp P, Bootman MD (2000) The versatility and universality of calcium signalling. Nat Rev Mol Cell Biol 1:11–21
2. Heizmann CW, Krebs J, Moss SE (2004) Calcium signal transduction and cellular control mechanisms. Biochim Biophys Acta 1742:1–206
3. Chazin WJ (2011) Relating form and function of EF-hand calcium binding proteins. Acc Chem Res 44:171–179
4. Kawasaki H, Nakayama S, Kretsinger RH (1998) Classification and evolution of EF-hand proteins. Biometals 11:277–295
5. Leclerc E, Stürchler E, Heizmann CW (2009) Calcium regulation by EF-hand proteins in the brain. In: Mikoshiba K (ed) Handbook of neurochemistry and molecular neurobiology. Springer, New York, pp 509–532
6. Bernstein HG, Blazejczyk M, Rudka T, Gundelfinger ED, Dobrowolny H, Bogerts B et al (2005) The Alzheimer disease-related calcium-binding protein Calmyrin is present in human forebrain with an altered distribution in Alzheimer's as compared to normal ageing brains. Neuropathol Appl Neurobiol 31:314–324
7. Donato R (2003) Intracellular and extracellular roles of S100 proteins. Microsc Res Tech 60:540–551
8. Dattilo BM, Fritz G, Leclerc E, Kooi CW, Heizmann CW, Chazin WJ (2007) The extracellular region of the receptor for advanced glycation end products is composed of two independent structural units. Biochemistry 46:6957–6970
9. Ostendorp T, Leclerc E, Galichet A, Koch M, Demling N, Weigle B et al (2007) Structural and functional insights into RAGE activation by multimeric S100B. EMBO J 26:3868–3878

Chapter 14

Phage Display Selection of Peptides that Target Calcium-Binding Proteins

Stefan W. Vetter

Abstract

Phage display allows to rapidly identify peptide sequences with binding affinity towards target proteins, for example, calcium-binding proteins (CBPs). Phage technology allows screening of 10^9 or more independent peptide sequences and can identify CBP binding peptides within 2 weeks. Adjusting of screening conditions allows selecting CBPs binding peptides that are either calcium-dependent or independent. Obtained peptide sequences can be used to identify CBP target proteins based on sequence homology or to quickly obtain peptide-based CBP inhibitors to modulate CBP-target interactions.

The protocol described here uses a commercially available phage display library, in which random 12-mer peptides are displayed on filamentous M13 phages. The library was screened against the calcium-binding protein S100B.

Key words: Phage display, Peptide library, Ph.D.-12 library, Filamentous phage M13, Calcium-binding proteins, S100B, Calcium-dependent peptide binding, Phage ELISA, EF-hand

1. Introduction

Phage display describes a methodology to produce large libraries of genetically engineered bacteriophages and to screen these phage libraries based on phenotypical properties (1–3). Because phage particles contain the genome that encodes the proteins that make up the phage capsid, a physical link between phenotype (displayed engineered capsid protein) and genotype (engineered phage capsid protein gene) is established (4–6). A key feature of phage display is the ability to exponentially amplify a single phage particle in *E. coli* and to produce very large numbers of phages (>10^{15}) within a few milliliters of bacterial growth culture within hours.

Claus W. Heizmann (ed.), *Calcium-Binding Proteins and RAGE: From Structural Basics to Clinical Applications*, Methods in Molecular Biology, vol. 963, DOI 10.1007/978-1-62703-230-8_14, © Springer Science+Business Media New York 2013

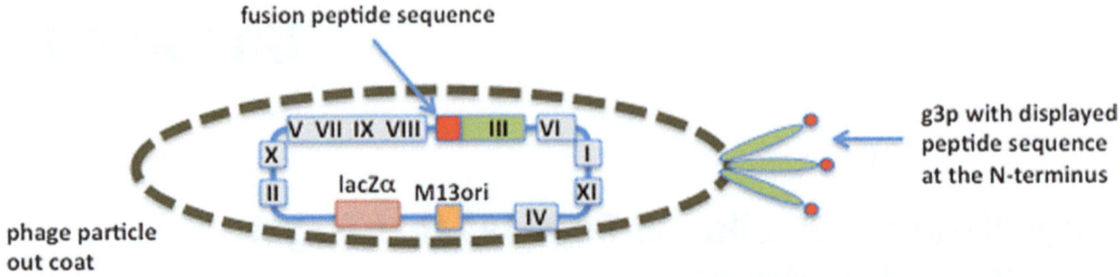

Fig. 1. Schematic presentation of a Ph.D. display phage based on M13KE. The M13 genome encodes 11 proteins (I-XI), of which five are incorporated into the phage coat (III, VI, VII, VIII, IX) and six are required for phage assembly. The display peptides (***red box***) are fused to the N-terminus of gene III (or g3) (***green box***). The M13 ori (***orange box***) ensure efficient packaging of the genome into the phage capsid. M13KE contains also the gene encoding lacZα fragment and thus allows for blue/white screening. The g3p protein is displayed at the tip of the phage particle in three to five copies, with the N-terminal fusion peptide being accessible from the bulk solvent.

The biology of phage display is very robust and withstands screening conditions that cannot be used with other display technologies, such as bacterial display (7–9), yeast display (10–12), or ribosome display (13–15).

Filamentous phage M13 is most frequently used for phage display. The genome of M13 encodes eleven proteins, of which five are incorporated into the phage coat (Fig. 1). The tube like body of the phage is formed by protein g8p (gVIIIp), while the tips of the phage body are made up of g3p and g6p at one end and g7p and g9p at the other end. A mature M13 phage contains about 2,700 copies of g8p, but only three to five copies of g3p. In order to display a nonnative peptide sequence on the surface of the phage it is necessary to engineer a fusion protein in which the desired peptide sequence is fused to one of the coat proteins. Depending on the size of the displayed peptide, such fusions may interfere with phage assembly, phage stability, or phage infectivity. The most commonly used fusion is a fusion to the N-terminus of g3p (16, 17). This protein contains a leader sequence, which directs export of the protein to the periplasmic space, where it is then proteolytically removed. Export to the periplasm allows formation of disulfide bonds within displayed proteins and peptides. Because functional g3p is needed for phage infectivity only short peptide sequences can be displayed on all g3p copies of a phage. Peptide sequences shorter than approximately 50 residues can be displayed in all copies of g3p (18). The display of longer peptide sequences on g3p requires the presence of both, engineered g3p-fusion protein and wild-type g3p, in order for the phage to remain infective. Such systems have been developed for the display of antibody fragments and other proteins and are generally phagemid based (19–22).

Commercial, premade phage display libraries are available and we use the Ph.D.-12 library from New England Biolabs in this protocol.

This library is part of a library system of three libraries (Ph.D.-12, Ph.D.-7, and Ph.D.-C7C), cloning vectors for custom library construction (23), and a vector for transfer of selected peptide sequences to monovalent display on maltose-binding protein (24).

The NEB Ph.D.-12 library displays a random 12-residue peptide, Ph.D.-7 displays a random 7-residue peptide, and Ph.D.-C7C displays a random 7-residues peptide that is conformation restrained by a disulfide bridge formed between two cysteine residues flanking the randomized heptapeptide. The peptide sequences are cloned between the g3p leader peptide sequence and the N-terminal start of mature g3p. Therefore, the library peptides are displayed as N-terminal fusions to g3p. The first residue of the displayed peptide is the N-terminus of mature g3p incorporated into the phage capsid. All copies of g3p present at the tip of the phage display the same peptide sequence, resulting in multivalent display of the peptide (Fig. 1). Multivalent display can increase effective binding affinity due to avidity effects and also be beneficial for the binding to homo-dimeric targets, such as S100B.

The Ph.D. libraries purchased from New England Biolabs should be used without in-house amplification prior to the first round of affinity panning to avoid a reduction of library diversity.

The principle of affinity-based selection (or panning) is simple: display phages are allowed to bind to an immobilized target protein and nonbinding phages are removed by repeated washing. The phages that did bind tightly to the immobilized protein are then released (eluted) using suitable conditions. Eluted phages are used to infect *E. coli* cells, which will subsequently produce hundreds of copies of the phage that originally infected the bacterial cells. The amplified phages are then isolated from the cell culture supernatant and used in the next round of affinity selection (Fig. 2). Typically three rounds or panning are sufficient to isolate high affinity binding peptides. Certain control experiments are necessary to perform to ensure that the selective pressure is directed towards isolating high affinity binding peptide specific to the target protein. M13KE phages have a tendency to eliminate additional genetic sequences and to revert to wild-type g3p. In addition, contamination of experiments with wild-type phage can occur and overgrow display phages quickly. In addition, the immobilization of the target protein, blocking nonspecific binding and conditions used for washing and elution of phages are important.

Finally, it is important to realize that the Ph.D.-7 library can provide almost complete coverage of the sequence space, whereas the Ph.D.-12 only can cover a fraction of the sequence space. The construction of phage libraries usually limits the number of independent clones that can be obtained after transformation of the vector into bacterial cells to about 10^9. A random 7mer peptide allows for $20^7 = 1.28 \times 10^9$ different sequences and can be fully represented by the Ph.D.-7 libraries. However, a random 12mer

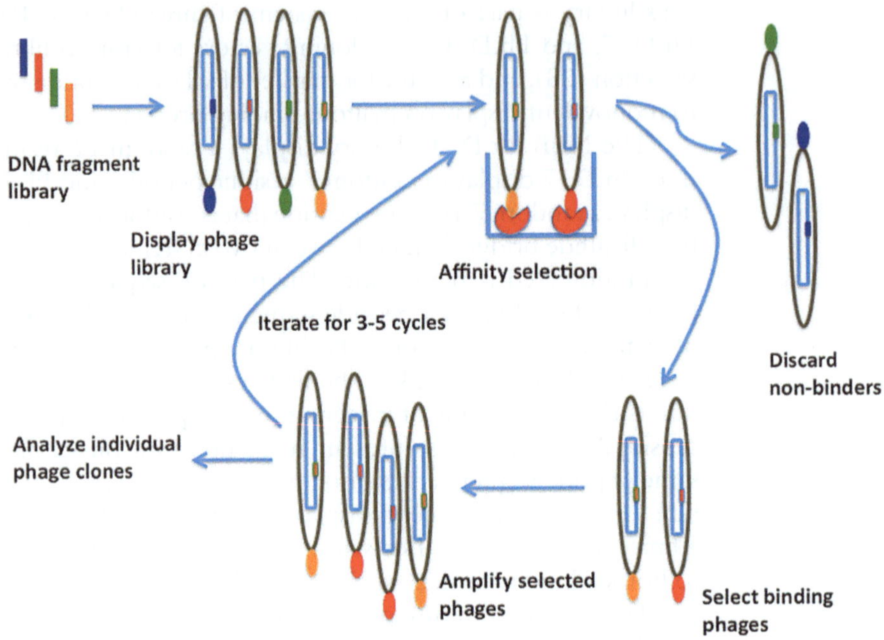

Fig. 2. The principle of phage display library panning. A gene library (***colored bars top left***) is cloned into the engineered M13KE vector and phage particles are produced, each displaying a unique peptide sequence fused to g3p at the tip of the phage (***colored elipsoids***). A minute fraction of the displayed peptides (***red and orange***) has binding affinity to an immobilized target protein and will remain bound to the target, while nonbinding phages (***blue and green***) can be washed out. Once released from the target protein, the binding peptides can be amplified and screened again in a second round of affinity selection (panning). After several rounds of panning, individual phage clones are analyzed for binding and the sequence of the displayed peptide is decoded by DNA sequencing.

peptide allows for $20^{12} = 4.096 \times 10^{15}$ sequences and thus only a minute fraction (less than 1 millionth) of sequences can be represented in the phage display library. However, while specific sequences will be absent from the Ph.D.-12 library, phenotypically dominant amino acid substitution in each position of the peptide can be identified. The Ph.D. libraries have been successfully screened against many targets and applications including epitope mapping (25–29), small molecule binders (30–34), and mapping of protein–protein interactions (35–39).

2. Materials

This protocol uses the Ph.D.-12 phage display peptide libraries available from New England Biolabs (NEB). The protocol can be used without modification for the other NEB phage display libraries (Ph.D.-7, Ph.D.-C7C). Commercial phage display libraries obtained from other vendors might need adjustment of the protocol.

Phagemid-based libraries need substantial changes to the protocol, particularly in regards to phage amplification.

1. Ph.D.-12 library kit (New England Biolabs Prod# E8110S).
2. *E. coli* strain ER2738 (supplied together with NEB-E8110S kit).
3. LB-Medium (Lennox): 10 g/L tryptone, 5 g/L yeast extract, 5 g/L NaCl, autoclave, store at ambient temperature.
4. 1,000× IPTG stock: Dissolve IPTG (isopropyl-β-d-1-thiogalactopyranoside, FW: 238.3 g/mol) in water to a final concentration of 200 mM (~48 mg/ml), sterile filter and store at −20°C.
5. 1,000×Xgal stock: Dissolve Xgal (5-bromo-4-chloro-3-indoyl-β-D-galactoside, FW: 408.6 g/mol) in DMF to a final concentration of 100 mM (~41 mg/ml). Store at −20°C.
6. 1,000× Tetracycline stock: Dissolve tetracycline in 75% ethanol to a final concentration of 20 mg/ml. Store at −20°C.
7. LB–Tet Plates: To 200 ml of LB add 3 g of agar, add a stir bar, autoclave, and let cool down with stirring until the flask can be touched (<60°C). Add 200 μl of the 1,000× tetracycline stock solution, stir and pour into 100×15 mm petridishes. 200 ml should yield about 10–15 plates.
8. LB/IPTG/Xgal plates: To 1 L of LB add 15 g agar, add a big stir bar, autoclave, and let cool down while stirring until the flask can be touched with bare hands (<60°C). Add 1 ml of the 1,000× IPTG, and 1 ml of 1,000× Xgal stock solution, mix and pour into 100×15 mm petridishes. 1 L should yield about 60–70 plates.
9. Top Agar: To 500 ml of LB add 3.5 g of agar, autoclave and dispense into 45 ml aliquots in 50 ml centrifuge tubes. Store at ambient temperature. To melt, heat in a microwave or boiling water, then keep liquid in 45–50°C water bath.
10. Immobilization buffer: 100 mM sodium bicarbonate ($NaHCO_3$, FW: 84.0 g/mol) in water, pH 8.6.
11. Tris-buffered saline (TBS): 50 mM Tris–HCl, 150 mM sodium chloride, adjust to pH 7.5 with NaOH or (HCl when using Tris-base). Autoclave and store at ambient temperature.
12. Tris-buffered saline with Tween 20 (TBST): add 0.1% Tween-20 to TBS.
13. Blocking buffer: Dissolve 50 mg/ml (5%w/v) of fat free milk powder in immobilization buffer (100 mM sodium bicarbonate, pH 8.6).
14. PEG/NaCl for phage precipitation: 20% (w:v) PEG-8000, 2.5 M NaCl. Add 20 g of polyethylene glycol (PEG) 8000 and 14.6 g of sodium chloride into an autoclavable, graduated

bottle. Add water to 100 mL final volume. The PEG most likely will not dissolve until heated (autoclaved). After autoclaving two layers will have separated, mix the layers while still warm, otherwise the PEG phase will solidify. Store at ambient temperature.

15. HRP/Anti-M13 monoclonal antibody (GE Healthcare prod# 27-9421-01).
16. HRP ELISA substrate kit using either ABTS (2,2′-azinobis (3-ethylbenzothiazoline-6-sulfonic acid)-diammonium salt) (Pierce prod# 34026) or OPD (*o*-phenylenediamine) (Pierce prod # 34006) as substrate.
17. Multiwell plate reader capable to read at 405 nm (ABTS) or 490 nm (OPD).
18. The target protein. In this example we use recombinant human S100B expressed in *E. coli.*

3. Methods

The following step-by-step protocol describes three rounds of panning performed during day 2, day 3, and day 4. If more rounds of panning are desired, simply repeat all steps described for day 3.

3.1. Protocol: Panning of Ph.D.-12 Against S100B

Day 1:

1. Streak ER2738 cells onto LB/tet agar plate (see Note 1).
2. Coat one ELISA well with 100 μg/ml S100B in immobilization buffer 0.1 M $NaHCO_3$, pH 8.6, incubate at 4°C overnight (see Note 2).

Day 2 (panning round 1):

1. Block the S100B-coated ELISA well (1.2.) with 5% milk in TBST for 3 h at room temperature (see Note 3). After blocking, wash with TBST containing 2 mM $CaCl_2$ (see Note 4).
2. Inoculate 2 ml LB with a single colony of ER2738 (1.1.), once cells start to grow visible, inoculate 50 ml LB medium (see Note 5).
3. Dilute the Ph.D.-12 phage stock provided by NEB to 10^{12} pfu/ml in TBS containing 2 mM $CaCl_2$.
4. Add 100 μl of diluted phages to the blocked S100B well and incubate for 30 min at room temperature (see Note 6).
5. Wash the well six times with TBST containing 2 mM $CaCl_2$ (see Note 7).
6. Elute phages by adding 100 μl TBS containing 5 mM EDTA and incubate for 30 min (see Note 8).

7. Use 10 μl of the eluted phages (2.6.) to determine the phage titer (see Note 9).
8. Use 50 μl phage eluate (2.6.) for phage amplification. Infect 20 ml of early to mid-log ER2738 cells (2.2.) and grow for 4–5 h (see Note 10).
9. Clarify the cell supernatant from (2.8.) by centrifugation at 10,000 × *g* for 20 min. Carefully remove the majority of the supernatant without disturbing the cell pellet and add 1/5 volume of 20% PEG, 5 M NaCl for precipitation (see Note 11), keep at 4°C overnight.
10. Coat one ELISA well with 100 μg/ml S100B as in (same as Day 2; 1.).

Day 3 (panning round 2):

1. Block the S100B-coated ELISA well (2.10.) with 5% milk in TBS (same as 2.1.)
2. Inoculate 2 ml LB with a single colony of ER2738 (1.1.), once cells start to grow visible, inoculate 50 ml LB medium (same in day 2; 2.).
3. Isolate amplified phages that were precipitated overnight (2.9.) by centrifugation at 10,000 g for 10 min. Decant the supernatant carefully, resuspend the pellet in 1 ml of TBS. Transfer to a microcentrifuge tube and centrifuge at maximal speed for 5 min and transfer the supernatant into a clean 2 ml centrifuge tube. Add 1/5 volume of PEG/NaCl, mix and incubate on ice for 10 min. Pellet the phages by centrifugation and resuspend in 200 μl TBS. This is your stock of amplified phages (see Note 12). Resuspend in TBS and determine phage count by UV spectrometry and dilute to 10^{12} phages/ml in TBS, 2 mM $CaCl_2$ (see Note 13).
4. Check the phage titer plates from (2.7.) and determine the number of phages eluted in panning round 1. Ensure that almost all plaques are blue (see Note 14).
5. Add 100 μl of diluted amplified phages (3.3.) to the blocked S100B well and incubate for 30 min (same as in day 2; 4.).
6. Wash the well six to ten times with TBST, 2 mM $CaCl_2$ (same as in 2.5.).
7. Elute phages by adding 100 μl TBS containing 5 mM EDTA and incubate for 30 min (same as in day 2; 6.).
8. Use 10 μl of the eluted phages (3.7.) to determine the phage titer (same as in day 2; 7.).
9. Use 50 μl of phage eluate (3.7.) to infect 20 ml of early to mid-log ER2738 cells (3.2.). Let cell grow for 4–5 h. (same as in day 2; 8.).
10. Clarify the cell supernatant from (3.9.) by centrifugation and add PEG/NaCl for precipitation, keep at 4°C overnight (same as in day 2; 9.).

11. Coat one ELISA well with 100 μg/ml S100B (same as in day 1; 2.).

Day 4 (panning round 3):

1. Block the S100B-coated ELISA well (3.11.) with 5% milk in TBS (same as in day 2; 1.).
2. Inoculate 2 ml LB with a single colony of ER2738 (1.1.), once cells start to grow visible, inoculate 50 ml LB medium (same as in day 2; 2.).
3. Isolate amplified phages (3.10.), resuspend in TBS and determine phage count by UV spectrometry and dilute to 10^{12} phages/ml (same as in day 3; 3.).
4. Check the phage titer plates from the second round of panning (day 3; 8.) to determine the number of phages elute in panning round 2. Ensure that the vast majority of plaques are blue (same as in day 3; 4.).
5. Add 100 μl of diluted phages to the blocked S100B well and incubate for 20 min (same as in day 2; 4.).
6. Wash the well ten times with TBST, 2 mM $CaCl_2$ (same as in day 2; 5.).
7. Elute phages with 100 μl TBS, 5 mM EDTA for 10 min (same as in day 2; 6.).
8. Use 10 μl of the eluted phages (day 4; 7.) to determine the phage titer (same as in day 2; 7.).
9. Store the remaining eluted phages at 4°C (see Note15).

Day 5 (amplifying individual phage clones):

1. Inoculate 2 ml LB with a single colony of ER2738 (1.1.), once cells start to grow visible, inoculate 100 ml LB medium (same as day 2; 2.).
2. Check the phage titer plates of the plates from the third round of panning (4.8.) and determine the number of phages eluted in panning 3. Ensure that the vast majority of plaques is blue (same as day 3; 4.).
3. Choose a phage titer plate that has well separated phage plaques. Pick 48 single, blue colonies off the titer plates (4.8) (see Note 16) and amplify single phage clones in 2 ml LB for 5 h (see Note 17). Spin samples down briefly, transfer supernatant to fresh tube and centrifuge at maximum speed for 5 min. Remove the upper 80% of the supernatant. These are your amplified phage clones.
4. Precipitate amplified phages from clarified supernatants (day 5; 3.) by adding 1/5 volume PEG/NaCl for precipitation, keep at 4°C until next day (same as day 2; 9.).

5. Coat 48 wells of an ELISA plate with 100 μl S100B (100 μg/ml) (same as day 1; 2.).

Day 6 (Phage ELISA; Note 18):

1. Block all wells of the ELISA plate (5.5) with 5% milk (same as day 2; 1.).
2. Isolate the precipitated phages (5.4), resuspend in 200 μl TBS and measure phage count by UV spectrometry, adjust to 10^{12} phages/ml (same as day 3; 3.).
3. Add 100 μl of each of the 48 single clone phage solutions (day 6; 2.) to an ELISA plate well coated with S100B and blocked and one well only blocked, but not coated with S100B. Allow phages to bind for 30 min.
4. Remove the phages and wash six times with TBST containing 2 mM $CaCl_2$.
5. Dilute the HRP/Anti-M13 antibody 1: 5,000 in TBS and add 100 μl to each well of the plate. Incubate for 30 min and wash the plate six times with TBST.
6. Prepare 20 ml of ODP substrate according the manufacturer's recommendations.
7. Add 200 μl ODP substrate solution to each well and allow the HRP reaction to proceed until a suitable color intensity has developed (see Note 19).
8. Measure the absorbance in each well with a plate reader at 490 nm.
9. Analyze ELISA data and identify phage clones that show high absorbance in S100B-coated wells and low absorbance in wells not coated with S100B (see Note 20).
10. Inoculate 2 ml LB with a single colony of ER2738 (same as day 2; 2.).

Day 7 (ssDNA preparation for sequencing):

1. Dilute the ER2738 overnight culture (day 6; 10) 100-fold and let grow for about 2 h until OD_{600} between 0.1 and 0.5.
2. Use the single phage clones that were identified in phage ELISA as having specific binding activity for S100B (day 6; 9). Use these phages to infect 3–5 ml of ER2738 cells. Let cells grow for 5 h (see Note 21).
3. Use a commercial kit to isolate ssDNA isolation from M13 and isolate ssDNA, determine ssDNA concentration by UV (see Note 22).
4. Submit ssDNA samples for sequencing using the -96gIII sequencing primer (see Note 23).

3.2. Next Steps: Analyze the Returned DNA Sequences

1. Check the quality of your sequencing data by looking at the sequencing chromatograms in the region of the peptide insert and eliminate all sequences that suggest multiple sequences in the region of the peptide insert (see Note 24).
2. Align the DNA sequences and identify the region that represents the inserted peptide sequence.
3. Identify and flag identical nucleotide sequences that were picked up multiple times (see Note 25).
4. Reverse translate these region into amino acid sequences.
5. Analyze amino acid distribution in corresponding positions of the peptide. Identify identical and homologous amino acids in corresponding positions. Identify sequences with identical amino acid sequence and analyze codon usage in identical positions (see Note 26).
6. Deduce one or several consensus peptide sequences either manually or with the help of bioinformatic tools (recently reviewed by Huang et al. (40)).

3.3. Possible Further Experiments

Monovalent display of selected peptide on MBP.

In the Ph.D. phage display system, three to five copies of peptide are displayed on the tip of the phage. This can lead to avidity effects when measuring binding affinities. The NEB system allows straightforward cloning of the inserted peptide sequence to the N-terminus of maltose-binding protein (MBP) (24). The peptide–MBP fusion protein is usually highly expressed and highly soluble and ensures monovalent display of the peptide. It is thus suitable for determination of binding affinities by ELISA, fluorescence polarization, surface plasmon resonance, or any other technique.

Chemically synthesis selected peptides

Selected peptide sequences can be synthesized chemically for affinity measurements or for use as peptide-based inhibitors.

4. Notes

1. Strain maintenance

   The recommended strain ER2738 provided together with the NEB Ph.D.—phage library kit should be used. The F′ strain F′ *proA*+*B*+ *lacI*^{q} *Δ(lacZ)M15 zzf::Tn10*(Tet^R)*/fhuA2 glnV Δ(lac-proAB) thi-1 Δ(hsdS-mcrB)5* grows rapidly and is well suited for M13 amplification. Other F′ strains (e.g., XL-1 blue, TG1) may be suitable as well, but should not be the first choice.

   M13 phages only infect F+ *E. coli* cells, ER2738 cells contain an F-factor that also carries a tetracycline resistance gene. Therefore,

ER2837 cells should be grown on selective plates containing 20 μg/ml tetracycline. Tetracycline can be included in liquid cultures as well, but it is not necessary. For long-term storage, freeze ER2738 cells as glycerol stocks (14% glycerol).

2. Immobilizing target protein

   Any protocol used for immobilizing antigens for ELISA assays can be used to immobilize target proteins for phage display affinity panning. It is convenient to use 96-well, high-binding ELISA plates. Plates wells should be sealed with sealant tape (clear packaging tape works fine too) to prevent evaporation.

3. Blocking plate wells

   Other blocking buffers, including 3% bovine serum albumin (BSA) in TBS or PBS, with or without 0.1% Tween 20, or any of several commercially available ELISA blocking buffers may also be used.

   After overnight immobilization, the target protein is removed with a pipettor and the wells are washed three times with TBS (see also Note 6). To block, fill the wells to the top with blocking solution, do not seal the plate. After blocking, wash the plate six times with TBST to remove the blocking buffer proteins.

4. Washing plate wells

   Wells are being wash by filling the well entirely with washing buffer (e.g., TBST) and then flicking the plate upside down into a sink. The time between filling the well and emptying it can be varied when phages are being eluted. After the bulk of the wash buffer is poured out, the plate is padded onto a stack of dry paper towels. It is important to remove liquid completely from the well and it is sometimes advisable to let the plates rest on the paper towels for a few seconds to allow drainage of residual fluid from the plate. Plates are usually padded onto the paper towels about five times. Do not pad onto wet areas of the paper towel stack.

5. Preparing ER2738 cells for infection with phages

   Tetracycline does not need to be included in liquid cultures. ER2738 cultures for infection with phages are needed either for phage amplification after rounds of panning (usually in the afternoon) or for the isolation of single phage clones (usually in the morning). ER2738 cells ready for phage infection in the afternoon can be grown from a single colony on an LB-agar plate. Use a sterile pipettor tip to pick a single colony off a LB-agar-Tet plate and drop the tip into a 16 × 160 mm glass tube containing 2 ml LB. Incubate the culture on a rotating wheel or shaking incubator at 37°C for several hours until an OD_{600} between 0.1 and 0.5. To have fresh ER2738 culture ready for infection in the morning it is possible to dilute an overnight culture 100-fold (200 μl into 20 ml LB in a 100 ml sterile flask on a shaker). The culture will have grown to OD_{600} between 0.1 and 0.5 within about 2 h.

6. Affinity binding of phages to immobilized target protein

   100 μl of phages are added to the blocked well containing the target protein and allowed to incubate at room temperature with mild rocking motion (if possible). The length of the time allowed for phages to bind to the plate is a parameter in the panning protocol and typical incubation times vary between 5 min and 1 h.

7. Washing after phage binding

   The conditions used for washing after binding of phages determines the stringency of the panning procedure. Parameters that should be considered are the composition of the wash buffer (e.g., calcium vs. no calcium, detergent concentration, ionic strength, etc.), as well as the time allowed between filling the well with wash buffer and removing it (the longer the incubation time during washes, the more phages will dissociate from the target). Of course, the number of washes is critical too. Typical conditions use a few seconds between filling the well and emptying it out and approximately six to ten wash steps. The stringency of the wash protocol can be increased in subsequent rounds of panning.

8. Eluting bound phages

   Addition of EDTA to the elution buffer can be used to selectively release phages that display peptides with calcium-dependent binding to CBPs. Phages displaying calcium-independent binders will remain bound to the plate (including phages bound to blocking proteins).

   If calcium-independent binding peptides are desired, then EDTA should be included in the washing buffer. Calcium-independent binding phages can be eluted under acidic conditions (0.1 M glycine, pH 2.2). After elution of the phages, the pH needs to be adjusted back to neutral by adding 1 M Tris–HCl pH 9.

9. Phage titer

   Linearity between the number of infectious phages and plaques obtained requires that more cells are available for infection than phages. Therefore, phages must be serially diluted and then added to cells for infection in order to determine the number of plaque forming units (pfu) correctly.

9.1 Start an ER2738 culture from a plate and incubate with shaking until OD600 ~0.5 has been reached.

9.2. Pre-warm LB/IPTG/Xgal plates at 37°C for 1 h or longer.

9.3. Melt top-agar in a hot water bath (or microwave) and then maintain liquid at 45–48°C.

9.4. Prepare four to six serial tenfold dilution of phages to be tittered in LB. Dilution can be done in 1.5 ml microcentrifuge tubes using 1 ml LB.

9.5. Dispense 3 ml top agar into sterile glass tubes and keep liquid at 45°C.

9.6. Dispense 200 µl of ER2738 cell in mid-log phase into microcentrifuge tubes

9.7. Infect the cells from step 5 with 10 µl of each of the dilution from step 4, vortex and let sit at ambient temperature for 5 min. Do infections in dublicate.

9.8. Transfer all 210 µl of cells from step 7 into the 45–48°C warm top agar, vortex and poor immediately onto a pre-warmed LB/IPTG/Xgal plate. Tilt the plate to spread the top-agar evenly.

9.9. Once the top agar has solidified, invert plates and incubate at 37°C overnight.

9.10. Count phage plaques on plates that have approximately 100 plaques. Plaques will appear blue.

9.11. Calculate the plaque forming units (pfu) per 10 µl of original stock solution of phages using the appropriate dilution factor.

9.12. Amplified phages (*see* below) usually have 10^8–10^{11} pfu/µl; typical panning eluates have 10–10^4 pfu/µl.

10. Phage amplification

Phages eluted in an affinity capture experiment need to be amplified. The amplified phages will be used in the next round of panning.

Grow 20 ml of ER2738 in a 100 ml flask to early or mid log growth phase (OD_{600} <0.5). Add the phages and let grow for 4–5 h. It is important to have more cells then phages, otherwise some phages will not be able to infect cells and will not be amplified. As a rough estimate, an *E. coli* culture with an OD_{600} of 1 has a cell density close to 10^9 cells/ml. Do not grow the culture for longer then 5 h. The Ph.D. phage display system has a tendency to be genetically instable and eliminate the inserted peptide sequence from the phage genome, thus producing phages that do not display the correct peptide sequences any longer.

11. Precipitating amplified phages

Phage particles are isolated from the cell culture supernatant by precipitation with PEG and NaCl. The higher the phage dilution, the longer the time needed for effective precipitation. Allow the phages to precipitate from cell culture supernatants to for at least 2 h on ice, or better, let the sample sit at 4°C overnight.

12. Phage isolation

    Addition of PEG/NaCl to the resuspended phage pellet should result in visible clouding. The phage pellet should be white. Contamination with cell debris can be removed by repeating the PEG/NaCl precipitation step.
    The phage stock can be stored at 4°C for several weeks.

13. Phage count by UV

    The UV absorbance of phage preparation is dominated by the ssDNA absorbance and can be used as a quick way to approximate the phage concentration. The approximate absorbance coefficient of wild Fd and M13 phages per nucleotide is $\varepsilon_{269} = 6750 \pm 300\ M^{-1}\ cm^{-1}$ (41). This can be used to estimate the number of phages using the following formula:

$$\text{phages} / l = \frac{E_{269} * N}{b * \varepsilon_{269}}$$

    E = measured absorbance at 269 nm

    N = Avogadro number = $6.022 \times 10^{23}\ mol^{-1}$

    b = number of base pairs in the packaged phage genome = 7,325 for Ph.D.-12

    ε = average molar extinction coefficient per base, use $\varepsilon_{269} = 6{,}750\ M^{-1}\ cm^{-1}$

$$\text{phages} / ml = \frac{E_{269} * 6.022 * 10^{23}}{7325 * 6750 * 1000} \approx E_{269} * 1.2 * 10^{13}$$

14. Checking for wild-type phage contamination
    Phage plaques should be blue. Wild-type phages will result in clear plaques.

    The M13KE phage genome has been engineered to include the lacZα-peptide, thus allowing blue/white screening of phage-infected colonies or plaques. However, these alterations also reduce infectivity of the phages and increase the time needed for replication. Wild-type M13 phages will overgrow M13KE phages quickly once a contamination has occurred. Wild-type phage lack the lacZα-peptide and result in colorless colonies/plaques, which are often also larger and less well defined at the edges than M13KE-infected colonies/plaques. Contamination of display phage with wild-type phage can be monitored by plating on LB/ITPG/Xgal plates, were blue colonies indicate display phage-infected colonies.

    If infection with wild-type phages has occurred or is suspected it is necessary to thoroughly clean all contaminated equipment and to sanitize and to discard all infected reagents. Incubators should be cleaned with 70% ethanol, diluted bleach, or heat sterilized is possible. Reagents and media should be sanitized with diluted bleach prior to disposal.

Media bottles and flasks should be heat sterilized in a drying oven at 150–200°C overnight if possible or sanitized with bleach. Pipettors should be cleaned, soaked in 70% ethanol or isopropanol, or autoclaved.

To prevent wild-type phage contaminations from the beginning it is important to work cleanly, to sanitize the work bench areas, and to discard ER2738 liquid cultures promptly. Bacterial liquid cultures should be sanitized with bleach and discarded. Plates containing ER2738 cells should be discard as soon as they are no longer needed. All plates should be sealed tightly with parafilm at all times during storage. Stacks of parafilm sealed plates from phage titer determinations should be taped to avoid plate lids from falling off and discard promptly. Biological and biohazards waste containers should be emptied routinely to eliminate them as a potential source of contaminations. Also, to prevent cross-contaminations and spreading wild-type phages all pipette tips should be aerosol resistant filter tips.

15. Number of panning rounds

    Additional round of panning can be performed, however, usually three round of panning yield sufficient enrichment of target-specific peptides. Eluted phages can be stored as 50% glycerol stock at −20°C for several years and can be retrieved for amplification is necessary.

16. How many clones to pick

    Processing 48 samples for phage isolation is time consuming and it might be easier to practice with fewer samples initially. The total number of clones depends on the results of the phage ELISA (*see* below).

17. Picking phage clones

    The output plate from panning round 3 of panning should not be incubated for longer than 18 h at 37°C and then stored at 4°C for not longer than 3 days. With a sterile pipette tip, pick into the center of a blue plaque and drop the tip into 2 ml of LB medium in a sterile tube.

18. Phage ELISA

    Phage ELISA are used to screen individual phage clones for binding to the target protein. A phage ELISA is similar to a regular ELISA in which the primary antibody is replaced by the display phage and the secondary antibody is a labeled antibody directed against g8p of the M13 coat. A good anti-M13 antibody labeled with horseradish peroxidase (HRP) is available from GE Healthcare (HRP/Anti-M13 monoclonal conjugate prod # 27-9421-01). Typically, a phage ELISA is done after the last round of affinity selection. It is also possible to perform phage ELISA on amplified phage pools obtained after each round of panning.

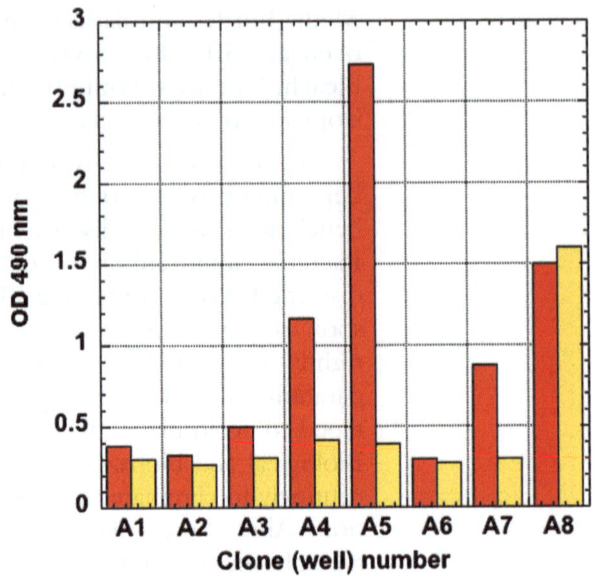

Fig. 3. Example results from a phage ELISA assay. OPD was used as HRP substrate and absorbance reading for clones A1 to A8 are shown. The two bars for each clone correspond to the absorbance reading in wells coated with S100B and blocked (***left bar, red***) and wells blocked without coating of S100B (***right bars***, ***yellow***). Clones A1, A2, A3, and A6 do not show specific binding of phages for S100B. Clones A4, A5, and A7 show better binding to wells coated with S100B then to wells without S100B. Clone A5 shows the best binding properties for S100B. Clone A8 is an example of phages that bind equally well to wells coated with S100B or without S100B. This suggests binding of the phages to the blocking protein.

19. HRP substrates in ELISA

    ODP is more sensitive as an HRP substrate compared to ABTS. If color development is too quick, or background is too high wash the plate a few times with TBST, 2 mM $CaCl_2$ and apply the HRP substrate again. Alternatively repeat the phage ELISA and increase washing steps. Color development can be stopped by adding 100 μl of a 1% SDS solution (ABTS substrate) or 50 μl 2.5 M sulfuric acid (OPD substrate).

20. ELISA screen hit-rate

    It is normal to have a certain number of phage clones that do not show specific binding affinity for the target protein (low absorbance reading in S100B-coated wells) or that show binding to the protein used for blocking, or the plastic of the ELISA plate (high absorbance in S100B-coated and no-coated wells). The percentage of phage clones that show specific binding can vary between 20 and 80%. Figure 3 shows an example of typical phage ELISA results.

21. Growing many ER2738 cultures in parallel

    It is convenient to grow cell in heat sterilized glass tube (16 × 160 mm) on a rotating wheel placed in a 37°C incubator.

Table 1
Analysis of selected phages: DNA sequencing results

Sequencing outcome	Primers used	Number of sequences
Sequences reliable in the randomized region	-96gIII -26gIII	15 0
Sequences not reliable in the randomized region	-96gIII -28gIII	2 9
Sequences without read	-96gIII -28gIII	4 1
Sequences indicating multiple sequences in the randomized region	-96gIII -28gIII	4 included in "not reliable"
Sequences without randomized insert	-96gIII	1
Total number of clones sequenced	-96gIII -28gIII	26 10

36 samples of single-stranded phage DNA were prepared (see day 7) and sequenced using either sequencing primers, -28gIII and -96gIII. Sequences that reliably covered the region of the displayed peptide were only obtained with primer -96gIII. Sequences obtained with -26gIII started very close to the displayed peptide coding region and were considered not-reliable. A total of five sequences gave no usable sequencing trace and four sequences indicated multiple sequences in the region encoding the displayed peptide. One sequence was found to not contain the display peptide sequence inserted

However, it is not advised to use more than 2.5 ml of medium per tube to ensure sufficient aeration.

22. Isolation of M13 ssDNA

    The kit from Qiagen (Qiaprep Spin M13Kit prod# 27704) works well, but products from other companies are acceptable too. Yields for ssDNA from M13 are generally lower then yields expected for high copy number plasmids isolated from an equal volume of cell culture. Also ssDNA yields can vary two- to tenfold between phage clones.

    Determine the concentration of isolated ssDNA by UV spectrometry (OD_{260} of 1 = 33 ng/μl) or the agarose gel electrophoresis. Yields are frequently in the 20–100 ng/μl range when using 2.5 ml culture and eluting ssDNA in 50 μl elution buffer.

23. Sequencing of M13 ssDNA

    Depending on the sequencing technology, sequencing of ssDNA requires less template (100 ng) than sequencing of double-stranded plasmids. M13 phages contain the (+)DNA strand. New England Biolabs provides two reverse primers, -28gIII and -96gIII. Depending on the sequencing instrument, the -28gIII primers can lead to incomplete reads within the randomized peptide region (see Tables 1 and 2). The -96gIII primer yields better sequencing results because the regions flanking the randomized sequence are readable.

Table 2
Analysis of selected phages: Amino acid sequences

		Frequency
Sequences reliable in the randomized region obtained using primer -96gIII		
>Seq1	NVKKLLFAIPLVVPFYSHS HLNILSTLWKYR GGGSAETVESCLAKSHTENS	(6/36)
>Seq2	NVKKLLFAIPLVVPFYSHS KLLTLSNWPNIR GGGSAETVESCLAKSHTENS	(4/36)
>Seq3	NVKKLLFAIPLVVPFYSHS HLPTSSLFDTTH GGGSAETVESCLAKSHTENS	(2/36)
>Seq4	NVKKLLFAIPLVVPFYSHS HLRLQDALPTRP GGGSAETVESCLAKSHTENS	(2/36)
>Seq5	NVKKLLFAIPLVVPFYSHS HLSWRDLQVQSY GGGSAETVESCLAKSHTENS	(1/36)
Sequences starting within or very close the randomized region obtained using primer -28gIII		
>Seq6	NVKKLLFAIPLVVPFYSHS HLPQNFTKLSP-	(1/36)
>Seq7	NVKKLLFAIPLVVPFYSHS HLPQNFTKLS--	(1/36)
>Seq8	NVKKLLFAIPLVVPFYSHS HLPQNFTKL---	(1/36)
>Seq9	NVKKLLFAIPLVVPFYSHS GSSHRLSWSQ--	(1/36)
>Seq10	NVKKLLFAIPLVVPFYSHS GHSQFASPSTAL	(1/36)
Sequences not reliable in the randomized region obtained using primer -96gIII		
>Seq11	NVKKLLFAIPLVVPFYSHS WH*GKISPLQVF	(1/36)

Alignment of peptide sequences obtained after sequencing 36 clones after three rounds of panning the Ph.D.12 library against S100B. The region underlined in yellow is part of the leader peptide sequence and removed in the periplasmic space. The region in green represents the sequences from the random 12-mer peptide library and peptides fused to the N-terminal part of g3p (blue) via a short GGG linker

Using primer -96gIII, 15 sequences were obtained that allow to reliably deduce the amino acid sequence of the displayed peptide. Within this group of five distinct sequences, some occur multiple times. For example, Seq1 was found six times, whereas sequence 5 was found only once among the total of 36 samples sequenced

Using primer -28gIII sequences staring within or very close to the displayed peptide coding region were obtained. Sequences 6, 7, and 8 seem to fit the consensus pattern seen in sequences obtained with primer -96gIII. However, sequences 9 and 10 do not fit this pattern and do not represent reliable sequences. Sequence 11 is a case in which a stop codon (*) appears within the sequence. Sequences that cannot be considered reliable should not be used when deducing a consensus sequences

The majority of reliable sequences (11/16) shows the His-Leu dipeptide sequence at the N-terminus, which suggests strong phenotypic selection for this sequence. Further analysis of codon usage revealed that while all His residues were encoded by CAT, the leucine residues were encoded by CTT as well as by CTG codon

The sequences isolated in this screen are in agreement with, but not identical to, previously reported S100B binding peptide consensus sequences of the format +OXO*XOO (+: positive charge residue, O: hydrophobic residue, X: variable residue, *: hydrophilic residue) (42)

24. Contaminated DNA sequences

    It can happen that a phage clone is contaminated with a second (minor) phage clone. Such contaminations are easy to recognize in the sequencing chromatograms but will be missed when exclusively relying on the sequence file in text format. Figure 4 shows an example of a contaminated DNA sequence.

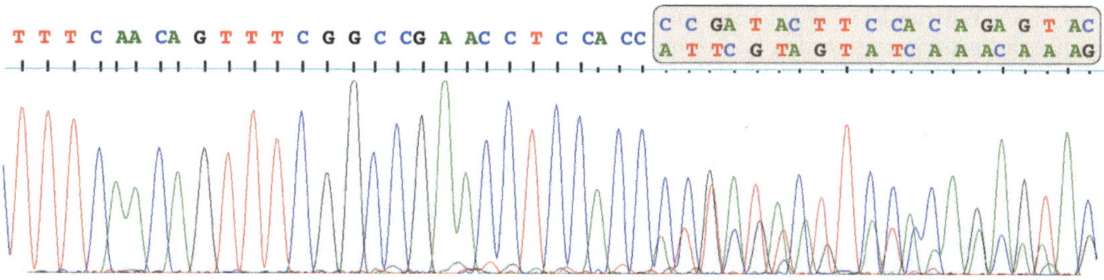

Fig. 4. Sequencing chromatogram of a phage clone contaminated with a second clone. A unambiguous sequence can be read initially, however, then a second sequence appears (highlighted area) and it becomes impossible to unambiguously assign a nucleotide sequence. For example, the first three nucleotides pairs (C or A), (C or T), (G or T) in the highlighted area can form 2 x 2 x 2 = 16 possible codons, encoding either of proline, threonine, leucine, methionine, or isoleucine. This problem cannot be detected if only the text based sequencing results is considered. Manual checking of the sequencing chromatogram to ensure contamination free sequencing results is important.

25. Multiple identical DNA sequences

    Multiple phage clones with identical nucleotide sequence originate from the same single phage clone. This clone may be represented multiple times because of superior infectivity, faster replication times compared to other clones, or possibly because of tighter binding to the target protein. If the same nucleotide sequence is picked up multiple times, one should consider that the phage clone has overgrown other binding clones at some stage of the panning and amplification scheme.

26. Codon usage

    Phenotypic selection pressure will select homologous or identical peptide sequences, independent of codon usage. In other words, finding the same amino acid sequence to be encoded by different gene sequences is very strong evidence for successful phenotypic selection. Similarly, homologous amino acid substitutions indicate phenotypic selection (see Table 2). Finding the same nucleotide sequence multiple times may indicative of overrepresentation of one dominant clone.

References

1. Katz BA (1997) Structural and mechanistic determinants of affinity and specificity of ligands discovered or engineered by phage display. Annu Rev Biophys Biomol Struct 26:27–45
2. Smith GP, Petrenko VA (1997) Phage display. Chem Rev 97:391–410
3. Burritt JB, Bond CW, Doss KW, Jesaitis AJ (1996) Filamentous phage display of oligopeptide libraries. Anal Biochem 238:1–13
4. Rodi DJ, Makowski L (1999) Phage-display technology—finding a needle in a vast molecular haystack. Curr Opin Biotechnol 10:87–93
5. Sidhu SS (2001) Engineering M13 for phage display. Biomol Eng 18:57–63
6. Sidhu SS, Koide S (2007) Phage display for engineering and analyzing protein interaction interfaces. Curr Opin Struct Biol 17:481–487
7. Samuelson P, Gunneriusson E, Nygren PA, Stahl S (2002) Display of proteins on bacteria. J Biotechnol 96:129–154
8. Hansson M, Samuelson P, Gunneriusson E, Stahl S (2001) Surface display on gram positive bacteria. Comb Chem High Throughput Screen 4:171–184
9. Desvaux M, Dumas E, Chafsey I, Hebraud M (2006) Protein cell surface display in Gram-positive bacteria: from single protein to macromolecular protein structure. FEMS Microbiol Lett 256:1–15

10. Gai SA, Wittrup KD (2007) Yeast surface display for protein engineering and characterization. Curr Opin Struct Biol 17:467–473
11. Pepper LR, Cho YK, Boder ET, Shusta EV (2008) A decade of yeast surface display technology: where are we now? Comb Chem High Throughput Screen 11:127–134
12. Feldhaus MJ, Siegel RW (2004) Yeast display of antibody fragments: a discovery and characterization platform. J Immunol Methods 290:69–80
13. Schaffitzel C, Hanes J, Jermutus L, Pluckthun A (1999) Ribosom display: an in vitro method for selection and evolution of antibodies from libraries. J Immunol Methods 231:119–135
14. Pluckthun A, Amstutz P, Forrer P, Zahnd C (2001) In vitro display technologies: novel developments and applications. Curr Opin Biotechnol 12:400–405
15. Pluckthun A (2012) Ribosome display: a perspective. Methods Mol Biol 805:3–28
16. Makowski L (1994) Phage display: structure, assembly and engineering of filamentous bacteriophage M13. Curr Opin Struct Biol 4:225–230
17. Georgieva Y, Konthur Z (2011) Design and screening of M13 phage display cDNA libraries. Molecules 16:1667–1681
18. Cwirla SE, Peters EA, Barrett RW, Dower WJ (1990) Peptides on phage: a vast library of peptides for identifying ligands. Proc Natl Acad Sci USA 87:6378–6382
19. Hust M, Dubel S (2005) Phage display vectors for the in vitro generation of human antibody fragments. Methods Mol Biol 295:71–96
20. Lowman HB, Wells JA (1991) Monovalent phage display: a method for selecting variant proteins from random libraries. Methods 3:205–216
21. Dunn IS (1996) Phage display of proteins. Curr Opin Biotechnol 7:547–553
22. Soderlind E, Simonsson AC, Borrebaeck CA (1992) Phage display technology in antibody engineering: design of phagemid vectors and in vitro maturation systems. Immunol Rev 130:109–124
23. Noren KA, Noren CJ (2001) Construction of high-complexity combinatorial phage display peptide libraries. Methods 23:169–178
24. Zwick MB, Bonnycastle LL, Noren KA, Venturini S, Leong E, Barbas CF 3rd, Noren CJ, Scott JK (1998) The maltose-binding protein as a scaffold for monovalent display of peptides derived from phage libraries. Anal Biochem 264:87–97
25. Birkenmeier G, Osman AA, Kopperschlager G, Mothes T (1997) Epitope mapping by screening of phage display libraries of a monoclonal antibody directed against the receptor binding domain of human alpha-2-macroglobulin. FEBS Lett 416:193–196
26. Dore JM, Morard F, Vita N, Wijdenes J (1998) Identification and location on syndecan-1 core protein of the epitopes of B-B2 and B-B4 monoclonal antibodies. FEBS Lett 426:67–70
27. Youn JH, Myung HJ, Liav A, Chatterjee D, Brennan PJ, Choi IH, Cho SN, Shin JS (2004) Production and characterization of peptide mimotopes of phenolic glycolipid-I of Mycobacterium leprae. FEMS Immunol Med Microbiol 41:51–57
28. Rowley MJ, O'Connor K, Wijeyewickrema L (2004) Phage display for epitope determination: a paradigm for identifying receptor-ligand interactions. Biotechnol Annu Rev 10: 151–188
29. Spillner E, Deckers S, Grunwald T, Bredehorst R (2003) Paratope-based protein identification by antibody and peptide phage display. Anal Biochem 321:96–104
30. Rozinov MN, Nolan GP (1998) Evolution of peptides that modulate the spectral qualities of bound, small-molecule fluorophores. Chem Biol 5:713–728
31. Miura Y, Sasao Y, Kamihira M, Sakaki A, Iijima S, Kobayashi K (2004) Peptides binding to a Gb3 mimic selected from a phage library. Biochim Biophys Acta 1673:131–138
32. Zhao SW, Shen PP, Zhou Y, Wei Y, Xin XB, Hua ZC (2005) Selecting peptide ligands of microcystin-LR from phage displayed random libraries. Environ Int 31:535–541
33. Pustowka A, Dietz J, Ferner J, Baumann M, Landersz M, Konigs C, Schwalbe H, Dietrich U (2003) Identification of peptide ligands for target RNA structures derived from the HIV-1 packaging signal psi by screening phage-displayed peptide libraries. Chembiochem 4: 1093–1097
34. Sugimura Y, Hosono M, Wada F, Yoshimura T, Maki M, Hitomi K (2006) Screening for the preferred substrate sequence of transglutaminase using a phage-displayed peptide library: identification of peptide substrates for TGASE 2 and Factor XIIIA. J Biol Chem 281:17699–17706
35. Dintilhac A, Bernues J (2002) HMGB1 interacts with many apparently unrelated proteins by recognizing short amino acid sequences. J Biol Chem 277:7021–7028

36. Bitto E, McKay DB (2003) The periplasmic molecular chaperone protein SurA binds a peptide motif that is characteristic of integral outer membrane proteins. J Biol Chem 278:49316–49322
37. Carter DM, Gagnon JN, Damlaj M, Mandava S, Makowski L, Rodi DJ, Pawelek PD, Coulton JW (2006) Phage display reveals multiple contact sites between FhuA, an outer membrane receptor of Escherichia coli, and TonB. J Mol Biol 357:236–251
38. Hu B, Gilkes DM, Chen J (2007) Efficient p53 activation and apoptosis by simultaneous disruption of binding to MDM2 and MDMX. Cancer Res 67:8810–8817
39. Nelson TJ, Alkon DL (2007) Protection against beta-amyloid-induced apoptosis by peptides interacting with beta-amyloid. J Biol Chem 282:31238–31249
40. Huang J, Ru B, Dai P (2011) Bioinformatics resources and tools for phage display. Molecules 16:694–709
41. Day LA (1969) Conformations of single-stranded DNA and coat protein in fd bacteriophage as revealed by ultraviolet absorption spectroscopy. J Mol Biol 39:265–277
42. IvnenkovV V, Jamieson GA, Gruenstein E, Dimlich RVW (1995) Characterization of S-100b binding epitopes. J Biol Chem 270: 14651–14658

Part IV

RAGE-Meditated Cell Signaling and Pathology

Chapter 15

RAGE-Mediated Cell Signaling

Ari Rouhiainen, Juha Kuja-Panula, Sarka Tumova, and Heikki Rauvala

Abstract

RAGE (receptor for advanced glycation end products) is a multi-ligand receptor that belongs to the immunoglobulin superfamily of transmembrane proteins. RAGE binds AGEs (advanced glycation end products), HMGB1 (high-mobility group box-1; also designated as amphoterin), members of the S100 protein family, glycosaminoglycans and amyloid β peptides. Recent studies using tools of structural biology have started to unravel common molecular patterns in the diverse set of ligands recognized by RAGE. The distal Ig domain (V1 domain) of RAGE has a positively charged patch, the geometry of which fits to anionic surfaces displayed at least in a proportion of RAGE ligands. Association of RAGE to itself, to HSPGs (heparan sulfate proteoglycans), and to Toll-like receptors in the cell membrane plays a key role in cell signaling initiated by RAGE ligation. Ligation of RAGE activates cell signaling pathways that regulate migration of several cell types. Furthermore, RAGE ligation has profound effects on the transcriptional profile of cells. RAGE signaling has been mainly studied as a pathogenetic factor of several diseases, where acute or chronic inflammation plays a role. Recent studies have suggested a physiological role for RAGE in normal lung function and in neuronal signaling.

Key words: RAGE, HSPG, Heparin, Toll-like receptors, AGE, HMGB1, Amphoterin, S100 proteins, EF-hand, Calcium, β amyloid, Inflammation

1. Introduction

RAGE (receptor for advanced glycation end products) was initially isolated as the receptor of AGEs (advanced glycation end products) that accumulate in diabetes and during senescence as a result of nonenzymatic glycation and oxidation of proteins and lipids (1). Since then, RAGE has been shown to bind a diverse set of ligands in addition to the AGEs, including HMGB1 (high-mobility group box-1; also designated as amphoterin), several members of the S100 protein family and β amyloid peptides. Currently, RAGE can be viewed as a multi-ligand or pattern recognition receptor playing

Claus W. Heizmann (ed.), *Calcium-Binding Proteins and RAGE: From Structural Basics to Clinical Applications*, Methods in Molecular Biology, vol. 963, DOI 10.1007/978-1-62703-230-8_15, © Springer Science+Business Media New York 2013

key roles in innate and adaptive immunity. The role of RAGE as a pattern recognition receptor involved in inflammatory diseases thus resembles the roles of Toll-like receptors. The goal of this review is to give an overview of the RAGE ligands, to consider how these seemingly different molecules could be recognized by RAGE, to outline the cell signaling events induced by RAGE ligation, and to summarize the known physiological and pathophysiological outcomes of RAGE signaling.

2. Structure of RAGE

A gene coding for human RAGE is located to the major histocompatibility complex in chromosome 6 (2). The RAGE gene has 11 exons and 10 short introns. RAGE has been described in mammalia but not in other species, whereas the closest homologue, ALCAM, occurs also in other species (3, 4). The regulatory 5′UTR of the RAGE gene has binding sites for nuclear factor-κB (NF-κB), specificity protein 1 (SP1), activator protein 1 (AP1, (5)), E-twenty-six 1 (Ets-1, (5)), hypoxia-inducible factor-1 (HIF-1, (6)), early growth response gene 1 (Egr-1), and thyroid transcription factor-1 (TTF-1) that regulate the transcription of the RAGE gene (7, 8). Further, additional transcription factor binding sites on the RAGE promoter have been described, including an activator protein 2 (AP-2) site, and epigenetic regulation of the RAGE gene occurs via methylation of cytosines in the AP-2 and SP-1 binding sites of the RAGE promoter (9).

The 3′UTR of the RAGE mRNA is differentially polyadenylated, which regulates stability of the mRNA (10). Further, the 3′UTR has a target site for mir-30 miRNA that is able to downregulate the RAGE mRNA expression (11). The RAGE transcript is also extensively modulated by alternative splicing to yield altered exons and introns, and thus different translation products (12). One mechanism that regulates splicing involves the binding of heterogeneous nuclear ribonucleoprotein H to G-rich cis-elements (13).

The most abundant protein form coded by the RAGE mRNA in humans consists of 404 amino acids. It has a signal sequence for secretion, three immunoglobulin domains (a V-type, a C1-type, and a C2-type domain), a transmembrane helix and a short cytoplasmic domain (12, 14). The cytoplasmic domain does not have homology to any known signaling domains (14). The extracellular domain binds several ligands, and the cytoplasmic domain mediates intracellular signal transduction (Fig. 1).

The RAGE polypeptide can be posttranslationally modified by disulfide bond formation, phosphorylation, proteolysis, and

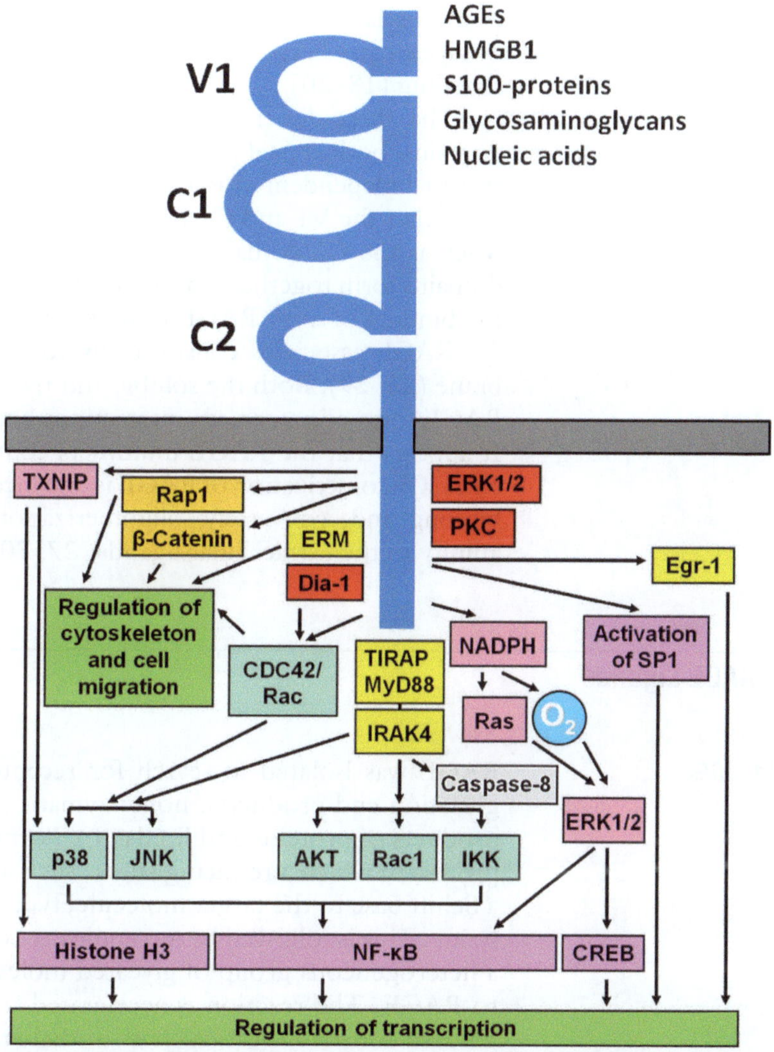

Fig. 1. RAGE extracellular ligands, cytoplasmic interaction partners, and signaling pathways. Extracellular ligands bind to the V1 domain that is the major ligand-binding domain of RAGE. Ligand binding induces cytosolic signaling cascades. Signaling partners in *red boxes* have been shown to bind directly to the RAGE cytoplasmic domain using biochemical approaches, and signaling molecules in *yellow boxes* have been shown to associate with RAGE using immunoprecipitation. For additional details, see the text.

glycosylation. The disulfide bonds are formed between the conserved cysteines within the Ig domains (14). Proteolysis of membrane-bound RAGE can produce a soluble ectodomain, a transmembrane form of the cytoplasmic domain, or a soluble form of the cytoplasmic domain. The soluble ectodomain can be localized to the extracellular space or intracellular vesicles, whereas the soluble C-terminus localizes to cytoplasm or nucleus (15–17). The transmembrane forms are predicted to be associated with membranes. Glycosylation occurs in the V1 domain that has two

N-glycosylation sites (14, 18), and modification by carboxylated *N*-glycans has been shown to promote RAGE ligand binding and signaling (18–20).

The three-dimensional structure of the RAGE extracellular domains, both crystal and solution structures, have been solved in several independent studies (4, 21–26). These structural studies show that the V1 and C1 domains form a compact unit together, whereas the C2 domain occurs as a single unit. The V1 and C1 domains form together a large cationic surface area which mediates the binding of most RAGE ligands (reviewed in refs. 27, 28).

RAGE associates constitutively to itself in the plasma membrane (25, 29). Both the soluble and the transmembrane forms of RAGE form oligomers via interactions between their V1 domains. It appears that the RAGE multimers contain four or more molecules. Oligomerization of RAGE is enhanced by extracellular ligand binding and, conversely, oligomerization increases the binding affinity to the extracellular ligands (27, 30).

3. RAGE Ligands

3.1. AGEs

RAGE was isolated in search for receptors of AGEs (advanced glycation end products), nonenzymatic glycation, and oxidation products of proteins and lipids (1). Formation of AGEs is a slow process in which a reducing sugar, such as glucose, initially forms a Schiff base to the target molecule, typically to ε-amino group of lysine. This is followed by a complex set of reactions ending up in a heterogeneous group of glycated molecules that display affinity to RAGE. The reaction is accelerated under conditions of high glucose concentration, such as in diabetes. It appears that AGE formation is also accelerated by hypoxia and ischemia/reperfusion, which is probably due to oxidative stress and tissue damage (31). AGEs and other RAGE ligands bind to the distal V1 domain of RAGE (Fig. 1). Excessive activation of RAGE by the AGEs is suggested to perturb cell functions in diseases where inflammation is a pathogenetic factor, such as in atherosclerotic complications of diabetes.

3.2. HMGB1 (Amphoterin)

Search for endogenously occurring RAGE-binding proteins from tissues led to isolation of amphoterin (32). "Amphoterin" (heparin-binding p30) refers to a Janus-faced molecule both in structure (the polycationic N-terminal part followed by a polyanionic C-terminal sequence) and cellular expression (extracellular and intracellular localization). It was initially isolated from developing brain as a heparin-binding protein that enhances neurite extension in brain neurons (33). Amphoterin turned out to have the same amino acid sequence as the DNA-binding protein HMGB1 (high-mobility group box-1; (34, 35)). The designation "HMG" was used to refer specifically to nuclear proteins, whereas "amphoterin" refers to an extracellularly acting protein. Extracellular functions of HMGB1/amphoterin are currently generally accepted within the HMG field and have been extensively studied during the last few years (reviewed in ref. (36)). Therefore, we use the currently recommended designation HMGB1 in this chapter.

HMGB1 is expressed at the extracellular side of the plasma membrane, which was initially shown by lactoperoxidase-catalyzed cell surface iodination in vitro (33). Furthermore, several anti-protein and anti-synthetic peptide antibodies were shown to detect HMGB1 in the cytoplasm and at the cell surface (35, 37). Accumulation of HMGB1 in the extracellular space in vivo was shown in plasma samples of patients with sepsis (38, 39). It appears that secretion from monocytes is a major source of extracellular HMGB1 in the context of inflammation (39, 40).

A high extracellular level of HMGB1 can be found as a consequence of cell damage, which is predictably important in pathophysiological contexts, and HMGB1 is currently included in the group of molecules designated as "DAMPs" (damage-associated molecular patterns). However, a physiologically regulated release of HMGB1 to the extracellular space also occurs upon many types of cell stimulation, such as activation by contact with the extracellular matrix (41, 42). A similar shift to a peripheral location within the cells, followed by export to the extracellular space, is induced by LPS (lipopolysaccharide) and by several cytokines (43–45). Many posttranslational modifications have been suggested to guide HMGB1 to a nonconventional secretory route, including acetylation (46), phosphorylation (47), methylation (48), and poly(ADP)-ribosylation (49). Very recently, an autophagy-based unconventional secretory pathway has been suggested that facilitates export of a subset of cytosolic proteins to the extracellular space (50). IL-1β and HMGB1 were proposed to be secreted through this novel secretion pathway.

HMGB1 binds to RAGE at nanomolar concentrations (K_D 5–10 nM) that appear reasonable to explain, at least partially, binding of the extracellularly occurring HMGB1 to the cell surface. The major RAGE-binding area of HMGB1 has been mapped to the C-terminal area of the B box (51), in particular to the V1 domain binding (Fig. 1).

Regulation of cell motility is the most widely studied consequence of HMGB1 binding to the cell surface, and has been in most, if not in all cases, ascribed to HMGB1/RAGE interactions. Neurite outgrowth is the initial form of migration response found for HMGB1 (33, 37, 52–54). Furthermore, HMGB1/RAGE has been shown to regulate migration and invasion of tumor cells in several studies (35, 41, 42, 51, 55). It also affects migration of other cell types, including endothelial cells (56–58), granulocytes (59), monocytes (40), dendritic cells (60–62), smooth muscle cells (63), and stem cells (64–66). Very recently, endothelial cell sprouting due to HMGB1/RAGE interaction was shown to require the presence of cell surface heparan sulfate as the coreceptor of RAGE (67). Whether this is the case for other migration responses induced by HMGB1/RAGE in various cell types, requires further studies.

3.3. S100 Proteins

The S100 protein family consists of small EF-hand calcium-binding proteins that interact with a diverse set of target proteins (for review, see ref. 68). Currently, more than 20 family members are known. Most S100 proteins are localized in the cytoplasm where they have important functions as calcium sensors. Many S100 proteins can be also secreted although they lack a classic-type secretion signal. As in the case of HMGB1, the mechanism of secretion is still unclear. The S100 proteins form dimers and oligomers which is important for their functions, especially for their extracellular cytokine-like functions.

RAGE was initially identified as a cell surface receptor for S100B and S100A12 (69) which can compete for RAGE binding with the AGEs and HMGB1. It appears that most, if not all, S100 proteins bind to RAGE, and RAGE is currently regarded as a general S100 receptor. The different S100 proteins bind to RAGE with varying affinities, ranging from nanomolar to micromolar K_D values. The V1 domain plays a crucial role in S100 binding, with varying contributions by the C1 and C2 domains depending on the specific S100 protein. Interactions of the different S100 proteins with RAGE have been recently reviewed in detail (70) and are not further discussed here.

Consequences of S100 binding to RAGE appear somewhat similar to those of HMGB1. Neurite outgrowth and other forms of migration responses have been documented for S100/RAGE interactions. Many S100 proteins have prominent extracellular roles in inflammation and cancer that are, at least partially, mediated through binding to RAGE.

3.4. Amyloid β

Accumulation of amyloid β in the extracellular space is a hallmark and possible pathogenetic factor in Alzheimer's disease. RAGE binds amyloid β (71) and may be involved in neurotoxic effects of the amyloid β peptides within the brain. Furthermore, RAGE has been shown to transport amyloid β from plasma through the blood–brain barrier into the brain (72), possibly contributing to accumulation in the brain tissue. Amyloid β oligomers that are suggested to perturb neuronal functions bind to the V1 domain of RAGE, whereas the amyloid β aggregates interact with the C1 domain (73). HMGB1 and S100 proteins are upregulated at sites of inflammation and may contribute, together with amyloid β, to neurodegenerative changes through binding to RAGE.

4. How Does Rage Bind Many Different Types of Ligands?

Similar to the Toll-like receptors, RAGE can be viewed as a pattern recognition receptor (74). However, no obvious common molecular patterns exist that RAGE would recognize in the diverse set of ligands that have been shown to bind to RAGE and to activate RAGE-dependent signaling cascades.

Recent structural studies have started to unravel the common molecular patterns recognized by RAGE. The unusually high contents of lysine and arginine residues form a large positively charged patch on the surface of V1 and C1 domains of RAGE (25, 29). It therefore appears conceivable that the V1-C1 Ig-domains of RAGE would bind negatively charged ligands. In AGEs, N_ε-carboxymethyl-lysine (CML) and N_ε-carboxy-ethyl-lysine (CEL) constitute two major AGE structures found in tissues. Recent studies on the solution structure of a CEL peptide-RAGE V1 domain complex have shown that the carboxyethyl moiety of CEL fits inside a positively charged cavity of the V1 domain, and the peptide backbone makes specific contacts with the V1 domain (26). The geometry found in these NMR studies is compatible with many CML/CEL-modified sites of plasma proteins.

The structural studies cited above explain how the diverse CML/CEL-patterned structures of AGEs are recognized by RAGE. HMGB1 and its different posttranslationally modified forms constitute another major set of RAGE ligands, and it seems reasonable to consider whether HMGB1 could follow the concept shown for AGE/RAGE interactions. We have previously mapped the RAGE-binding area of HMGB1 to the C-terminal area of the B box (amino acid residues 150–183) using biochemical and cell biological approaches (51). This 33-amino acid stretch contains 12 cationic and 5 anionic residues and could provide electrostatic interactions with the V1 domain. Furthermore, the 33-amino acid RAGE-binding area is followed by a 30-amino acid stretch containing only glutamate and aspartate residues. This polyanionic tail of HMGB1 could enhance HMGB1 binding to the V1-C1 domain of RAGE. In support of this model, the polyanionic tail of HMGB1 has been shown to be involved in RAGE interactions in efferocytosis assays (75). Analysis of solution structures of the HMGB peptide-V1 domain is however warranted to reveal the exact binding characteristics of HMGB1 and RAGE.

The S100 family consists of acidic proteins, and it appears likely that the highly positively charged areas at the surfaces of the RAGE V1 and C1 domains participate in their binding. Furthermore, the 33-amino acid RAGE-binding area of HMGB1 bears some sequence homology with N-terminal regions of the S100 proteins (76). Tools of structural biology are currently required to elucidate binding characteristics of the S100 proteins to RAGE and differences and similarities compared to AGE and HMGB1 binding.

As the other RAGE ligands, the amyloid β peptides bind the V1 domain (Fig. 1) with the positively charged patch being the likely binding site. Short versions of theV1 domain-binding amyloid β peptide have been recently indentified using RAGE tryptophan fluorescence and mass spectrometry analysis of non-covalent ligand-receptor complexes. These studies have mapped the RAGE-binding area of the amyloid β peptide to residues 17–23, a hydrophobic sequence flanked by anionic residues at the C-terminal end. Some similarity compared to the C-terminal sequence of the RAGE-binding area of HMGB1 was found (77). Further structural studies on ligand–receptor complexes are required to understand the exact mode of binding.

5. Signaling Pathways Activated By RAGE Ligation

Transfection studies showed that ligation of full-length RAGE induces neurite outgrowth, whereas the cytosolic domain-deleted RAGE is inactive (52), demonstrating that RAGE-activated cytosolic signaling is required for cell motility regulation, but the identity of the cytoplasmic binding partners has been unclear. The formin homology protein Diaphanous-1 (Dia-1) has been recently shown to bind directly to the cytosolic domain of RAGE (78). Furthermore, Dia-1 has been shown to activate the Rho family GTPases Cdc42 and Rac that were previously shown to be critical for the cell motility response induced by RAGE ligation (52, 76). In addition to Dia-1, ERK1/2 has been reported to use the cytosolic tail of RAGE as the docking site (79).

The RAGE/Dia/GTPase/cytoskeleton axis appears sufficient to explain the migration control by HMGB1 and possibly by other RAGE ligands in most cell types (Fig. 1). Another recently found feature in the migration control through RAGE is the role of integrin activation (56, 59, 80). RAGE ligation has been shown to induce the active conformation of the β_1 and β_2 integrins and to polarize integrin and CD44 localization for migration function. The small GTPase Rap1has been found to function as a cytosolic mediator of integrin activation due to RAGE ligation (Fig. 15.1). Very recently, β-catenin (81) and ezrin/radexin/moesin (ERM) proteins (82) have been also implicated in cytoskeletal regulation through RAGE (Fig. 1).

RAGE activates transcription through several routes (Fig. 1). RAGE-mediated activation of the transcription factor NF-κB depends on the classical mitogen-activated protein kinase (MAPK) pathway involving the small GTPase Ras and ERK1/2 (83). In addition, two other MAPK kinases, p38 MAP kinase and the stress-activated protein kinase/c-Jun-NH_2-terminal kinase (SAPK/JNK) are activated by RAGE (55). Very recently, RAGE was reported to signal to the TIRAP-MyD88 pathway (84) that is a well-known route in the signaling cascade from ligation of Toll-like receptors to NF-κB activation (Fig. 1). The finding that RAGE is able to activate the transcription factors NF-κB, CREB, and Sp1 ((7, 51), see Fig. 1) suggests that in agreement with the current literature, RAGE signaling regulates widely cell behavior.

In addition to the transcription factor pathways discussed above, RAGE has been reported to affect transcription through an epigenetic mechanism involving activation of TXNIP (thioredoxin interacting protein) that regulates chromatin remodeling (85). The RAGE–TXNIP pathways is suggested to regulate acetylation and trimethylation of histone H3 (Fig. 1), which affects chromatin structure and inflammation in diabetic retinopathy.

Sustained RAGE activation due to a positive feedback loop is a characteristic feature in RAGE signaling (86, 87). Ligation of RAGE results in activation of the transcription factor NF-κB. Since the RAGE promoter has an NF-κB binding site, this results in enhanced RAGE expression and signaling. This phenomenon is for example reflected in the expression pattern: RAGE is expressed at low levels in many tissues but accumulates on areas where its ligands are expressed.

One open question in RAGE signaling is to what extent the different RAGE ligands are able to preferably activate a certain signaling pathway. It seems reasonable that some specificity in the signaling pathway(s) activated would be imparted by the differences in ligand binding to the different domains of RAGE; however, supporting evidence is currently lacking. Furthermore, it is conceivable that the degree of oligomerization would impact the cytosolic signaling pathways, for example, by enhancing recruitment of adaptor molecules, such as Dia-1. Further work is however required to understand how the ligand binding may affect the conformation and oligomerization of RAGE, and how these changes are transmitted to the cytosolic signaling pathways (reviewed in refs. 27, 28).

6. Cooperation of RAGE and HSPGS

Some RAGE ligands as well as RAGE itself bind to strongly to heparin. In fact, heparin-binding has been exploited in the isolation of HMGB1 (33) and of RAGE (71). In binding assays, where the ectodomain of RAGE or its different domains are immobilized on ELISA wells, the V1 domain binds heparin with the same K_D as

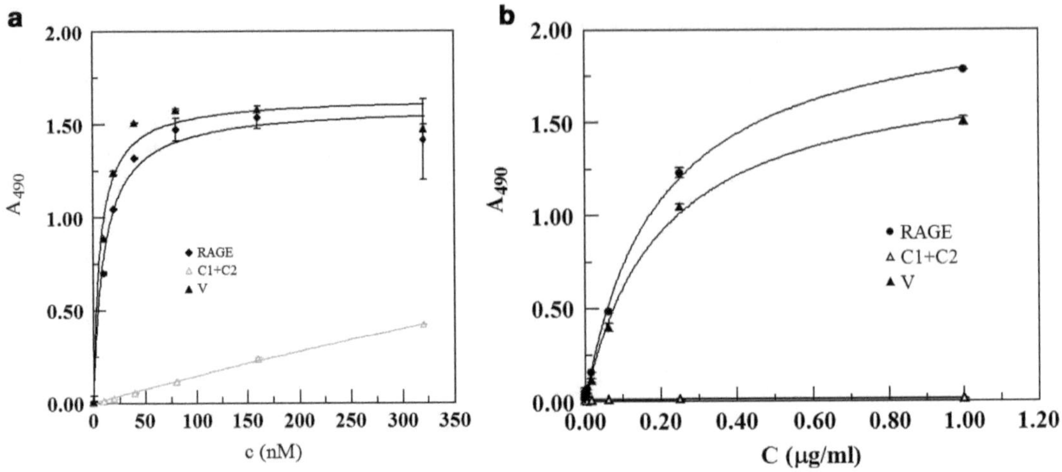

Fig. 2. Binding of amyloid β 1-42 peptide (A) and heparin-BSA (B) to RAGE and its domains. Binding assays were carried out as described (147) with minor modifications. Briefly, Fc-fusion proteins of sRAGE (40) or RAGE domains (40) in a concentration of 5 μg/ml were bound to protein A (1 μg/ml, Sigma-Aldrich, St. Louis, MO, USA) coated wells (Maxisorb ELISA plates, Nunc A/S, Roskilde, Denmark). Unbound proteins were washed away, and the wells were blocked with BSA (Sigma-Aldrich, St. Louis, MO, USA). Biotinylated amyloid β 1-42 peptide (1–320 nM, American Peptide, Sunnyvale, CA, USA; panel (**a**)) or biotinylated heparin-BSA (0–1 μg/ml, Sigma-Aldrich; panel (**b**)) were bound to the wells, and unbound ligands were washed away. Bound biotinylated proteins were detected with streptavidin-HRP (Sigma-Aldrich) and peroxidase substrate (Sigma-Aldrich), and an absorbance at 490 nm was measured. Results were analyzed using PSI-Plot (Poly Software International, Pearl River, NY, USA). The full-length ectodomain and the V1 domain display very similar binding characteristics. Both amyloid β 1-42 and heparin-BSA bound to the V domain of RAGE with $K_D = 6.9 \pm 1.6$ nM and 67 ± 7 nm, respectively. K_D values for full-length RAGE for amyloid β and heparin-BSA were 8.9 ± 1.3 nM and 67 ± 7 nm, respectively. RAGE lacking the V1 domain (C1C2 RAGE) bound to amyloid β 1-42 with $K_D = 2.2 \pm 1.1$ μM, however, it did not bind to heparin. Amyloid β 1-42 rapidly forms fibrils in an aqueous solution and the ability of C1C2 RAGE to bind amyloid β 1-42 may be due to the binding of the C1 domain of RAGE to amyloid β aggregates (73).

the whole ectodomain, whereas the ectodomain lacking the V1 domain does not display heparin-binding activity (Fig. 2). This finding is to expected based on the positively charged patch on the V1 domain. Furthermore, in similar assays heparin/heparan sulfate (HS) inhibits ligand binding to immobilized RAGE ectodomain, as demonstrated for amyloid β and heparin-BSA in Table 1. Binding of heparin to the V1 domain (Fig. 2) and inhibition of ligand binding to the ectodomain of RAGE by polyanionic carbohydrates is consistent with the idea that the positively charged patch at the surface of the V1 domain plays a key role in ligand–RAGE interactions. However, some degree of specificity is required, since other negatively charged carbohydrates including chondroitin sulfate C and colominic acid (polysialic acid, the type of sialic acid polymer found in the embryonic form of N-CAM) display little or no activity (Table 1). Interestingly, the ectodomain of the transmembrane HSPG (heparan sulfate proteoglycan) syndecan-3 (*N*-syndecan) isolated from brain (88) is as potent or even more potent inhibitor than heparin in RAGE binding. Of heparin-derived fragments,

Table 1
Inhibition of amyloid β 1-42 and heparin-BSA binding to RAGE by glycosamoniglycans and charged molecules

Compound	IC50(μg/ml) for inhibition of Aβ binding to RAGE	IC50(μg/ml) for inhibition of heparin-BSA binding to RAGE
Heparin	2	5
Syndecan-3	1	n.d.
Heparan sulfate	4	15
Enoxaparin	4	5
2-O-desulfated heparin	3	4
6-O-desulfated heparin	10	8
N-desulfated heparin	>40	>70
Chondrotin sulfate A	>40	>150
Chondrotin sulfate C	20	11
Chondrotin sulfate E	2	5
Dextran sulfate	5	5
Sucrose octasulfate	>100	>200
Colominic acid	>100	>100
Heparin dodecasaccharide	10	5
Heparin decasaccharide	12	6
Heparin hexasaccharide	45	10
Heparin tetrasaccharide	>40	14

The inhibitory effect of glycosamoniglycans and charged molecules was tested using the microwell-binding assay as described in Fig. 2. Biotinylated amyloid β 1-42 peptide (20 nM, American Peptide, Sunnyvale, CA, USA) or (B) biotinylated heparin-BSA(0.2 μg/ml, Sigma-Aldrich) were bound to wells coated with the ectodomain of RAGE in the presence or absence of inhibitors, and unbound ligands were washed away. Heparin oligosaccharides were from Iduron (Manchester, UK) and soluble Syndecan-3 was isolated as described (88). The IC50 values for different glycosaminoglycans and charged molecules were calculated using PSI-Plot. The soluble ectodomain of syndecan-3 and heparin were the best inhibitors, with the lowest IC50 values. In addition, some other sulfated oligosaccharides had an inhibitory effect. N-desulfation of heparin strongly reduced its inhibitory activity, whereas 2-O and 6-O desulfation only had milder effects. The length of the heparin oligosaccharide chain also plays a role as it directly correlated to the ability of the glycan to inhibit both amyloid β 1-42 and heparin-BSA binding to RAGE

oligosaccharides with eight to ten monosaccharide units still retain significant inhibitory activity (Table 1).

The role of HS in RAGE signaling has been recently explored using biochemical and cell biological assays in vitro (67). HMGB1-induced ERK1/2 and p38 phosphorylation were shown to be abolished when cell surface HS was removed or pharmacologically blocked, which resulted in decreased HMGB1-induced cell motility that was analyzed using endothelial cell sprouting assays.

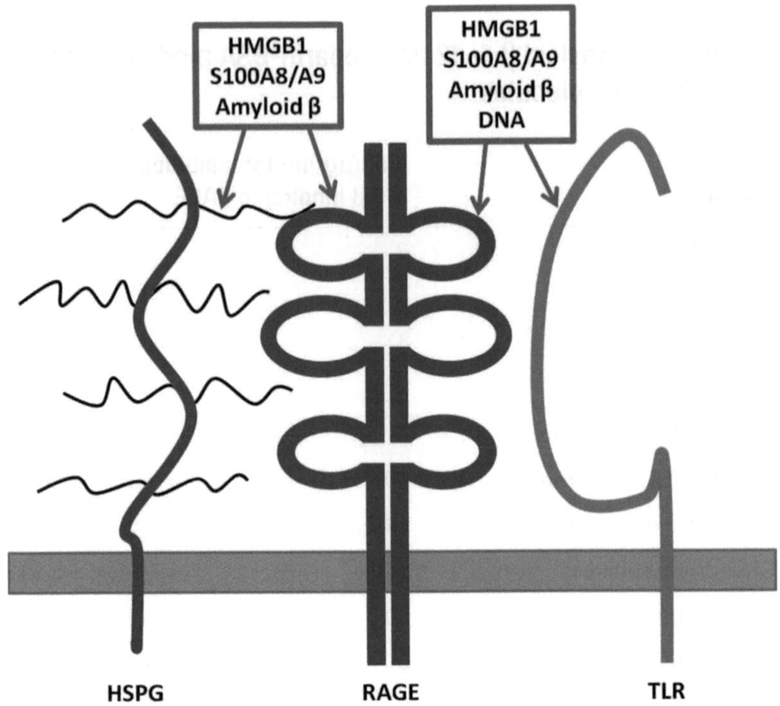

Fig. 3. RAGE associates to itself, to HSPGs, and to Toll-like receptors in the plasma membrane. RAGE, HSPGs, and TLRs share common ligands that may enhance association of the receptors. Some of the common ligands and interaction partners of RAGE, HSPGs, and TLRs are presented in the model shown in the figure. In addition to interactions mediated by soluble ligands, RAGE and HSPGs display constitutive interaction on cell surface mediated by the RAGE V1 domain and heparan sulfate side chains of membrane-associated HSPGs. RAGE-HSPG interaction is required for RAGE signaling into cells (67). In addition to RAGE, HMGB1 binds to both transmembrane (148) and matrix-associated proteoglycans (149). S100A8/A9 binds to heparan sulfate proteoglycans on the endothelial cell surface by direct interaction that is mediated by S100A9 (94). Amyloid-β binds to both matrix-associated (150) and transmembrane forms (151) of HSPGs. Further, TLR4 signaling can be induced by soluble heparan sulfates (152), and TLR4 can bind to the secreted proteoglycan biglycan that induces TLR4 signaling (153). Therefore, HSPGs may associate to TLRs in the plasma membrane in addition to their association to RAGE. RAGE–TLR complex formation induced by soluble ligands may either activate or inhibit the receptor complex (91, 95, 154). Whether amyloid-β binds directly to TLR4 is currently unknown, but a receptor complex of TLR4/TLR6/CD36 mediates amyloid-β-induced activation of macrophages (96). Cell signaling may be also affected by soluble ectodomains of the receptors. The ectodomains of HSPGs, RAGE, and TLR4 can exist in soluble forms that are still able to bind extracellular ligands (1, 149, 155).

Interestingly, loss of the HS-RAGE interaction, but not the HS-HMGB1 interaction, was shown to be essential in RAGE-signaling that promotes endothelial cell sprouting. Furthermore, HSPG and RAGE were shown to form constitutively complexes at the cell surface, which was rapidly enhanced by the addition of the ligand. In contrast to the role of the HSPGs in FGF (fibroblast growth factor) signaling, the requirement of HS appeared absolute in the sense that it could not be compensated by increasing the concentration of the ligand.

The current data suggest a model in which RAGE oligomers associate to HSPGs at the cell surface (Fig. 3). In particular, the transmembrane HSPGs, syndecans, fulfill the requirements to complex with RAGE at the cell surface. However, the current data do not exclude the possibility that the PI-linked HSPGs, glypicans, would play a similar role. Whether other RAGE ligands would also enhance formation of HSPG-RAGE complexes, is an open question. It seems probable to us that other heparin-HS-binding ligands, such as S100A/A9 and amyloid β, would act in an analogous manner to HMGB1 (Fig. 3). It would be also interesting to learn whether other cell types in addition to endothelial cells require HSPG in RAGE-dependent migratory responses and whether all or only some RAGE-dependent signaling pathways are affected by HSPGs.

7. Cooperation of RAGE and Toll-Like Receptors

In addition to RAGE, Toll-like receptors (TLR2 and TLR4) have been suggested to bind HMGB1 and to mediate its effects in inflammatory conditions (89). The proinflammatory activity of HMGB1 is strongly enhanced when HMGB1 is complexed with DNA/lipids (90, 91). This type of complexes enhance inflammation in a manner that depends on both RAGE and Toll-like receptors. Furthermore, the oxidation status of HMGB1 has been reported to strongly influence on the receptor binding and inflammatory activity (92, 93). As in the case of HSPG-RAGE cooperation, HMGB1 alone or in complex with DNA/lipids binds to both receptors and could recruit Toll-like receptors (such as TLR4) to RAGE multimers at the cell surface (Fig. 3). Other ligands displaying dual binding activity to the receptors, such as S100 proteins (94, 95), amyloid β (71, 96), and DNA fragments released upon tissue injury (23, 91) could act in a similar manner (Fig. 3). Experimental evidence of the Toll-like receptor/RAGE assemblies at the cell surface has been recently shown (95).

Besides regulation of inflammatory responses, RAGE and Toll-like receptors may cooperate in other pathophysiological mechanisms. Excessive extracellular expression of HMGB1 was recently shown to cause an amnesic effect (97). The HMGB1-induced amnesia was shown to require both RAGE and TLR4.

In addition to their cooperation as cell surface receptors, RAGE and TLR 2/4, have been recently found to cooperate in cytosolic signaling (84). Ligand binding to RAGE results in phosphorylation of Ser391 in the cytosolic moiety of the receptor by PKCζ. The phosphorylated cytosolic tail then binds TIRAP and MyD88, which are known to act as adaptor proteins for TLR2 and TLR4 and transduce the signal to downstream molecules

(Fig. 1). In this scenario, the functional interaction of RAGE and Toll-like receptors would explain why engaging either of the two receptors causes similar outcomes in various physiological and pathophysiological settings.

8. Pathophysiological Outcomes of RAGE Signaling

The signaling pathways stimulated by RAGE ligation (Fig. 1), such as the pronounced activation of NF-κB, are consistent with a role in inflammatory responses. Furthermore, many RAGE ligands, such as the S100 proteins and HMGB1, have been shown to play key roles in inflammation. Polymorphisms of the RAGE gene that have been recently described (see below) provide additional evidence for involvement in human inflammatory conditions.

The key finding suggesting a role for RAGE signaling in acute inflammation comes from search of mediators of sepsis in human (39). HMGB1 was found to accumulate at high levels in human plasma after the early secretion of proinflammatory cytokines, and neutralization of HMGB1 was reported to confer protection even at the late stage of sepsis (98). Studies using RAGE null mice in sepsis models displayed a significant protection compared to the wild-type mice, and use of soluble RAGE as a decoy receptor also lead to enhanced survival rate (99). The studies in animal models clearly implicate RAGE in injurious hyperinflammatory mechanisms and suggest that the effects of HMGB1 are, at least partially, mediated by RAGE signaling.

Stroke is another acute state where HMGB1-induced RAGE signaling has been shown to play a pathogenetic role. Ischemic injury of stroke causes tissue damage, and inflammatory mechanisms cause further injury in the penumbra around the ischemic core. Tissue injury is strongly reduced by RNAi-mediated down-regulation of HMGB1 expression (100) or by HMGB1-neutralizing antibodies (101). RAGE involvement is suggested by the finding that disruption of the RAGE gene confers protection in an animal model of stroke (102).

In addition to acute inflammatory conditions, sustained activation of RAGE has been implicated in several diseases where chronic inflammation plays a pathogenetic role, such as diabetes and its vascular complications (86, 87). In particular, chronic RAGE activation caused by the positive feedback loop in the signaling mechanism (see above) has been suggested as a pathogenetic factor in atherosclerosis and its complications (86, 87, 103, 104). Several RAGE ligands, such as the AGEs, S100 proteins and HMGB1, are likely to act as ligands inducing sustained RAGE activation in chronic inflammatory conditions.

Search for amyloid β receptors identified RAGE as a binding component and mediator of neurotoxicity of amyloid β (71, 73). Furthermore, RAGE has been reported to mediate inhibition of long-term potentiation (LTP) by amyloid β (105), suggesting a role for RAGE signaling in memory impairment in Alzheimer's disease. Furthermore, inflammation is an important component in neurodegenerative diseases where RAGE and its ligands are suggested to play important roles (28).

RAGE signaling has been also implicated in multiple sclerosis using the mouse model of EAE (experimental allergic encephalomyelitis). In these studies, S100-induced RAGE signaling has been shown to recruit damage-provoking T cells into brain (31, 106). RAGE signaling is therefore suggested to play key roles in T cell-mediated adaptive immune response.

HMGB1 is highly expressed in tumor cells, regulates their motility, localizes to leading edges of cells, and promotes plasmin-based proteolytic activity (35, 41, 42). HMGB1 has been therefore suggested to regulate invasive migration (107), which apparently depends on RAGE signaling (52). Furthermore, HMGB1-induced RAGE signaling has been found to enhance cancer progression in implanted and spontaneously occurring tumors (55). Furthermore, S100 proteins are implicated in cancer in a plethora of studies (70). In addition to regulating tumor cells, HMGB1 is suggested to be important for angiogenesis that enhances tumor spread (57, 58). Involvement of RAGE appears probably since HMGB1 induces homing of endothelial cells in a RAGE-dependent manner (56). Chronic inflammation is known to predispose to cancer, and tumor development is suggested to become enhanced due to inflammation induced by RAGE signaling (108).

9. What Is the Physiological Function of RAGE Signaling?

RAGE signaling has been extensively described as a malefactor in various diseases, as briefly outlined above. Only very few studies have addressed the physiological function and possible beneficial roles of RAGE signaling.

The RAGE knockout mice appear healthy without any overt phenotype (99, 109). Hyperactivity and increased sensitivity to auditory stimuli have been recently reported in RAGE knockout mice (110), suggesting that RAGE is required in neuronal signaling in brain. In the peripheral nervous system, RAGE appears beneficial in neurite outgrowth required for regeneration after injury, as has been shown for the sciatic nerve (53, 54). Further, bone remodeling is altered in RAGE knockout mice due to decreased bone resorption by osteoclasts (111).

RAGE is expressed at low levels in normal healthy tissues except for lung where high expression levels are detected, which would be compatible with a role in tissue homeostasis (see also Subheading 10 below). Blockade of RAGE has been reported to result in disturbed cell–matrix contacts in lung cells (112). Based on the apparent ability of RAGE to bind glycosaminoglycans via the positively charged cluster in the V1-C1 domain (see above), the role of RAGE as a receptor for HSPGs and other glycosaminoglycans present in the extracellular matrix appears reasonable and warrants further studies. Furthermore, HMGB1 binds to HSPGs, is expressed in cell matrices, and promotes cell migratory responses as a matrix-bound ligand (107).

It has become clear that RAGE-signaling regulates the innate immune system (see above), which acts as the first-line defense against foreign invaders. It seems reasonable that RAGE does not only participate in harmful chronic inflammation, but is also required for the protective function of inflammation, for example, in migration of immune cells to the sites of infection and injury.

10. Polymorphisms of RAGE

The RAGE gene has several single nucleotide polymorphisms (SNPs) that affect the structure, expression, or functions of RAGE. Studies on the SNPs of the RAGE gene have provided important clues to physiological and pathophysiological roles of RAGE signaling. We will summarize below the data reporting association of the RAGE SNPs to biological functions and diseases.

The highest expression level of RAGE in adults is in the lungs. Two genome-wide studies were recently reported associating six genetic loci to natural variation of lung function estimated by spirometric measurements (113, 114). One of these loci is RAGE (designated as AGER and AGER-PPT2). The G82S SNP of the RAGE gene associated to lung function in these studies promotes glycosylation of the second N-glycosylation site in the V1 domain; glycosylation at this site has been previously shown to affect RAGE ligand binding and signaling (18, 23, 115). RAGE thus appears to play a role in normal lung biology, perhaps affecting development/remodeling of lung (see also above, Subheading 9).

In addition to the normal variation of lung function, the RAGE polymorphism discussed above has been associated to susceptibility to develop chronic obstructive pulmonary disease (116). A polymorphism in the promoter of RAGE has been associate to lung cancer (117). Another lung disease linked to RAGE SNPs is -374T/A in the promoter region that is a risk for invasive aspergillosis (118). This SNP enhances the promoter activity of the RAGE gene in vitro and increases the expression of RAGE in monocytes in vivo (118, 119).

Since RAGE was identified as a receptor for AGEs produced in hyperglycaemic conditions, the RAGE SNPs have been intensively studied in the field of diabetes. In these studies, multiple RAGE SNPs have been associated to diabetes or diabetic complications. Two SNPs, -429T/C in the promoter and G82S that cause enhanced RAGE expression and signaling, respectively, have been described as risk factors of diabetes and diabetes related complications (119–126). However, Balasubby et al. (127) showed an opposite association of G82S with diabetic retinopathy in Indian patients.

In addition to diabetes and diabetes associated diseases, the G82S SNP is a risk factor in many other diseases, where autoimmune reactions or chronic inflammation play a significant role (128–133). One mechanism that may explain the effects of the G82S polymorphism is increased inflammation that is predicted by increased expression of proinflammatory cytokines and other inflammatory biomarkers (129, 134–136). Furthermore, plasma sRAGE levels are decreased in individuals carrying the G82S allele (133, 137–139). Soluble RAGE that is normally present in plasma is a biomarker of inflammation, for example, its levels are decreased in severe pancreatitis without organ failure (140). Whether the downregulation of plasma sRAGE in individuals carrying the G82S allele mediates inflammation or whether it is just an indicator of inflammatory reactions is currently poorly understood. The concentrations of sRAGE do not necessarily correlate with the levels of the transmembrane form of RAGE (118). Theoretically, sRAGE may act in vivo as a decoy receptor suppressing the inflammation-enhancing activity of the transmembrane RAGE. Alongside these studies, there are some reports that were unable to show any association of the G82S SNP to diseases, or even showed an opposite association (127, 141–143).

One of the first physiological ligands for RAGE, HMGB1, was isolated from brain and shown to induce neurite outgrowth in cortical neurons (33) in a RAGE-dependent manner (32). Furthermore, HMGB1 is highly expressed in brain during organogenesis and affects forebrain development (144). Of brain diseases, the RAGE G82S SNP has been described to associate to Alzheimer's disease and schizophrenia (128, 145), which may relate to the regulation of development/plasticity or inflammation in brain.

The association of the -374A/T SNP in the RAGE promoter to vascular complications was recently studied in a meta-analysis by Lu and Feng (146). The meta-analysis revealed that the allele -374A is protective for vascular complications in diabetes as suggested by earlier studies (142), whereas it is a risk factor for infections (118). Further molecular and cellular studies on the consequences of the polymorphism are required to understand these associations.

11. Concluding Remarks

Studies using tools of structural biology have recently shown that the positively charged surface in the V1 domain of RAGE binds a pattern of negatively charged residues in AGEs. The V1 domain of RAGE binds in addition several other ligands, and one key question is whether other RAGE ligands are recognized in a similar manner compared to the AGEs. It seems probable that the charged surface of the V1 domain participates in binding of most if not all RAGE ligands. However, other types of interactions specific for each ligand are likely to be found that are important to initiate signaling cascades due to RAGE ligation. Studies using approaches of structural biology, biochemistry, and cell biology will be important to understand the details of ligand binding. These studies are likely to identify novel RAGE antagonists to be tested for the treatment of diseases where RAGE-signaling plays a role.

RAGE associates to itself and to other membrane proteins, such as HSPGs and Toll-like receptors, in the plasma membrane. Such receptor assemblies are likely to have a key role in the signaling mechanism of RAGE but have been so far addressed only in very few studies. It will be important to learn what ligands promote such receptor assemblies and, vice versa, how association of RAGE to itself or other membrane receptors affects ligand binding. Further, recruitment of adaptor proteins, such as Dia-1, and the signaling pathways activated by the different ligands are likely to be affected by the configuration of the receptor assembly used by the different cell types. As the details of ligand binding to RAGE itself, understanding the receptor assemblies used in RAGE signaling is likely to pave the way for novel pharmaceutical interventions.

RAGE has been studied for many years as a key cell surface receptor involved in disease mechanisms but its physiological function has been unclear. Recent extensive genome-wide association studies on RAGE polymorphisms for lung function and the role of RAGE in maintaining normal lung cells have implicated RAGE in lung biology and respiratory health. Furthermore, variation of the RAGE gene is currently leading to novel mechanistic insights into susceptibility to chronic pulmonary diseases.

It is probable that RAGE signaling is required for physiological regulation in other tissues in addition to lung. One obvious candidate would be nervous system development/plasticity but very few studies have so far addressed the role of RAGE in the nervous system. Behavioral alterations found in the RAGE knockout mice and the roles of the RAGE ligands HMGB1and S100 proteins in brain development/plasticity provide clues to such functions that currently warrant further studies.

References

1. Schmidt AM, Vianna M, Gerlach M et al (1992) Isolation and characterization of 2 binding-proteins for advanced glycosylation end-products from bovine lung which are present on the endothelial-cell surface. J Biol Chem 267:14987–14997
2. Sugaya K, Fukagawa T, Matsumoto K et al (1994) Three genes in the human MHC class-III region near the junction with the class-II—gene for receptor of advanced glycosylation end-products, PBX2 homeobox gene and a notch homolog, human counterpart of mouse mammary-tumor gene int-3. Genomics 23:408–419
3. Bowen MA, Patel DD, Li X et al (1995) Cloning, mapping, and characterization of activated leukocyte-cell adhesion molecule (ALCAM), a CD6 ligand. J Exp Med 181: 2213–2220
4. Koch M, Chitayat S, Dattilo BM et al (2010) Structural basis for ligand recognition and activation of RAGE. Structure 18: 1342–1352
5. Tsuji A, Wakisaka N, Kondo S et al (2008) Induction of receptor for advanced glycation end products by EBV latent membrane protein 1 and its correlation with angiogenesis and cervical lymph node metastasis in nasopharyngeal carcinoma. Clin Cancer Res 14: 5368–5375
6. Pichiule P, Chavez JC, Schmidt AM et al (2007) Hypoxia-inducible factor-1 mediates neuronal expression of the receptor for advanced glycation end products following hypoxia/ischemia. J Biol Chem 282: 36330–36340
7. Li JF, Schmidt AM (1997) Characterization and functional analysis of the promoter of RAGE, the receptor for advanced glycation end products. J Biol Chem 272: 16498–16506
8. Reynolds PR, Kasteler SD, Cosio MG et al (2008) RAGE: developmental expression and positive feedback regulation by Egr-1 during cigarette smoke exposure in pulmonary epithelial cells. Am J Physiol Lung Cell Mol Physiol 294:L1094–L1101
9. Tohgi H, Utsugisawa K, Nagane Y et al (1999) Decrease with age in methylcytosines in the promoter region of receptor for advanced glycated end products (RAGE) gene in autopsy human cortex. Brain Res Mol Brain Res 65: 124–128
10. Caballero JJ, Girón MD, Vargas AM et al (2004) AU-rich elements in the mRNA 3' -untranslated region of the rat receptor for advanced glycation end products and their relevance to mRNA stability. Biochem Biophys Res Commun 319:247–255
11. Shi SL, Yu LP, Chiu C et al (2008) Podocyte-selective deletion of dicer induces proteinuria and glomerulosclerosis. J Am Soc Nephrol 19:2159–2169
12. Hudson BI, Carter AM, Harja E et al (2008) Identification, classification, and expression of RAGE gene splice variants. FASEB J 22: 1572–1580
13. Ohe K, Watanabe T, Harada S et al (2010) Regulation of alternative splicing of the receptor for advanced glycation endproducts (RAGE) through G-rich cis-elements and heterogenous nuclear ribonucleoprotein H. J Biochem 147:651–659
14. Neeper M, Schmidt AM, Brett J et al (1992) Cloning and expression of a cell-surface receptor for advanced glycosylation end-products of proteins. J Biol Chem 267:14998–15004
15. Galichet A, Weibel M, Heizmann CW (2008) Calcium-regulated intramembrane proteolysis of the RAGE receptor. Biochem Biophys Res Commun 370:1–5
16. Raucci A, Cugusi S, Antonelli A et al (2008) A soluble form of the receptor for advanced glycation endproducts (RAGE) is produced by proteolytic cleavage of the membrane-bound form by the sheddase a disintegrin and metalloprotease 10 (ADAM10). FASEB J 22: 3716–3727
17. Zhang L, Bukulin M, Kojro E et al (2008) Receptor for advanced glycation end products is subjected to protein ectodomain shedding by metalloproteinases. J Biol Chem 283: 35507–35516
18. Srikrishna G, Huttunen HJ, Johansson L et al (2002) N-glycans on the receptor for advanced glycation end products influence amphoterin binding and neurite outgrowth. J Neurochem 80:998–1008
19. Srikrishna G, Nayak J, Weigle B et al (2010) Carboxylated N-Glycans on RAGE promote S100A12 binding and signaling. J Cell Biochem 110:645–659
20. Turovskaya O, Foell D, Sinha P et al (2008) RAGE, carboxylated glycans and S100A8/A9 play essential roles in colitis-associated carcinogenesis. Carcinogenesis 29:2035–2043
21. Dattilo BM, Fritz G, Leclerc E et al (2007) The extracellular region of the receptor for advanced glycation end products is composed of two independent structural units. Biochemistry 46:6957–6970

22. Matsumoto S, Yoshida T, Murata H et al (2008) Solution structure of the variable-type domain of the receptor for advanced glycation end products: new insight into AGE-RAGE interaction. Biochemistry 47:12299–12311
23. Park H, Adsit FG, Boyington JC (2011) The 1.5 angstrom crystal structure of human receptor for advanced glycation endproducts (RAGE) ectodomains reveals unique features determining ligand binding (vol 285, pg 40762, 2010). J Biol Chem 286: 19178–19178
24. Sárkány Z, Ikonen TP, Ferreira-da-Silva F et al (2011) Solution structure of the soluble receptor for advanced glycation end products (sRAGE). J Biol Chem 286:37525–37534
25. Xie JJ, Reverdatto S, Frolov A et al (2008) Structural basis for pattern recognition by the receptor for advanced glycation end products (RAGE). J Biol Chem 283:27255–27269
26. Xue J, Rai V, Singer D et al (2011) Advanced glycation end product recognition by the receptor for AGEs. Structure 19:722–732
27. Fritz G (2011) RAGE: a single receptor fits multiple ligands. Trends Biochem Sci 36:625–632
28. Leclerc E, Sturchler E, Vetter SW et al (2009) Crosstalk between calcium, amyloid beta and the receptor for advanced glycation endproducts in Alzheimer's disease. Rev Neurosci 20:95–110
29. Xie J, Burz DS, He W et al (2007) Hexameric calgranulin C (S100A12) binds to the receptor for advanced glycated end products (RAGE) using symmetric hydrophobic target-binding patches. J Biol Chem 282:4218–4231
30. Ostendorp T, Leclerc E, Galichet A et al (2007) Structural and functional insights into RAGE activation by multimeric S100B. EMBO J 26:3868–3878
31. Ramasamy R, Yan SF, Schmidt AM (2009) RAGE: therapeutic target and biomarker of the inflammatory response-the evidence mounts. J Leukoc Biol 86:505–512
32. Hori O, Brett J, Slattery T et al (1995) The receptor for advanced glycation end-products (RAGE) is a cellular-binding site for amphoterin—mediation of neurite outgrowth and coexpression of rage and amphoterin in the developing nervous-system. J Biol Chem 270:25752–25761
33. Rauvala H, Pihlaskari R (1987) Isolation and some characteristics of an adhesive factor of brain that enhances neurite outgrowth in central neurons. J Biol Chem 262:16625–16635
34. Bianchi ME, Beltrame M, Paonessa G (1989) Specific recognition of cruciform DNA by nuclear protein HMG1. Science 243: 1056–1059
35. Merenmies J, Pihlaskari R, Laitinen J et al (1991) 30-kDa heparin-binding protein of brain (amphoterin) involved in neurite outgrowth: amino acid sequence and localization in the filopodia of the advancing plasma membrane. J Biol Chem 266:16722–16729
36. Rauvala H, Rouhiainen A (2010) Physiological and pathophysiological outcomes of the interactions of HMGB1 with cell surface receptors. Biochim Biophys Acta 1799:164–170
37. Rauvala H, Merenmies J, Pihlaskari R et al (1988) The adhesive and neurite-promoting molecule p30: analysis of the amino-terminal sequence and production of antipeptide antibodies that detect p30 at the surface of neuroblastoma cells and of brain neurons. J Cell Biol 107:2293–2305
38. Sunden-Cullberg J, Norrby-Teglund A, Rouhiainen A et al (2005) Persistent elevation of high mobility group box-1 protein (HMGB1) in patients with severe sepsis and septic shock. Crit Care Med 33:564–573
39. Wang HC, Bloom O, Zhang MH et al (1999) HMG-1 as a late mediator of endotoxin lethality in mice. Science 285:248–251
40. Rouhiainen A, Kuja-Panula J, Wilkman E et al (2004) Regulation of monocyte migration by amphoterin (HMGB1). Blood 104:1174–1182
41. Fages C, Nolo R, Huttunen HJ et al (2000) Regulation of cell migration by amphoterin. J Cell Sci 113:611–620
42. Parkkinen J, Raulo E, Merenmies J et al (1993) Amphoterin, the 30-kDa protein in a family of HMG1-type polypeptides: enhanced expression in transformed cells, leading edge localization and interactions with plasminogen activation. J Biol Chem 268:19726–19738
43. Lotze MT, Tracey KJ (2005) High-mobility group box 1 protein (HMGB): nuclear weapon in the immune arsenal. Nat Rev Immunol 5:331–342
44. Muller S, Ronfani L, Bianchi ME (2004) Regulated expression and subcellular localization of HMGB1, a chromatin protein with a cytokine function. J Intern Med 255: 332–343
45. Ulloa L, Messmer D (2006) High-mobility group box 1 (HMGB1) protein: friend and foe. Cytokine Growth Factor Rev 17:189–201
46. Bonaldi T, Talamo F, Scaffidi P et al (2003) Monocytic cells hyperacetylate chromatin protein HMGB1 to redirect it towards secretion. EMBO J 22:5551–5560

47. Youn JH, Shin JS (2006) Nucleocytoplasmic shuttling of HMGB1 is regulated by phosphorylation that redirects it toward secretion. J Immunol 177:7889–7897
48. Ito I, Fukazawa J, Yoshida M (2007) Post-translational methylation of high mobility group box 1 (HMGB1) causes its cytoplasmic localization in neutrophils. J Biol Chem 282:16336–16344
49. Ditsworth D, Zong WX, Thompson CB (2007) Activation of poly(ADP)-ribose polymerase (PARP-1) induces release of the pro-inflammatory mediator HMGB1 from the nucleus. J Biol Chem 282:17845–17854
50. Dupont N, Jiang S, Pilli M et al (2011) Autophagy-based unconventional secretory pathway for extracellular delivery of IL-1β. EMBO J 30:4701–4711
51. Huttunen HJ, Fages C, Kuja-Panula J et al (2002) Receptor for advanced glycation end products-binding COOH-terminal motif of amphoterin inhibits invasive migration and metastasis. Cancer Res 62:4805–4811
52. Huttunen HJ, Fages C, Rauvala H (1999) Receptor for advanced glycation end products (RAGE)-mediated neurite outgrowth and activation of NF-kappa B require the cytoplasmic domain of the receptor but different downstream signaling pathways. J Biol Chem 274:19919–19924
53. Rong LL, Trojaborg W, Qu W et al (2004) Antagonism of RAGE suppresses peripheral nerve regeneration. FASEB J 18:1812–1817
54. Rong LL, Yan SF, Wendt T et al (2004) RAGE modulates peripheral nerve regeneration via recruitment of both inflammatory and axonal outgrowth pathways. FASEB J 18:1818–1825
55. Taguchi A, Blood DC, del Toro G et al (2000) Blockade of RAGE-amphoterin signalling suppresses tumour growth and metastases. Nature 405:354–360
56. Chavakis E, Hain A, Vinci M et al (2007) High-mobility group box 1 activates integrin-dependent homing of endothelial progenitor cells. Circ Res 100:204–212
57. Schlueter C, Weber H, Meyer B et al (2005) Angiogenetic signaling through hypoxia—HMGB1: an angiogenetic switch molecule. Am J Pathol 166:1259–1263
58. van Beijnum JR, Dings RP, van der Linden E et al (2006) Gene expression of tumor angiogenesis dissected: specific targeting of colon cancer angiogenic vasculature. Blood 108:2339–2348
59. Orlova VV, Choi EY, Xie CP et al (2007) A novel pathway of HMGB1-mediated inflammatory cell recruitment that requires Mac-1-integrin. EMBO J 26:1129–1139
60. Dumitriu IE, Baruah P, Manfredi AA et al (2005) HMGB1: guiding immunity from within. Trends Immunol 26:381–387
61. Dumitriu IE, Bianchi ME, Bacci M et al (2007) The secretion of HMGB1 is required for the migration of maturing dendritic cells. J Leukoc Biol 81:84–91
62. Yang D, Chen Q, Yang H et al (2007) High mobility group box-1 protein induces the migration and activation of human dendritic cells and acts as an alarmin. J Leukoc Biol 81:59–66
63. Porto A, Palumbo R, Pieroni M et al (2006) Smooth muscle cells in human atherosclerotic plaques secrete and proliferate in response to high mobility group box 1 protein. FASEB J 20:2565–2566
64. Germani A, Limana F, Capogrossi MC (2007) Pivotal advance: high-mobility group box 1 protein—a cytokine with a role in cardiac repair. J Leukoc Biol 81:41–45
65. Limana F, Germani A, Zacheo A et al (2005) Exogenous high-mobility group box 1 protein induces myocardial regeneration after infarction via enhanced cardiac c-kit(+) cell proliferation and differentiation. Circ Res 97:E73–E83
66. Palumbo R, Sampaolesi M, De Marchis F et al (2004) Extracellular HMGB1, a signal of tissue damage, induces mesoangioblast migration and proliferation. J Cell Biol 164:441–449
67. Xu D, Young J, Song D et al (2011) Heparan sulfate is essential for high mobility group protein-1 (HMGB1) signaling by the receptor for advanced glycation end products (RAGE). J Biol Chem 268:41736–41744
68. Leclerc E, Fritz G, Vetter SW et al (2009) Binding of S100 proteins to RAGE: an update. Biochim Biophys Acta 1793:993–1007
69. Hofmann MA, Drury S, Fu CF et al (1999) RAGE mediates a novel proinflammatory axis: a central cell surface receptor for S100/calgranulin polypeptides. Cell 97:889–901
70. Leclerc E, Heizmann C (2011) The importance of Ca2+/Zn2+ signaling S100 proteins and RAGE in translational medicine. Front Biosci S3:1232–1262
71. Yan SD, Chen X, Fu J et al (1996) RAGE and amyloid-beta peptide neurotoxicity in Alzheimer's disease. Nature 382:685–691
72. Deane R, Yan SD, Submamaryan RK et al (2003) RAGE mediates amyloid-beta peptide transport across the blood–brain barrier and accumulation in brain. Nat Med 9:907–913

73. Sturchler E, Galichet A, Weibel M et al (2008) Site-specific blockade of RAGE-V-d prevents amyloid-beta oligomer neurotoxicity. J Neurosci 28:5149–5158
74. Bopp C, Bierhaus A, Hofer S et al (2008) Bench-to-bedside review: the inflammation-perpetuating pattern-recognition receptor RAGE as a therapeutic target in sepsis. Crit Care 12:201. doi:210.1186/cc6164
75. Banerjee S, Friggeri A, Liu G et al (2010) The C-terminal acidic tail is responsible for the inhibitory effects of HMGB1 on efferocytosis. J Leukoc Biol 88:973–979
76. Huttunen HJ, Rauvala H (2004) Amphoterin as an extracellular regulator of cell motility: from discovery to disease. J Intern Med 255:351–366
77. Gospodarska E, Kupniewska-Kozak A, Goch G et al (2011) Binding studies of truncated variants of the A beta peptide to the V-domain of the RAGE receptor reveal A beta residues responsible for binding. Biochim Biophys Acta 1814:592–609
78. Hudson BI, Kalea AZ, Arriero MD et al (2008) Interaction of the RAGE cytoplasmic domain with diaphanous-1 is required for ligand-stimulated cellular migration through activation of Rac1 and Cdc42. J Biol Chem 283:34457–34468
79. Ishihara K, Tsutsumi K, Kawane S et al (2003) The receptor for advanced glycation end-products (RAGE) directly binds to ERK by a D-domain-like docking site. FEBS Lett 550:107–113
80. Chavakis T, Bierhaus A, Al-Fakhri N et al (2003) The pattern recognition receptor (RAGE) is a counterreceptor for leukocyte integrins: a novel pathway for inflammatory cell recruitment. J Exp Med 198:1507–1515
81. Xiong F, Leonov S, Howard AC et al (2011) Receptor for advanced glycation end products (RAGE) prevents endothelial cell membrane resealing and regulates F-actin remodeling in a beta-catenin-dependent manner. J Biol Chem 286:35061–35070
82. Buckley ST, Medina C, Kasper M et al (2011) Interplay between RAGE, CD44, and focal adhesion molecules in epithelial-mesenchymal transition of alveolar epithelial cells. Am J Physiol Lung Cell Mol Physiol 300:L548–L559
83. Lander HM, Tauras JM, Ogiste JS et al (1997) Activation of the receptor for advanced glycation end products triggers a p21(ras)-dependent mitogen-activated protein kinase pathway regulated by oxidant stress. J Biol Chem 272:17810–17814
84. Sakaguchi M, Murata H, Yamamoto K et al (2011) TIRAP, an adaptor protein for TLR2/4, transduces a signal from RAGE phosphorylated upon ligand binding. PLoS One 6. doi:10.1371/journal.pone.0023132
85. Perrone L, Devi TS, Hosoya KI et al (2009) Thioredoxin interacting protein (TXNIP) induces inflammation through chromatin modification in retinal capillary endothelial cells under diabetic conditions. J Cell Physiol 221:262–272
86. Bierhaus A, Humpert PM, Morcos M et al (2005) Understanding RAGE, the receptor for advanced glycation end products. J Mol Med 83:876–886
87. Schmidt AM, Yan SD, Yan SF et al (2000) The biology of the receptor for advanced glycation end products and its ligands. Biochim Biophys Acta 1498:99–111
88. Raulo E, Chernousov MA, Carey DJ et al (1994) Isolation of a neuronal cell surface receptor of heparin binding growth-associated molecule (HB-GAM): identification as N-syndecan (syndecan-3). J Biol Chem 269:12999–13004
89. Park JS, Svetkauskaite D, He QB et al (2004) Involvement of toll-like receptors 2 and 4 in cellular activation by high mobility group box 1 protein. J Biol Chem 279:7370–7377
90. Rouhiainen A, Tumova S, Valmu L et al (2007) Pivotal advance: analysis of proinflammatory activity of highly purified eukaryotic recombinant HMGB1 (amphoterin). J Leukoc Biol 81:49–58
91. Tian J, Avalos AM, Mao SY et al (2007) Toll-like receptor 9-dependent activation by DNA-containing immune complexes is mediated by HMGB1 and RAGE. Nat Immunol 8:487–496
92. Yang H, Lundbäck P, Ottosson L et al (2011) Redox modification of cysteine residues regulates the cytokine activity of HMGB1. Mol Med. doi:10.2119/molmed.2011.00389, Epub ahead of print
93. Yang HA, Hreggvidsdottir HS, Palmblad K et al (2010) A critical cysteine is required for HMGB1 binding to toll-like receptor 4 and activation of macrophage cytokine release. Proc Natl Acad Sci U S A 107:11942–11947
94. Robinson MJ, Tessier P, Poulsom R et al (2002) The S100 family heterodimer, MRP-8/14, binds with high affinity to heparin and heparan sulfate glycosaminoglycans on endothelial cells. J Biol Chem 277:3658–3665
95. Sorci G, Giovannini G, Riuzzi F et al (2011) The danger signal S100B integrates pathogen- and danger-sensing pathways to restrain

inflammation. PLoS Pathog 7:E1001315. doi:10.1371/journal.ppat.1001315
96. Stewart CR, Stuart LM, Wilkinson K et al (2010) CD36 ligands promote sterile inflammation through assembly of a toll-like receptor 4 and 6 heterodimer. Nat Immunol 11:155–161
97. Mazarati A, Maroso M, Iori V et al (2011) High-mobility group box-1 impairs memory in mice through both toll-like receptor 4 and receptor for advanced glycation end products. Exp Neurol 232:143–148
98. Yang H, Ochani M, Li JH et al (2004) Reversing established sepsis with antagonists of endogenous high-mobility group box 1. Proc Natl Acad Sci U S A 101:296–301
99. Liliensiek B, Weigand MA, Bierhaus A et al (2004) Receptor for advanced glycation end products (RAGE) regulates sepsis but not the adaptive immune response. J Clin Invest 113:1641–1650
100. Kim JB, Choi JS, Yu YM et al (2006) HMGB1, a novel cytokine-like mediator linking acute neuronal death and delayed neuroinflammation in the postischemic brain. J Neurosci 26:6413–6421
101. Liu K, Mori S, Takahashi HK et al (2007) Anti-high mobility group box 1 monoclonal antibody ameliorates brain infarction induced by transient ischemia in rats. FASEB J 21:3904–3916
102. Muhammad S, Barakat W, Stoyanov S et al (2008) The HMGB1 receptor RAGE mediates ischemic brain damage. J Neurosci 28:12023–12031
103. Naka Y, Bucciarelli LG, Wendt T et al (2004) RAGE axis—animal models and novel insights into the vascular complications of diabetes. Arterioscler Thromb Vasc Biol 24: 1342–1349
104. Wautier JL, Schmidt AM (2004) Protein glycation—a firm link to endothelial cell dysfunction. Circ Res 95:233–238
105. Origlia N, Capsoni S, Cattaneo A et al (2009) A beta-dependent inhibition of LTP in different intracortical circuits of the visual cortex: the role of RAGE. J Alzheimers Dis 17:59–68
106. Yan SS, Wu ZY, Zhang HP et al (2003) Suppression of experimental autoimmune encephalomyelitis by selective blockade of encephalitogenic T-cell infiltration of the central nervous system. Nat Med 9:287–293
107. Rauvala H, Huttunen HJ, Fages C et al (2000) Heparin-binding proteins HB-GAM (pleiotrophin) and amphoterin in the regulation of cell motility. Matrix Biol 19:377–387
108. Vakkila J, Lotze MT (2004) Opinion—inflammation and necrosis promote tumour growth. Nat Rev Immunol 4:641–648
109. Yonekura H, Yamamoto Y, Sakurai S et al (2005) Roles of the receptor for advanced glycation endproducts in diabetes-induced vascular injury. J Pharmacol Sci 97:305–311
110. Sakatani S, Yamada K, Homma C et al (2009) Deletion of RAGE causes hyperactivity and increased sensitivity to auditory stimuli in mice. PLoS One 4:E8309. doi:10.1371/journal.pone.0008309
111. Zhou Z, Immel D, Xi CX et al (2006) Regulation of osteoclast function and bone mass by RAGE. J Exp Med 203:1067–1080
112. Queisser MA, Kouri FM, Konigshoff M et al (2008) Loss of RAGE in pulmonary fibrosis—molecular relations to functional changes in pulmonary cell types. Am J Respir Cell Mol Biol 39:337–345
113. Hancock DB, Eijgelsheim M, Wilk JB et al (2010) Meta-analyses of genome-wide association studies identify multiple loci associated with pulmonary function. Nat Genet 42:45–52
114. Repapi E, Sayers I, Wain LV et al (2010) Genome-wide association study identifies five loci associated with lung function. Nat Genet 42:36–44
115. Hudson BI, Stickland MH, Grant PJ (1998) Identification of polymorphisms in the receptor for advanced glycation end products (RAGE) gene—prevalence in type 2 diabetes and ethnic groups. Diabetes 47:1155–1157
116. Castaldi P, Cho M, Litonjua A et al (2011) The association of genome-wide significant spirometric loci with chronic obstructive pulmonary disease susceptibility. Am J Respir Cell Mol Biol 45:1147–1153
117. Schenk S, Schraml P, Bendik I et al (2001) A novel polymorphism in the promoter of the RAGE gene is associated with non-small cell lung cancer. Lung Cancer 32:7–12
118. Cunha C, Giovannini G, Pierini A et al (2011) Genetically-determined hyperfunction of the S100B/RAGE axis is a risk factor for aspergillosis in stem cell transplant recipients. PLoS One 6:e27962. doi:10.1371/journal.pone.0027962
119. Hudson BI, Stickland MH, Futers TS et al (2001) Effects of novel polymorphisms in the RAGE gene on transcriptional regulation and their association with diabetic retinopathy. Diabetes 50:1505–1511
120. Forbes JM, Soderlund J, Yap FYT et al (2011) Receptor for advanced glycation end-products (RAGE) provides a link between genetic

susceptibility and environmental factors in type 1 diabetes (vol 54, pg 1032, 2011). Diabetologia 54:1586–1587

121. Kucukhuseyin O, Yilmaz-Aydogan H, Isbir C et al (2011) Is there any association between GLY82 ser polymorphism of rage gene and Turkish diabetic and non diabetic patients with coronary artery disease? Mol Biol Rep 39:4423–4428. doi:1007/s11033-011-1230-3, Epub ahead of print
122. Kumaramanickavel G, Ramprasad VL, Sripriya S et al (2002) Association of Gly82Ser polymorphism in the RAGE gene with diabetic retinopathy in type II diabetic Asian Indian patients. J Diabetes Complications 16:391–394
123. Laki J, Kiszel P, Vatay A et al (2007) The HLA 8.1 ancestral haplotype is strongly linked to the C allele of -429 T > C promoter polymorphism of receptor of the advanced glycation endproduct (RAGE) gene. Haplotype-independent association of the -429C allele with high hemoglobin(A1C) levels in diabetic patients. Mol Immunol 44:648–655
124. Prasad P, Tiwari AK, Kumar KMP et al (2010) Association analysis of ADPRT1, AKR1B1, RAGE, GFPT2 and PAI-1 gene polymorphisms with chronic renal insufficiency among Asian Indians with type-2 diabetes. BMC Med Genet 11:52. doi:10.1186/1471-2350-11-52
125. Prevost G, Fajardy I, Besmond C et al (2005) Polymorphisms of the receptor of advanced glycation endproducts (RAGE) and the development of nephropathy in type 1 diabetic patients. Diabetes Metab 31:35–39
126. Sullivan C, Futers T, Barrett J et al (2005) RAGE polymorphisms and the heritability of insulin resistance: the Leeds family study. Diab Vasc Dis Res 2:42–44
127. Balasubbu S, Sundaresan P, Rajendran A et al (2010) Association analysis of nine candidate gene polymorphisms in Indian patients with type 2 diabetic retinopathy. BMC Med Genet 11:158. doi:10.1186/1471-2350-11-158
128. Daborg J, von Otter M, Sjolander A et al (2010) Association of the RAGE G82S polymorphism with Alzheimer's disease. J Neural Transm 117:861–867
129. Gao JX, Shao YH, Lai WY et al (2010) Association of polymorphisms in the RAGE gene with serum CRP levels and coronary artery disease in the Chinese Han population. J Hum Genet 55:668–675
130. Hofmann MA, Drury S, Hudson BI et al (2002) RAGE and arthritis: the G82S polymorphism amplifies the inflammatory response. Genes Immun 3:123–135
131. Kanková K, Záhejský J, Márová I et al (2001) Polymorphisms in the RAGE gene influence susceptibility to diabetes-associated microvascular dermatoses in NIDDM. J Diabetes Complications 15:185–192
132. Li K, Dai D, Zhao B et al (2010) Association between the RAGE G82S polymorphism and Alzheimer's disease. J Neural Transm 117:97–104
133. Li KS, Zhao B, Dai DW et al (2011) A functional p. 82G > S polymorphism in the RAGE gene is associated with multiple sclerosis in the Chinese population. Mult Scler 17:914–921
134. Jang Y, Kim JY, Kang SM et al (2007) Association of the Gly82Ser polymorphism in the receptor for advanced glycation end products (RAGE) gene with circulating levels of soluble RAGE and inflammatory markers in nondiabetic and nonobese Koreans. Metabolism 56:199–205
135. Kim OY, Jo SH, Jang Y et al (2009) G allele at RAGE SNP82 is associated with proinflammatory markers in obese subjects. Nutr Res 29:106–113
136. Mokbel A, Rashid L, Al-Harizy R (2011) Decreased level of soluble receptors of advanced glycated end products (sRAGE) and glycine82serine (G82S) polymorphism in Egyptian patients with RA. The Egyptian Rheumatologist 33:53–60
137. Boor P, Celec P, Klenovicsova K et al (2010) Association of biochemical parameters and RAGE gene polymorphisms in healthy infants and their mothers. Clin Chim Acta 411: 1034–1040
138. Gaens KHJ, Ferreira I, van der Kallen CJH et al (2009) Association of polymorphism in the receptor for advanced glycation end products (RAGE) gene with circulating RAGE levels. J Clin Endocrinol Metab 94:5174–5180
139. Peng WH, Lu L, Wang LJ et al (2009) RAGE gene polymorphisms are associated with circulating levels of endogenous secretory RAGE but not with coronary artery disease in Chinese patients with type 2 diabetes mellitus. Arch Med Res 40:393–398
140. Lindström O, Tukiainen E, Kylänpää L et al (2009) Circulating levels of a soluble form of receptor for advanced glycation end products, and high-mobility group box chromosomal protein 1 in patients with acute pancreatitis. Pancreas 38:E215–E220
141. Caillier SJ, Briggs F, Cree BAC et al (2008) Uncoupling the roles of HLA-DRB1 and HLA-DRB5 genes in multiple sclerosis. J Immunol 181:5473–5480

142. Kalea AZ, Schmidt AM, Hudson BI (2009) RAGE: a novel biological and genetic marker for vascular disease. Clin Sci 116:621–637
143. Lindholm E, Bakhtadze E, Cilio C et al (2008) Association between LTA. TNF and AGER polymorphisms and late diabetic complications. PLoS One 3:e2546. doi:10.1371/journal.pone.0002546
144. Zhao X, Kuja-Panula J, Rouhiainen A et al (2011) High mobility group box-1 (HMGB1; amphoterin) is required for zebrafish brain development. J Biol Chem 286:23200–23213
145. Suchankova P, Klang J, Cavanna C et al (2011) Is the Gly82Ser polymorphism in the RAGE gene relevant to schizophrenia and the personality trait psychoticism? J Psychiatry Neurosci 36. doi:10.1503/jpn.110024, Epub ahead of print
146. Lu W, Feng B (2010) The -374A allele of the RAGE gene as a potential protective factor for vascular complications in type 2 diabetes: a meta-analysis. Tohoku J Exp Med 220: 291–297
147. Raulo E, Tumova S, Pavlov I et al (2005) The two thrombospondin type I repeat domains of the heparin-binding growth-associated molecule bind to heparin/heparan sulfate and regulate neurite extension and plasticity in hippocampal neurons. J Biol Chem 280: 41576–41583
148. Salmivirta M, Rauvala H, Elenius K et al (1992) Neurite growth-promoting protein (amphoterin, p30) binds syndecan. Exp Cell Res 200:444–451
149. Milev P, Chiba A, Häring M et al (1998) High affinity binding and overlapping localization of neurocan and phosphacan protein-tyrosine phosphatase-zeta/beta with tenascin-R, amphoterin, and the heparin-binding growth-associated molecule. J Biol Chem 273: 6998–7005
150. NarindrasorasakS,LoweryD,Gonzalezdewhitt P et al (1991) High-affinity interactions between the alzheimers beta-amyloid precursor proteins and the basement-membrane form of heparan-sulfate proteoglycan. J Biol Chem 266:12878–12883
151. Kanekiyo T, Zhang JA, Liu QA et al (2011) Heparan sulphate proteoglycan and the low-density lipoprotein receptor-related protein 1 constitute major pathways for neuronal amyloid-beta uptake. J Neurosci 31:1644–1651
152. Johnson GB, Brunn GJ, Kodaira Y et al (2002) Receptor-mediated monitoring of tissue well-being via detection of soluble heparan sulfate by toll-like receptor 4. J Immunol 168: 5233–5239
153. Schaefer L, Babelova A, Kiss E et al (2005) The matrix component biglycan is proinflammatory and signals through toll-like receptors 4 and 2 in macrophages. J Clin Invest 115:2223–2233
154. Popovic PJ, DeMarco R, Lotze MT et al (2006) High mobility group B1 protein suppresses the human plasmacytoid dendritic cell response to TLR9 agonists. J Immunol 177:8701–8707
155. Iwami K, Matsuguchi T, Masuda A et al (2000) Cutting edge: naturally occurring soluble form of mouse toll-like receptor 4 inhibits lipopolysaccharide signaling. J Immunol 165:6682–6686

Chapter 16

RAGE Splicing Variants in Mammals

Katharina Anna Sterenczak, Ingo Nolte, and Hugo Murua Escobar

Abstract

The receptor for advanced glycation end products (RAGE) is a multiligand receptor of environmental stressors which plays key roles in pathophysiological processes, including immune/inflammatory disorders, Alzheimer's disease, diabetic arteriosclerosis, tumorigenesis, and metastasis. Besides the full-length RAGE protein in humans nearly 20 natural occurring RAGE splicing variants were described on mRNA and protein level. These naturally occurring isoforms are characterized by either N-terminally or C-terminally truncations and are discussed as possible regulators of the full-length RAGE receptor either by competitive ligand binding or by displacing the full-length protein in the membrane. Accordingly, expression deregulations of the naturally occurring isoforms were supposed to have significant effect on RAGE-mediated disorders. Thereby the soluble C-truncated RAGE isoforms present in plasma and tissues are the mostly focused isoforms in research and clinics. Deregulations of the circulating levels of soluble RAGE forms were reported in several RAGE-associated pathological disorders including for example atherosclerosis, diabetes, renal failure, Alzheimer's disease, and several cancer types.

Regarding other mammalian species, the canine *RAGE* gene showed high similarities to the corresponding human structures indicating RAGE to be evolutionary highly conserved between both species. Similar to humans the canine RAGE showed a complex and extensive splicing activity leading to a manifold pattern of RAGE isoforms. Due to the similarities seen in several canine and human diseases—including cancer—comparative structural and functional analyses allow the development of RAGE and ligand-specific therapeutic approaches beneficial for human and veterinary medicine.

Key words: RAGE, Isoforms, Mammalian variants, Comparative genetics, Receptor for advanced glycation end products

1. Introduction: The Receptor for Advanced Glycation End Products

The receptor for advanced glycation end products (RAGE) is a ~50 kDa cell-membrane bound protein that was firstly characterized in 1992 (1). This receptor has the ability to specifically bind molecules (ligands) which are present in the surrounding extracellular environment. RAGE was named after one of its first discovered

Claus W. Heizmann (ed.), *Calcium-Binding Proteins and RAGE: From Structural Basics to Clinical Applications*, Methods in Molecular Biology, vol. 963, DOI 10.1007/978-1-62703-230-8_16, © Springer Science+Business Media New York 2013

ligands—the advanced glycation end products (AGEs). Following the characterization of RAGE further additional ligands were found including High Mobility Group 1 protein (HMGB1, formerly known as amphoterin), proteins of the S100/calgranulin family, amyloid-beta, and Mac-1 (1–5) and thus the protein was referred to as a multiligand receptor of environmental stressors. In humans, RAGE is a cell surface molecule belonging to the immunoglobulin superfamily and being encoded on genomic level on chromosome HSA 6p21.3 (1, 6, 7). The human *RAGE* gene consists of 11 exons and 10 introns resulting in the 404 aa RAGE protein. The protein is structurally subdivided into three domains: the extracellular domain (ED), the highly hydrophobic α-helix transmembrane domain (TM), and the cytosolic domain (CD). The extracellular domain itself is subdivided into the variable (V) domain—that is essential for ligand binding—and two constant (C) domains. The terminal highly acidic cytosolic domain (CD) is essential for RAGE signaling (8) which is induced by specific binding of a RAGE ligand to its receptor. This interaction leads to signal transduction pathways within the cell activating, i.e., mitogen-activated protein (MAP) kinases, regulating cell growth, proliferation, and inflammation (1, 3, 9–15). The RAGE protein is basically expressed in a wide range of cell types and organs and with the exception of the lung, where RAGE expression is constantly high, the expression levels of RAGE remain low during lifetime. However, in terms of inflammatory-related pathophysiological states high levels of RAGE ligands were described which in turn increase RAGE expression levels due to a positive feedback loop thereby intensifying RAGE-mediated cellular effects. Thus, RAGE levels were associated with various diseases including diabetic arteriosclerosis (16, 17), impaired wound healing (18), Alzheimer's disease (5, 19, 20), immune and inflammatory disorders (3, 8, 21, 22), and various cancers (23–27).

Besides the above-mentioned RAGE protein which is composed of the three domains (ED, TM, CD)—also called full-length RAGE—further naturally occurring RAGE isoforms were detected on mRNA and protein level. These isoforms are characterized by mRNA transcripts with missing or additional exons/introns or parts of them resulting from alternative splicing of the *RAGE* pre-mRNA. In general, the alternative splicing of pre-mRNA transcripts is a central mechanism in the regulation of gene expression through the generation of multiple different transcripts from a single gene enabling diversity of proteins found in organisms, highly exceeding the number of genes encoding them. The first human RAGE isoforms were found in 1999 and further studies which followed during the last decade resulted in identification of altogether 19 human *RAGE* transcript isoforms (28–34). The variants were discussed to function as possible receptor regulators by acting as competitive inhibitors of the full-length receptor, either by ligand binding or displacing the full-length protein in the membrane (32, 35). Deregulation of the naturally occurring protein isoforms was supposed to have significant effect on RAGE-mediated diseases.

Structurally, these isoforms are characterized by either N-terminally or C-terminally truncated proteins and were detected in different types of tissues and cell lines initially without a compliant nomenclature containing the following isoforms: hRAGEsec (28), sRAGE1, sRAGE2, sRAGE3 (29), N-truncated and Secretory (30), Δ8-RAGE (31), RageΔ, NtRAGEΔ, sRAGEΔ (32, 33), and RAGE_v4-RAGE_v13 (34). Hudson et al. summarized these human isoforms and introduced a uniform nomenclature according to the Human Gene Nomenclature Committee (HGNC) (34). In so doing, the RAGE variants were renamed into RAGE_v1 to RAGE_v19 (34) and we will refer to this nomenclature.

Regarding the biological relevance of RAGE isoforms, the soluble C-truncated RAGE isoforms, which circulate in plasma and tissues and competitively bind RAGE ligands, represent the splicing variants mostly focused on in research and clinics. Generally two mechanisms were reported for the generation of soluble RAGE variants by either alternative splicing or shedding of the full-length receptor from the membrane (36–39) leading to soluble RAGE showing the ability to bind ligands. Recombinant soluble RAGE successfully prevented or reversed RAGE-mediated effects in several disorders such as diabetic and nondiabetic atherothrombosis, ischemia/reperfusion injury, collagen-induced arthritis, neurodegeneration and in terms of cancer, tumor cell migration, and invasion (23, 40). Deregulations of the circulating levels of soluble RAGE forms were reported in several RAGE-associated pathological disorders including for example chronic inflammatory diseases such as atherosclerosis, diabetes, renal failure, and the aging process (41), Alzheimer's disease (42), oesophageal cancer, lung cancer, chondrosarcoma, prostate tumor, and brain tumor (43–46).

Regarding other mammalian species, the canine and murine *RAGE* genes were also found to undergo extensive alternative splicing and discussed as a possible mechanism for receptor regulation (47, 48). Within the following chapter, we concentrate on the canine *RAGE* transcript isoforms that have been characterized in several tissues and cell lines of different origins.

2. The Canine *RAGE* Gene

2.1. Characterization of the Canine RAGE Gene and Protein

The characterization of the canine RAGE gene and protein showed high similarities to the corresponding human structures indicating the receptor to be highly evolutionary conserved between mammalians. Similar to the human *RAGE* gene, the canine counterpart

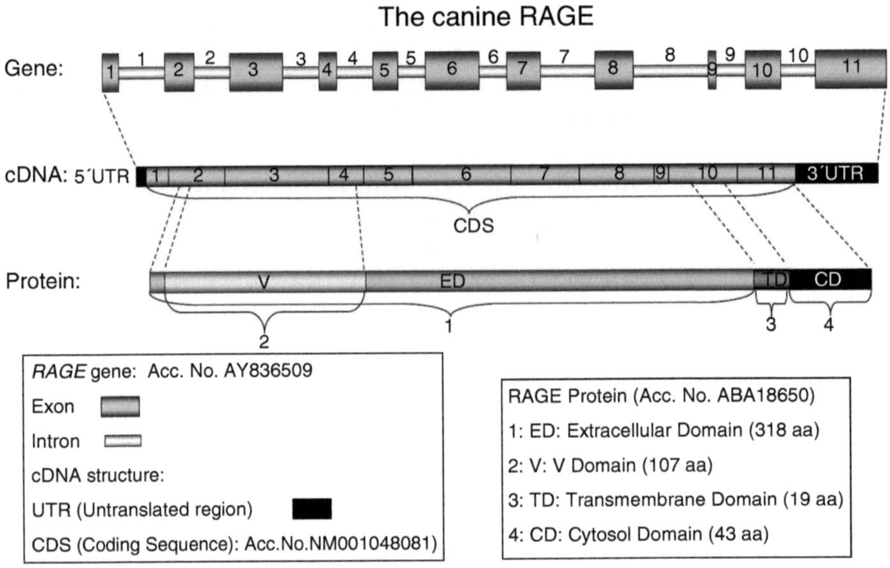

Fig. 1. Characterization of the canine receptor for advanced glycation end products (RAGE) (according to (49)). The canine RAGE was characterized on genomic, cDNA, and protein level. The canine *RAGE* gene consists of 11 exons and 10 introns with a total genomic length of 2,835 bp. The complete *RAGE* cDNA spans 1384 bp in total and is composed of an 18 bp 5′UTR, a 1215 bp CDS, and a 151 bp 3′UTR. The canine RAGE protein is composed of 404 aa and with a molecular weight of 43 kDa. The full-length protein is composed of the extracellular domain (ED), the transmembrane domain (TM), and the cytosol domain (CD). The ED includes the ligand-binding V domain and two C domains. Acc. No. = Accession Number.

consists of 11 exons and 10 introns encoding a 404 aa full-length protein which is composed of the three domains ED, TM, and CD resembling the human full-length RAGE protein (49) (Fig. 1). The chromosomal assignment localized the canine *RAGE* gene on the canine chromosome CFA12 that shares structural similarities with the human chromosome HSA 6 (49–51) which itself bears the human *RAGE* gene. Moreover, the short 5′UTR showed 100% identity to the human corresponding part of the gene indicating possibly similar mechanisms in the regulation of the translation and protein biosynthesis between both species (49). Comparison of human RAGE protein with other mammalians including dog, rat, mouse, cattle showed high identities between all species analyzed with the highest index between humans and dogs when comparing the amino acid sequences (49). Consequently, RAGE-mediated pathological effects, which were reported in humans, could show the same underlying mechanisms in other mammalians especially the dog. Regarding the regulation of the full-length receptor by alternative splicing, similar mechanisms were reported in dogs and mice (47, 48, 52) underlying the evidence of similar function and pathology of RAGE between mammalians.

2.2. Characterization of Canine RAGE Transcript Isoforms

Similar to the human RAGE, the canine *RAGE* gene undergoes extensive alternative splicing leading to the generation of a wide range of structurally differing mRNA transcript isoforms (48). These variants were characterized by a systematic screening of different canine tissues for *RAGE* transcript isoforms including non-neoplastic tissues, tumor tissues, and various cell lines derived from primary tumors. The non-neoplastic tissues included pancreas, spleen, testis, and thyroid tissue samples. The tumor samples included neoplasias such as histiocytoma, malignant lymphoma, malignant melanoma, mastocytoma, and thyroid carcinoma. The canine cell lines used within the study included CT1258 (prostate adenocarcinoma), MTH53A (non-neoplastic tissue of the mammary gland), MTH52C (malignant small-cell carcinoma of the mammary gland), and ZMTH3 (pleomorphic adenoma of the mammary gland). For the characterization, RNA isolations were performed followed by copy DNA (cDNA) syntheses, gene-specific polymerase chain reactions (PCRs), gel electrophoreses, cloning, and sequencing of the generated clones. The PCR assay used specifically amplified the complete coding sequence of the canine *RAGE* gene beginning at the starting codon and ending with the termination codon for translation. In total, 24 previously not described *RAGE* isoforms could be characterized showing different structural characteristics within the different sample groups (non-neoplastic, neoplastic, cell lines) (48).

2.3. Structural Setup of the Canine RAGE Isoforms

Analyses of the sequence data showed a manifold set of different structural variances between the *RAGE* transcript variant clones initially caused by alternative splicing of *RAGE* pre-mRNA. Generally, modifications such as insertions of introns, deletion of exons, or parts of them could be found.

One recurrent noticeable structural character was the insertion of intron 1, which was found in non-neoplastic samples and in neoplastic and cell line samples as well. The solely insertion of intron 1 was found in the following samples: cell lines CT1258 and ZMTH3, histiocytoma, malignant lymphoma, mastocytoma (Acc. Nos.: EU428803, EU428809, EU428793, EU428795, EU428800) (Fig. 2). The translation into a protein would lead to a membrane bound RAGE protein isoform without the ability to bind ligands due to a missing V domain within the ED. The insertion of intron 1 leads to a frameshift within the coding sequence. As a result, a premature stop codon in exon 2 is generated leading to termination of translation. Further downstream the mRNA sequence an alternative start codon located within exon 6 could be potentially used for the translation into an alternative protein. As the ligand-binding V domain is encoded within the exons 2–4, this alternative protein isoform would miss this domain within its ED. The resulting protein would exist in form of a truncated ED, a viable TM and CD, thus representing a truncated RAGE protein at

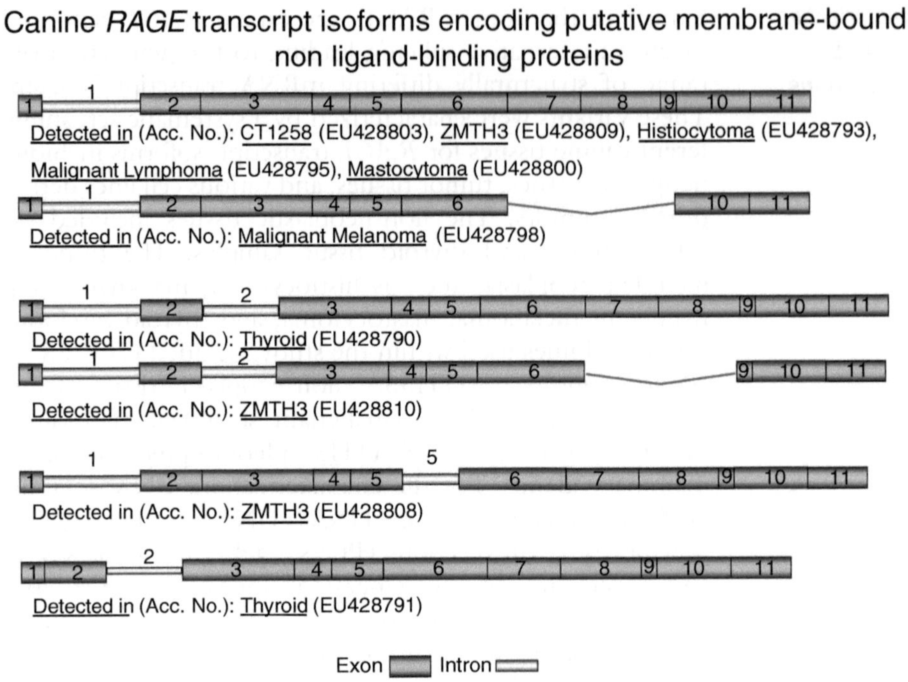

Fig. 2. Characterization of canine RAGE transcript isoforms encoding putative membrane-bound proteins (according to (48, 49)). These isoforms encode, if translated, putative N-truncated membrane-bound non-ligand-binding RAGE protein isoforms. The putative proteins could be characterized by truncated extracellular domains (ED)—missing the ligand binding V domain—followed by viable transmembrane domains (TM) and cytosol domains (CD). These isoforms could act as competitive inhibitors of the full-length RAGE by displacing the latter in the cell membrane. The isoforms show insertions of intron 1 and/or intron 2 and/or intron 5. Further modifications include deletions of exon 7 and/or exon 8 and/or exon 9. The variants were detected in non-neoplastic thyroid tissue, the cell lines CT1258 and ZMTH3 and in histiocytoma, malignant melanoma and mastocytoma tumor samples. Acc. No. = Accession Number.

its N-terminus when compared to the full-length RAGE. This protein, if expressed, could act as competitive inhibitor of the full-length RAGE through the displacement of the latter within the cell membrane. Importantly, this protein isoform would not be able to induce cellular pathways, although it possesses a viable CD due to the missing interaction with extracellular ligands. Interestingly, this isoform corresponds to human RAGE_v2 isoform (formerly known as N-truncated RAGE, Nt-RAGE, N-RAGE) showing similar mechanisms in splicing of *RAGE* within both species.

Besides the sole insertion of intron 1, additional modifications would lead, if translated, into N-truncated membrane-bound non ligand-binding isoforms. These modifications include insertions of intron 2 and/or intron 5 and/or deletions of exons 7, 8, and 9. All of these isoforms would, if translated, show a truncated extracellular domain with a viable TM and CD. Such isoforms were found in non-neoplastic thyroid, the cell line ZMTH3, histiocytoma, malignant lymphoma, malignant melanoma, mastocytoma

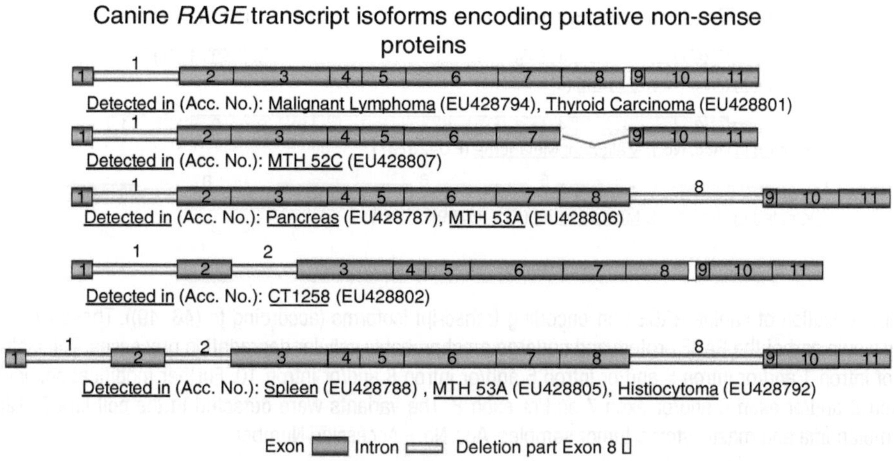

Fig. 3. Characterization of canine RAGE transcript isoforms encoding putative nonsense proteins (according to (48, 49)). These isoforms encode, if translated, putative N-truncated nonsense RAGE protein isoforms with no distinct cellular localization and function. The putative proteins could be characterized by truncated extracellular (ED) and transmembrane domains (TM) followed by viable cytosol domains. The isoforms show insertions of intron 1 and/or intron 2 and/or intron 8. Further modifications include deletion of exon 8 or a 16 bp long part of exon 8. The variants were detected in non-neoplastic pancreas and spleen tissue, the cell lines CT1258, MTH 53A, and MTH52C and in histiocytoma, malignant lymphoma, and thyroid carcinoma samples. Acc. No. = Accession Number.

(Acc. Nos.: EU428790, EU428791, EU428808, EU428810, EU428798) (Fig. 2).

In addition to the already mentioned modifications, other combinations of insertions of introns such as 1, 2, 8 and/or deletions of exon 8 or of a part of it encode proteins, when translated, which represent N-terminally truncated nonsense-mutated isoforms showing an incomplete TM and a viable CD. These nonsense isoforms indicate proteins with no distinct cellular localization or apparent function and were detected in non-neoplastic pancreas and spleen, the cell lines CT1258, MTH53A, MTH52C, in histiocytoma, malignant lymphoma, and thyroid carcinoma (Acc. Nos.: EU428787, EU428788, EU428802, EU428805, EU428806, EU428807, EU428792, EU428794, EU428801) (Fig. 3).

Generally, the more alterations within a transcript, the more likely this isoform does not encode a viable protein. Due to fatal frame shifts and the alteration of the original genetic information and structure, premature stop codons or instability of the respective mRNA molecule impede translation into native protein. It seems more likely that these extremely altered mRNAs undergo cellular degradation processes. Further *RAGE* isoforms could be detected which did not encode any putative viable proteins or parts of them in the cell line CT1258, malignant melanoma, and mastocytoma (Acc. Nos.: EU428804, EU428797, EU428799) (Fig. 4). Consequently, these variants most likely represent mRNAs undergoing cellular degradation without expression on protein level.

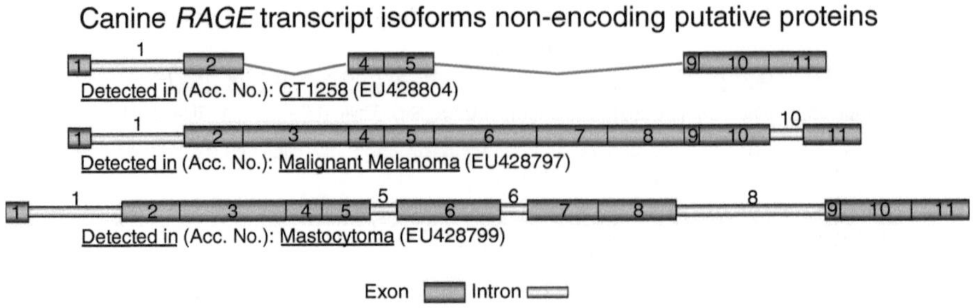

Fig. 4. Characterization of canine RAGE non-encoding transcript isoforms (according to (48, 49)). These isoforms do not encode any viable part of the RAGE protein and undergo most probably cellular degradation processes. The isoforms show insertions of intron 1 and/or intron 5 and/or intron 6 and/or intron 8 and/or intron 10. Further modifications include deletions of exon 3 and/or exon 6 and/or exon 7 and/or exon 8. The variants were detected in the cell line CT1258, and in malignant melanoma and mastocytoma tumor samples. Acc. No. = Accession Number.

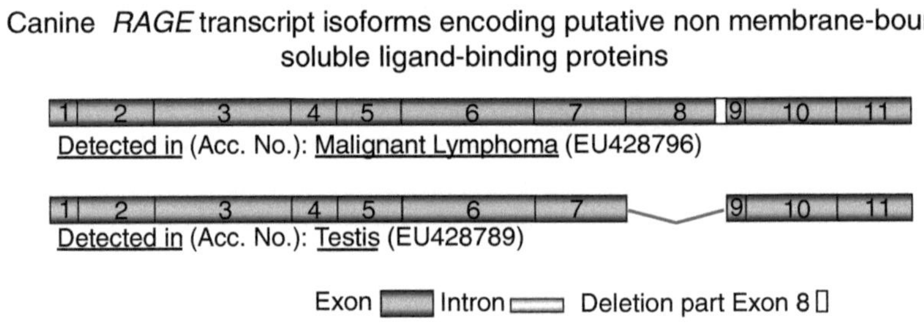

Fig. 5. Characterization of canine RAGE transcript isoforms encoding putative non-membrane-bound soluble proteins (according to (48, 49)). These isoforms encode, if translated, putative C-truncated non-membrane-bound soluble ligand-binding RAGE protein isoforms. The putative proteins could be characterized by a viable extracellular domain and missing transmembrane (TM) and cytosol domain (CD). These soluble RAGE isoforms could act as competitive inhibitors of full-length RAGE by binding of RAGE ligands—due to a viable V domain within the ED—without induction of cellular pathways. The isoforms show modifications including deletion of exon 8 or 16 bp of exon 8. The variants were detected in non-neoplastic testis tissue and in malignant lymphoma tumor samples. Acc. No. = Accession Number.

Besides the above-mentioned modifications leading to proteins truncated at the N-terminus, also *RAGE* isoforms could be detected which, when translated into viable proteins, would result in C-truncated isoforms. Such modifications include deletion of exon 8 or 16 bp of exon 8 which were found in non-neoplastic testis tissue and malignant lymphoma (Acc. Nos.: EU428789, EU428796) (Fig. 5). Due to a frame shift within the mRNA, premature stop codons would be generated in either exon 7 or exon 10 terminating the downstream translation of exons 10 and 11. As the TM and CD are encoded in those exons this modifications would lead, when translated, into C-truncated RAGE isoforms with viable EDs but missing TM and CDs. Such isoforms represent soluble, non-membrane-bound isoforms. Soluble RAGE acts as competitive inhibitor of the full-length RAGE by binding of RAGE ligands due to a viable V-domain within the ED without initiating cellular effects. Thus, soluble C-truncated RAGE isoforms are able to reverse RAGE-mediated effects and represent a valuable therapeutic tool in the impairment of RAGE-mediated pathophysiological processes. Interestingly the isoform found in non-neoplastic testis tissue resembles the human RAGE_v3 isoform (formerly known as Δ8-RAGE) showing similar splicing within both species.

3. Comparison of Human and Canine RAGE Splicing Variants

The 24 previously not described characterized canine splicing variants showed a wide set of modifications including insertions and/or deletions of exons and introns or parts of them. The resulting transcripts encode, if translated, C- or N-terminally truncated protein isoforms and nonsense isoforms. A small number of detected transcripts did not encode any viable part of the RAGE protein and represent, most likely, targets for RNA degradation. Generally, the insertion of intron 1 was a remarkable recurrent modification, which was found in all samples analyzed with exclusion of non-neoplastic testis and thyroid tissue samples. The presence of the respective transcripts varied in the different samples which was comparable to findings in humans showing a different range of variants depending on the sample source (34).

A closer look at specific canine variants showed two transcript isoforms that resemble human RAGE_v2 and RAGE_v3 splicing variants. The canine intron 1 containing isoform thereby resembles human RAGE_v2 (formerly known as nt-RAGE, N-RAGE, N-RAGE, N-truncated RAGE) and was found in both cell line and tumor samples (see Chapter 2 for Acc. Nos.). The canine transcript missing exon 8 corresponds to human RAGE_v3 (formerly known as Δ^8-RAGE) and was found in non-neoplastic testis tissue (Acc. No.: EU428789).

These resembling structures between human and canine RAGE transcript variants indicate similar mechanisms in *RAGE* splicing and thus the biology of the receptor. However, bioinformatic analyses by Hudson et al. revealed that the human RAGE_v2 and RAGE_v3 variants represent targets of the nonsense-mediated decay (NMD) machinery for degradation of mRNA inhibiting the expression on protein level (34, 53). If this is also true for the respective canine splicing variants further studies will be needed showing their expression on protein level.

4. Conclusion

Taken together, the analysis of the canine *RAGE* gene and its transcript isoforms showed high similarities to the corresponding human structures indicating RAGE to be evolutionary highly conserved between both species. Similar to humans, canine *RAGE* showed a complex and extensive splicing activity leading to a manifold pattern of transcript isoforms. The elucidation of the biological impact of this isoforms and their role in the regulation of the full-length receptor needs further analyses. Until now, reports are rare due to missing specific canine antibodies or ELISA assays for detection of specific viable alternatively spliced RAGE isoforms.

References

1. Neeper M et al (1992) Cloning and expression of a cell surface receptor for advanced glycosylation end products of proteins. J Biol Chem 267:14998–15004
2. Chavakis T et al (2003) The pattern recognition receptor (RAGE) is a counterreceptor for leukocyte integrins: a novel pathway for inflammatory cell recruitment. J Exp Med 198: 1507–1515
3. Hofmann MA et al (1999) RAGE mediates a novel proinflammatory axis: a central cell surface receptor for S100/calgranulin polypeptides. Cell 97:889–901
4. Hori O et al (1995) The receptor for advanced glycation end products (RAGE) is a cellular binding site for amphoterin. Mediation of neurite outgrowth and co-expression of rage and amphoterin in the developing nervous system. J Biol Chem 270:25752–25761
5. Yan SD et al (1996) RAGE and amyloid-beta peptide neurotoxicity in Alzheimer's disease. Nature 382:685–691
6. Sugaya K et al (1994) Three genes in the human MHC class III region near the junction with the class II: gene for receptor of advanced glycosylation end products, PBX2 homeobox gene and a notch homolog, human counterpart of mouse mammary tumor gene int-3. Genomics 23:408–419
7. Vissing H et al (1994) Localization of the human gene for advanced glycosylation end product-specific receptor (AGER) to chromosome 6p21.3. Genomics 24:606–608
8. Schmidt AM et al (2001) The multiligand receptor RAGE as a progression factor amplifying immune and inflammatory responses. J Clin Invest 108:949–955
9. Deora AA et al (1998) A redox-triggered ras-effector interaction. Recruitment of phosphatidylinositol 3′-kinase to Ras by redox stress. J Biol Chem 273:29923–29928
10. Huang JS et al (2001) Role of receptor for advanced glycation end-product (RAGE) and the JAK/STAT-signaling pathway in AGE-induced collagen production in NRK-49F cells. J Cell Biochem 81:102–113
11. Huttunen HJ, Fages C, Rauvala H (1999) Receptor for advanced glycation end products (RAGE)-mediated neurite outgrowth and activation of NF-kappaB require the cytoplasmic domain of the receptor but different downstream signaling pathways. J Biol Chem 274: 19919–19924
12. Huttunen HJ, Kuja-Panula J, Rauvala H (2002) Receptor for advanced glycation end products (RAGE) signaling induces CREB-dependent chromogranin expression during neuronal differentiation. J Biol Chem 277: 38635–38646
13. Kislinger T et al (1999) N(epsilon)-(carboxymethyl)lysine adducts of proteins are ligands for receptor for advanced glycation end products that activate cell signaling pathways and modulate gene expression. J Biol Chem 274:31740–31749
14. Lander HM et al (1997) Activation of the receptor for advanced glycation end products triggers a p21(ras)-dependent mitogen-activated protein kinase pathway regulated by oxidant stress. J Biol Chem 272:17810–17814
15. Yan SD et al (1994) Enhanced cellular oxidant stress by the interaction of advanced glycation end products with their receptors/binding proteins. J Biol Chem 269:9889–9897
16. Cipollone F et al (2003) The receptor RAGE as a progression factor amplifying arachidonate-dependent inflammatory and proteolytic response in human atherosclerotic plaques: role of glycemic control. Circulation 108: 1070–1077
17. Park L et al (1998) Suppression of accelerated diabetic atherosclerosis by the soluble receptor for advanced glycation endproducts. Nat Med 4:1025–1031
18. Goova MT et al (2001) Blockade of receptor for advanced glycation end-products restores effective wound healing in diabetic mice. Am J Pathol 159:513–525
19. Lue LF et al (2001) Involvement of microglial receptor for advanced glycation endproducts (RAGE) in Alzheimer's disease: identification of a cellular activation mechanism. Exp Neurol 171:29–45

20. Lue LF et al (2005) Preventing activation of receptor for advanced glycation endproducts in Alzheimer's disease. Curr Drug Targets CNS Neurol Disord 4:249–266
21. Hofmann MA et al (2002) RAGE and arthritis: the G82S polymorphism amplifies the inflammatory response. Genes Immun 3: 123–135
22. Chavakis T, Bierhaus A, Nawroth PP (2004) RAGE (receptor for advanced glycation end products): a central player in the inflammatory response. Microbes Infect 6:1219–1225
23. Taguchi A et al (2000) Blockade of RAGE-amphoterin signalling suppresses tumor growth and metastases. Nature 405:354–360
24. Huttunen HJ et al (2002) Receptor for advanced glycation end products-binding COOH-terminal motif of amphoterin inhibits invasive migration and metastasis. Cancer Res 62:4805–4811
25. Bartling B et al (2005) Down-regulation of the receptor for advanced glycation end-products (RAGE) supports non-small cell lung carcinoma. Carcinogenesis 26:293–301
26. Bhawal UK et al (2005) Association of expression of receptor for advanced glycation end products and invasive activity of oral squamous cell carcinoma. Oncology 69:246–255
27. Ishiguro H et al (2005) Receptor for advanced glycation end products (RAGE) and its ligand, amphoterin are overexpressed and associated with prostate cancer development. Prostate 64:92–100
28. Malherbe P et al (1999) cDNA cloning of a novel secreted isoform of the human receptor for advanced glycation end products and characterization of cells co-expressing cell-surface scavenger receptors and Swedish mutant amyloid precursor protein. Brain Res Mol Brain Res 71:159–170
29. Schlueter C et al (2003) Tissue-specific expression patterns of the RAGE receptor and its soluble forms—a result of regulated alternative splicing? Biochim Biophys Acta 1630:1–6
30. Yonekura H et al (2003) Novel splice variants of the receptor for advanced glycation endproducts expressed in human vascular endothelial cells and pericytes, and their putative roles in diabetes-induced vascular injury. Biochem J 370:1097–1109
31. Park IH et al (2004) Expression of a novel secreted splice variant of the receptor for advanced glycation end products (RAGE) in human brain astrocytes and peripheral blood mononuclear cells. Mol Immunol 40:1203–1211
32. Ding Q, Keller JN (2005) Evaluation of rage isoforms, ligands, and signaling in the brain. Biochim Biophys Acta 1746:18–27
33. Ding Q, Keller JN (2005) Splice variants of the receptor for advanced glycosylation end products (RAGE) in human brain. Neurosci Lett 373:67–72
34. Hudson BI et al (2008) Identification, classification, and expression of RAGE gene splice variants. FASEB J 22:1572–1580
35. Geroldi D et al (2005) Decreased plasma levels of soluble receptor for advanced glycation endproducts in patients with essential hypertension. J Hypertens 23(9):1725–1729
36. Parkin E, Harris B (2009) A disintegrin and metalloproteinase (ADAM)-mediated ectodomain shedding of ADAM10. J Neurochem 108:1464–1479
37. Raucci A et al (2008) A soluble form of the receptor for advanced glycation endproducts (RAGE) is produced by proteolytic cleavage of the membrane-bound form by the sheddase a disintegrin and metalloprotease 10 (ADAM10). FASEB J 22:3716–3727
38. Zhang L et al (2008) Receptor for advanced glycation end products is subjected to protein ectodomain shedding by metalloproteinases. J Biol Chem 283:35507–35516
39. Santilli F et al (2009) Soluble forms of RAGE in human diseases: clinical and therapeutical implications. Curr Med Chem 16:940–952
40. Vazzana N et al (2009) Soluble forms of RAGE in internal medicine. Intern Emerg Med 4:389–401
41. Maillard-Lefebvre H et al (2009) Soluble receptor for advanced glycation end products: a new biomarker in diagnosis and prognosis of chronic inflammatory diseases. Rheumatology (Oxford) 48:1190–1196
42. Emanuele E et al (2005) Circulating levels of soluble receptor for advanced glycation end products in Alzheimer disease and vascular dementia. Arch Neurol 62:1734–1736
43. Jing R et al (2010) Receptor for advanced glycation end products (RAGE) soluble form (sRAGE): a new biomarker for lung cancer. Neoplasma 57:55–61
44. Jing RR et al (2010) Tissue-specific expression profiling of receptor for advanced glycation end products and its soluble forms in esophageal and lung cancer. Genet Test Mol Biomarkers 14:355–361
45. Takeuchi A et al (2007) Endogenous secretory receptor for advanced glycation endproducts as a novel prognostic marker in chondrosarcoma. Cancer 109:2532–2540
46. Kalea AZ et al (2010) Alternatively spliced RAGEv1 inhibits tumorigenesis through suppression of JNK signaling. Cancer Res 70: 5628–5638

47. Kalea AZ et al (2009) Alternative splicing of the murine receptor for advanced glycation end-products (RAGE) gene. FASEB J 23: 1766–1774
48. Sterenczak KA et al (2009) Cloning, characterisation, and comparative quantitative expression analyses of receptor for advanced glycation end products (RAGE) transcript forms. Gene 434:35–42
49. Murua Escobar H et al (2006) Cloning and characterization of the canine receptor for advanced glycation end products. Gene 369:45–52
50. Breen M et al (1999) Reciprocal chromosome painting reveals detailed regions of conserved synteny between the karyotypes of the domestic dog (Canis familiaris) and human. Genomics 61:145–155
51. Yang F et al (1999) A complete comparative chromosome map for the dog, red fox, and human and its integration with canine genetic maps. Genomics 62:189–202
52. Harashima A et al (2006) Identification of mouse orthologue of endogenous secretory receptor for advanced glycation end-products: structure, function and expression. Biochem J 396:109–115
53. Khajavi M, Inoue K, Lupski JR (2006) Nonsense-mediated mRNA decay modulates clinical outcome of genetic disease. Eur J Hum Genet 14:1074–1081

Part V

Ca^{2+}-Binding Proteins in Human Diseases and As Drug Targets

Chapter 17

Screening for Novel Calcium-Binding Proteins that Regulate Cardiac Hypertrophy: CIB1 as an Example

Joerg Heineke

Abstract

Calcium-binding proteins have a crucial function in the regulation of cardiac contractility as well as in the regulation of cardiac signal-transduction. Because they sense calcium concentrations and at the same time bind specific signaling molecules, some of these proteins are critically involved in the establishment of signaling microdomains, which are insulated from the large cytosolic calcium fluctuations involved in cardiac excitation–contraction coupling. In this regard, we have recently identified the calcium-binding protein CIB1 as an important regulator of pathological cardiac hypertrophy and transition to heart failure. It is almost certain that more, currently unknown calcium-binding proteins with similar regulatory function in cardiac signaling exist. Here, I suggest screening strategies to identify these calcium-binding proteins with impact on cardiac hypertrophy and provide a detailed protocol for the identification of protein interaction partners. I also describe cell culture-based models for cardiomyocyte hypertrophy as well as mouse models for pathological or physiological hypertrophy and strategies to analyze the impact of candidate genes on the development of hypertrophy.

Key words: Calcium, Heart, Hypertrophy, Screening, CIB1, Calcium-binding proteins, EF-hand

1. Introduction

Cardiac hypertrophy describes a postnatal growth process of cardiac myocytes, which leads to enlargement of the heart. Depending on whether growth is induced by either pathological stimulation or physiologically occurring stimuli, this hypertrophy is termed "pathological" or "physiological." Pathological stimuli that trigger hypertrophy are on the one hand those leading to pressure overload (like chronic arterial hypertension, aortic valve stenosis) as well as those leading to volume overload (like myocardial infarction or aortic valve insufficiency) of the heart (1). Pathological hypertrophy is

Claus W. Heizmann (ed.), *Calcium-Binding Proteins and RAGE: From Structural Basics to Clinical Applications*, Methods in Molecular Biology, vol. 963, DOI 10.1007/978-1-62703-230-8_17, © Springer Science+Business Media New York 2013

often associated with myocardial fibrosis and reactivation of the so-called embryonic gene program (e.g., re-expression of ANP and BNP) in ventricular cardiac myocytes. Importantly, pathological cardiac growth is coupled with increasing cardiac dysfunction and development of heart failure. In contrast, physiological hypertrophy is triggered by regular strenuous exercise (e.g., in athletes) or during pregnancy (1). Physiological hypertrophy is purely adaptive and not associated with the development of heart failure, cardiac dysfunction, cardiac fibrosis, or embryonic myocardial gene-expression (1). Pathological cardiac hypertrophy is viewed as an important treatment target during heart failure and various signaling pathways within cardiomyocytes have already been identified, although specific molecular treatment strategies remain elusive.

Within cardiac myocytes calcium participates in multiple crucial cellular functions, such as the regulation of signal-transduction as well as the induction of muscle contraction. Entrance of calcium through the voltage-dependent L-type channel upon membrane depolarization triggers a massive release of calcium from the sarcoplasmatic reticulum (SR) and its subsequent binding to troponin C to trigger interaction of myosin and actin filaments and cardiac contraction during systole (2). In that regard, cytosolic calcium levels between systole (cardiac contraction) and diastole (cardiac relaxation) fluctuate between 1 μM in systole and 100 nM in diastole with every heartbeat. Therefore, targeted activation of signaling molecules, for example during pathological loading of the heart, has to be tightly controlled and insulated from these constant variations in cytosolic contraction-related calcium. In fact, it has even been suggested that “contractile calcium” is not at all contributing to the activation of signaling molecules regulating cardiac hypertrophy upon pathological cardiac overload (e.g., chronic arterial hypertension) (3, 4). Indeed, enhancement of the contractile calcium pool by genetic means (resulting in increased activity of the SR calcium pump SERCA) did not result in cardiac hypertrophy (3). This implies the existence of a separate calcium pool that is responsible for activation of hypertrophy-related signaling molecules and that has therefore been termed “signaling calcium” (4). It has been suggested that signaling calcium acts in microdomains in which calcium-dependent kinases or phosphatases are exposed to exceedingly high local concentrations of calcium in addition to their respective molecular targets. Calcium-binding proteins play a critical role within signaling microdomains. One such molecule is calmodulin, which is a prototypical calcium sensor with four canonical calcium-binding EF-hand domains (5). With its relatively low affinity, calcium only binds to calmodulin at high calcium concentrations and consequently induces a conformational switch within the molecule that enables binding of target molecules. Calcium/calmodulin-binding partners with hypertrophy-inducing properties include the calmodulin-dependent kinase II (CamKII), the transcriptional co-activator CAMTA2, as well as the phosphatase calcineurin. CamKII is a serine–threonine kinase with a broad spectrum of substrates. Upon calcium release from the nuclear envelope CamKII becomes activated by calcium/calmodulin to phosphorylate HDAC5, leading to its nuclear export and derepression of pro-hypertrophic gene-expression (6). Therefore, CamKII overexpression in the mouse heart induces, while genetic deletion of CamKIIδ (the main isoform in the heart) protects from, cardiac hypertrophy and failure (7–10). Calmodulin-binding transcription activator 2 (CAMTA2) is highly expressed in heart and brain and activates hypertrophic gene-expression by association with the cardiac homeodomain protein Nkx2-5 in cardiac myocytes (11). While cardiomyocyte-specific overexpression of CAMTA2 leads to severe cardiac hypertrophy, homozygous deletion of CAMAT2 in mice blunts hypertrophy due to cardiac pressure overload or stimulation by angiotensin II infusion (11). Finally, calcium/calmodulin binding to the phosphatase calcineurin shifts the C-terminal auto-inhibitory domain off the catalytic site of calcineurin, rendering the molecule active to dephosphorylate its substrates (12). The most prominent calcineurin substrates belong to the family of nuclear factor of activated T cells (NFAT). Upon their dephosphorylation by calcineurin, NFATs translocate to the nucleus to activate pro-hypertrophic gene-expression (12). In that regard it was shown that constitutive overexpression of activated calcineurin or activated NFAT leads to cardiac hypertrophy in mice, while in turn genetic deletion of the stress-responsive isoform of calcineurin (calcineurin Aβ) inhibits the development of myocardial hypertrophy (13, 14). Calcineurin consists of two subunits: Calcineurin A (CnA), which is bound by calcium/calmodulin and contains the catalytic center in its N-terminal region, and calcineurin B (CNB), a small 4 EF-hands containing molecule with high affinity for calcium. CnA is not stable without CNB and even constitutively active calcineurin depends on calcium-bound CNB (15, 16). We recently discovered CIB1 as another calcium-binding protein with crucial function for CnA activation (16). CIB1 is ubiquitously expressed, although a particularly high expression has been observed in the heart. It was originally identified to bind to the cytoplasmic domain of the integrin αIIb in platelets, where it inhibits activation of the αIIbβ3 integrin complex (17). CIB1 is a 22-kDa protein containing four EF-hand domains, but only the third and fourth C-terminal EF hand domains bind calcium at physiological concentrations (18). It can be myristoylated at the N-terminus and plasma-membrane localization of the protein has been demonstrated in multiple cell types (19, 20). Indeed, CIB1 binds and influences multiple signaling proteins in vicinity of the plasma-membrane, including sphingosine kinase 1, focal adhesion kinase, p21-activated kinase, and calcineurin (17, 19, 21–23).

We have recently shown that CIB1 directly interacts with CNB and supports the recruitment of calcineurin to the sarcolemma (the plasma-membrane of cardiac myocytes) as well as its activation (16). CIB1 also binds to the L-type calcium channel during pathological hemodynamic overload and thereby sets up a microdomain, in which calcineurin is anchored close to its local calcium source and its main target NFAT at the membrane (16). Importantly, ablation of CIB1 protected the mouse heart from pathological hypertrophy and the development of heart failure, thus rendering CIB1 a potential target of future therapeutic strategies (16).

1.1. Screening for Novel Calcium-Binding Proteins

Because the existence of more calcium-binding regulatory proteins with crucial functions in cardiomyocyte (microdomain) signaling and induction of myocardial hypertrophy or heart failure is very likely, adequate strategies need to be employed to enable fast, efficient, and reliable identification of these proteins. Most likely this will contribute not only to our understanding of calcium-dependent signal-transduction in cells with large variations in cytosolic calcium concentrations but also to the development of new medical therapies counteracting hypertrophy and heart failure. As an unbiased approach, multiple different screening procedures could be used for this endeavor:

1. Gene-expression profiling in order to identify calcium-binding proteins that are up- or down-regulated in isolated cardiac cells or whole hearts under different conditions. For example, one could sample myocardial gene-expression in different mouse models of cardiac hypertrophy (e.g., mouse models of myocardial infarction, physiological hypertrophy, or transverse aortic constriction to induce cardiac pressure overload, which will lead to pathological hypertrophy and transition to heart failure as described in detail below) and compare it between various disease stages and the un-diseased state. Gene-expression analysis can be applied in this regard to myocardial tissue or specific cells isolated from the heart (e.g., cardiomyocytes, endothelial cells, fibroblasts) or of course to cultured myocardial cells. Besides analyzing different disease states, one could also analyze the impact of reduced or enhanced activation of certain pro- or anti-hypertrophic signaling molecules (e.g., transcription factors, signaling kinases, or phosphatases). For this, one could over-express or down-regulate the respective molecule in cell culture using adenoviral or lenti-viral overexpression or si/shRNA-based techniques for downregulation. For mouse models, one could sample tissue or cells from systemic or cell-specific knockout mice or transgenes with cell-specific overexpression. Expressed genes with significant down- or upregulation (e.g., by twofold or more) in hypertrophy or induced by pro-hypertrophic signaling molecules should be

screened for calcium-binding protein motifs like EF-hand motifs, calmodulin-binding domains, or calcium-binding C2 domains (for example on the NCBI Web site: http://www.ncbi.nlm.nih.gov/cdd?term), in order to identify novel calcium-binding signaling proteins with impact on cardiac hypertrophy. By this approach, we previously identified CIB1 as an important regulator of pathological hypertrophy (16): We over-expressed the pro-hypertrophic transcription factor GATA4 in primary isolated neonatal rat cardiomyocytes (NRCM, a well-established, reliable in vitro model of cardiomyocyte hypertrophy, described in detail below) and compared the gene-expression with NRCMs over-expressing a control protein (lac-Z). Transcriptional profiling by micro-array revealed 2–2.5-fold upregulation of CIB1 mRNA, which we later independently confirmed by Northern-Blot and on the protein level by Western-blot (16). We extended our expression analysis to show that CIB1 was also up-regulated in mouse hearts during pathological hypertrophy due to aortic constriction (a good model of pathological hypertrophy and heart failure in vivo, described in detail below), but not in physiological hypertrophy (induced by chronic swimming exercise in mice, also described in detail below) (16). Besides micro-array-based transcriptome profiling, one could also employ next-generation sequencing approaches, or proteomic methods (e.g., DIGE). Since all these techniques require very specialized, expensive equipment and knowledge they are almost always performed by core-facilities and the only thing the researcher has to do is design the experiment and then submit the samples to the facility. Because of this fact, I describe possible experimental setups here, instead of giving detailed protocols of the transcriptional profiling methods itself.

Of course, lack of expression changes would not at all mean that the respective calcium-binding protein is not important during cardiac hypertrophy; therefore two other screening approaches are subsequently briefly described.

2. One could screen for protein interactions to identify (novel) calcium-binding proteins with impact on cardiac hypertrophy. In this regard the "interactome" (all interaction partners of a given protein) of hypertrophy-inducing transcription factors or signal-transduction proteins (e.g., particular kinases or phosphatases) could be of high interest. For example, a yeast-two hybrid screen with the pro-hypertrophic transcription factor GATA4 as bait revealed the transcription factor NFAT3 as previously unknown interaction partner of GATA4 (14). Because NFAT was known at that time already as being activated by the phosphatase calcineurin in T-cells, the importance of calcineurin for NFAT activation and hypertrophy in cardiomyocytes was discovered as a result of that initial yeast-

two hybrid screen (14). In addition, the small non-catalytic subunit of calcineurin (CNB) was identified as an important binding partner of CIB1 when CIB1 was used as bait in a yeast-two hybrid screen (16). Since yeast-two hybrid screens are commercially available now (for example the "Matchmaker" system from Clontech), I will not describe the yeast-two hybrid screen in detail here. As alternative approach, one could use a GST-fusion protein pulldown analysis to identify new interaction partners. For this procedure, a GST-fusion protein of the pro-hypertrophic molecule of interest needs to be cloned, synthesized, and purified. Subsequently, this GST-fusion protein as well as GST (both coupled to sepharose beads) are separately incubated with cardiac protein lysates (either from hearts or cardiomyocytes). After washing, the proteins bound to the beads are released by a denaturing step and analyzed side by side with SDS-PAGE gel analysis and subsequent Coomassie staining. Specific interaction partners will show up as a specific band as the results of the GST-fusion protein pulldown, but not of the GST pulldown. Bands of interest are excised and sent for mass-spectrometry analysis to identify the protein. The method is described in detail below. To confirm the results of the screen, one needs to perform independent immunoprecipitations or GST-pulldown experiments to confirm the interaction.

3. In order to identify novel calcium-binding proteins with impact on hypertrophy one could also conduct a "functional" screen. For this approach, a large amount of clones (e.g., a particular cDNA library) are tested for their ability to induce cardiomyocyte hypertrophy or hypertrophic signaling. As an example, pools of cDNA clones were tested for their ability to activate the ANF (atrial natriuretic factor) promoter, which typically becomes activated in cardiomyocyte hypertrophy (11). In detail, 2000 cDNA pools (each containing around 100 cDNA clones) were co-transfected into COS cells with an ANF promoter construct, which was coupled to a luciferase cassette (11). One clone that was found to activate the ANF promoter by more than 20-fold was CAMTA2, a calcium/calmodulin-binding protein, which was subsequently identified as a strong inducer of cardiomyocyte hypertrophy in vitro and in mouse hearts in vivo (11). Besides screening in COS cells one could directly use neonatal rat cardiac myocytes to perform a functional screen. However, because of the low transfection efficiency of cardiomyocytes (usually 1–2%), one has to deliver the cDNA clones for example via lentiviral gene-transfer. In addition to cDNA libraries, also shRNA libraries could be used to target cardiac myocytes in a screening procedure. In that case, a loss of function approach could reveal calcium-binding proteins that are required to mount a hypertrophy response.

Besides using ANF-luciferase as a readout in cardiac myocytes, cell size could also be used as an alternative, although an automated software-based approach for cell-size measurements needs to be in place for that.

In the following sections, I give a detailed overview on models of cardiomyocyte hypertrophy in vitro and in vivo and I also suggest a protocol for a GST-pulldown protein interaction screen.

2. Materials

2.1. Neonatal Rat Cardiac Myocytes as a Model of Cardiac Hypertrophy In Vitro

Course, small 90° angled forceps (2×) (Roboz).
Dressing forceps (Roboz).
Spring scissors (Roboz).
10×ADS buffer: 1.164 M NaCl, 54 mM KCl, 56 mM dextrose, 109 mM NaH_2PO_4, 4.057 mM $MgSO_4$, 200 mM HEPES, adjust pH to 7.3 with 1 M NaOH. Dilute 1:10 with H_2O and adjust pH again for 1×ADS.
1×ADS buffer (*red*): Add 2% phenol red to 1×ADS.
Enzyme solution (200 ml, for the preparation of 85–120 hearts): 88 mg collagenase, type 2 (Worthington Biochemical Corporation). 20 mg pancreatin (Sigma-Aldrich), fill up to 200 ml with 1×ADS, filter sterile.
Percoll Gradient (for 10 gradients): Percoll stock: 40.5 ml Percoll reagent (GE Healthcare) + 4.5 ml 10×ADS, top Percoll: 16 ml Percoll stock + 24 ml 1×ADS, bottom Percoll: 26 ml Percoll stock + 14 ml 1×ADS (red).
Plating medium: 400 ml DMEM high Glucose, 100 ml Medium 199, 10% Horse serum, 5% FBS, 1% Glutamine, 1% Penicillin/Streptomycin.
Maintenance Medium: 400 ml DMEM high Glucose, 100 ml Medium 199, 1% Glutamine, 1% Penicillin/Streptomycin.

2.2. Transverse Aortic Constriction and Swimming Exercise as Mouse Models of Hypertrophy In Vivo

MiniVent (ventilator, Harvard Apparatus).
Isoflurane vaporizer (Vetequip Inc.).
Face mask for mice (Vetequip Inc.).
Electric Shaver (GE).
Operating table (equipped to be heated, Harvard Apparatus).
Buprenorphine (Merck).
Bandaging tape.
1½ 20 gauge angiocatheter, blunted (BD).
Blunt scissors (Roboz).
Spring scissors (Roboz).
Course, small 90° angled forceps (2×) (Roboz).
Dressing forceps (Roboz).
Chest retractor (Fine Science Tools).

Needle holder (Roboz).
27 Gauge needle 90° curved, blunted (BD).
7–0 silk suture material for constriction (Johnson & Johnson).
7–0 prolene suture material for chest closure (Johnson & Johnson).
40 cm (width) × 60 cm (length) × 40 cm (height) plastic basin.
Water heater.
Thermometer.

2.3. GST-Pulldown Screen for the Identification of Protein-Binding Partners

100 mM IPTG (Sigma-Aldrich).
Cold Buffer: 50 mM Tris–HCl, pH 7.5, 150 mM NaCl, 1 mM EDTA, 10% glycerol.
100 mM PMSF in Isopropanol.
Glutathione Sepharose 4B beads (GE Healthcare).
Binding Buffer: 20 mM HEPES (pH 7.6), 75 mM NaCl, 0.05% NP40.
Protease Inhibitor Tablet (complete, from Roche).
SDS-PAGE Resolving Gel (12%, 80 ml).
26.4 ml H_2O, 32.0 ml 30% Acrylamide/Bis (37.5:1), 20 ml 1.5 M Tris–HCl (pH 8.8), 0.8 ml 10%SDS, 0.8 ml 10% ammonium persulfate, 0.032 ml TEMED.
SDS-PAGE Stacking Gel (10 ml).
6.8 ml H_2O, 1.7 ml 30% Acrylamide/Bis (37.5:1), 1.25 ml 1.5 M Tris–HCl (pH 6.8), 0.1 ml 10%SDS, 0.1 ml 10% ammonium persulfate, 0.01 ml TEMED.
40% Acrylamide (Roth).
5×Western Sample Buffer: 10% SDS (w/v), 10 mM DTT or β-mercapto-ethanol, 0.2 M Tris–HCl, pH6.8, 0.05% Bromophenol blue (w/v).
Coomassie staining solution: 0.2 g Coomassie Brilliant Blue R-250, 30% methanol, 10% acetic acid, H_2O to a total volume of 200 ml.
Coomassie destaining solution: 30% methanol, 10% acetic acid in H_2O.

3. Methods

3.1. Neonatal Rat Cardiac Myocytes as a Model of Cardiac Hypertrophy In Vitro

For gene-expression profiling, GST-pulldown screens, as well as for the conduction of a functional screen it is crucial to establish a good cell culture-based model of cardiac hypertrophy in the lab. In addition, this will also be indispensable for the examination of candidate genes obtained from screens, for example to test their impact on cardiac hypertrophy.

Preparation of neonatal rat hearts (a typical preparation will use a maximum of 100 rat hearts, although the prep can easily be downscaled to isolate hearts from only 50 rats):

1. Newborn rats (1–3 days old) are held with large forceps and are quickly decapitated and transferred to lie on their back onto a clean paper towel. 4–10 neonates are processed at one time.
2. The neonate is fixed with large forceps and the thorax is cut along the sternum with a small scissors, but care has to be taken not to open the abdominal cavity.
3. The tweezers are used to spread open the thorax and by carefully pushing downwards release the heart, which is captured by a small curved forceps and immediately transferred to a Petri dish containing cold 1×ADS buffer on ice (between 30 and 50 hearts can be collected in one dish like this).
4. After all hearts have been collected, the atria and vessels are removed from the base of the heart and the ventricles are transferred to a fresh Petri dish with cold 1×ADS on ice.
5. The hearts are cut into about eight pieces each within the Petri dish and then transferred to a 50 ml Falcon (max. 50 hearts/Falcon).

   Predigestion of the hearts
6. The superfluous 1×ADS is removed and ice-cold enzyme solution is added to the Falcons containing the hearts (0.3 ml/heart).
7. The Falcons are rotated (6–7 revolutions/min) at 37°C for 10 min. FBS is thawed during this time.
8. After the heart pieces settle on the ground of the Falcon, the supernatant is removed and discarded.

   Digestion of hearts
9. Fresh enzyme solution is given into the Falcons containing the heart pieces. Incubation is conducted at 37°C with rotation for 20 min.
10. After the heart pieces settle, the supernatant (containing the cells) is removed and transferred into a 15 ml Falcon containing 2 ml of FBS. New enzyme solution is added to the heart pieces and the digestion is repeated. Together, the digestion procedure should be performed five to six times. See also Note 1.
11. The Falcon containing the supernatant with cells and FBS is centrifuged in a tabletop centrifuge at 700 rpm (94 × *g*) for 5 min at room temperature. The supernatant is removed and discarded and the cells are resuspended in 1 ml FBS using a blue (1 ml) pipette tip. Subsequently, the cells are transferred to a 50 ml Falcon, in which all cells are collected and stored in a tissue culture incubator at 37°C (with the lid only loosely closed) during the digestion procedure.
12. After the last digest, the 50 ml Falcon with all collected cells is centrifuged at 700 rpm for 7 min. The cells are then resuspended

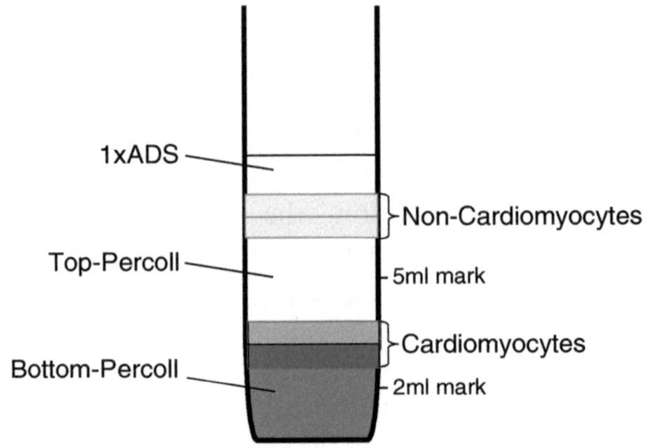

Fig. 1. Schematic representation of the Percoll gradient after centrifugation.

using a pipette tip with 1 ml 1×ADS. Add 1×ADS to obtain a total volume of 20 ml 1×ADS (if only 50 hearts were processed, ADS should be added only to a total volume of 10 ml; the total volume required in this step can be calculated as the amount of Percoll gradients × 2 ml).

13. During the incubation times of the digest, one should gelatinize the tissue culture plates (with 1% Gelatine, which is incubated briefly on the plates before it is removed and the plates are dried under the cell culture hood) and set up the Percoll gradients (see next paragraph).

    A Percoll gradient is used in order to separate the cardiomyocytes from the non-myocyte cells (i.e., fibroblasts and endothelial cells).

14. The Percoll gradients have to be carefully prepared (preferably during the digestion procedures). The number of Percoll gradients depends on the amount of rat hearts used in the beginning. Roughly 1 gradient is needed per 10 hearts (and therefore 10 gradients are needed, when 100 hearts are used in the beginning). For each gradient 4 ml of top-Percoll (white) are dispensed at the bottom of a 15 ml Falcon. Then 3 ml of bottom-Percoll solution are carefully placed under the top-Percoll by bringing the pipette tip to the bottom of the Falcon and slowly releasing the pipette's content (see Fig. 1).

15. 2 ml of cells are very carefully loaded on top of each gradient before the gradients are centrifuged at 3,000 rpm (1,730 × g) for 30 min at room temperature. It is important to inactivate the brakes of the centrifuge to enable a very slow reduction of speed by the centrifuge and therefore keep the gradients and the cell separation intact.

16. Two main layers of cells can be observed (Fig. 1). The top layer (between 1×ADS and top-Percoll) is formed by non-

myocytes. When these cells are not needed, this layer is removed by a sterile glass pipette with suction (approximately until the 5 ml mark of the Falcon, see Note 2). The next cell layer contains the cardiomyocytes and should be taken up with a 5 ml pipette and transferred to a 50 ml Falcon (up to a maximum of 5 gradients per Falcon). The 50 ml Falcons need to be filled up until the 40 ml mark with 1×ADS before they are centrifuged at 700 rpm for 7 min. The supernatant is then discarded, while the cells are resuspended in 1 ml 1×ADS with a blue (1 ml) tip and Falcons are subsequently filled up to the 30 ml mark. Another centrifugation is performed at 700 rpm, for 7 min, after which the supernatant is discarded and the cardiomyocyte pellet resuspended in 1 ml plating medium, before more plating medium is added to a total of 0.3 ml/heart.

17. In order to count the cells, two 10 μl aliquots of the cell suspension are each supplemented with 10 μl trypan blue and the non-blue cells (the dead cells are stained blue by trypan blue) are counted by a hemocytometer. In order to determine the cell concentration, the result from the hemocytometer still needs to be multiplied by 2 (because of the dilution by trypan blue). The cells are then seeded onto gelatinized cell-culture plates at a cell density of roughly (212 cells/mm^2). For the analysis of cardiac hypertrophy, cardiomyocyte density should be between 60 and 70% (see also Note 3).
18. The cells are kept in plating medium overnight, before they are twice washed in PBS and switched to serum-free maintenance medium on day 2 of culture.

Hypertrophy can be induced in these primary cardiomyocytes by adding growth factors or serum (see Table 1) to the culture medium on day 3 of culture (i.e., after 24 h without any serum).

Table 1
Different pro-hypertrophic agents, their final concentration in cell culture media, and potential suppliers are listed

Pro-hypertrophic agent	Supplier	Final concentration in culture media
Phenylephrine	Sigma-Aldrich	20 μM
Endothelin-1	Sigma-Aldrich	100nM
Isoproterenol	Sigma-Aldrich	10 μM
Fetal bovine serum (FBS)	Invitrogen/Life technologies	2%
IGF-1	R&D systems	20 ng/ml

Typically, cardiomyocyte hypertrophy emerges 24–48 h in growth factor-stimulated versus unstimulated cells (i.e., cells kept in maintenance medium without added serum or growth factors). Hypertrophy can be quantified by measuring cardiomyocyte cell size using an imaging software (e.g., Image J from the NIH Web site), preferably after immunostaining for a cardiomyocyte-specific marker (like α-actinin). As another characteristic of hypertrophy the protein/DNA ratio is increased and can be measured. Finally one should analyze the typical changes in gene-expression that are associated with cardiomyocyte hypertrophy (increased mRNA expression of ANP, BNP, α-skeletal Actin, and β-MHC and decreased expression of α-MHC and SERCA2a); this is typically done by standard Northern Blot or quantitative real-time PCR. In order to test whether any identified calcium-binding protein participates in the regulation of cardiac hypertrophy, one could start by either over-expressing or down-regulating this gene and evaluate the effect on hypertrophy. In our lab, overexpression is typically achieved by infecting the cells with a recombinant adenovirus (generated with the help of the Adeasy system from Stratagene), while downregulation is achieved by transfecting specific siRNAs (obtained for example from Sigma-Aldrich) with the use of a standard transfection reagent (e.g., Fugene6 from Roche or Lipofectamine 2000 from Invitrogen). As an example, specific downregulation of CIB1 by siRNA reduced the hypertrophic response after phenylephrine (PE) stimulation in neonatal rat cardiac myocytes, while adenoviral CIB1 overexpression increased hypertrophy during PE stimulation (16).

3.2. Transverse Aortic Constriction and Swimming Exercise as Mouse Models of Hypertrophy In Vivo

In order to enable screening or analysis of a given candidate gene in vivo, mouse models for pathological as well as physiological hypertrophy need to be in place. Importantly, these protocols first need to be approved by the local animal care facilities before they can be applied.

3.2.1. Transverse Aortic Constriction (TAC)

This method was originally described by Rockmann et al. and is illustrated here in Fig. 2 (Fig. 2a contains a schematic representation of the procedure) (24). For best results 8–12-week-old mice should be used. Before starting the procedure, mice are depilated at the ventral thoracic skin and also receive an intraperitoneal injection of buprenorphine (0.1 mg/kg) for intra- and postoperative analgesia.

1. Mice are anesthetized by 3% isoflurane and 600 ml/min oxygen flow in an induction chamber for 3–5 min, before being transferred onto their back on the operating table while staying connected to the mask anesthesia apparatus. The forelegs are held down by bandaging tape. In order to expose the trachea, a midline incision is made just cranial to the manubrium

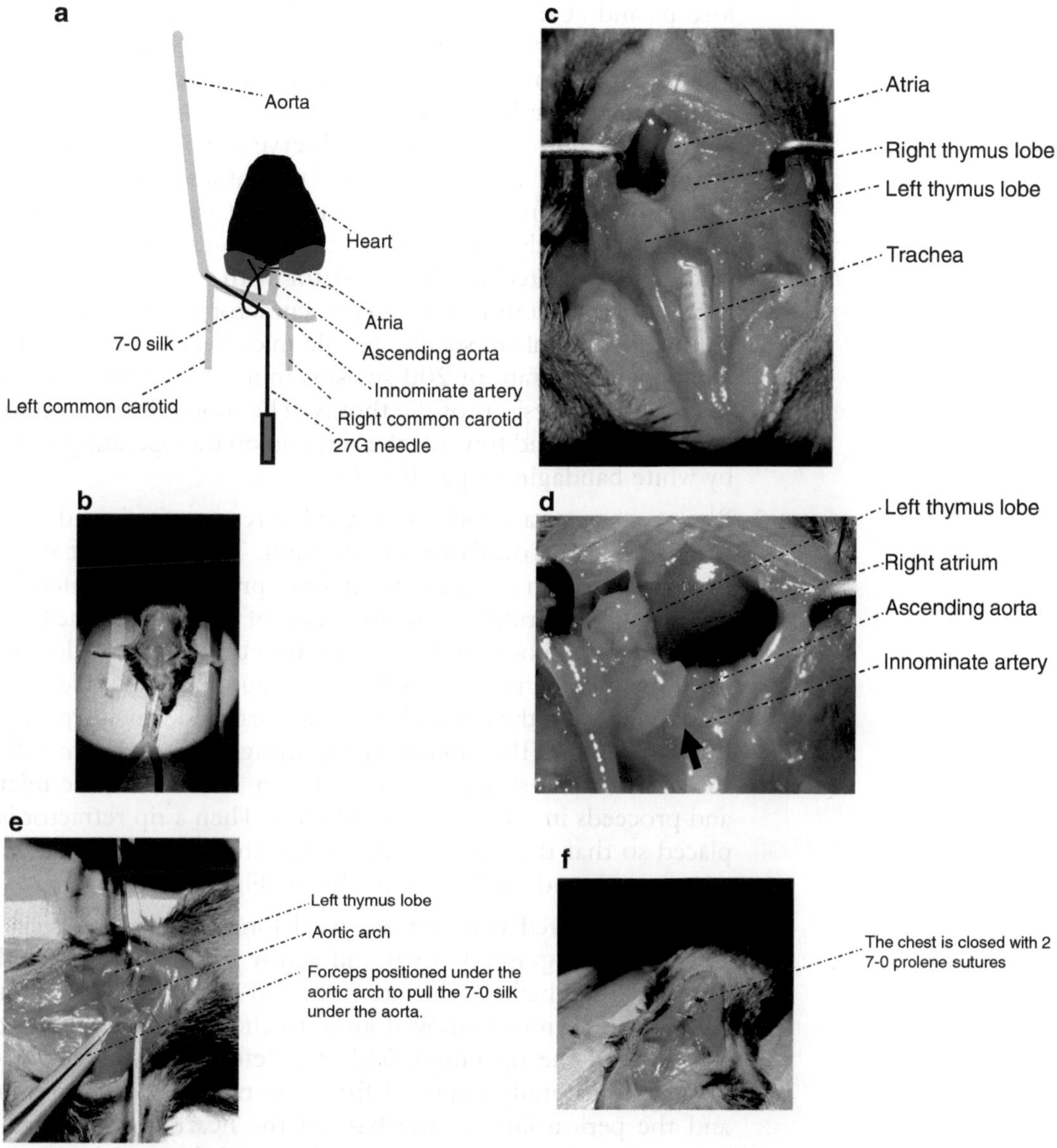

Fig. 2. (**a**) Schematic representation of the TAC procedure. (**b**) Positioning of the intubated mouse on the operating table for TAC surgery. (**c–f**) Anatomical and procedural details during TAC surgery. The *black arrow* demonstrates where a passage under the aortic arch should be initiated. See also text for details.

with a small scissors and the skin is subcutaneously undermined towards the symphysis of the mandible. The skin is incised and the muscles overlying the trachea are moved away from the midline with small tweezers. The mouse is then quickly moved away from the mask anesthesia apparatus into a position with a good light source still on its back with the head to the right. The trachea is fixed with a toothed curved forceps and at the same time, a 1½ 20 gauge angio-catheter with a blunted tip is inserted into the trachea through the mouth with the mouse's head bent far backwards towards the table. A slight resistance is felt when the catheter passes the vocal cords and one will clearly observe the catheter moving into the trachea. The plastic end of the catheter at the mouth is held by two fingers, while the metal rot part of the angio-catheter is quickly removed and the angio-catheter tubing is carefully connected to the ventilator with the mouse on its back on the operating table. The ventilator is set to a stroke volume of 0.2 ml (in smaller female mice 0.175 ml is recommended) and a rate of 200 strokes/min. One should clearly see regular excursions of the thorax. The mouse is fixed on its back with the head towards the surgeon on the operating table by white bandaging tape (Fig. 2b).

2. Blunt scissors or a scapel can be used to remove fascia and subcutaneous tissue overlying the sternum. The skin is cut open over the sternum towards the xiphoid process. Subsequently, the sternum is carefully split along 2/3 of its length to the base of the heart and one has to be extremely careful not to deviate from the midline more than 2 mm, because this would lead to the laceration of the internal mammary arteries, which typically leads to death of the mouse. The splitting of the sternum with the use of small spring scissors is begun at the thoracic inlet and proceeds in four to five small bites. Then a rip retractor is placed so that the surgeon can see the thymus overlaying the great vessels and the base of the heart (Fig. 2c).
3. Small 90° curved tweezers are used for the following steps. The thymus is grasped slowly and gently at its caudal end and is gently lifted before split into right and left halves. The right half of the thymus is moved away to the right side where it stays out of the operating field. The left half of the thymus should be carefully removed from the transverse aortic arch and the pericardium at the base of the heart to which it is attached to. One should then clearly see the base of the heart, the ascending aortic arch, as well as the innominate and right carotid arteries (see Fig. 2d).
4. Next, small 90° curved tweezers are used to create a passage dorsal of the ascending aortic arch (starting at the tip of the black arrow as depicted in Fig. 2d). For this, the closed tweezers are moved downwards at the left side of the trachea towards the base of the heart and are repeatedly moved under the ascending aortic arch between the innominate and left carotid arteries. When the aortic arch is completely tunnelled, the tweezers are inserted under the aorta and carefully opened to grasp a 7–0 silk to place it along under the aorta (Fig. 2e).

5. A 90° bent blunted 27-gauge needle is put on top of the aorta and a fine needle holder is used to place a triple looped surgeons knot on the aorta including the 27-gauge needle (see Note 4). The knot is tied and quickly another single knot is added to secure the first knot. This should ideally be done within 10 s and then the needle is carefully pulled out of the knot.
6. Before closing the chest wall, trapped air should be removed from the thoracic cavity by massaging it with thumb and index finger, in order to avoid pneumothorax. The chest wall is then closed with two to three simple interrupted 7–0 prolene sutures and great care should again be applied not to lacerate the internal mammary arteries (Fig. 2f). The skin and subcutaneous tissue are closed with surgical glue after the edges of the wound are apposed.
7. The isoflurane can then be discontinued so that ventilation of the mouse occurs with 100% room air; the fixation of the mouse can be released and care should be taken not to extubate the mouse during that process. Respiration needs to be carefully monitored as the mouse will increasingly try to breathe on its own. The mouse can be briefly disconnected from the ventilator and if respiratory efforts progressively slow down, the mouse can be reconnected to the ventilator. If not, the mouse is transferred to an oxygen-enriched, 37°C incubator for full recovery.

During sham operation the exact same procedure is performed, but no ligature is placed. With some practice, mortality rates for TAC surgery in wild-type mice can be as low as 10–20%. In general, cardiac hypertrophy will develop within 1–2 weeks and cardiac dysfunction within 6–8 weeks after TAC surgery, although the development of cardiac dysfunction strongly depends on the genetic background of the mice: while FVBN mice never develop cardiac dysfunction after TAC, C57/Bl6 mice do. In order to test the impact of certain calcium-binding proteins, one should generate or obtain general or preferable cell-specific (e.g., cardiomyocyte specific) knockout mice or over-expressing transgenic mice of the respective gene. If one wants to analyze the development of cardiac hypertrophy, TAC and sham surgery have to be performed in the respective knockout or transgenic mice as well as littermate control/wild-type mice; mice have to be sacrificed 2 or 3 weeks after TAC in order to analyze cardiac hypertrophy as described below. For the analysis of cardiac dysfunction, one should follow left ventricular dimensions and function by weekly echocardiography (starting 2 weeks after surgery) and mice should be sacrificed between 6 and 12 weeks after TAC or sham operation. For example, we analyzed cardiac hypertrophy in *Cib1* homozygous (−/−) and heterozygous (+/−) knockout mice in comparison to wild-type mice (+/+) and discovered that the *Cib1* mutant mice develop significantly less cardiac hypertrophy after 2 weeks of TAC. In addition, weekly echocardiography revealed that the *Cib1* −/− mice were significantly protected against cardiac dysfunction between 3 weeks and 6 weeks after TAC (16).

3.2.2. Chronic Swimming as a Model of Physiological Hypertrophy

Physiological hypertrophy is a purely adaptive response and is associated with activation of adaptive signaling cascades within cardiac myocytes and other cell types in the heart (1). It is therefore recommended, once a novel calcium-binding protein with impact on pathological hypertrophy has been identified, to also carefully test whether this protein might also influence the development of physiological hypertrophy. For example, if inhibition of the protein of interest does also inhibit physiological hypertrophy, interference with adaptive signaling could render such a protein unsuitable as a future therapeutic target of cardiac hypertrophy. *Cib1* −/− mice, while being protected from pathological hypertrophy, developed normal physiological hypertrophy and therefore CIB1 could be a good treatment target for the therapy of pathological hypertrophy and heart failure (16).

1. Fresh water needs to be filled in the water tank for mouse swimming for each session and the water depth should be 15–20 cm, while the water temperature needs to be brought to 30–32°C. Importantly, mice have to be constantly supervised during swimming to be able to rescue them from drowning, although that is rarely necessary (see Note 5).
2. In order to slowly accustom the mice to swimming, the swimming time is gradually increased: on the first day, mice swim for 10 min twice a day, which is gradually increased by 10 min per session each day; therefore on day 2, mice swim twice for 20 min, day 3 twice for 30 min, until on the 9th day mice swim twice for 90 min (see also Note 6).
3. From the 9th day until the 21st day, mice swim twice per day for 90 min. Mice are sacrificed on day 22 and hypertrophy is quantified.

3.2.3. Quantification of Hypertrophy In Vivo

The most important measure of cardiac hypertrophy is also very easy to obtain: the heart weight. After removal of the heart from the mouse, it is washed in sterile, cold 1×PBS, and forceps are used to carefully squeeze the heart to remove blood from its chambers. The heart is then transferred to a clean paper towel and excess tissue (e.g., epicardial fat) and the big vessels are removed from the base of the heart. It is then washed again, before being carefully dried on a paper towel and weighed. The heart weight is usually well in proportion with body weight or other measures of body size. Therefore, heart weight is expressed as a ratio with body weight (heart weight/body weight, HW/BW in mg/g) or tibia length (heart weight/tibia length, HW/TL in mg/mm). While the normal HW/BW ratio is about 4.5–5.0, it increases after TAC typically by 30–60%, while during chronic swimming exercise it increases only by 10–20%. In addition to the whole heart, one can also weigh the left or right ventricle or the atria separately and also normalize these measures to body weight or tibia length. It is important to note that hypertrophy can occur as either eccentric or concentric hypertrophy. During eccentric hypertrophy cardiomyocytes preferably grow in length (leading at the whole heart level to ventricular dilation), while during concentric hypertrophy cardiomyocytes preferably become thicker (leading at the whole heart level to increased ventricular wall thickness and reduced chamber size). In order to quantify these differential cardiomyocyte growth responses one can measure the maximal cardiomyocyte transverse (thickness) or longitudinal (length) diameter at the level of the nucleus in transverse cardiac, histological sections, in which the cardiomyocyte plasma-membrane is marked either by immunostaining for a membrane protein (e.g., dystrophin) or by wheat-germ-agglutinin—FITC/TRITC (labelling myocyte membranes through interaction with the glycocalix; Sigma-Aldrich)—and nuclei are stained by a nuclear dye (e.g., DAPI). This in situ detection of cardiomyocyte size should preferably be done after in situ fixation and relaxation of hearts (by injecting a 4% paraformaldehyde, 50 mM KCl solution in PBS into the apex at a rate of 1 ml/min for 1 min while carefully holding and thereby closing the big vessels at the base of the heart, then releasing the heart at the aorta, and repeating injection before the whole heart is placed into that solution), subsequent embedding in paraffin, and preparation of 10 μm thick transverse paraffin sections. One could also isolate cardiomyocytes from adult hearts and measure the dimensions of the isolated cells, although the description of this method is beyond the scope of this chapter (see for example (25)).

In addition to cardiac size, pathological hypertrophy (but not physiological hypertrophy) is characterized by reactivation of the embryonic gene-program, which includes upregulation of ANP, BNP, α-skeletal actin, and β-myosin heavy chain (MHC) mRNAs and downregulation of α-MHC as well as SERCA2a mRNAs. We typically quantify these in total heart RNA with quantitative real-time PCR.

3.3. GST-Pulldown Screen for the Identification of Protein-Binding Partners

In order to find an interaction partner of a given cardiac signaling protein or a calcium-binding protein, one could generate a recombinant GST-fusion protein, for example by cloning the cDNA of interest into a GST containing cloning vector (e.g., pGEX-4T-1 from Amersham). After generation of the GST-fusion protein it is coupled to Sepharose beads and is incubated with cardiac protein lysate. Subsequently, after careful washing, the interacting proteins are resolved by SDS-PAGE gel electrophoresis. As a control, a

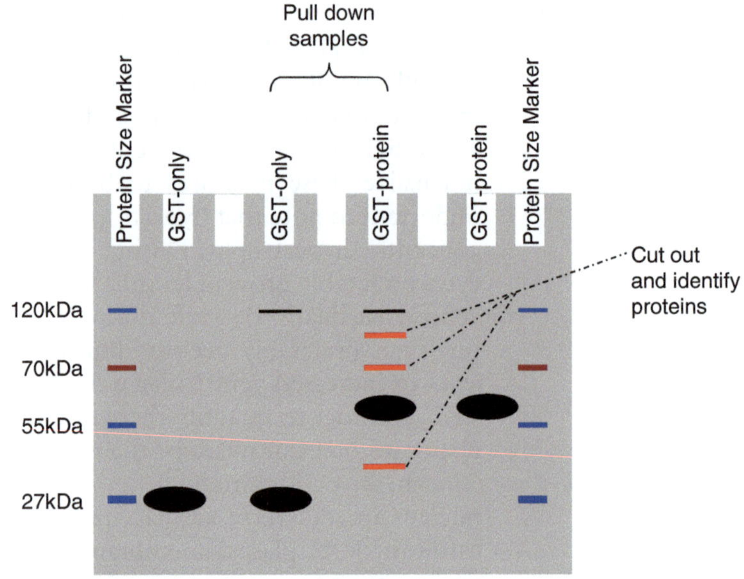

Fig. 3. Schematic representation of the SDS-PAGE run to identify novel interaction partner in a GST-pulldown screen. The strong black bands represent the GST-only or GST-fusion proteins, respectively.

pulldown with recombinant GST-only protein is performed. Proteins that interact with the GST-protein of interest fusion protein, but not with the GST-only protein, are sent for identification to mass-spectrometry analysis (see also Fig. 3). One drawback of this method is that proteins with low cellular expression cannot easily be identified. Therefore, it is recommended to incubate the GST-fusion proteins with high amounts of protein lysate (1–3 mg or even more) and also, if possible, to use lysate of a particular subcellular fraction (e.g., nuclear, mitochondrial, membrane, cytoskeletal, cytosolic), in which the particular interacting partner is suspected.

Generation of a GST-fusion protein:

1. The plasmid encoding for the GST-fusion protein is transformed into bacterial cells, which allow for high-level protein expression (e.g., BL21-cells from Invitrogen).
2. On the next day, a single bacterial colony is picked and transferred into 5 ml LB medium with the appropriate antibiotic (depending on the plasmid that is used to generate the GST-fusion protein) and incubated overnight in a 37°C incubator (with shaking at 225 rpms).
3. In the morning of the following day, the 5 ml culture is transferred into 50–200 ml of pre-warmed LB medium (including antibiotic) and is grown until the optical density of 0.6–0.8 is reached. Then expression of the protein should be

induced. The substance used for induction also depends on the plasmid that is used to express the GST-fusion protein. For plasmids containing the T7-lac operator or the trc promoter, IPTG (final concentration range 0.05–2 mM, typical concentration 0.2 mM) is added to the bacterial culture. The culture is then incubated at 37°C on the shaker (225 rpm) for 3 h. See Note 7.

4. The culture is then centrifuged for 10 min at 4,500 rpm (3,893 × *g*). The pellet can be stored at −80°C, if necessary.
5. The pellet is resuspended in 10 ml Cold buffer (after addition of DTT, PMSF, and protease inhibitors). 1 ml lysozyme (from a 10 mg/ml stock, made fresh) is then added to the culture followed by a 30-min incubation on ice. This is typically done in 50 ml Falcons.
6. Sonification is performed for four intervals (30 s each, we use the Omni Ruptor 250 from Omni International Inc. at 1/3 of total intensity). Keep the samples on ice between intervals.
7. 500 μl of 20% Triton X100 are added and incubated with rotation at 4°C for 30–60 min. Lysozyme, sonification, and TritonX serve to break up the bacterial cells and release the recombinant protein.
8. A centrifugation is conducted at 12,000 × *g*, 4°C for 20 min and the supernatant is transferred into a new tube (e.g., 15 ml Falcon). 800 μl of 50% Glutathione Sepharose 4B beads are added to the supernatant and incubated first for 20 min at room temperature, and then for 2 h at 4°C with rotation. During this step the GST-fusion protein is bound to the beads. See Note 8.
9. Another centrifugation at 500 × *g*, 4°C for 1 min is performed. The supernatant is removed and the beads are washed twice with 1×PBS. In the end, the beads are taken up in 1,500 μl 1×PBS.
10. In order to check purity and expression level of the GST-(fusion) proteins, 20 μl of the suspension with beads is mixed with 5 μl 5× Western sample buffer and heated at 95°C for 5 min. The samples are then loaded on an SDS-PAGE gel. GST-only and GST-(fusion)-proteins should be run side by side for comparison.

GST-pulldown screen

11. 1ml protein extract (1–3 mg of protein, preferably of a subcellular protein fraction), if possible in binding buffer (but other buffers used for protein isolation will also work, see also Note 9), are incubated in separate tubes with 40 μl of beads (50% slurry) bound to either GST-only or GST-protein overnight at 4°C with rotation. During this incubation, potential

interaction partners from the lysate bind to GST-protein but not to the GST-only beads.

12. The beads are subsequently washed three times with 1.5 ml Binding Buffer. The centrifugations are conducted at 500 × *g* for 5 min. After the last centrifugation, the supernatant is completely removed (e.g., with the help of gel-loading tips). The beads are taken up in 50 μl 2× Western sample buffer and heated at 95°C for 5 min. After the samples are cooled down, 2 μl of 40% Acrylamide are added and incubated for 20–30 min at room temperature.
13. The samples are subsequently loaded on a large SDS-PAGE gel (e.g., gel size of roughly 20 × 20 cm, see also Fig. 3 and Note 10). GST-only and GST-protein pulldown samples are run side by side, with a blank line in between. In addition, a 40 μl sample (50% slurry) of GST-only as well as GST-protein beads are run on the same gel, in order to be able to find out the pattern of protein bands caused by the GST proteins.
14. The gel is stained in Coomassie staining solution for 1–2 h.
15. The gel is destained with Coomassie destaining solution.
16. Protein bands that selectively appear in the GST-protein pulldown lane but not in the GST-only pulldown lane or the control lanes in which either GST-only or GST-protein was loaded should be identified and cut out (see also Fig. 3). The proteins are then sent for identification by mass spectrometry to a proteomic core facility. It is important to work very carefully and wear gloves, in order to avoid contamination of the protein samples.

When new interaction partners of the protein of interest are identified, one should then verify this interaction in independent experiments. For example, another GST-pulldown experiment could be performed, in which cardiac lysate is incubated either with GST-only coupled Sepharose beads (as control) or GST-protein beads. Subsequently, after careful washing of the samples, both are run on an SDS-PAGE gel, and the interaction partner can be detected specifically by antibody. Alternatively, the interaction could be confirmed by co-immuno-precipitation.

4. Notes

1. Digestion of the heart pieces is performed until they are (almost) completely gone. The amount of enzyme solution that is used can be progressively reduced as the amount of heart tissue diminishes.
2. The non-cardiomyocytes (mostly fibroblasts, but also endothelial cells, inflammatory cells, and progenitor cells)

can also be washed with 1×ADS (like described for the cardiomyocytes later in the protocol) and cultured or used for other applications.

3. If the neonatal rat cardiomyocytes grow too dense (>80% density), analysis and manipulation of the hypertrophic response are not possible. For best results, the cardiomyocytes should be grown on the culture plate as isolated cells (no bigger cell conglomerates with a lot of cardiomyocytes sticking to each other).
4. If one wants a milder TAC, one could also use a 25G needle instead of the 27G. Some strains of mice will not at all tolerate TAC with the 27G needle, which gives a severe TAC.
5. Occasionally, mice would let themselves float for extended time periods, instead to actively swim. In this case, the water basin should be irritated a little bit, in order to motivate the mice to swim.
6. At least 4–5 h need to be there in between the two swim sessions each day.
7. We typically use 50 ml of bacterial culture here, but for plasmids that give a low expression one can scale the volume up to 200 ml (or even more). In addition, one should carefully test which incubation time and which culture conditions work best for any given protein produced in bacterial cells after induction. Usually, the culture is incubated for 3–4 h at 37°C. However, if this does not give good results, incubation of the culture (after induction) at room temperature (6 h to overnight) could also be tested. It is important in this regard to cool down the bacterial culture (on ice) to the required temperature right before induction is performed.
8. Preparation of 50% Slurry with Glutathione Sepharose 4B: 1 ml of beads is briefly centrifuged at 500×*g* and washed twice with 1×PBS. After the last centrifugation, the supernatant is completely removed and 800 μl of PBS are added to the beads.
9. One could make the screen more stringent by increasing the salt concentration (various salt concentrations between 0 and 1 M can be tried) as well as increasing the concentrations of NP40 as nonionic detergent (concentrations between 0 and 2% can be tested).
10. It has to be decided in which range of protein size one wants to look for an interaction partner, in order to decide which percentage gel to run. In addition, the size of the GST-only (around 27 kDa) as well as GST-protein needs to be taken into account for this decision. For example, proteins between 12 and 45 kDa are best separated on a 15% SDS-PAGE gel, between 15 and 100 kDa on a 10% gel, and between 25 and 200 kDa on an 8% gel.

References

1. Heineke J, Molkentin JD (2006) Regulation of cardiac hypertrophy by intracellular signaling pathways. Nat Rev Mol Cell Biol 7:589–600
2. Bers DM (2002) Cardiac excitation-contraction coupling. Nature 415:198–205
3. Kiriazis H, Sato Y, Kadambi VJ, Schmidt AG, Gerst MJ, Hoit BD, Kranias EG (2002) Hypertrophy and functional alterations in hyperdynamic phospholamban-knockout mouse hearts under chronic aortic stenosis. Cardiovasc Res 53:372–381
4. Molkentin JD (2006) Dichotomy of Ca2+ in the heart: contraction versus intracellular signaling. J Clin Invest 116:623–626
5. Schaub MC, Heizmann CW (2008) Calcium, troponin, calmodulin, S100 proteins: from myocardial basics to new therapeutic strategies. Biochem Biophys Res Commun 369:247–264
6. Wu X, Zhang T, Bossuyt J, Li X, McKinsey TA, Dedman JR, Olson EN, Chen J, Brown JH, Bers DM (2006) Local InsP(3)-dependent perinuclear Ca signaling in cardiac myocyte excitation-transcription coupling. J Clin Invest 116:675–682
7. Backs J, Backs T, Neef S, Kreusser MM, Lehmann LH, Patrick DM, Grueter CE, Qi X, Richardson JA, Hill JA, Katus HA, Bassel-Duby R, Maier LS, Olson EN (2009) The delta isoform of CaM kinase II is required for pathological cardiac hypertrophy and remodeling after pressure overload. Proc Natl Acad Sci USA 106:2342–2347
8. Ling H, Zhang T, Pereira L, Means CK, Cheng H, Gu Y, Dalton ND, Peterson KL, Chen J, Bers D, Heller Brown J (2009) Requirement for Ca2+/calmodulin-dependent kinase II in the transition from pressure overload-induced cardiac hypertrophy to heart failure in mice. J Clin Invest 119:1230–1240
9. Zhang T, Johnson EN, Gu Y, Morissette MR, Sah VP, Gigena MS, Belke DD, Dillmann WH, Rogers TB, Schulman H, Ross J Jr, Brown JH (2002) The cardiac-specific nuclear delta(B) isoform of Ca2+/calmodulin-dependent protein kinase II induces hypertrophy and dilated cardiomyopathy associated with increased protein phosphatase 2A activity. J Biol Chem 277:1261–1267
10. Zhang T, Maier LS, Dalton ND, Miyamoto S, Ross J Jr, Bers DM, Brown JH (2003) The deltaC isoform of CaMKII is activated in cardiac hypertrophy and induces dilated cardiomyopathy and heart failure. Circ Res 92:912–919
11. Song K, Backs J, McAnally J, Qi X, Gerard RD, Richardson JA, Hill JA, Bassel-Duby R, Olson EN (2006) The transcriptional coactivator CAMTA2 stimulates cardiac growth by opposing class II histone deacetylases. Cell 125: 453–466
12. Bueno OF, van Rooij E, Molkentin JD, Doevendans PA, De Windt LJ (2002) Calcineurin and hypertrophic heart disease: novel insights and remaining questions. Cardiovasc Res 53:806–821
13. Bueno OF, Wilkins BJ, Tymitz KM, Glascock BJ, Kimball TF, Lorenz JN, Molkentin JD (2002) Impaired cardiac hypertrophic response in calcineurin Abeta -deficient mice. Proc Natl Acad Sci USA 99:4586–4591
14. Molkentin JD, Lu JR, Antos CL, Markham B, Richardson J, Robbins J, Grant SR, Olson EN (1998) A calcineurin-dependent transcriptional pathway for cardiac hypertrophy. Cell 93:215–228
15. Yang SA, Klee CB (2000) Low affinity Ca2+-binding sites of calcineurin B mediate conformational changes in calcineurin A. Biochemistry 39:16147–16154
16. Heineke J, Auger-Messier M, Correll RN, Xu J, Benard MJ, Yuan W, Drexler H, Parise LV, Molkentin JD (2010) CIB1 is a regulator of pathological cardiac hypertrophy. Nat Med 16:872–879
17. Naik UP, Patel PM, Parise LV (1997) Identification of a novel calcium-binding protein that interacts with the integrin alphaIIb cytoplasmic domain. J Biol Chem 272: 4651–4654
18. Gentry HR, Singer AU, Betts L, Yang C, Ferrara JD, Sondek J, Parise LV (2005) Structural and biochemical characterization of CIB1 delineates a new family of EF-hand-containing proteins. J Biol Chem 280: 8407–8415
19. Naik MU, Naik UP (2003) Calcium-and integrin-binding protein regulates focal adhesion kinase activity during platelet spreading on immobilized fibrinogen. Blood 102:3629–3636
20. Stabler SM, Ostrowski LL, Janicki SM, Monteiro MJ (1999) A myristoylated calcium-binding protein that preferentially interacts with the Alzheimer's disease presenilin 2 protein. J Cell Biol 145:1277–1292
21. Haataja L, Kaartinen V, Groffen J, Heisterkamp N (2002) The small GTPase Rac3 interacts with the integrin-binding protein CIB and promotes integrin alpha(IIb)beta(3)-mediated adhesion and spreading. J Biol Chem 277: 8321–8328

22. Jarman KE, Moretti PA, Zebol JR, Pitson SM (2010) Translocation of sphingosine kinase 1 to the plasma membrane is mediated by calcium and integrin binding protein 1. J Biol Chem 285:483–492
23. Leisner TM, Liu M, Jaffer ZM, Chernoff J, Parise LV (2005) Essential role of CIB1 in regulating PAK1 activation and cell migration. J Cell Biol 170:465–476
24. Rockman HA, Ross RS, Harris AN, Knowlton KU, Steinhelper ME, Field LJ, Ross J Jr, Chien KR (1991) Segregation of atrial-specific and inducible expression of an atrial natriuretic factor transgene in an in vivo murine model of cardiac hypertrophy. Proc Natl Acad Sci USA 88:8277–8281
25. Heineke J, Ruetten H, Willenbockel C, Gross SC, Naguib M, Schaefer A, Kempf T, Hilfiker-Kleiner D, Caroni P, Kraft T, Kaiser RA, Molkentin JD, Drexler H, Wollert KC (2005) Attenuation of cardiac remodeling after myocardial infarction by muscle LIM protein-calcineurin signaling at the sarcomeric Z-disc. Proc Natl Acad Sci USA 102:1655–1660

Chapter 18

In Vivo Screening of S100B Inhibitors for Melanoma Therapy

Danna B. Zimmer, Rena G. Lapidus, and David J. Weber

Abstract

S100 proteins are markers for numerous cancers, and in many cases high S100 protein levels are a prognostic indicator for poor survival. One such case is S100B, which is overproduced in a very large percentage of malignant melanoma cases. Elevated S100B protein was more recently validated to have causative effects towards cancer progression via down-regulating the tumor suppressor protein, p53. Towards eliminating this problem in melanoma, targeting S100B with small molecule inhibitors was initiated. This work relies on numerous chemical biology technologies including structural biology, computer-aided drug design, compound screening, and medicinal chemistry approaches. Another important component of drug development is the ability to test compounds and various molecular scaffolds for their efficacy in vivo. This chapter briefly describes the development of S100B inhibitors, termed SBiXs, for melanoma therapy with a focus on the inclusion of in vivo screening at an early stage in the drug discovery process.

Key words: In vivo screening, Preclinical testing, Intratumoral delivery, Systemic delivery, Pharmacokinetics, Pharmacodynamics, Maximum tolerated dose, Therapeutic window, Genetically modified mouse models, S100 proteins, EF-hand

1. Introduction

In malignant melanoma (MM), the tumor marker S100B binds directly to wild-type p53, dissociates the p53 tetramer, enhances hdm2-dependent ubiquitination of p53, and down-regulates p53-dependent tumor suppression functions (1–3). As a proof of principle for drug design, inhibiting S100B with small interfering antisense RNA (siRNAS100B) or with several *S*100*B* *i*nhibitor*s* (SBiXs; X = compound number) was achieved and shown to restore wild-type p53 at the protein level. Inhibiting S100B production was also found to cause an increase in the levels of p53 gene

Claus W. Heizmann (ed.), *Calcium-Binding Proteins and RAGE: From Structural Basics to Clinical Applications*, Methods in Molecular Biology, vol. 963, DOI 10.1007/978-1-62703-230-8_18, © Springer Science+Business Media New York 2013

products as necessary to induce cell growth arrest and apoptosis in malignant melanoma (2, 4–7). Also encouraging from a drug development standpoint is that ablation of S100B expression in mice via gene targeting produces unremarkable phenotypes with few, if any, problematic physiological consequences (8–13).

With any drug-design program, there is a need to obtain physiological data at an early stage in the process to help determine whether a compound or a series of compounds induce off-target effects and/or cause other unanticipated toxicities. This is particularly important for S100 inhibitors since there are over 20 structurally similar proteins in the S100 protein family, and they each regulate several physiologically important pathways in a cell-specific manner (14, 15). Thus, an S100 inhibitor could have multiple phenotypes depending on the number of S100 proteins it blocked and the S100 status of the cell-type targeted. In the case of blocking the S100B–p53 protein–protein interaction in malignant melanoma, there exists a separate issue with regard to a feedback loop that is initiated when S100B is inhibited since the gene for S100B itself is up regulated by the tumor suppressor protein, when p53 levels are restored (2). To address these and other issues, this chapter describes the discovery/development of SBiXs for melanoma therapy with a focus on the importance of doing in vivo screening of lead compounds at an early stage in the drug development process. It is likely that such an approach of performing early in vivo screening could benefit many drug discovery programs.

Small molecule inhibitors have been reported for numerous S100 family members, some of which are in clinical trials. Inhibitors of the S100A10–annexin A2 interaction have not been validated in vivo but their predicted clinical uses include angiogenesis and cancer metastasis therapy (16, 17). The anti-allergic drug cromolyn disrupts S100P–RAGE interaction and reduces pancreatic tumor formation in animal models (18). Interestingly, cromolyn also binds other S100 family members (S100A1, S100A12, and S100A13), but its effects on the interaction of these family members with their target proteins have not been fully investigated. Other small molecules that bind S100A1 include pentamidine and propanolol (19). In the case of S100A4, several phenothiazines block S100A4-mediated depolymerization of myosin-IIA filaments (20, 21). S100A4 also binds anti-allergic drugs and a modified version of azaxanthone, which is in clinical trials for treatment of metastatic disease states (21, 22). Two SBiXs developed by our group disrupt S100B–p53 complex formation and prevent unregulated melanoma cell growth and are currently being tested in human and veterinary clinical trials as potential melanoma therapeutics.

For the development of SBiXs, a combination of computer-aided drug design (CADD), high-throughput screening (HTS), structural biology, medicinal chemistry, and in vivo biology/drug testing approaches is employed. The benefits of a structure-based

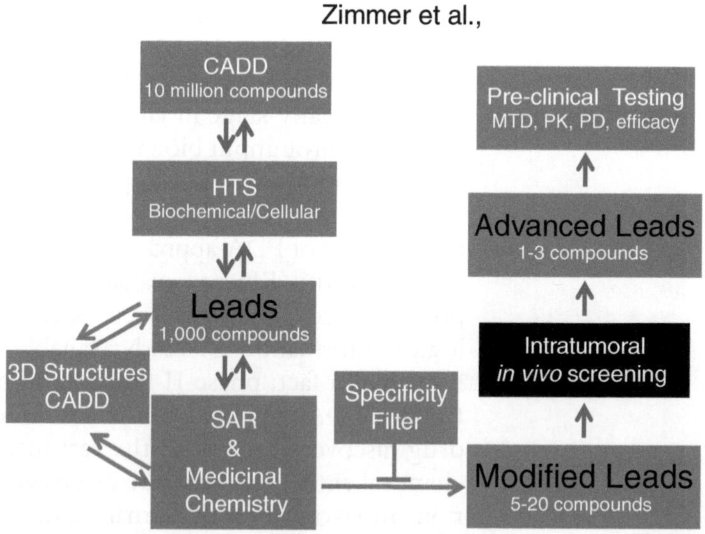

Fig. 1. Strategy for the discovery and development of SBiXs. An iterative process that includes computer-aided drug design (CADD), high-throughput screening (HTS), structure/activity (SAR) and medicinal chemistry, 3D structure (NMR and X-ray), and a specificity filter is used to engineer modified SBiXs for intratumoral in vivo screening and subsequent advanced leads suitable for preclinical evaluation.

approach are taken advantage of for developing new compounds as well as addressing issues with regard to target specificity since the 3D structures for several S100 and S100–target complexes are published (4, 15, 19, 23–27). Likewise, structure/activity relationships (SAR) and SAR by NMR approaches are more efficiently directed when the structures are available (28, 29). This is especially important for the difficult hurdle of inhibiting a protein–protein interaction such as the S100B–p53 complex (4, 7, 27, 30). More specifically, modifications on a particular scaffold that depend on the structure of a lead compound bound to S100B are achieved iteratively by using CADD-directed medicinal chemistry; likewise, structural studies can be used to identify multiple sites within the p53-binding cleft on S100B via structural biology approaches (NMR, X-ray), so compounds and combinations of compounds can be rationally linked together synthetically to obtain tight and highly specific S100B binders (i.e., relative to binding S100A1 for example). Such a drug development process is very dependent on the latest new results, so this process occurs in a very iterative manner (Fig. 1). Once promising leads are identified, then it is important that they are screened for their in vivo efficacy and toxicity. Such leads are chosen based on binding affinities (K_Ds) and/or specificity both in biochemical assays (i.e., S100B thermodynamic/kinetic data versus other S100s) and in cellular assays (i.e., IC_{50}s). For the cellular assays, specificity and other potential off-target effects are evaluated by comparing IC_{50}

values of isogenic cell lines plus/minus S100B present (31, 32). Of the approximately 100–200 lead compounds that undergo extensive in vitro testing each year, eight to ten typically meet the criteria for early-stage in vivo screening.

High-throughput biochemical and cellular screens identify two types of lead SBiXs: FDA-approved compounds and new chemical entities. Since there are no effective treatments for melanoma, repurposing of FDA-approved lead SBiXs for melanoma therapy is a high priority. Efficacy evaluation is the primary focus for repurposing approved drugs, since extensive pharmacological/toxicological information is already available for many species, including humans. In fact, phase II human and phase I canine clinical trials for two FDA-approved SBiXs are underway. However, like most drug discovery programs, the vast majority of lead SBiXs are new chemical entities that require extensive development and optimization. In vivo testing in animal models plays a fundamental role in developing new anticancer drugs. Prior to human testing, preclinical tolerability (maximum tolerated dose, MTD), pharmacokinetics (PK), pharmacodynamic (PD), and efficacy assays provide key information that is used to improve advanced leads. For example, a compound may need to be more lipophilic to pass through the cell membrane and reach its target or side groups may need to be added to allow for oral delivery or brain penetration.

In the case of accessible tumors such as melanoma, drugs can be delivered directly to the tumor (intratumoral) without optimization for systemic delivery and/or minimization of toxic effects on normal cells. In addition, intratumoral delivery can achieve significantly higher drug concentrations at the site of action than systemic delivery. In the clinical setting, intratumoral administration has been used to deliver gene therapy constructs, complex biologics, and small molecules to a variety of cancers including adenoviral based p53 genes in head and neck cancer (33); TNFalpha genes in rectal cancer (34); interleukin-2 in melanoma (35); immunostimulant CpG in brain cancers (36); single treatment of a metastatic squamous cell carcinoma lesion (37); BCNU in combination with radiotherapy in glioma (38); and para-toluenesulfonamide in non-small cell lung cancer (39). Intratumoral delivery has been used in the preclinical evaluation of siRNAs (40) and immune modulators (41, 42) as well as to reduce the toxicity of approved agents such as melphalan (43); 2-deoxy-D-glucose alone and in combination with carboplatin (44); and paclitaxel/docetaxel in mammary, bladder, prostate, and head and neck cancers (45–48). Our S100B inhibitor drug development program includes an intratumoral in vivo screen of lead SBiXs before preclinical testing of advanced leads. While in vivo testing prior to medicinal chemistry optimization and ADME testing is atypical, the availability of in vivo screening data early in the drug discovery process focuses resources on SBiXs with the greatest probability of clinical success.

A major consideration for any in vivo trial is the selection of an appropriate animal model. Mouse models are a mainstay in animal testing because it is feasible to perform studies in a short period of time. However, no single mouse model recapitulates the complex genetics and biology of human melanoma (49–52). Therefore, the biological question being asked as well as the advantages and limitations of the various types of models were important criteria in selecting an animal model for intratumoral in vivo screening of SBiXs. Syngeneic transplantation models involve the implantation of well-characterized melanoma cell lines into a syngeneic host and do not recapitulate many aspects of the human disease because characterized mouse lines do not reflect the heterogeneity observed in human tumors. Xenogeneic transplantation models involve the implantation of cell lines or patient-derived cells into an immunocompromised host. Both of these transplantation models can be flank models in which subcutaneous tumors are grown on the backs of mice or orthotopic models in which tumors are grown at the site of origin (i.e., breast cancer cells in the mammary fat pad). Orthotopic models have a better chance to metastasize but are more difficult to monitor. One disadvantage of human xenograft models is the fact that the host is immunocompromised and immunosurvelliance has been shown to play a role in limiting metastasis (53). In the case of SBiXs, S100B's effects on immune responses in the CNS are well documented although little is known about its role in peripheral immune responses. Nonetheless, human xenograft models are usually considered to be superior to genetically modified mouse models because the composition of the resulting tumor mimics the heterogeneity observed in patients. However, the development of multi-allelic genetically engineered mouse models that mimic spontaneous tumorigenesis and heterogeneity as well as target validation in simulated clinical trials using standard of care chemotherapeutics confirm the utility of genetically modified mouse models in developing cancer therapeutics (54).

The RAS-induced INK4a/ARF$^{-/-}$ mouse melanoma model (55) was chosen for in vivo screening of lead SBiXs because it has (1) an intact S100B–p53 signaling pathway (elevated S100B and wild-type p53), (2) an intact immune system, (3) tumors which are amenable to intratumoral delivery, and (4) a proven record in developing new melanoma therapies (55, 56). The Tyr::RAS^{G12V}/INK4a/ARF$^{-/-}$ line is bigenic and contains two genomic mutations on an FVB background: a mutated H-ras^{G12V} transgene on the Y chromosome and inactivated INK4a/ARF alleles on chromosome 4. This model is not commercially available and we maintain a breeding colony that generates experimental animals as well as breeders for the individual Tyr::RAS^{G12V} and INK4a/ARF$^{-/-}$ lines. Founders for both lines were obtained from the National Cancer Institute Mutant Mouse Resource (Frederick, MD). At 2–3 months of age experimental Tyr::RAS^{G12V}/INK4a/

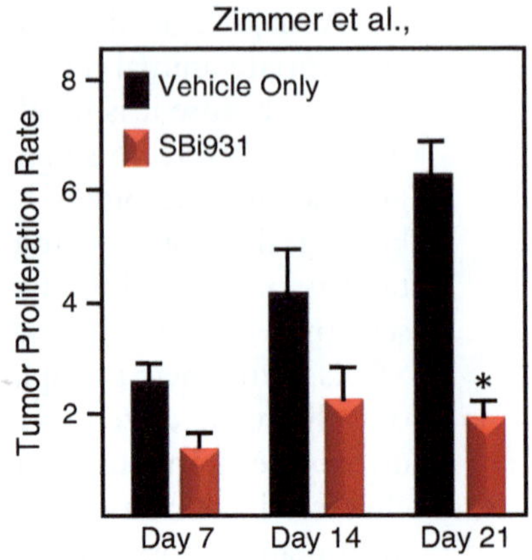

Fig. 2. Intratumoral in vivo screening trial. The histogram depicts the mean tumor responsiveness for intratumoral injection of SBi931 (*red bars*) or vehicle only (*black bars*) every other day for 21 days. *Asterisk* denotes $p < 0.05$ when compared to vehicle only.

ARF$^{-/-}$ males develop spontaneous cutaneous melanomas in the pinna of the ears (30%), torso (23%), and tail (20%) without distant metastasis (55). Our intratumoral in vivo screening protocol for modified SBiXs is a longitudinal design with a study period of 3–8 weeks and uses the relative tumor proliferation rate (the tumor volume at a particular treatment interval/tumor volume at the time of treatment initiation) as the primary outcome (Fig. 2). Although this trial design is not optimized for garnering tolerability, PK or PD information, the gross/histological pathology, SBiX levels, and p53 pathway reactivation in the tumors are monitored and this information is useful in selecting advanced leads that will proceed to preclinical testing.

In the case of modified leads with predicted ADME properties favorable for systemic administration, concurrent tolerability (MTD) and pharmacokinetic (PK) assays are also conducted. These trials provide valuable pharmacological information that cannot be garnered from local administration trials including potential effects of the route of administration on tumor responsiveness (57). MTD and PK studies are conducted in the same species/strain and employ the same dosing scheme that will be used in subsequent efficacy studies. For example, if the expected SBiX dosing scheme is 2 weeks of consecutive intraperitoneal (IP) administration in female nu/nu mice, then the MTD experiment should be dosed IP daily × 14 in female nu/nu mice. The route of administration, oral (PO), intraperitoneal (IP), intravenous (IV), or subcutaneous (SC), is determined by the predicted chemical properties of the SBiX. In MTD experiments, animals are monitored post cessation of dosing for delayed toxicity for a minimum of 1 week and optimally 2 weeks. In mouse models, the best indicator for toxicity is

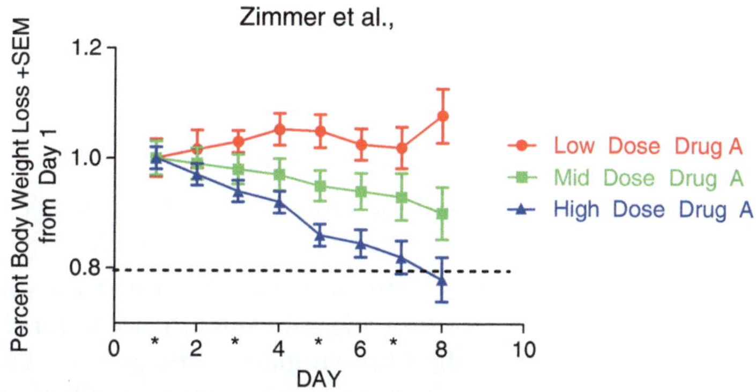

Fig. 3. Systemic tolerability testing/MTD. The graph depicts the percent body weight loss verses time for groups ($n = 3$ mice) receiving daily IP doses (mid, low, or high) of a drug for 7 consecutive days. The decrease in mean body weight in the high-dose group was greater than 20% indicating that this dose exceeds the MTD.

weight. MTDs for novel cancer drugs are considered an LD_{10}, the dose at which 10% of mice in one group die or lose 20% of their body weight (58) (Fig. 3). The therapeutic index for cancer drugs is very small in that the MTD can be equivalent to the effective dose. The goal of PK studies is to select a dose and route of administration that allow the plasma/target tissue drug levels to exceed the IC_{50} value from cellular assays and/or K_D from biochemical screens. SBiX levels in the tumor tissue provide useful information about tumor penetrance, half-life in target tissue, and the compound's ability to leave the bloodstream. Also, SBiX levels in brain tissue assess brain penetration and potential for use in metastatic disease therapy and primary brain cancers. The data from the intratumoral in vivo screening, MTD, and PK trials described below determines if modified leads are suitable for preclinical testing, require additional medicinal chemistry optimization and further testing prior to preclinical testing, or are eliminated.

2. Materials

1. SBiX stock: 100 mM in sterile dimethylsulfoxide (DMSO). Dissolve appropriate amount of SBiX in sterile-filtered DMSO to achieve final concentration of 100 mM. Prepare 100 μl aliquots and store at −20°C.
2. SBiX Injection Solution: For intratumoral in vivo screening, the SBiX injection solution composition is dependent upon the tumor volume and the K_D/IC_{50} of the SBiX, i.e., injection volume ≤ 20% of tumor volume and $[SBiX]_{intratumor}$ = fivefold

over K_D or IC_{50}, whichever is greater. For systemic tolerability studies, SBiX injection solution composition is dependent upon the dose and appropriate volume range for the route of administration. Dilutions are made in sterile phosphate-buffered saline (PBS).

3. Vehicle-Only Control: Dilute sterile DMSO into PBS in the same ratios used to prepare the SBiX injection solution in item 2.
4. Avertin solution: Avertin stock solution (100%) prepared by dissolving tribromoethanol in tertiary amyl alcohol and stored in a brown glass bottle at 4°C. The avertin working solution (2.5%) is prepared fresh weekly by diluting 1.25 ml stock solution (100%) to 48.75 ml sterile filtered PBS. After the pH is adjusted to 7.4, the solution is filtered through a 0.22 micron filter and stored in a brown bottle at 4°C. All containers should be labeled in accordance with institutional and regulatory policies regarding the use of animals.
5. Syringes: 20 G×38 mm gastric gavage needles (plastic or metal). 1 cc tuberculin syringes with 25G, 26G, 27G, or 30G needle.
6. Calipers: TTC Digital Caliper 0–150 mm.

3. Methods

All experiments involving animals described in this section were approved by the institutional IACUC and performed in accordance with the NIH guidelines for the treatment of animals. Appropriate institutional and/or regulatory agency approval is required for these procedures.

3.1. Intratumoral In Vivo Screening

Ten or more experimental Tyr::RAS^{G12V}/INK4a/ARF$^{-/-}$ males are obtained from our in-house breeding colony and randomly assigned to a cohort (experimental or control) when tumors ≥20 mm^3 in size develop.

1. Assess and record the general health status using the following criteria: body weight, grooming, eating, drinking, alterness/lethargy, mobility, coat (smooth vs. ruffled), and posture (hunch vs. upright) (see Note 1).
2. Assess and record the gross pathology of the tumor using the following criteria: erosion/invasion, redness, inflammation, and weeping (edema and presence of fluids) (see Note 1).
3. Restrain the animal and measure the length (l) and width (w) of the tumor (mm) with calipers (see Note 2).
4. Slowly inject the SBiX injection solution or vehicle-only control directly into the tumor.

5. Monitor and record the overall health status, gross tumor pathology, and tumor size daily.
6. Administer SBiX or vehicle-only control every other day. Rotate the injection site to minimize inflammation (see Note 3).
7. At the conclusion of the study period (3–8 weeks), euthanize animals using anesthetic overdose followed by decapitation. Administer 0.03 ml/g body weight of sterile Avertin (2.5% w/v) by ip injection. Administer additional doses of 0.15 ml as needed to attain a deep plane of anesthesia as determined by pedial and breathing reflexes (see Notes 4–6).
8. Examine and record the gross pathology of the tumor. Excise and weigh the tumor. Slice the tumor in half along the long axis. Process one-half for histological analysis (formalin fixation and paraffin-embedding) and the other half for biochemical/pharmacological analyses (snap frozen).
9. Examine internal organs for gross pathology and remove any abnormal organs for processing and histological analysis.
10. Tumor response is expressed as the mean tumor proliferation rate ± the SEM at 7, 14, and 21 days, i.e., the tumor volume ($V = \frac{1}{2}ab^2$) on day 7, 14, or 21 divided by the tumor volume at day 0 (Fig. 2).

3.2. Systemic Tolerability Testing (MTD and PK)

MTD studies are preferably carried out in efficacy species/strain but can be performed in an outbred mouse followed by a small bridging study in efficacy species/strain. Mice should be acclimated for at least 3 days prior to start of experiment. Three mice are the absolute minimum required and more than 5 are utilized if fine-tuning of the dose is required. The dosing schedule and route of administration (PO, IP, SC, IV, or IM) should be chosen based on the desired dosing schedule/route of administration in the proposed efficacy experiment. Mice should be monitored for 1–2 weeks post cessation of dosing to monitor for long-term toxicities.

1. Divide mice into four groups of three to ten mice. Each group will receive a different dose, i.e., 10×, 5×, 2.5×, and 1× dosing. If the vehicle is known to be nontoxic it may be excluded from the MTD study.
2. Novel compounds should be dosed IP, IM, SC, or IV using dosing schedule planned for future efficacy experiment. Depending on the route of administration, mice are either restrained by hand (IP, IM, and PO) or immobilized in a mouse restrainer and the tail is warmed by infrared light, hot water, or warm ethanol (IV). Usually mice are not anesthetized for dosing (see Note 7 for dosing volumes and needle size).
3. Mice should be monitored or observed daily and weighed 3 times per week.

4. Unacceptable toxicity is the inability of a mouse to ambulate in order to drink and eat in a 24-h period or 20% body weight loss at any time or 15% body weight loss over 72 h as compared to day 1 of the experiment. Animals that meet these criteria should be removed from the study and euthanized (see Note 8) (Fig. 3).
5. When dosing is complete, mice are monitored for 2 weeks to detect delayed toxicities.

Pharmacokinetic studies are carried out in naïve mice that may be tumor bearing or not. PK experiments are usually carried out with 18 mice (unless the desire is to test more time points).

1. Mice are divided into six groups of three mice.
2. Three mice are in the control group (time zero—no drug or vehicle) and 15 mice are in the treated groups (see Note 9) if mice will be implanted with tumor cells.
3. SBiXs are usually dosed ONCE by the route of administration planned for the efficacy experiment. The dose can range from 10 to 50 mg/kg.
4. Mice are euthanized (e.g., CO_2 inhalation) at specific time points post dosing. For SBiXs, typical time points are 15 min, 30 min, 1 h, 4 h, and 8 h post dosing (see Note 6).
5. Blood is collected by cardiac puncture (27G needle with 1 ml disposable syringe) and plasma is isolated after centrifugation in BD microtainer heparinized tubes. Mice are placed in left hand and 1 cc syringe with 27 G 1/2 ″ needle is inserted under the sternum towards the heart. The plunger is slowly pulled from the barrel. If no blood is observed, the needle is pulled out and reinserted. The needle is removed from barrel before injecting blood into microtainer tube. The tube is immediately capped and inverted several times to prevent clotting. Samples are centrifuged in a tabletop centrifuge for 10 min at max speed. Plasma is poured off into labeled microfuge tube and stored at −20°C until analysis.
6. Tumor, brain, or other tissue may be collected at time points noted. Tissue is snap frozen in an ethanol/dry ice bath and stored at −80°C until analysis.

4. Notes

1. Any animals exhibiting clinically significant health issues should be evaluated for removal from the study.
2. If a scruff hold blocks tumor access, then isoflurane anesthesia can be used for immobilization.

3. The trial schedule can be adjusted to match the predicted pharmacokinetics of the SBiX. Trials for SBiXs that are predicted to concentrate in tumor cells have used a 5-day on and 2-day off schedule. Trials for SBiXs that are not predicted to concentrate in tumors have used everyday and every other day schedules.
4. The trial duration for a typical SBiX in vivo screen is 3 weeks. However, the trial duration can be shortened or lengthened. If tumors progress to complete remission and/or unexpected toxicities are encountered, then trial duration can be shortened. Trial duration for the Tyr::RAS^{G12V}/INK4a/ARF$^{-/-}$ model should not exceed 8 weeks because animals will be approaching the age (5.5 months) at which 50% of the mice succumb to tumor growth.
5. Avertin has been reported to be an irritant and the following steps should be taken to minimize decomposition: (1) store and use under sterile conditions; (2) store in the dark at 4°C; (3) adjust pH of solution to >5.0; (4) discard stock solution after 4 months; (5) discard working solution after 7 days; and (6) discard any solutions that become discolored or have precipitate.
6. Other approved methods of euthanasia can be used.
7.

Route	Needle size and volume
Oral (PO)	20G × 38 mm gastric gavage needles (plastic or metal) Dosing volume should be 10 ml/kg and no more than 5 ml/kg
Intraperitoneal (IP)	26 or 27G needle with 1 ml disposable syringe Dosing volume should be 10 ml/kg and no more than 5 ml/kg
Subcutaneous (SC)	26 or 27G needle with 1 ml disposable syringe Dosing volume should not exceed 300 ul
Intravenous (IV)	27G to 30G needle with 1 ml disposable syringe Dosing volume should not exceed 10 ml/kg
Intramuscular (IM)	27G needle with 1 ml disposable syringe 0.05 ml can be injected into calf muscle

8. If one mouse in a group of three loses 20% body weight loss, this may be past the MTD since the number of animals is very small. The study would have to be repeated using the reported MTD and lower doses.
9. The cell line of choice will be grown in culture using optimal conditions. Usually, 1×10^6 to 5×10^6 cells are injected with 33% to 50% Matrigel™ (BD Biosciences) on the right flank of mice. Cells are injected in 0.15–0.2 ml of volume with a

27G tuberculin syringe. Tumors should be monitored 3 times per week with electronic calipers as described earlier. When tumors reach approximately 400–500 mm^3, the mice can be divided into six groups of three mice and the PK study may be initiated.

Acknowledgments

These studies were supported by NIH grant CA107331 (DJW) and the Center for Biomolecular Therapeutics (CBT), The University of Maryland School of Medicine, and the Institute for Bioscience and Biotechnology Research (IBBR).

References

1. Lin J, Blake M, Tang C, Zimmer D, Rustandi RR, Weber DJ, Carrier F (2001) Inhibition of p53 transcriptional activity by the S100B calcium-binding protein. J Biol Chem 276: 35037–35041
2. Lin J, Yang Q, Yan Z, Markowitz J, Wilder PT, Carrier F, Weber DJ (2004) Inhibiting S100B restores p53 levels in primary malignant melanoma cancer cells. J Biol Chem 279: 34071–34077
3. Lin J, Yang Q, Wilder PT, Carrier F, Weber DJ (2010) The calcium-binding protein S100B down-regulates p53 and apoptosis in malignant melanoma. J Biol Chem 285:27487–27498
4. Markowitz J, Mackerell AD Jr, Carrier F, Charpentier TH, Weber DJ (2005) Design of Inhibitors for S100B. Current Top Med Chem 5:1093–1108
5. Markowitz J, MacKerell AD Jr, Weber DJ (2007) A search for inhibitors of S100B, a member of the S100 family of calcium-binding proteins. Mini Rev Med Chem 7:609–616
6. Smith J, Stewart BJ, Glaysher S, Peregrin K, Knight LA, Weber DJ, Cree IA (2010) The effect of pentamidine on melanoma ex vivo. Anticancer Drugs 21:181–185
7. Wilder PT, Charpentier TH, Liriano MA, Gianni K, Varney KM, Pozharski E, Coop A, Toth EA, MacKerell AD Jr, Weber DJ (2010) In vitro screening and structural characterization of inhibitors of the S100B-p53 interaction. Intl J High Throughput Screen 1:109–126
8. Xiong Z, O'Hanlon D, Becker LE, Roder J, MacDonald JF, Marks A (2000) Enhanced calcium transients in glial cells in neonatal cerebellar cultures derived from S100B null mice. Exp Cell Res 257:281–289
9. Roltsch E, Holcomb L, Young KA, Marks A, Zimmer DB (2010) PSAPP mice exhibit regionally selective reductions in gliosis and plaque deposition in response to S100B ablation. J Neuroinflammation 7:78
10. Tsoporis JN, Overgaard CB, Izhar S, Parker TG (2009) S100B modulates the hemodynamic response to norepinephrine stimulation. Am J Hypertens 22:1048–1053
11. Schulte-Herbruggen O, Hortnagl H, Ponath G, Rothermundt M, Hellweg R (2008) Distinct regulation of brain-derived neurotrophic factor and noradrenaline in S100B knockout mice. Neurosci Lett 442:100–103
12. Kim HS, Seto-Ohshima A, Nishiyama H, Itohara S (2011) Normal delay eyeblink conditioning in mice devoid of astrocytic S100B. Neurosci Lett 489:148–153
13. Dyck RH, Bogoch II, Marks A, Melvin NR, Teskey GC (2002) Enhanced epileptogenesis in S100B knockout mice. Brain Res Mol Brain Res 106:22–29
14. Leclerc E, Heizmann CW (2011) The importance of Ca2+/Zn2+ signaling S100 proteins and RAGE in translational medicine. Front Biosci (Schol Ed) 3:1232–1262
15. Zimmer DB, Weber DJ (2010) The calcium-dependent interaction of S100B with its protein targets. Cardiovasc Psychiatry Neurol pii:728052
16. Li C, Reddy TRK, Fischer PM, Dekker LV (2010) A Cy5-labeled S100A10 tracer used to identify inhibitors of the protein interaction with annexin A2. Assay Drug Dev Technol 8:85–95

17. Reddy TR, Li C, Guo X, Myrvang HK, Fischer PM, Dekker LV (2011) Design, synthesis, and structure-activity relationship exploration of 1-substituted 4-aroyl-3-hydroxy-5-phenyl-1 H-pyrrol-2(5 H)-one analogues as inhibitors of the annexin A2-S100A10 protein interaction. J Med Chem 54:2080–2094
18. Arumugam T, Ramachandran V, Logsdon CD (2006) Effect of cromolyn on S100P interactions with RAGE and pancreatic cancer growth and invasion in mouse models. J Natl Cancer Inst 98:1806–1818
19. Wright NT, Cannon BR, Zimmer DB, Weber DJ (2009) S100A1: structure, function, and therapeutic potential. Curr Chem Biol 3: 138–145
20. Malashkevich VN, Dulyaninova NG, Ramagopal UA, Liriano MA, Varney KM, Knight D, Brenowitz M, Weber DJ, Almo SC, Bresnick AR (2010) Phenothiazines inhibit S100A4 function by inducing protein oligomerization. Proc Natl Acad Sci USA 107: 8605–8610
21. House RP, Garrett SC, Bresnick AR (2012) Moving aggressively: S100A4 and tumor invasion. In: Fatatis A (ed) Signaling pathways and molecular mediators in mestastasis. Springer, New York, Chapter 4
22. Mack GS, Marshall A (2010) Lost in migration. Nat Biotechnol 28:214–229
23. Bhattacharya S, Bunick CG, Chazin WJ (2004) Target selectivity in EF-hand calcium binding proteins. Biochim Biophys Acta 1742:69–79
24. Fritz G, Botelho HM, Morozova-Roche LA, Gomes CM (2010) Natural and amyloid self-assembly of S100 proteins: structural basis of functional diversity. FEBS J 277:4578–4590
25. Santamaria-Kisiel L, Rintala-Dempsey AC, Shaw GS (2006) Calcium-dependent and -independent interactions of the S100 protein family. Biochem J 396:201–214
26. Weber DJ, Rustandi R, Carrier F, Zimmer DB (2000) Interaction of dimeric S100B(bb) with the tumor suppressor protein p53: A model for Ca^{2+} dependent S100-target protein interactions. In: Pochet R (ed) CALCIUM: the molecular basis of calcium action in biology and medicine. Kluwer Academic, Dordrecht, The Netherlands, pp 521–539
27. Charpentier TH, Wilder PT, Liriano MA, Varney KM, Zhong S, Coop A, Pozharski E, MacKerell AD Jr, Toth EA, Weber DJ (2009) Small molecules bound to unique sites in the target protein binding cleft of calcium-bound S100B as characterized by nuclear magnetic resonance and X-ray crystallography. Biochemistry 48:6202–6212
28. Shuker SB, Hajduk PJ, Meadows RP, Fesik SW (1996) Discovering high-affinity ligands for proteins: SAR by NMR. Science 274: 1531–1534
29. Huang Q, Petros AM, Virgin HW, Fesik SW, Olejniczak ET (2002) Solution structure of a Bcl-2 homolog from Kaposi sarcoma virus. Proc Natl Acad Sci USA 99:3428–3433
30. Wilder PT, Lin J, Bair CL, Charpentier TH, Yang D, Liriano M, Varney KM, Lee A, Oppenheim AB, Adhya S, Carrier F, Weber DJ (2006) Recognition of the tumor suppressor protein p53 and other protein targets by the calcium-binding protein S100B. Biochium Biophys Acta 1763:1284–1297
31. Vitolo MI, Weiss MB, Szmacinski M, Tahir K, Waldman T, Park BH, Martin SS, Weber DJ, Bachman KE (2009) Deletion of PTEN promotes tumorigenic signaling, resistance to anoikis, and altered response to chemotherapeutic agents in human mammary epithelial cells. Cancer Res 69:8275–8283
32. Weiss MB, Vitolo MI, Mohseni M, Rosen DM, Denmeade SR, Park BH, Weber DJ, Bachman KE (2010) Deletion of p53 in human mammary epithelial cells causes chromosomal instability and altered therapeutic response. Oncogene 29:4715–4724
33. Nemunaitis J (2011) Head and neck cancer: response to p53-based therapeutics. Head Neck 33:131–134
34. Citrin D, Camphausen K, Wood BJ, Quezado M, Denobile J, Pingpank JF, Royal RE, Alexander HR, Seidel G, Steinberg SM, Shuttack Y, Libutti SK (2010) A pilot feasibility study of TNFerade biologic with capecitabine and radiation therapy followed by surgical resection for the treatment of rectal cancer. Oncology 79:382–388
35. Weide B, Eigentler TK, Pflugfelder A, Leiter U, Meier F, Bauer J, Schmidt D, Radny P, Pfohler C, Garbe C (2011) Survival after intratumoral interleukin-2 treatment of 72 melanoma patients and response upon the first chemotherapy during follow-up. Cancer Immunol Immunother 60:487–493
36. Carpentier A, Metellus P, Ursu R, Zohar S, Lafitte F, Barrie M, Meng Y, Richard M, Parizot C, Laigle-Donadey F, Gorochov G, Psimaras D, Sanson M, Tibi A, Chinot O, Carpentier AF (2010) Intracerebral administration of CpG oligonucleotide for patients with recurrent glioblastoma: a phase II study. Neuro Oncol 12:401–408
37. Akeda T, Yamanaka K, Kitagawa H, Kawabata E, Tsuda K, Kakeda M, Omoto Y, Habe K, Isoda K, Kurokawa I, Mizutani H (2011)

Intratumoral injection of OK-432 suppresses metastatic squamous cell carcinoma lesion inducing interferon-gamma and tumour necrosis factor-alpha. Clin Exp Dermatol 37:193–194. doi:10.1111

38. Jenkinson MD, Smith TS, Haylock B, Husband D, Shenoy A, Vinjamuri S, Walker C, Pietronigro D, Warnke PC (2010) Phase II trial of intratumoral BCNU injection and radiotherapy on untreated adult malignant glioma. J Neurooncol 99:103–113
39. He J, Ying W, Yang H, Xu X, Shao W, Guan Y, Jiang M, Wu Y, Zhong B, Wang D, Tucker S, Zhong N (2009) Gemcitabine plus cisplatin chemotherapy with concurrent para-toluene-sulfonamide local injection therapy for peripherally advanced nonsmall cell lung cancer larger than 3 cm in the greatest dimension. Anticancer Drugs 20:838–844
40. Dharmapuri S, Peruzzi D, Marra E, Palombo F, Bett AJ, Bartz SR, Yong M, Ciliberto G, La Monica N, Buser CA, Toniatti C, Aurisicchio L (2011) Intratumor RNA interference of cell cycle genes slows down tumor progression. Gene Ther 18:727–733
41. Moyer JS, Li J, Wei S, Teitz-Tennenbaum S, Chang AE (2008) Intratumoral dendritic cells and chemoradiation for the treatment of murine squamous cell carcinoma. J Immunother 31:885–895
42. Houot R, Levy R (2009) T-cell modulation combined with intratumoral CpG cures lymphoma in a mouse model without the need for chemotherapy. Blood 113:3546–3552
43. Matsuo M, Yamaguchi K, Feril LB Jr, Endo H, Ogawa K, Tachibana K, Nakayama J (2011) Synergistic inhibition of malignant melanoma proliferation by melphalan combined with ultrasound and microbubbles. Ultrason Sonochem 18:1218–1224
44. Pina Y, Houston SK, Murray TG, Boutrid H, Celdran M, Feuer W, Shi W, Hernandez E, Lampidis TJ (2010) Focal, periocular delivery of 2-deoxy-D-glucose as adjuvant to chemotherapy for treatment of advanced retinoblastoma. Invest Ophthalmol Vis Sci 51:6149–6156
45. Al-Ghananeem AM, Malkawi AH, Muammer YM, Balko JM, Black EP, Mourad W, Romond E (2009) Intratumoral delivery of Paclitaxel in solid tumor from biodegradable hyaluronan nanoparticle formulations. AAPS PharmSciTech 10:410–417
46. Shikanov A, Shikanov S, Vaisman B, Golenser J, Domb AJ (2008) Paclitaxel tumor biodistribution and efficacy after intratumoral injection of a biodegradable extended release implant. Int J Pharm 358:114–120
47. Yoo GH, Subramanian G, Piechocki MP, Ensley JF, Kucuk O, Tulunay OE, Lonardo F, Kim H, Won J, Stevens T, Lin HS (2008) Effect of docetaxel on the surgical tumor microenvironment of head and neck cancer in murine models, Archives Otolaryngology. Arch Otolaryngol Head Neck Surg 134:735–742
48. Shikanov S, Shikanov A, Gofrit O, Nyska A, Corn B, Domb AJ (2009) Intratumoral delivery of paclitaxel for treatment of orthotopic prostate cancer. J Pharm Sci 98:1005–1014
49. Sausville EA, Burger AM (2006) Contributions of human tumor xenografts to anticancer drug development. Cancer Res 66:3351–3354, discussion 3354
50. Becker JC, Houben R, Schrama D, Voigt H, Ugurel S, Reisfeld RA (2010) Mouse models for melanoma: a personal perspective. Exp Dermatol 19:157–164
51. Heyer J, Kwong LN, Lowe SW, Chin L (2010) Non-germline genetically engineered mouse models for translational cancer research. Nature Rev Cancer 10:470–480
52. de Jong M, Maina T (2010) Of mice and humans: are they the same?—implications in cancer translational research. J Nuc Med 51:501–504
53. Eyles J, Puaux AL, Wang X, Toh B, Prakash C, Hong M, Tan TG, Zheng L, Ong LC, Jin Y, Kato M, Prevost-Blondel A, Chow P, Yang H, Abastado JP (2010) Tumor cells disseminate early, but immunosurveillance limits metastatic outgrowth, in a mouse model of melanoma. J Clin Invest 120:2030–2039
54. Singh M, Lima A, Molina R, Hamilton P, Clermont AC, Devasthali V, Thompson JD, Cheng JH, Bou Reslan H, Ho CCK, Cao TC, Lee CV, Nannini MA, Fuh G, Carano RAD, Koeppen H, Yu RX, Forrest WF, Plowman GD, Johnson L (2010) Assessing therapeutic responses in Kras mutant cancers using genetically engineered mouse models. Nat Biotechnol 28:585–593
55. Chin L, Pomerantz J, Polsky D, Jacobson M, Cohen C, Cordon-Cardo C, Horner JW 2nd, DePinho RA (1997) Cooperative effects of INK4a and ras in melanoma susceptibility in vivo. Genes Dev 11:2822–2834
56. Larue L, Beermann F (2007) Cutaneous melanoma in genetically modified animals. Pigment Cell Res 20:485–497
57. Atkinson JM, Shelat AA, Carcaboso AM, Kranenburg TA, Arnold LA, Boulos N, Wright K, Johnson RA, Poppleton H, Mohankumar

KM, Feau C, Phoenix T, Gibson P, Zhu L, Tong Y, Eden C, Ellison DW, Priebe W, Koul D, Yung WK, Gajjar A, Stewart CF, Guy RK, Gilbertson RJ (2011) An integrated in vitro and in vivo high-throughput screen identifies treatment leads for ependymoma. Cancer Cell 20:384–399

58. Workman P, Aboagye EO, Balkwill F, Balmain A, Bruder G, Chaplin DJ, Double JA, Everitt J, Farningham DA, Glennie MJ, Kelland LR, Robinson V, Stratford IJ, Tozer GM, Watson S, Wedge SR, Eccles SA (2010) Guidelines for the welfare and use of animals in cancer research. Br J Cancer 102:1555–1577

Chapter 19

Arrayed Primer Extension Microarray for the Analysis of Genes Associated with Congenital Stationary Night Blindness

Kadri Vaidla, Janne Üksti, Christina Zeitz, and Eneli Oitmaa

Abstract

Arrayed primer extension (APEX) is a microarray-based genotyping method that enables to simultaneously analyze hundreds of known mutations in the genome. APEX-based microarrays are successfully used for molecular diagnostics of various genetic disorders.

Congenital stationary night blindness (CSNB) is a rare retinal disease caused by mutations in genes involved in phototransduction cascade and signaling from photoreceptors to adjacent neurons in the retina. As CSNB is clinically and genetically heterogeneous, the identification of the underlying cause of the disease can be challenging. In this chapter, we describe an APEX-based method for the analysis of genes associated with CSNB.

Key words: APEX, CSNB, Microarray, Mutation analysis, Night blindness, Calcium

1. Introduction

Congenital stationary night blindness (CSNB) is a rare heterogeneous retinal disease that is present at birth. Clinically the disease is characterized by vision impairment under dim light conditions and can be associated with other ocular defects as myopia, nystagmus, strabismus, and reduced visual acuity (1). Different forms of CSNB are classified according to their mode of inheritance, phenotype, and mutated genes (2). CSNB is caused by mutations in genes encoding different components of the phototransduction cascade or proteins involved in signaling from photoreceptors to adjacent neurons (1).

Claus W. Heizmann (ed.), *Calcium-Binding Proteins and RAGE: From Structural Basics to Clinical Applications*, Methods in Molecular Biology, vol. 963, DOI 10.1007/978-1-62703-230-8_19, © Springer Science+Business Media New York 2013

To date, more than 10 different genes have been associated with CSNB (3, 4). Most of the patients with mutations in these genes show a typical electrophysiological phenotype characterized by an electronegative waveform of the dark-adapted bright flash electroretinogram (ERG), in which the amplitude of the b-wave is smaller than that of the a-wave (5). According to this so-called Schubert-Bornschein type of ERG response CSNB can be divided into two subtypes—incomplete (ic) and complete (c) CSNB (2). icCSNB has been characterized by both a reduced rod b-wave and substantially reduced cone responses due to both ON- and OFF-bipolar cell dysfunction, while the complete type is associated with a drastically reduced rod b-wave response due to ON-bipolar cell dysfunction, but largely normal cone b-wave amplitudes (6). icCSNB has been associated with mutations in *CACNA1F* (7, 8), *CABP4* (9), and *CACNA2D4* (10), which encode for proteins forming the alpha subunit of a calcium channel, for a calcium-binding protein which interacts with this subunit, and for an auxiliary subunit of this channel, respectively. The latter one is most likely important for the correct localization of the alpha subunit. These proteins are located at the synapses of photoreceptor cells. cCSNB has been associated with mutations in *NYX* (11, 12), *GRM6* (13, 14), and *TRPM1* (3, 15, 16)—proteins expressed in bipolar cells and which are important for the further signaling in the retina (1, 3).

The majority of mutations associated with CSNB have been identified in *CACNA1F* and *NYX* genes (1). Still, a number of patients lack mutations in these genes indicating that new gene defects need to be discovered.

Because of the heterogeneity of CSNB, the identification of the underlying cause of the disease is complicated (17). Clinical data does not always direct to the disease-causing mutated gene. Thus, in some cases, direct sequencing of all known genes associated with CSNB might be requested. However, this is time-consuming and cost-ineffective. In order to overcome the challenges to confirm the diagnosis of CSNB and to determine the genetic defect an arrayed primer extension (APEX)-based microarray for the analysis of genes associated with CSNB has been developed (17).

APEX is a genotyping method that allows detection of hundreds of known variations in the genome in a single multiplexed reaction (18, 19). The APEX reaction involves the target DNA hybridization to the sequence-specific oligonucleotides immobilized on a glass slide followed by an enzymatic single base extension reaction in which DNA polymerase incorporates a dye-labelled terminator nucleotide to the hybridized oligonucleotide.

For the mutation detection by APEX method the DNA regions of interest are first amplified by polymerase chain reaction (PCR).

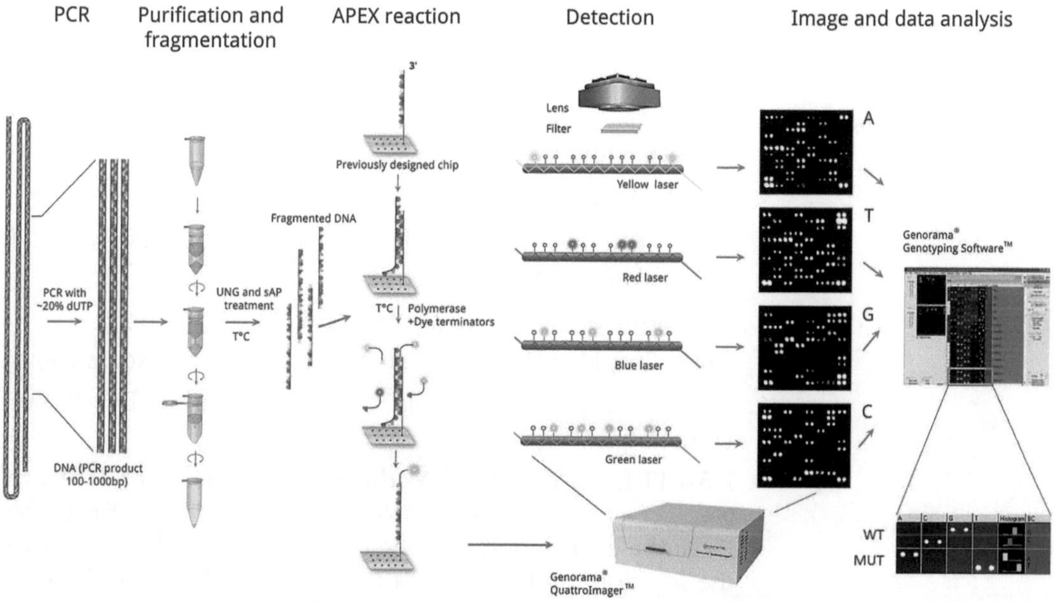

Fig. 1. Schematic overview of the APEX method. The DNA regions of interest are first amplified by PCR. In order to achieve the fragmentation of the PCR products with uracile-DNA glycosylase a fraction of dTTPs are substituted with dUTPs in the reaction mixture. The fragmented DNA is mixed with a thermostable DNA polymerase and fluorescently labelled ddNTPs for APEX reaction. After the single base extension reaction and washing steps, the microarray signals will be detected and analyzed. *WT* wild type, *MUT* mutation.

The amplified products are concentrated and purified with PCR purification columns. The fragmentation of purified products and functional inactivation of the unincorporated residual dNTPs are achieved by shrimp alkaline phosphatase (sAP) and Uracil DNA-Glycosylase (UNG) treatment. Fragmented and denatured PCR products are used for the primer extension reaction on the microarray. A schematic overview of the APEX-based mutation detection on a microarray is presented in Fig. 1.

APEX array for the analysis of CSNB-associated genes is a robust, cost-effective, and easily modifiable tool for a first-pass genetic testing (17). Current Asper Biotech CSNB microarray covers 159 mutations from 11 genes—*RHO* (MIM 180380), *PDE6B* (MIM 180072), *GNAT1* (MIM 139330), *CABP4* (MIM 608965), *GRM6* (MIM 604096), *SAG* (MIM 181031), *NYX* (MIM 300278), *CACNA1F* (MIM 300110), *CACNA2D4* (MIM 608171), *GRK1* (MIM 180381), and *TRPM1* (MIM 603576). This microarray is regularly updated to cover as many known mutations associated with CSNB as possible.

2. Materials

2.1. Polymerase Chain Reaction

1. 10× PCR buffer: 750 mM Tris–HCl (pH 8.8 at 25°C), 200 mM $(NH_4)_2SO_4$, 0.1% Tween 20.
2. 25 mM $MgCl_2$.
3. dNTP mix: 2.5 mM of dATP, dCTP, dGTP, 2 mM of dTTP, and 0.5 mM of dUTP.
4. PCR primers (Metabion GmbH, Germany).
5. Smart-Taq Hot DNA Polymerase (10 U/μl) (Naxo, Estonia).
6. 1× Tris–borate–EDTA (TBE) Buffer: 89 mM Tris–Borate, 10 mM EDTA (pH 8.3).
7. 1.5% TBE agarose gel with ethidium bromide: For a 1.5% agarose gel, add 1 gram of agarose to 100 ml of 1× TBE buffer. Place the gel solution into the microwave oven, boil, and swirl the solution until agarose particles are dissolved. Cool the gel to 60°C, add ethidium bromide (final concentration 0.5 μg/ml). Pour the gel into the gel casting tray, insert the comb into the gel, and let the gel set about 30 min.
8. DNA loading dye.
9. Thermal cycler.
10. ddH_2O.

2.2. Column Purification

1. GenJet™ PCR Purification Kit (Fermentas, Lithuania).

2.3. Fragmentation

1. UNG 1 U/μl.
2. sAP 1 U/μl.
3. 10× UNG buffer: 500 mM Tris–HCl (pH 9.0), 200 mM $(NH_4)_2SO_4$.
4. 1.5% TBE agarose gel with ethidium bromide and 1× TBE Buffer.

2.4. APEX

1. CSNB microarray slides (Asper Biotech Ltd., Estonia).
2. Thermo Sequenase™ DNA Polymerase 32 U/μl (GE Healthcare, UK).
3. Thermo Sequenase™ Reaction Buffer: 260 mM Tris–HCl (pH 9.5) and 65 mM $MgCl_2$.
4. Thermo Sequenase™ Dilution Buffer: 10 mM Tris–HCl (pH 8.0), 1 mM 2-mercaptoethanol, 0.5% Tween 20 (v/v), 0.5% Nonidet P-40 (v/v).
5. Fluorescently labelled ddNTPs: 50 μM Cy3-ddCTP, 50 μM Texas Red-5-ddATP, 50 μM Fluorescein-12-ddGTP, 50 μM Cy5-ddUTP.

6. LifterSlip™ cover slides 22 × 32 mm^2 (Erie Scientific Company, NH, USA).
7. 0.3% Alcanox detergent solution (Alcanox Inc., NY, USA).
8. Atlas Antifade (BioAtlas, Estonia).
9. Genorama QuattroImager™ (Genorama Ltd., Estonia).
10. Thermal plate with humidified chamber.
11. dH_2O.

3. Methods

3.1. Polymerase Chain Reaction

1. Amplify the DNA regions flanking CSNB mutations by PCR in singleplex reactions. Prepare the PCR reaction mix for each PCR primer pair in a total volume of 25 μl containing 30 ng of genomic DNA, 1X PCR buffer, 2.5 mM $MgCl_2$, 0.25 mM dNTP mix with dUTP (see Note 1), 5 pmol of forward and reverse primer, 1 U Smart Taq Hot DNA Polymerase, and ddH_2O.
2. Conduct the PCR amplification in a thermal cycler under the following conditions: 95°C for 15 min; 9 cycles at 95°C for 20 s, 68°C for 20 s (−1°C per cycle), and 72°C for 40 s; 18 cycles at 95°C for 20 s, 58°C for 20 s, and 72°C for 40 s; 9 cycles at 95°C for 20 s, 56°C for 20 s, and 72°C for 40 s; and final extension at 72°C for 7 min.
3. Control the amplification efficiency by loading the samples on a 1.5% agarose gel in 1× TBE Buffer.

3.2. Concentration and Purification of the Amplified Products

1. Pool the PCR amplification products into 1.5 ml microcentrifuge tubes. For 160 μl of pooled amplification product add 160 μl of Binding Buffer mix briefly by pipetting.
2. Transfer the solution from step 2 into the GeneJET™ purification column. Centrifuge the column at 12,000 × g (10,000–14,000 rpm, depending on the rotor type) for 1 min, and discard the flow-through.
3. Add 700 μl of Wash Buffer to the column. Centrifuge the column at 12,000 × g for 1 min. Discard the flow-through and place the purification column back into the collection tube.
4. Centrifuge the empty column for 2 min at 12,000 × g to remove any residual wash buffer (see Note 2).
5. Insert the column into a new 1.5 ml microcentrifuge tube. Add 30 μl of Elution Buffer in the center of the column and incubate at room temperature for 1 min.

6. Centrifuge the column at 12,000 × g for 2 min. Discard the purification column.
7. Store the purified PCR products at −20°C.

3.3. Fragmentation and Inactivation of dNTPs

1. Prepare the UNG–sAP reaction mix containing 4 μl of 10× UNG Buffer, 1.2 U of UNG, and 1 U of sAP and add the mixture to 30 μl of purified PCR products.
2. Incubate the sample at 37°C for 1 h. For product fragmentation denature the sample for 10 min at 95°C.
3. Control the fragmentation efficiency by loading the samples on a 1.5% agarose gel in 1× TBE Buffer. Use non-denatured samples as controls. Fragmentation is successful if no intact denatured PCR product is visible in the gel.

3.4. APEX Reaction and Slide Imaging

1. Wash the CSNB APEX slides 1× in 95°C dH_2O. Lift the slides slowly out of the water in order not to leave any droplets of water onto the slides (see Note 3).
2. Place the slides on a prewarmed thermoplate.
3. Prepare the APEX reaction mixture as follows:

   Tube A: 35 μl of UNG–sAP-treated and fragmented amplification products.

   Tube B: 5 μl of Reaction Buffer and 1.2 μl of each fluorescently labelled 50 μM ddNTP.

   Tube C: Dilute 3.2 U of Thermo Sequenase™ DNA Polymerase with 0.9 μl of dilution buffer.
4. Denature the PCR products in the tube A for 10 min at 95°C.
5. Meanwhile mix the content of tube C with tube B.
6. Centrifuge the tube A briefly and add the prepared reaction mixture (tube B + C) to the sample. Vortex and centrifuge briefly and apply the mixture immediately to the pre-warmed slides on a heated plate (see Note 4).
7. Cover the slide quickly with LifterSlip™ (see Notes 5 and 6) and incubate for 20 min at 58°C in the dark and humid chamber.
8. To terminate the reaction wash slides after incubation with 95°C dH_2O, then in 0.3% warm Alcanox solution for 3 min, and twice with 95°C dH_2O.
9. Remove the slides from water; apply a droplet of Atlas Antifade Reagent to minimize bleaching, and cover the slide with a coverslip.
10. Image the slides with Genorama QuattroImager™. Mutations are determined by using Genorama Genotyping Software™ (Genorama Ltd., Estonia).

4. Notes

1. 20% of dTTPs are substituted with dUTPs in order to achieve the fragmentation of the PCR products during UNG–sAP treatment. Fragmentation is necessary for the APEX reaction efficiency.
2. This step is essential as the presence of residual ethanol in the DNA sample may inhibit subsequent reactions.
3. The surface of silanized slides is hydrophobic and no droplets of water should stay on the glass surface if the slides are slowly lifted out of the water.
4. The reaction mixture should be immediately applied onto the microarray slide to assure that the DNA is denatured.
5. The reaction mixture applied onto microarray slide should be covered with LifterSlip™ as fast as possible to avoid evaporation.
6. In order to remove the bubbles that may have formed, gently bend the slides.

References

1. Zeitz C (2007) Molecular genetics and protein function involved in nocturnal vision. Expert Rev Ophthalmol 2:467–485
2. Miyake Y, Yagasaki K, Horiguchi M et al (1986) Congenital stationary night blindness with negative electroretinogram. A new classification. Arch Ophthalmol 104:1013–1020
3. Audo I, Kohl S, Leroy BP et al (2009) TRPM1 is mutated in patients with autosomal-recessive complete congenital stationary night blindness. Am J Hum Genet 85:720–729
4. Riazuddin SA, Shahzadi A, Zeitz C et al (2010) A mutation in SLC24A1 implicated in autosomal-recessive congenital stationary night blindness. Am J Hum Genet 87:523–531
5. Schubert G, Bornschein H (1952) Analysis of the human electroretinogram. Ophthalmologica 123:396–413
6. Audo I, Robson AG, Holder GE et al (2008) The negative ERG: clinical phenotypes and disease mechanisms of inner retinal dysfunction. Surv Ophthalmol 53:16–40
7. Bech-Hansen NT, Naylor MJ, Maybaum TA et al (1998) Loss-of-function mutations in a calcium-channel alpha1-subunit gene in Xp11.23 cause incomplete X-linked congenital stationary night blindness. Nat Genet 19:264–267
8. Strom TM, Nyakatura G, Apfelstedt-Sylla E et al (1998) An L-type calcium-channel gene mutated in incomplete X-linked congenital stationary night blindness. Nat Genet 19: 260–263
9. Zeitz C, Kloeckener-Gruissem B, Forster U et al (2006) Mutations in the calcium-binding protein 4 (CABP4) cause autosomal recessive night blindness. Am J Hum Genet 79: 657–667
10. Wycisk KA, Zeitz C, Feil S et al (2006) Mutation in the auxiliary calcium-channel subunit CACNA2D4 causes autosomal recessive cone dystrophy. Am J Hum Genet 79: 973–977
11. Bech-Hansen NT, Naylor MJ, Maybaum TA et al (2000) Mutations in NYX, encoding the leucine-rich proteoglycan nyctalopin, cause X-linked complete congenital stationary night blindness. Nat Genet 26:319–323
12. Pusch CM, Zeitz C, Brandau O et al (2000) The complete form of X-linked congenital stationary night blindness is caused by mutations in a gene encoding a leucine-rich repeat protein. Nat Genet 26:324–327
13. Dryja TP, McGee TL, Berson EL et al (2005) Night blindness and abnormal cone electroretinogram ON responses in patients with mutations in the GRM6 gene encoding mGluR6. Proc Natl Acad Sci USA 102:4884–4889
14. Zeitz C, van Genderen M, Neidhardt J et al (2005) Mutations in GRM6 cause autosomal

recessive congenital stationary night blindness with a distinctive scotopic 15-Hz flicker electroretinogram. Invest Ophthalmol Vis Sci 46:4328–4335

15. van Genderen MM, Bijveld MM, Claassen YB et al (2009) Mutations in TRPM1 are a common cause of complete congenital stationary night blindness. Am J Hum Genet 85:730–736
16. Li Z, Sergouniotis PI, Michaelides M et al (2009) Recessive mutations of the gene TRPM1 abrogate ON bipolar cell function and cause complete congenital stationary night blindness in humans. Am J Hum Genet 85:711–719
17. Zeitz C, Labs S, Lorenz B et al (2009) Genotyping microarray for CSNB-associated genes. Invest Ophthalmol Vis Sci 50:5919–5926
18. Kurg A, Tõnisson N, Georgiou I et al (2000) Arrayed primer extension: solid-phase four-color DNA resequencing and mutation detection technology. Genet Test 4:1–7
19. Tõnisson N, Kurg A, Kaasik K et al (2000) Unravelling genetic data by arrayed primer extension. Clin Chem Lab Med 38:165–170

Chapter 20

Nesfatin-1: Its Role in the Diagnosis and Treatment of Obesity and Some Psychiatric Disorders

Hiroyuki Shimizu and Masatomo Mori

Abstract

We discovered a new anorexigenic protein, nesfatin/nucleobindin-2 (NUCB2), which includes an EF-hand, calcium-binding motif. Nesfatin/NUCB2 is converted to nesfatin-1, which may be a physiologically active form in the body. Centrally and systemically administered nesfatin-1 inhibits appetite and body weight gain in rodents. The mid-segment of nesfatin-1 appears to be important in the inhibition of food intake. Intranasal administration of the mid-segment inhibits appetite. Nesfatin-1 may also be involved in the regulation of gastrointestinal function and insulin secretion. We have summarized the recent progress in the research of nesfatin-1.

Key words: Nesfatin-1, Nucleobindin-2, Calcium, EF-hand, Obesity

1. Introduction

Adipocytes produce various kinds of bioactive substances (e.g., adipokines), which produce inflammatory changes in the adipose tissue, and mediate humoral information to the brain. We have proposed the concept of a brain–adipose axis that involves an endogenous molecule that is expressed in both the hypothalamus and adipose tissue, exists in the general circulation, and is involved in the regulation of the feeding behavior and the determination of size and/or number of adipocytes (1). Adipokines play important roles in the transmission of the adiposity signal to the brain. Several adipokines, which modify the set-point for appetite, have been recently discovered. The clinical importance of these adipokines has been investigated in people with obesity. We have discussed the recent progress in the research of nesfatin/nucleobindin-2 (NUCB2) with a calcium-binding motif.

Claus W. Heizmann (ed.), *Calcium-Binding Proteins and RAGE: From Structural Basics to Clinical Applications*, Methods in Molecular Biology, vol. 963, DOI 10.1007/978-1-62703-230-8_20, © Springer Science+Business Media New York 2013

2. Nucleobindin-2: From Nuclear Factor to Bioactive Peptide

Troglitazone, a peroxisome proliferator-activated receptor-γ (PPAR γ) agonist, improves insulin sensitivity, and might modify satiety in type 2 diabetes patients (2). We attempted to identify a new PPAR γ-regulated protein, which modulates the feeding behavior of rodents. We performed a subtraction-cloning assay, and found that the expression of a new anorexigenic protein increased after troglitazone addition; we named this protein NEFA/NUCB2-encoded satiety- and fat-influencing protein (nesfatin) (3), similar to NEFA/NUCB2, which is a secreted protein with unknown function with 2 units of sequence similar to the EF-hand calcium-binding motif. Nesfatin/NUCB2 is composed of a signal peptide of 24 amino acids and a protein structure containing 396 amino acids (Fig. 1). The amino acid sequence of nesfatin/NUCB2 is highly conserved between humans, mice, and rats. Nesfatin/NUCB2 possess 2 or 3 potential cleavage sites for prohormone convertase (PC) that are flanked by a pair of basic amino acids, Lys (K)-Arg (R) or Arg (R)-Arg (R). The possible processed fragments are designated as nesfatin-1, nesfatin-2, and nesfatin-3 (Fig. 1). Nesfatin-3 includes a calcium-binding motif. In addition, we have recently found that

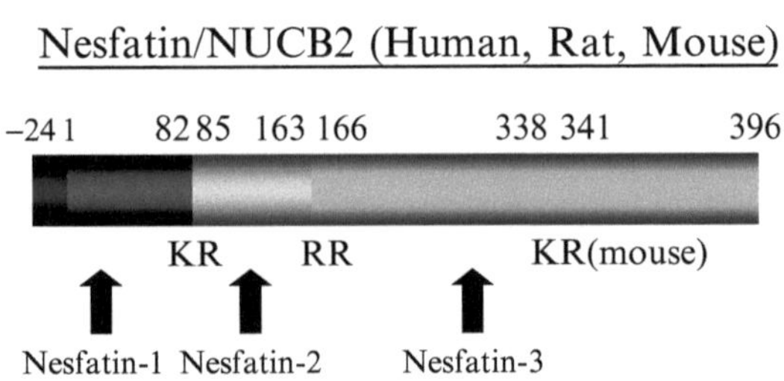

Amino acid sequence of Nesfatin-1

	1 … 42
human	VPIDIDKTKVQNIHPVESAKIEPPDTGLYYDEYLKQVIDVLE
rat	VPIDVDKTKVHNVDPVESARIEPPDTGLYYDEYLKQVIEVLE
mouse	VPIDVDKTKVHNTEPVENARIEPPDTGLYYDEYLKQVIEVLEE
	43 … 82
human	-TDKHFREKLQKADIEEIKSGRLSKELDLVSHHVRTKLDEL
rat	-TDPHFREKLQKADIEEIRSGRLSQELDLVSHKVRTRLDEL
mouse	-TDPHFREKLQKADIEEIRSGRLSQELDLVSHKVRTRLDEL

Fig. 1. Molecular structure of nesfatin/nucleobindin-2 (NUCB2), and comparison of the amino acid sequences of human, rat, and mouse nesfatin-1.

an increase in NUCB2 expression by troglitazone addition could be explained by the stabilization of NUCB2 mRNA levels through the activation of ERK1/2 pathway (4).

3. Possible Involvement in the Regulation of Feeding Behavior

Nesfatin/NUCB2 is expressed in the appetite-control hypothalamic nuclei such as the paraventricular nucleus (PVN), arcuate nucleus (ARC), supraoptic nucleus (SON) of the hypothalamus, lateral hypothalamic area (LHA), and zona incerta of rats (3). In addition, immunoreactivity of nesfatin-1 was found in the brainstem nuclei such as the nucleus of the solitary tract (NTS) and the dorsal nucleus of vagus (3, 5). Selective reduction of nesfatin/NUCB2 mRNA expression in the PVN of rat was found under starved conditions, and refeeding restored the expression levels, while no significant changes were observed because of starvation in other hypothalamic nuclei. Intracerebroventricular injection of nesfatin/NUCB2 to male Wistar rats showed a dose-dependent reduction in food consumption for 6 h after the injection. Immunohistochemical analysis showed that nesfatin/NUCB2 is present in the cytoplasm of the neuronal cell, but not in the nucleus, where it was co-localized with PC-1/3 and PC-2. Intracerebroventricular infusion of nesfatin-1, an amino-terminal fragment derived from NUCB2, showed a dose-dependent reduction in food intake by 6 h. In contrast, intracerebroventricular infusion of other possible fragments processed from nesfatin/NUCB2, nesfatin-2, nesfatin-3, and nesfatin-2/3 showed no promotion of satiety. These data indicates that the calcium-binding motif, present in nesfatin-3, is not involved in the inhibition of appetite induced by nesfatin/NUCB2. Furthermore, intracerebroventricular infusion of the antibody against nesfatin-1 increased food intake during the light period. In addition, the cerebrospinal fluid of rat showed nesfatin-1-like immunoreactivity, which indicated that nesfatin/NUCB2 is physiologically cleaved in the central nervous system. Conversion of nesfatin/NUCB2 to nesfatin-1 by PC is indispensable for its inhibition of feeding behavior in rats.

Chronic infusion of nesfatin-1 into the third ventricle of rat consistently reduced body weight gain, and the weight of white adipose tissue weight on sacrifice. On the other hand, rats gained body weight after the initiation of chronic intracerebroventricular administration of antisense morpholino oligonucleotide against the gene encoding nesfatin/NUCB2. These data strongly indicated that nesfatin/NUCB2 is involved in the physiological regulation of feeding behavior in rats.

In addition to its role as a satiety peptide, nesfatin-1 may mediate anxiety- and/or fear-related responses in the brain (6). Intracerebroventricular injection of nesfatin-1 showed a dose-dependent reduction in the percentage of time spent on the open arms of the elevated plus maze, increased latency to approach, and increased fear-potentiated startle response and the time spent freezing to both context and conditioned cues in a conditioned emotional response test. However, further studies are required to determine the effects of peripheral administration of nesfatin-1 or its analogues.

Results from two groups showed that nesfatin-1 crosses the blood–brain barrier of mice without saturation (7, 8). Thus, endogenous and peripherally administered exogenous nesfatin-1 was proposed to reach the brain, and inhibit feeding behavior. Intraperitoneal injection of nesfatin-1 in nonstarved male ICR mice showed a dose-dependent decrease in 3-h food intake during the dark cycle (IC_{50} = 0.35 nmoles/g body weight) (9). In addition, subcutaneous administration of nesfatin-1 inhibited food intake, and the anorexigenic action of nesfatin-1 continued for 14 h after the injection. Furthermore, the duration of anorexigenic action was longer after subcutaneous administration than after intraperitoneal administration. For determining the chronic effects of nesfatin-1, 10 nmoles of nesfatin-1 were intraperitoneally administered to male ICR mice twice a day. Repeated intraperitoneal injection of nesfatin-1 significantly inhibited body weight gain for 6 days (vehicle: 1.10 ± 0.24 g; 10 nmoles nesfatin-1/mouse/day, 0.11 ± 0.33 g; P = 0.03). These results suggested that peripherally administered nesfatin-1 could reduce food intake, and administration via the subcutaneous route appears to be the most plausible route for the administration of nesfatin-1-based antiobesity drugs.

In addition, we examined the possibility of a nasal administration of nesfatin-1 in male Wistar rats because the intranasal route provides an easy access to the brain, independent of the altered blood–brain barrier function such as "leptin resistance" (9). Nesfatin-1 (10 nmoles/rat) was administered into the right and left nostrils of rats. Intranasal administration of nesfatin-1 significantly inhibited food intake for 6 h (10). Anorexigenic action of nesfatin-1 disappeared after 6 h of administration. Observations indicated that the intranasal route may be one of the possible routes of administration of nesfatin-1 in the future for the treatment of morbidly obese people.

Next, we sought to identify the active segment of nesfatin-1 for the development of antiobesity drugs using nesfatin-1 analogues. Because nesfatin-1 has 3 distinct segments (Fig. 2), we examined the effect of each segment (1–23, 24–53, and 54–82 amino acid residues) on food intake in mice. Among the 3 segments of nesfatin-1, only the mid-segment (24–53 amino acid residues; M30), consisting of 30 amino acids, has an effect on appetite

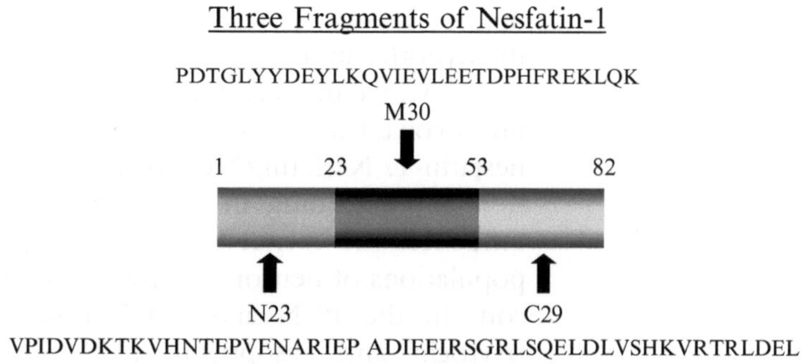

Fig. 2. Amino acid sequence of three distinct fragments of nesfatin-1.

(10). This mid-segment has a structure similar to that of an activation site of agouti gene-related protein (AgRP), and to the α-melanocyte-stimulating hormone (α-MSH). Intraperitoneally administered M30 had a dose-dependent (IC_{50} = 0.36 nmoles/g body weight) anorexigenic action similar to that of nesfatin-1. However, a single injection of M30 into the peritoneal spaces did not affect the concentrations of circulating glucose and lipids, 3 h after injection of this dosage, despite a significant reduction of food intake. Thus, a region with amino acid sequence similar to that of the active site of AgRP was indispensable for the anorexigenic action of nesfatin-1. This observation is of value for the future development of nesfatin-1 analogues.

4. Mechanism of the Anorexigenic Effect of Nesfatin-1

Nesfatin-1-induced anorexia occurred in Zucker fatty rats with a leptin receptor mutation. Pretreatment with anti-nesfatin-1 antibody did not block leptin-induced anorexia (3). The intraperitoneal injection of M30 decreased food intake under leptin-resistant conditions such as genetically obese (*ob/ob*) and genetically diabetic (*db/db*) mice and mice fed with a 45% high-fat diet for 28 days (11). These data indicate that the hypothalamic leptin signaling pathway does not exist downstream of the pathway by which nesfatin-1 causes anorexia. In contrast, central injection of α-MSH increased the expression of the nesfatin/NUCB2 gene in the PVN, and the reduction of food intake by nesfatin-1 was abolished by the administration of the melanocortin-3/4 receptor antagonist, SHU9119. However, intracerebroventricular administration of nesfatin-1 failed to show a significant change in pro-opiomelanocortin (*POMC*) gene expression in the ARC. On the basis of these observations, we concluded that nesfatin-1 signaling pathway might be associated with the melanocortin signaling pathway in the hypothalamus.

Next, we investigated the different regions of the brain possibly involved in the suppression of feeding behavior induced by nesfatin-1. Nesfatin/NUCB2 expression is modulated by starvation and refeeding in the PVN and SON (3, 12). Nesfatin-1 influences the excitability of a large proportion of different subpopulations of neurons located in the PVN (13). Nesfatin-1 neurons in the PVN and SON show concomitant expression of oxytocin and vasopressin (12). Double-labeling immunohistochemistry revealed a colocalization of nesfatin/NUCB2 with vasopressin and oxytocin in magnocellular neuroendocrine neurons, thyrotropin-releasing hormone (TRH), corticotropin-releasing hormone (CRH), somatostatin, neurotensin, and growth hormone-releasing hormone (GRH) in parvocellular neuroendorine neurons of the PVN (14). Magnocellular oxytocin neurons are activated by feeding, and intracerebroventricular administration of oxytocin antagonist increases food intake (14), which indicates a possible role of oxytocin in the regulation of feeding behavior. These data indicated that oxytocin may be involved in the nesfatin-1-induced anorexia in the hypothalamus.

Nesfatin/NUCB2-immunopositive neurons are also located in the ARC and are colocalized with POMC/cocaine- and amphetamine-regulated transcript (CART)-neurons but not with neuropeptide Y (NPY)/AgRP-neurons (3, 5). However, NPY neurons in the ARC are hyperpolarized by nesfatin-1, and glibenclamide, an ATP-sensitive potassium conductance antagonist, preventing nesfatin-1-induced hyperpolarization. This indicates that hyperpolarization of NPY neurons is important in the nesfatin-1-induced anorexia (15). Nesfatin-1, released from POMC/CART neurons, may directly affect NPY/AgRP neuronal activities. In addition, Inhoff et al. reported (16) that nesfatin-1 may be involved in the desacyl ghrelin-induced inhibition of the orexigenic effect of peripherally administered ghrelin in a freely fed rat because nesfatin/NUCB2-immunoreactive neurons in the ARC are activated by simultaneous injection of ghrelin and desacyl ghrelin. This newly proposed pathway suggests that peripherally administered desacyl ghrelin might block ghrelin-activated NPY/AgRP-neurons in the ARC via nesfatin-1-positive neurons (16).

Nesfatin/NUCB2 is also coexpressed with melanin-concentrating hormone (MCH) in the tuberal hypothalamic neurons of the rat, and approximately 80% of nesfatin-1-immunoreactive neurons were labeled for MCH (17). These data suggest that nesfatin/NUCB2, coexpressed in the MCH neurons may play a complex role not only in the regulation of food intake but also in other essential integrative brain functions involving MCH signaling, ranging from autonomic regulation, stress, mood, and cognition to sleep.

Nesfatin/NUCB2-immunoreactive cells were choline acetyltransferase positive in the Edinger–Westphal nucleus and dorsal motor nucleus of vagus, tyrosine hydroxylase positive in the NTS, and 5-hydroxytryptamine positive in the causal raphe nucleus (5). After intraperitoneal injection of M30, expression of c-*fos* was significantly increased in the brainstem and NTS accompanied by a significant inhibition of the 3-h food intake but not in the ARC. Injection of the mid-segment of nesfatin/NUCB2 significantly increased the expression of *POMC* and *CART* genes in the nucleus of the NTS but not in the ARC. Nesfatin/NUCB2 was colocalized with CART in almost all regions of the brain that expressed NUCB2, including brainstem nuclei (18). Our findings indicate that intraperitoneal injection of M30 causes anorexia, possibly, by activating the POMC and CART neurons in the NTS via a leptin-independent mechanism.

5. Peripheral Nesfatin/NUCB2

Reverse transcriptase-polymerase chain reaction showed that nesfatin/NUCB2 mRNA was expressed in various tissues of the body, including the brain and adipose tissue. We have shown that nesfatin/NUCB2 was also detected in the rat serum and cerebrospinal fluid by Western blot analysis. The origin of circulating nesfatin-1 in the serum remains to be determined because PC is also distributed in various tissues of the body. Thus, nesfatin-1 may be processed and released from various tissues expressing PC-1/3 or PC-2 in the entire body.

Recently, immunoreactivity of nesfatin/NUCB2 was identified in the rat gastric mucosa (19). Nesfatin/NUCB2-related peptides were thought to be involved in the physiological regulation of gastrointestinal function although whether nesfatin/NUCB2 is processed to nesfatin-1 in the gastric mucosa remains to be further investigated. Nesfatin-1 may be produced in the gastric mucosa because PC-1/3 and PC-2 mRNAs are abundant in the gastric mucosa (20). These results increased the possibility of nesfatin/NUCB2 gene expression being regulated by the nutritional status and indicate a regulatory role of peripheral nesfatin-1 in energy homeostasis.

In addition, the gastric vagal afferent is the major pathway that transmits the signal for starvation, derived from the ghrelin in the stomach (21). Some nesfatin/NUCB2-immunoreactive cells of the gastric mucosa of the rat coexpress ghrelin (19). Thus, the anorexigenic signal of peripherally administered nesfatin-1 and its active mid-segment, M30, may be transported to the brainstem via the vagal nerve. We tested the anorexigenic effects of M30 administered intraperitoneally in the mice pretreated with systemic capsaicin injection in order to examine the possible involvement of the vagal nerve in the anorexia. Intraperitoneal administration of M30 failed to show any anorexigenic action in capsaicin-pretreated mice (22). Therefore, the vagal nerve appears to play an important role in the induction of anorexia by intraperitoneally administered M30. Signal transmission from the vagal afferent nerves to the neurons in the NTS is known to involve neurons with muscarinic and nicotinic cholinergic fibers. Atropine methyl nitrate, an antagonist for the muscarinic cholinoceptor, and bretylium tosylate, an adrenergic ganglion blocker, did not affect suppression of food intake by intraperitoneally administered M30. In contrast, the nicotinic cholinergic blocker hexamethonium abolished the suppression of food intake induced by M30. These data indicate that the nicotinic cholinergic pathway is important in M30-induced anorexia.

Nesfatin/NUCB2-immunoreactive cells are colocalized with insulin in the pancreatic islets of both CD1 mice and Fischer 344 rats (23). Nesfatin-1 enhances glucose-induced insulin secretion by promoting Ca^{2+} influx through L-type channels (24, 25). Pancreatic β-cells are well known to express both PC-1/3 and PC-2, and thus it is conceivable that nesfatin/NUCB2 must be processed to nesfatin-1 in pancreatic β-cells. These results indicated the possibility that the abundant presence of nesfatin/NUCB2-immunoreactivity and its colocalization with insulin suggest a potential role for nesfatin/NUCB2-derived peptides in islet biology and glucose homeostasis. Furthermore, nesfatin-1 has recently been reported to enhance glucagon secretion from mouse islets (26).

Nesfatin-1 may affect insulin sensitivity in vivo. Insulin-stimulated glucose uptake in L6 muscle cells was inhibited by nesfatin-1 pretreatment, but basal and insulin-induced glucose uptake in adipocytes from nesfatin-1-treated rats was significantly increased (25).

We have hypothesized that nesfatin-1 from peripheral organs is transported to the brain by humoral and/or neural afferent pathways, and that central nesfatin-1 signal may modulate the balance of autonomic nervous system, which results in the modification of the different ways in which nesfatin-1 is metabolized in the peripheral organs (Fig. 3).

6. Future Direction of Clinical Research

On the basis of our results, we believe that nesfatin-1 could be useful for the diagnosis of some metabolic and psychiatric diseases. Furthermore, systemic or local administration of nesfatin-1, its fragments, or related analogues may improve metabolic abnormalities by reducing excessive body fat of the patients with obesity and metabolic syndrome.

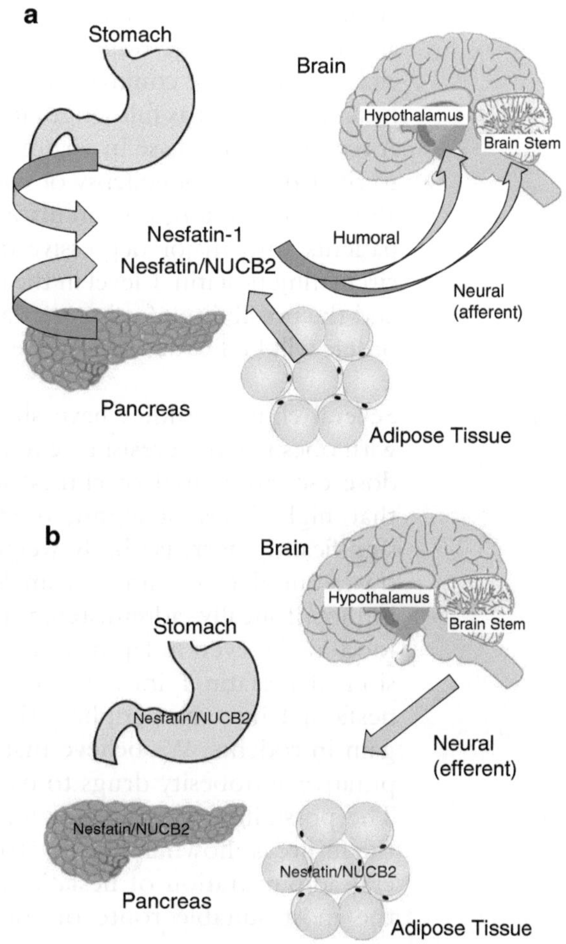

Fig. 3. Putative nesfatin-1 pathways in the brain and peripheral organs. (**a**) Afferent pathway: Nesfatin-1, released from peripheral organs, reaches the feeding-regulating center in the brain via humoral and neural pathways. (**b**) Efferent pathway: Nesfatin-1 signal from the brain is transported to the peripheral organs via neural pathway.

6.1. Diagnosis

We have detected nesfatin/NUCB2 in human serum by Western blot analysis and established a sensitive and specific enzyme-linked immunosorbent assay (ELISA) assay system for human nesfatin-1. In addition, we confirmed the existence of nesfatin-1 in human serum. By using our ELISA system, we have shown that circulating nesfatin-1 concentrations are negatively correlated with the body mass index (BMI) in men. This indicates the lack of "nesfatin-1" resistance in obese patients (27). Several studies have reported that circulating nesfatin concentrations are fluctuating in type 2 diabetes patients (25, 28, 29).

Serum nesfatin-1 levels, measured using commercially available assay system, are remarkably higher in patients who are newly diagnosed with primary generalized epilepsy (30). This increase of circulating nesfatin-1 levels might be useful to diagnose epilepsy and to monitor the response to antiepileptic treatment. It is not clear whether the commercially available assay system may detect nesfatin-1 as well as full-length nesfatin/NUCB2 in human serum. A remarkable increase in serum nesfatin levels may also be related to the existence of epilepsy or epileptic seizure already mentioned above. In addition, the circulating nesfatin-1 level is increased in patients with major depressive disorder (31). The fluctuation of circulating nesfatin-1 level in the patients with psychiatric disorders and the possibility of using nestfatin-1 as a biomarker of these disorders will be further investigated.

6.2. Treatment

Several clinical studies have shown that the majority of people with obesity have a resistance to leptin. A randomized, controlled, dose-escalation trial of clinical leptin treatment already showed that high doses of leptin, over a long time, are necessary to significantly decrease body weight in obese adults (32). In contrast, our data obtained from basic experiments indicated that intraperitoneally administered nesfatin-1 significantly inhibited food intake even in leptin-resistant animal models. Chronic infusion of nesfatin-1 into the brain and chronic administration of nesfatin-1 into the periphery significantly decreased body weight gain in rodents. We believe that nesfatin-1 and its analogues are putative antiobesity drugs to treat leptin-resistant obese patients. The possibility of treatment for leptin-resistant obese people in the future is shown in Fig. 4 (10). The data obtained from peripheral administration of nesfatin-1 and M30 in mice indicate that the most suitable route of administration is the subcutaneous

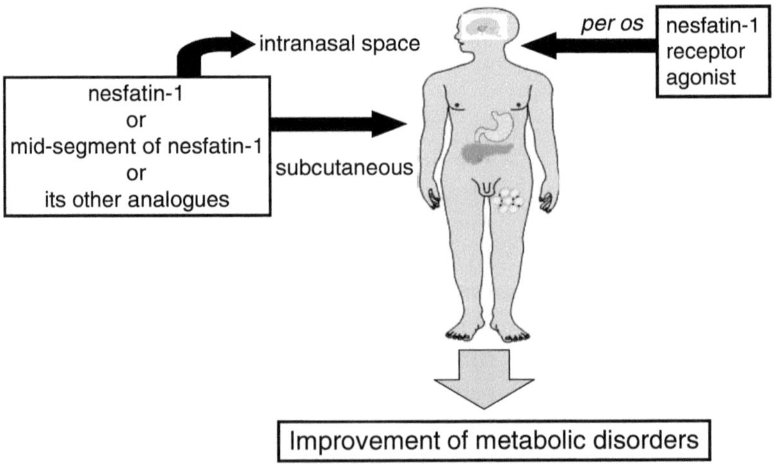

Fig. 4. Proposal for future clinical treatments of obesity using nesfatin-1-related drugs. Nesfatin-1-related peptides could be subcutaneously or intranasally administered. Nesfain-1 receptor agonists could be administered per os.

route similar to that used for daily administration of insulin in diabetes patients. However, some obese patients may have needle phobia to daily repeated subcutaneous only for weight reduction, and long-acting nesfatin-1 analogue has to be developed so that an injection has to be taken only once a day. Furthermore, results from another study indicated that the cerebrospinal fluid/plasma ratio of nesfatin-1/NUCB2 is decreased in obese humans, which indicates the existence of nesfatin-1/NUCB2 resistance in obese people (33). Because nesfatin-1 effectively reduced the food intake after intranasal administration in rats, this application may be another way for nesfatin-1 treatment. The nasal route is suitable for a frequent administration (before each meal) of nesfatin-1 to obese people. Although a specific receptor for nesfatin-1 has not been identified, a nesfatin-1 receptor agonist, administered per os, is an option for future treatment.

7. Conclusion

In this review, we have summarized the recent progress in the research of nesfatin-1, processed from nesfatin/NUCB2, which possesses a calcium-binding motif. In particular, the data obtained from basic research on nesfatin-1 and M30 will help to develop new antiobesity drugs for effective treatment of leptin-resistant obese people. However, further studies are required to determine the effects of nesfatin-1 on the endocrine and metabolic systems and its psychiatric effects. The influences of nesfatin-1, M30, and analogues, administered peripherally or intranasally, require further research before initiating clinical trials.

References

1. Shimizu H, Mori M (2005) The brain-adipose axis. Nutr Neurosci 8:7–20
2. Shimizu H, Tsuchiya T, Sato N et al (1998) Troglitazone reduces plasma leptin concentration but increases hunger in NIDDM patients. Diabetes Care 21:1470–1474
3. Oh-I S, Shimizu H, Satoh T et al (2006) Identification of nesfatin-1 as a satiety molecule in the hypothalamus. Nature 443:709–712
4. Yamada M, Horiguchi K, Umezawa R et al (2010) Troglitazone, a ligand of peroxisome proliferator-activated receptor-g, stabilizes NUCB2 (Nesfatin) mRNA by activating the ERK1/2 pathway: isolation and characterization of the human NUCB2 gene. Endocrinology 151:2494–2503
5. Brailoiu GC, Dun SL, Braoiu E et al (2007) Nesfatin-1: distribution and interaction with a G protein-coupled receptor in the rat brain. Endocrinology 148:5088–5094
6. Merali Z, Cayer C, Kent P et al (2008) Nesfatin-1 increases anxiety- and fear-related behaviors in the rat. Psychopharmacology (Berl) 201:115–123
7. Pan W, Hsuchou H, Kastin AJ (2007) Nesfatin-1 crosses the blood–brain barrier without saturation. Peptides 28:2223–2228
8. Price TO, Samson WK, Niehoff ML et al (2007) Permeability of the blood–brain barrier to a novel satiety molecule nesfatin-1. Peptides 28:2372–2381

9. Oh-I S, Shimizu H, Satoh T et al (2005) Molecular mechanisms associated with leptin resistance: n-3 polyunsaturated fatty acids induce alterations in the tight junction of the brain. Cell Metab 1:331–341
10. Shimizu H, Oh-I S, Okada S et al (2009) Nesfatin-1: an overview and future clinical application. Endocr J 56:537–543
11. Shimizu H, Oh-I S, Hashimoto K et al (2009) Peripheral administration of nesfatin-1 reduces food intake in mice: the leptin-independent mechanism. Endocrinology 150:662–671
12. Kohno D, Nakata M, Maejima Y et al (2008) Nesfatin-1 neurons in paraventricular and supraoptic nuclei of the hypothalamus coexpress oxytocin and vasopressin and are activated by refeeding. Endocrinology 149:1295–1301
13. Price CJ, Hoyda TD, Samson WK et al (2008) Nesfatin-1 influences the excitability of paraventricular nucleus neurons. J Neuroendocrinol 20:245–250
14. Douglas AJ, Johnstone LE, Leng G (2007) Neuroendocrine mechanism of change in food intake during pregnancy: a potential role for brain oxytocin. Physiol Behav 91:352–365
15. Price CJ, Samson WK, Ferguson AV (2008) Nesfatin-1 inhibits NPY neurons in the arcuate nucleus. Brain Res 1230:99–106
16. Inhoff T, Mönnikes H, Noetzel S et al (2008) Desacyl ghrelin inhibits the orexigenic effect of peripherally injected ghrelin in rats. Peptides 29:2159–2168
17. Fort P, Salvert D, Hanriot L et al (2008) The satiety molecule nesfatin-1 is co-expressed with melanin concentrating hormone in tuberal hypothalamic neurons of the rat. Neuroscience 155:174–178
18. Foo KS, Brismar H, Broberger C (2008) Distribution and neuropeptide coexistence of nucleobindin-2 mRNA/nesfatin-like immunoreactivity in the rat CNS. Neuroscience 156:563–579
19. Stengel A, Goebel M, Yakubov I et al (2009) Identification and characterization of nesfatin-1 immunoreactivity in endocrine cell types of the rat gastric oxyntic mucosa. Endocrinology 150:232–238
20. Macro JA, Dimaline R, Dockray GJ (1996) Identification and expression of prohormone-converting enzymes in the rat stomach. Am J Physiol 270:G87–G93
21. Date Y, Murakami N, Toshinai K et al (2002) The role of the gastric afferent vagal nerve in ghrelin-induced feeding and growth hormone secretion in rats. Gastroenterology 123: 1120–1128
22. Shimizu H, Ohsaki A, Oh-I S et al (2009) A new anorexigenic protein, nesfatin-1. Peptides 30:995–998
23. Gonzalez R, Tiwari A, Unniappan S (2009) Pancreatic beta cells colocalize insulin and pronesfatin immunoreactivity in rodents. Biochem Biophys Res Commun 381:643–648
24. Nakata M, Manaka K, Yamamoto S et al (2011) Nesfatin-1 enhances glucose-induced insulin secretion by promoting Ca^{2+} influx through L-type channels in mouse islet b-cells. Endocr J 58:305–313
25. Gonzalez R, Perry RL, Gao X et al (2011) Nutrient responsive nesfatin-1 regulates energy balance and induces glucose-stimulated insulin secretion in rats. Endocrinology 152:3628–3637
26. Riva M, Dekker M, Voss U et al (2011) Nesfatin-1 stimulates glucagon and insulin secretion and beta cell NUCB2 is reduced in human type 2 diabetic subjects. Cell Tissue Res 346:393–405
27. Tsuchiya T, Shimizu H, Yamada M et al (2010) Fasting concentrations of nesfatin-1 are negatively correlated with body mass index in non-obese males. Clin Endocrinol 73:484–490
28. Lo QC, Wang HY, Chen X et al (2010) Fasting plasma levels of nesfatin-1 in patients with type 1 and type 2 diabetes mellitus and the nutrient-related fluctuation of nesfatin-1 level in normal humans. Regul Pept 8:72–77
29. Zhang Z, Li L, Yang M et al (2012) Increased plasma levels of nesfatin-1 in patients with newly diagnosed type 2 diabetes mellitus. Exp Clin Endocrinol Diabetes 120:91–95
30. Aydin S, Dag E, Ozkan Y et al (2009) Nesfatin-1 and ghrelin levels in serum and saliva of epileptic patients: hormonal changes can have a major effect on seizure disorders. Mol Cell Biochem 32:49–56
31. Ari M, Ozturk OH, Bez Y et al (2010) High plasma nesfatin-1 level in patients with major depressive disorder. Prog Neuropsychopharmacol Biol Psychiatry 35:497–500
32. Heymsfield SB, Greeberg AS, Fujioka K et al (1999) Recombinant leptin for weight loss in obese and lean adults: a randomized, controlled, dose-escalation trial. JAMA 282:1568–1575
33. Tan BK, Hallschmid M, Kern W et al (2011) Decreased cerebrospinal fluid/plasma ratio of the novel satiety molecule, nesfatin-1/NUCB-2, in obese humans: evidence of nesfatin-1/NUCB-2 resistance and implications for obesity treatment. J Clin Endocrinol Metab 96:E669–E673

Part VI

Cell Line Authentication

Chapter 21

Human Cell Line Authentication: The Critical First Step in Any Project Using Human Cell Lines

Robert S. McLaren, Yvonne Reid, and Douglas R. Storts

Abstract

Short tandem repeat (STR) typing is a standard procedure used in many laboratories for the authentication of human cell lines. This technology, which is based on the informativeness of known polymorphism of numerous loci to uniquely identify a human cell line, has allowed for direct-amplification of human DNA stored on FTA® paper. We describe an application of this technology to create a unique STR profile by direct amplification of HCT 116 (ATCC® CCL-247™) cell line DNA, a cell line commonly used in colon research. The ability to perform direct-amplification of DNA opens up the possibility of using FTA® paper as a way to maintain long-term storage of DNA samples from a cell line and other human tissues, such as buccal cells.

Key words: STR, DNA profiling, Human cell lines, Authentication

1. Introduction

The propagation and culturing of human cell lines are essential for biomedical research. Over the decades, there has been an increased demand for cell lines as research tools and models by academic and industrial scientists. This demand has led to an increase in misidentified cell lines (1–6) and the inconsistencies in cell authenticity and quality that have resulted in false and invalid data.

Cellular cross-contamination of cultures is a persistent problem and occurs at frequencies ranging from 16 to 35% (7, 8). Contamination of cell lines continues as a result of mishandling and lack of best tissue culture practices (9, 10).

The discovery of DNA hypervariable regions (11, 12) in the human genome has made it possible to uniquely identify human cell lines (13, 14). The hypervariable regions, which are composed

Claus W. Heizmann (ed.), *Calcium-Binding Proteins and RAGE: From Structural Basics to Clinical Applications*, Methods in Molecular Biology, vol. 963, DOI 10.1007/978-1-62703-230-8_21,

of short tandem repeats (STRs), are distributed throughout the genome and represent some of the most informative markers for human cell line identification.

STR loci are polymorphic in the human population, in that multiple alleles exist that vary from one another in the number of repeats present at that locus. In a diploid cell line there are two copies of an allele at a given locus. These alleles have the same number of repeats (homozygous) or different number of repeats (heterozygous). With STR loci that segregate independently of each other it is possible to multiply the genotype frequencies obtained at multiple loci to estimate the DNA profile frequency. The greater the number of loci analyzed the greater the discrimination power and the more likely one is able to distinguish between two closely related human cell lines.

The length of the individual repeat in these STR loci is usually quite small, varying from three to six base pairs depending on the STR locus making them amenable to multiplex PCR amplification. Multiplexing of multiple STR markers is typically achieved by separating the amplified alleles on a capillary electrophoresis instrument based on size as well as on the basis of the fluorescent dye used to label the amplification product, In this way it is possible to distinguish amplification products from many different STR loci in one multiplex PCR reaction.

The PowerPlex® 18D System, which consists of 17 polymorphic STR markers plus amelogenin for gender determination, allows for direct amplification of human cell line DNA stored on FTA® paper. The FTA® paper enhances the long-term storage without prior purification of the DNA. We provide a detailed protocol on the use of the PowerPlex® 18D System for the direct amplification of human DNA spotted onto FTA® paper.

2. Materials

All reagents should be stored at the manufacturer's recommended temperature. FTA® paper permits the long-term storage (at least 14 years) of DNA at ambient temperature (20–25°C) following spotting of human samples as cell lines, blood, and buccal cells. Table 1 lists the reagents and storage temperatures used for the DNA typing human cells lines immobilized on FTA® cards. Table 2 lists the labware required.

Thermal cycling is typically performed using a GeneAmp® PCR System 9700 thermal cycler (Applied Biosystems/Life Technologies, Inc.) which can accept the MicroAmp® Optical 96-well reaction plate. Electrophoresis of the amplified products is performed using either an Applied Biosystems 4-capillalry 3130 Genetic Analyzer or an Applied Biosystems 16-capillary 3130*xl* Genetic Analyzer.

Table 1
Reagents for DNA typing of human cell lines

Reagent	Vendor	Recommended storage temperature
FTA® Micro Card	Fitzco, Inc. (now distributed by Whatman)	20–25°C
PowerPlex® 18D System (see Note 1)	Promega Corporation	-20°C
POP-4™ Performance Optimized Polymer (see Note 2)	Applied Biosystems/Life Technologies, Inc.	4°C
3130 & 3100-*Avant* Capillary Array 36 cm (see Note 3)	Applied Biosystems/Life Technologies, Inc.	20–25°C
3130*xl* and 3100 Capillary Array 36 cm (see Note 3)	Applied Biosystems/Life Technologies, Inc.	20–25°C
3730 Running Buffer, 10× (see Note 4)	Applied Biosystems/Life Technologies, Inc.	20–25°C
Hi-Di™ Formamide, 25 mL (see Note 5)	Applied Biosystems/Life Technologies, Inc.	-20°C

Table 2
Labware for DNA profiling of human cell lines

Labware	Vendor
MicroAmp® Optical 96-Well Reaction Plate (see Note 6)	Applied Biosystems/Life Technologies, Inc.
MicroAmp® Optical 8-Cap Strip	Applied Biosystems/Life Technologies, Inc.
96-Well Plate Base	Applied Biosystems/Life Technologies, Inc.
96-Well Plate Retainer	Applied Biosystems/Life Technologies, Inc.
96-Well Plate Septa	Applied Biosystems/Life Technologies, Inc.

Subsequent data analysis is performed using either the GeneMapper® *ID* Software, Version 3.2, or the GeneMapper *ID*-X Software, Version 1.2. The panels and bins text files designed for this chemistry and the appropriate software platform used to analyze data generated with the PowerPlex® 18D System are available for download from http://www.promega.com/resources/tools/genemapper-id-software-panels-and-bin-sets/. Detailed instructions on using these files in conjunction with either the GeneMapper® *ID* Software, Version 3.2, or the GeneMapper *ID*-X Software, Version 1.2, to analyze PowerPlex® 18D System data are available in the PowerPlex® 18D System Technical Manual (15).

3. Methods

The use of gloves and aerosol-resistant pipette tips is strongly recommended. To avoid contamination of the amplification setup area with allelic ladder and amplified product, post-amplification products/reactions are kept away from pre-amplification products and are stored in different rooms.

3.1. Spotting Cell Lines onto FTA® Card

1. Harvest cells (example cell line HCT 116, ATCC® CCL-247™) according to laboratory procedure.
2. Wash cells with PBS and resuspend cells in PBS at 10^6–10^7 cells/mL.
3. Spot 150 μL (10^6–10^7 cells/mL) of the cell suspension onto an FTA® card.
4. Distribute the cell suspension, evenly, within the circle.
5. Allow the FTA® card to dry completely at room temperature (20–25°C) before use. For long-term storage, cards may be stored at room temperature in a desiccator.

3.2. Direct-Amplification of DNA from 1.2 mm Diameter Punches from FTA® Card

1. Determine the number of amplification reactions to be performed and make sufficient reaction mix for the number of reactions plus 15% (to compensate for pipetting error and reagent loss due to multiple pipetting from the same tube of reaction mix to multiple reactions) using the reagents listed in Table 3 (see Note 1).
2. Completely thaw, at room temperature, the reagents from the PowerPlex® 18D System Pre-Amplification Components Box.
3. Centrifuge briefly to bring all contents to the bottom of each tube, and then vortex each tube, three times for 5 s each. *DO NOT* re-spin tubes of PowerPlex® 18D 5X Primer pair.

   Table 3 shows the reagent volumes per amplification reaction. To determine volumes for multiple reactions, multiply each volume by the number of reactions (including 15% extra) being performed.

Table 3
Reagents for reaction mix

Amplification reaction mix component	Volume per 25 μL reaction
PowerPlex® D 5× Master Mix	5 μL
PowerPlex® 18D 5× Primer pair Mix	5 μL
Water, Amplification Grade	15 μL

4. Dispense 25 μL of the amplification reaction mix to each well of a MicroAmp® Optical 96-Well Reaction Plate.
5. Use a 1.2 mm diameter Harris Punch to make a 1.2 mm diameter punch from the FTA® card and dispense one punch per well containing 25 μL of amplification reaction mix.

   It is acceptable to add the 1.2 mm diameter punch to the MicroAmp® Optical 96-Well Reaction Plate before the amplification reaction mix has been added.
6. For the two positive control amplification reactions, add 5 ng of the 2800 M control DNA supplied in the PowerPlex® 18D System Pre-Amplification Components Box to 25 μL of the amplification reaction mix in either the presence or absence of a 1.2 mm diameter punch from an FTA® card without cell DNA.
7. For the two negative controls, perform the reaction without DNA in either the presence or absence of a 1.2 mm diameter punch from an FTA® card without cell DNA.
8. Seal the wells of the MicroAmp® Optical 96-Well Reaction Plate with MicroAmp® Optical 8-Cap Strips and perform thermal cycling on a GeneAmp® PCR System 9700 system following the protocol below:

96°C for 2 min
Then 27 cycles of:
94°C for 10 s
60°C for 1 min
Then one final extension of:
60°C for 20 min
Ramp down to
4°C Soak

   After starting the GeneAmp® PCR System 9700 thermal cycler, the "Select Methods Options" screen appears. Select 25 μL for the reaction volume and "9600" for the ramp speed. The method described here is for 27 cycles. Cycling takes approximately 90 min. Depending upon the number of cells deposited it may be necessary to vary cycle number from 24 to 27 to obtain optimal signal on the Applied Biosystems 3130 or 3130*xl* Genetic Analyzer to prevent too low signal or saturating signal. It is recommended that preliminary experiments varying cycle number be performed to determine optimal signal.
9. After completion of thermal cycling, store the amplification products at –20°C in a light-protected container (see Note 7).

3.3. Detection of Amplified Fragments Using the Applied Biosystems 3130 or 3130xl Genetic Analyzer

Denature the amplified products prior to electrophoresis on the Applied Biosystems 3130 or 3130*xl* Genetic Analyzer. This is achieved by heat denaturation in a formamide-based loading cocktail.

1. Prepare a loading cocktail for one sample by combining 10 μL Hi-Di™ formamide with 1 μL CC5 ILS 500 (internal lane standards—ILS). For all the amplification reactions being analyzed multiply the volume of Hi-Di™ formamide and CC5 ILS 500 by the number of reactions being analyzed plus 15% (e.g., for 40 samples that would be 46 × 10 μL Hi-Di™ formamide and 46 × 1 μL CC5 ILS 500) to make a master mix loading cocktail.
2. Add 1 μL of amplified product from each well of the amplification reaction plate to a well of a new MicroAmp® Optical 96-Well Reaction Plate. Several wells (1–4 for a 96-well plate) should also contain 1 μL of the PowerPlex® 18D System Allelic Ladder Mix which allows the GeneMapper® *ID* Software, Version 3.2, or GeneMapper *ID*-X Software, Version 1.2, to assign allele calls to the PowerPlex® 18D System data.
3. Add 11 μL of loading cocktail to each well. Seal plate using 96-Well Plate Septa and using a centrifuge with a plate adapter, spin the plate briefly to remove bubbles from wells and ensure that all liquid is at the bottom of each well.
4. Denature samples by placing at 95°C for 3 min. This may be performed on a thermal cycler or on a dry block capable of accepting a 96-well PCR plate.
5. After heating for 3 min, quick chill the plate for 3 min on a crushed ice-slurry. Place the plate on the Applied Biosystems 3130 or 3130*xl* Genetic Analyzer.

3.4. Preparation of the Applied Biosystems 3130 or 3130xl Genetic Analyzer

1. Refer to the Applied Biosystems 3130 /3130*xl* Genetic Analyzer user's manual for instructions on cleaning, installing the capillary array, performing a spatial calibration, installing performance-optimized polymer, and changing running buffer.
2. Open the data collection software and select the "Module Manager" on the left-hand side of the software window. Select "New" and then in the Type drop-down list select "Regular".
3. In the Template drop-down list select "HIDFragmentAnalysis36_POP4". Confirm that the injection time is set to 5 s, the injection voltage is set at 3 kV, and the run time is 1,500 s.
4. Give a descriptive name to the run module (e.g., PowerPlex_18D_3kV_5Sec_1500) and select "OK".
5. Select "Protocol Manager" from the left-hand side of the data collection software window and select "New" on the right-hand side of the window.

6. Type in a name for the new protocol (e.g., PowerPlex_18D_3kV_5Sec_1500). Select "Regular" in the Type drop-down list.
7. In the Run Module drop-down list, select the Run Module created in the previous step. Select "G5" from the Dye Set drop-down list, and then select "OK".
8. Select "Plate Manager" from left-hand side of window of the data collection software and create a new plate record. Select "GeneMapper-Generic" in the Application drop-down list and select appropriate plate type (96-well).
9. Add entries to the "Owner" and "Operator" and select "OK".
10. In the GeneMapper plate record, enter sample names in the appropriate cells. In the "Results Group 1" column, select the desired Results Group location to save the data (see Note 8). In the "Instrument Protocol 1" column, select the Instrument Protocol created in steps 5–9, above. Ensure that this information is in each row containing a sample name. Select "OK".
11. Remove plate containing denatured samples from crushed ice-slurry (Subheading 3.3, step 5) and place in black 96-Well Plate Base.
12. Snap the white 96-Well plate retainer on top and place the plate in instrument. Close instrument doors.
13. Select "Spectral Viewer" from left-hand side of data collection software window and then select "G5" dye set. Confirm that the active spectral calibration for the G5 dye set is correct for the PowerPlex® 18D System (see Note 9). If not, select the correct spectral calibration from the list for the G5 dye set and then select "Set".
14. In the "Run Scheduler" window, locate the plate record created in steps 5 and 6, above. Click once on the name to highlight it and once on the image of the 96-well plate on the right of the screen. The plate should turn from yellow to green and the green run arrow at the top left of the data collection software window should become active.
15. Click on this arrow to begin the run. Each run takes approximately 40 min.

3.5. Analyzing Data with GeneMapper ID Software, Version 3.2

1. Download panels and bin text files for analysis of the data http://www.promega.com/resources/tools/genemapper-id-software-panels-and-bin-sets/. For analysis of PowerPlex® 18D System data using GeneMapper® *ID* software, version 3.2, the proper panels and bins text files are needed: PowerPlex_18D_Panels_vX.x.txt and PowerPlex_18D_Bins_vX.x.txt files, where "X.x" refers to the most recent version of the panels and bin text files.
2. Save the PowerPlex_18D_Panels_X.x.txt and PowerPlex_18D_Bins_X.x.txt files to a known location on your computer.

3. Open the GeneMapper® *ID* software, version 3.2. Select "Tools", and then "Panel Manager". Highlight the Panel Manager icon in the upper left navigation pane. Select "File", and then "Import Panels". Navigate to the panels text file you downloaded and select "Import". Go back to the navigation pane and highlight the PowerPlex_18D_Panels_X.x folder that you just imported. Select "File", and then "Import Bin Set". Navigate to the downloaded bins text file.
4. Select the file, and then "Import". At the bottom of the Panel Manager window, select "OK". The Panel Manager window will close automatically.

   A size standard .xml file (CC5_ILS_500.xml) is required to analyze data in the GeneMapper® *ID* software, version 3.2. This contains a listing of the sizes of all the size fragments in the CC5 ILS 500 ILS run with the amplified products and allelic ladder. This may also be downloaded from http://www.promega.com/resources/tools/genemapper-id-software-panels-and-bin-sets/ and imported into the software using the "GeneMapper Manager" tool.
5. Analyze data using an analysis method that can be created in the "GeneMapper Manager" tool. For full details see the PowerPlex® 18D System Technical Manual (15).
6. Analyze data with a 50–150 RFU analysis threshold. This value should be determined by individual laboratories from internal validation studies and is based on, among other things, the sensitivity of the capillary electrophoresis unit being used (sensitivities vary from instrument to instrument). In addition a 20% filter is typically assigned which removes labels from peaks that are less than 20% of the height of the tallest peak at a given locus.

During analysis the GeneMapper® *ID* software, version 3.2, determines the size of the amplified fragments in each sample by comparison of their mobility relative to the ILS. The same process is also performed with alleles in the allelic ladder. The software then compares the size of the peaks at a locus in an amplified sample to the sizes of alleles in the allelic ladder. If the size of an allele falls within a binning window around the size of a particular allele in the allelic ladder (typically the bin is ±0.5 bases) then the software assigns the same allele designation to the peak in the amplified sample as that allele in the allelic ladder (Fig. 1).

3.6. Analyzing Data with GeneMapper ID-X Software, Version 1.2

The process of analyzing data in GeneMapper *ID*-X Software, Version 1.2, is very similar to that of the GeneMapper® *ID* software, version 3.2, and is described in the PowerPlex® 18D System Technical Manual (15). One of the differences between the two software programs is that the configuration of the text files required

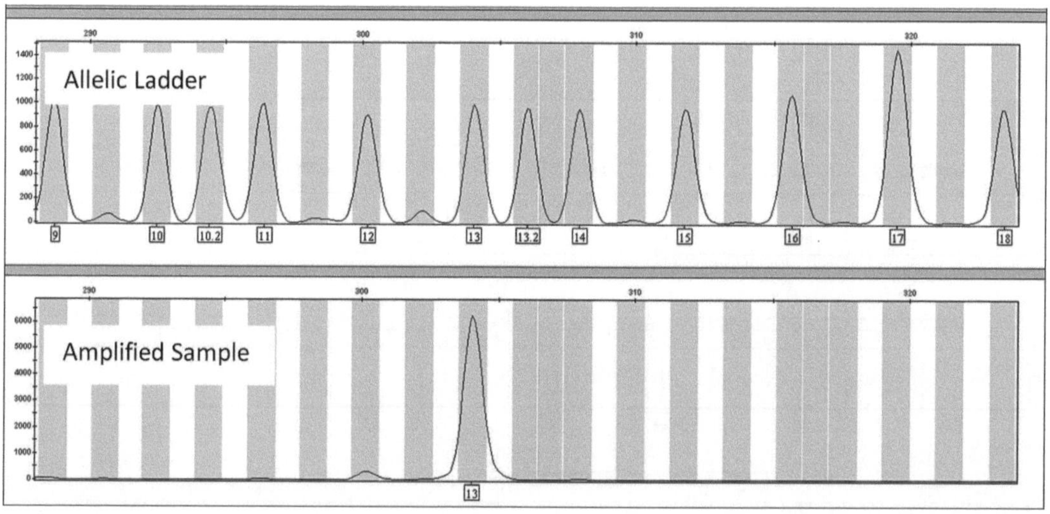

Fig. 1. Electropherogram of the amplified D18S51 locus from the PowerPlex® 18D System analyzed in GeneMapper® *ID* Software, Version 3.2, showing allelic ladder and amplified sample. The *grey bars* represent the ±0.5 base binning window centered on the alleles shown in the allelic ladder. The location of these bins is adjusted by the GeneMapper® *ID* Software, Version 3.2, based on sizing of the actual alleles in the allelic ladder. When a peak in the amplified sample aligns with one of these bins the software makes the corresponding allele assignment. In this case the sample contains allele 13 at D18S51.

for data analysis is changed to require three text files instead of two. The files that work for GeneMapper® *ID* software, version 3.2, will not work for GeneMapper *ID*-X Software. The three GeneMapper *ID*-X Software text files are the Panels, Bins, and Stutter text files.

1. Download the three GeneMapper *ID*-X Software text files from http://www.promega.com/resources/tools/genemapper-id-software-panels-and-bin-sets/ and import into GeneMapper *ID*-X Software following instructions in the PowerPlex® 18D System Technical Manual (15).
2. Import the size standard .xml file into the software using the "GeneMapper ID-X Manager" tool.

A separate size standard .xml file (CC5_ILS_500_IDX.xml) is required to analyze data in the GeneMapper *ID*-X Software, Version 1.2. This contains a listing of the sizes of all the size fragments in the CC5 ILS 500 ILS run with the amplified products and allelic ladder and is downloaded along with the panels, bins, and text files described above.

3. Analyze data in a similar manner to that previously described for GeneMapper® *ID* software, version 3.2, and is described in detail in the PowerPlex® 18D System Technical Manual (15).

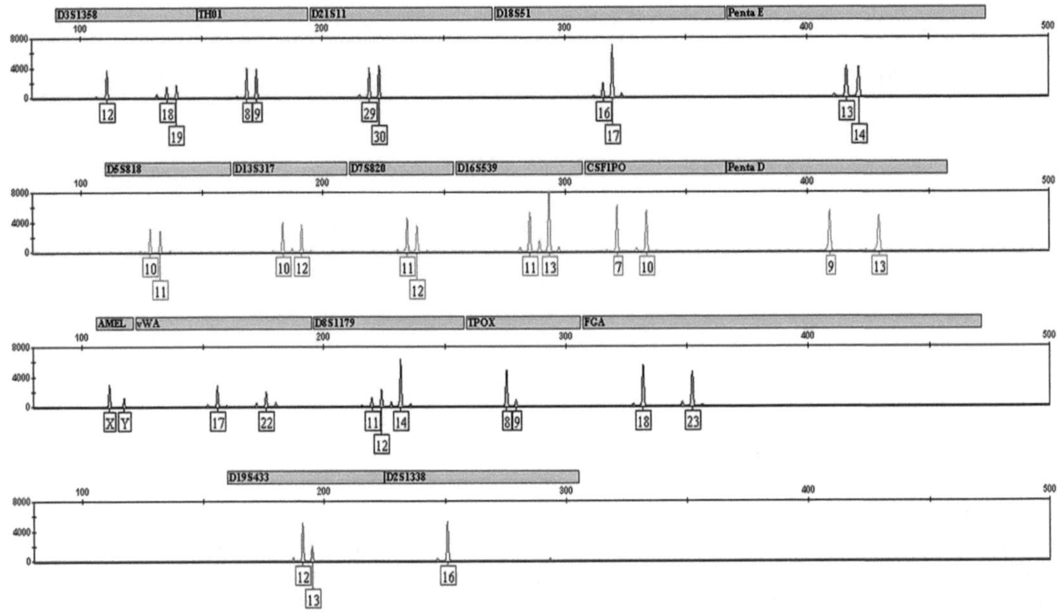

Fig. 2. PowerPlex® 18D System electropherogram of the colorectal carcinoma cell line HCT 116 (ATCC® CCL-247™). DNA was amplified from cells spotted onto FTA® paper. Data were analyzed in GeneMapper® *ID* Software, Version 3.2. Labels under each peak represent allele calls. This sample displays triallelic patterns at D3S1358 *(12, 18, 19)* and D8S1179 *(11, 12, 14)*. In addition there are several instances of peak height imbalance between alleles at a given locus (e.g., D18S51, TPOX, D8S1179, D19S433, and Amelogenin). Triallelic patterns at one or two loci and peak imbalance are not unusual in tumor cell lines displaying aneuploid genotype.

3.7. Example of Analyzed Data

Data are typically represented as a picture of the electropherogram showing all the amplified peaks present across each locus in each dye channel (Fig. 2). Both GeneMapper® *ID* Software, Version 3.2, and GeneMapper *ID*-X Software, Version 1.2, allow the user to choose what information is shown on peak labels (e.g., allele call, size in bases, height of peak, etc.). In addition, data may be exported from both pieces of software in a table format for reporting purposes.

4. Notes

1. The PowerPlex® 18D System is supplied with a pre-amplification components box (Blue label) and a post-amplification components box (Beige label). Keep the pre-component amplification box and post-component amplification box in separate rooms

to avoid contamination of the pre-amplification components with allelic ladder. The component boxes contain the following reagents:

Pre-amplification Components Box (Blue Label) from 200 Reaction PowerPlex® 18D System
1 mL PowerPlex® D 5× Master Mix
1 mL PowerPlex® 18D 5× Primer Pair Mix
25 μL 2800 M Control DNA, 10 ng/μL
5× 1,250 μL Water, Amplification Grade
Post-amplification Components Box (Beige Label) from 200 Reaction PowerPlex® 18D System
50 μL PowerPlex® 18D Allelic Ladder Mix
2× 150 μL CC5 ILS 500

2. The POP-4™ Performance Optimized Polymer should be stored at 4°C prior to installation on the Applied Biosystems 3130 or 3130*xl* Genetic Analyzer. Upon installation on the instrument it should be allowed to warm up to ambient room temperature for at least 30 min prior to installation. Once on the Applied Biosystems 3130 or 3130*xl* Genetic Analyzer the POP-4™ Performance Optimized Polymer is good for 7 days at room temperature. After that time it should be replaced. Use of POP-4™ Performance Optimized Polymer past this time may result in poor-quality data due to poor separation and/or an inability to keep the denatured amplicons in a denatured state.
3. Capillaries should be replaced after 100 injections. Using capillaries past this time may result in poor resolution and broader peaks, especially for the larger size amplified products.
4. Running buffer should be diluted to 1× with water and replaced fresh each day on the instrument.
5. Hi-Di™ formamide should be thawed, vortexed thoroughly to mix, and aliquoted into single use (e.g., 500 μL each). Aliquots should be stored at −20°C. Thaw enough single tube aliquots for each use and discard any unused formamide. Formamide breaks down to formic acid with repeat freeze/thaws. Formic acid is negatively charged and competes with the DNA for electrokinetic injection of the amplified DNA fragments into the capillary. This results in reduced peak heights.
6. A MicroAmp® Optical 96-Well Reaction Plate is used for the amplification reactions and for setting up the injection plate when loading samples onto the Applied Biosystems 3130 or

3130*xl* Genetic Analyzer. For each amplification reaction, two MicroAmp® Optical 96-Well Reaction Plates are used: one for pre-amplification and the other for post-amplification. It is recommended to keep these plates in separate areas to minimize chances for cross-contamination.

7. Amplification plates should be stored at −20°C upon completion of thermal cycling. Prolonged storage at 4°C may result in the appearance of degradation products, specifically peaks with a wide peak width relative to their height that migrate in the low size range (60–100 bases) of the electropherogram.
8. To create a new results group, select "New" in the drop-down menu for the "Results Group 1" columns on the plate record screen. Select the "General" tab and enter a name. Select the "Analysis" tab, and select "GeneMapper-Generic" in the Analysis type drop-down list.
9. In order to successfully generate data with the PowerPlex® 18D System on an Applied Biosystems 3130 or 3130*xl* Genetic Analyzer it is necessary to generate a PowerPlex® 5-Dye spectral calibration on the G5 dye set. This can be done using the PowerPlex® 5-Dye Matrix Standards, 3100/3130 (Cat.# DG4700), following the technical bulletin (16).

References

1. Schweppe R, Klopper JP, Korch C, Pugazhenthi U, Benezra M, Knauf JA et al (2008) Deoxyribonucleic acid profiling analysis of 40 human thyroid cancer cell lines reveals cross-contamination resulting in cell line redundancy and misidentification. J Clin Endocrinol Metab 93:4331–4341
2. Phuchareon J, Ohta Y, Woo JM, Eisele DW, Tetsu O (2009) Genetic profiling reveals cross-contamination and misidentification of 6 adenoid cystic carcinoma cell lines: ACC2, ACC3, ACCM, ACCNS, ACCs and CAC2. PLoS One 4:e6040
3. Thompson EW, Waltham M, Ramus SJ, Hutchins AM, Armes JE, Campbell IG et al (2004) LCC15-MB cells are MDA-MB-435: a review of misidentified breast and prostate cell lines. Clin Exp Metastasis 21:535–541
4. Boonstra JJ, van Marion R, Beer DG, Lin L, Chaves P, Ribeiro C et al (2010) Verification and unmasking of widely used human esophageal adenocarcinoma cell lines. J Natl Cancer Inst 102:271–274
5. Freshney RI (2010) Database of misidentified cell lines. Int J Cancer 26:302
6. MacLeod RAF, Dirks WG, Matsuo Y, Kaufmann M, Milch H, Drexler HG (1999) Widespread intraspecies cross-contamination of human tumor cell lines arising at source. Int J Cancer 83:555–563
7. Nelson-Rees WA (1978) The identification and monitoring of cell line specificity. Prog Clin Biol Res 26:25–79
8. Hukku B, Halton DM, Mally M, Peterson WD Jr (1984) Cell characterization by use of multiple genetic markets. In: Acton RT, Lynn JD (eds) Eukaryotic Cell Cultures. Plenum Publishing Co, New York, pp 13–31
9. Reid YA (2011) Characterization and authentication of cancer cell lines: an overview. Methods Mol Biol 731:35–43
10. Barallon R, Bauer SR, Butler J, Capes-Davis A, Dirks WG, Elmore E et al (2010) Recommendation of short tandem repeat profiling for authentication of human cell lines, stem cells, and tissues. In Vitro Cell Dev Biol Anim 46:727–732
11. Jeffreys AJ, Wilson V, Thein SL (1985) Hypervariable 'minisatellite' regions in human DNA. Nature 314:76–673
12. Nakamura Y, Leppert M, O'Connell P, Wolff R, Holm T, Culver M et al (1987) Variable number of tandem repeat (VNTR) markers for human gene mapping. Science 235:1616–1622
13. Gilbert DA, Reid YA, Gail MH, Pee D, White C, Hay RJ, O'Brien SJ (1990) Application of

DNA fingerprints for cell line individualization. Am J Hum Genet 47:499–514
14. Masters JR, Thompson JA, Daly-Burnes B, Reid YA, Dirks WG, Packer P et al (2001) Short tandem repeat profiling provides an international reference standard for human cell lines. Proc Natl Acad Sci USA 98:8012–8017
15. Technical Manual for PowerPlex® 18D System (Last revision, December 2011). Available on Promega's web page http://www.promega.com/resources/protocols/technical-manuals/101/powerplex-18d-system-protocol/
16. Technical Manual for PowerPlex 5-Dye Matrix Standards, 3100/3130. Available on Promega's webpage http://www.promega.com/resources/protocols/technical-bulletins/101/powerplex-5-dye-matrixstandards-3100-3130-protocol/

Part VII

Biomarkers, Diagnostic, Clinical Chemistry, and Gene Therapy

Chapter 22

Detection of Ca^{2+}-Binding S100 Proteins in Human Saliva by HPLC-ESI-MS

Massimo Castagnola, Tiziana Cabras, Federica Iavarone, Chiara Fanali, and Irene Messana

Abstract

High-performance liquid chromatography (HPLC) coupled with electrospray ionization (ESI) mass spectrometry (MS) is a relevant technique for the detection and relative quantitation of naturally occurring peptides and proteins. The peptide/protein mass is determined by deconvolution of the ESI-MS spectrum, and the resolution can be better than 1:10,000 with the instruments currently available. Accurate mass measurement, coupled with sufficient resolution, makes it possible to greatly restrict the enormous number of possible molecular formulas that might be represented by a specific molecular mass.

As soon as the protein mass has been unequivocally attributed to a specific structure by means of different enzymatic and chemical treatments, the m/z values detected in the ESI spectrum can be utilized to reveal the protein and to perform its relative quantitation, by the extracted ion current (XIC) procedure, in an unlimited number of samples. This chapter describes the HPLC-ESI-MS experimental conditions which allow detecting and quantifying—in human saliva—different S100 proteins and their isoforms.

Key words: Human saliva, HPLC-ESI-MS, Label-free quantification, XIC procedure, S100 proteins, EF-hand, Calcium, Post-translational modifications

1. Introduction

HPLC-ESI-MS is a technique very useful for peptide and small protein analysis. Electrospray ionization (ESI) is an atmospheric pressure ionization process and this makes it easy to use and ideally suited for online coupling HPLC with MS. The distinct characteristics of ESI are described in specific reference (1). However, it is relevant to outline that (a) ESI is a "soft" ionization technique, and thus it allows to measure the mass-to-charge ratio of intact molecular ions; (b) the ionization process is commonly performed

Claus W. Heizmann (ed.), *Calcium-Binding Proteins and RAGE: From Structural Basics to Clinical Applications*, Methods in Molecular Biology, vol. 963, DOI 10.1007/978-1-62703-230-8_22, © Springer Science+Business Media New York 2013

in acidic polar solvents, where the molecules of a given protein are differently positively charged for the interaction of the H^+ ions with the carboxylic groups of Asp, Glu, and the C-terminal amino acid residue; (c) during the ESI small peptides usually generate mono- and bi-charged ions, while larger peptides and proteins display statistical distribution of multiply-charged ions, with a mean charge increasing according to the molecular dimension, often reaching more than 20 positive charges; (d) this results in the ability of analyzing very high molecular weight species using the most available mass analyzers (2). As a consequence of the distinctive features of the ESI process, the ESI spectrum of proteins shows a characteristic gaussian distribution of multiply-charged ions with different mass-to-charge (m/z) values, as shown in Fig. 1 (ESI spectrum of the S100A8 protein).

The measurement of these multiply-charged ions gives the inherent molecular mass of the analyte molecules, reaching resolution values from 1:5,000 (low-middle resolution MS apparatus; i.e., ion trap (3)) to 1:1,000,000 (very high resolution MS apparatus; i.e., magnetic sector MS (4) or Fourier transform ion cyclotron resonance MS (5)). Different automated software perform the deconvolution, a mathematical manipulation, which allows calculating the protein average mass from the m/z values (Fig. 1).

Knowledge of the protein mass is not enough for its identification, and various references report the experimental approaches necessary for the definitive characterization (6–8). However, once the ESI spectrum corresponding to a chromatographic peak has been unequivocally attributed to a protein (or any isoform of it), the m/z values of the ions can be utilized to selectively extract from the total ion current HPLC profile the ionic current (XIC procedure) peak of the protein, as shown in the following sections. This chapter describes the detection and identification of several S100 proteins in human saliva by HPLC-ESI-MS (Table 1):

- *S100A7 D_{27} isoform.* It is a variant of the protein reported in Swiss-Prot data-bank (http://us.expasy.org/. Accession Number P31151) with an Asp residue instead of Glu in position 27. It has lost the N-terminal Met residue and it is N-terminally acetylated (6).
- *S100A7 E_{27} isoform.* It is the variant with the Glu residue in position 27. It has lost the N-terminal Met residue and it is N-terminally acetylated. It is detectable in pre-term newborn saliva, but not in adult saliva (6).
- *S100A8* (Accession Number P08109). It has not lost the N-terminal Met residue and it is not N-terminally acetylated (6).

Nine isoforms of S100A9 (Accession Number P06702):

- S100A9 short. It has lost the MTCKM N-terminal pentapeptide and it is N-terminally acetylated (6).

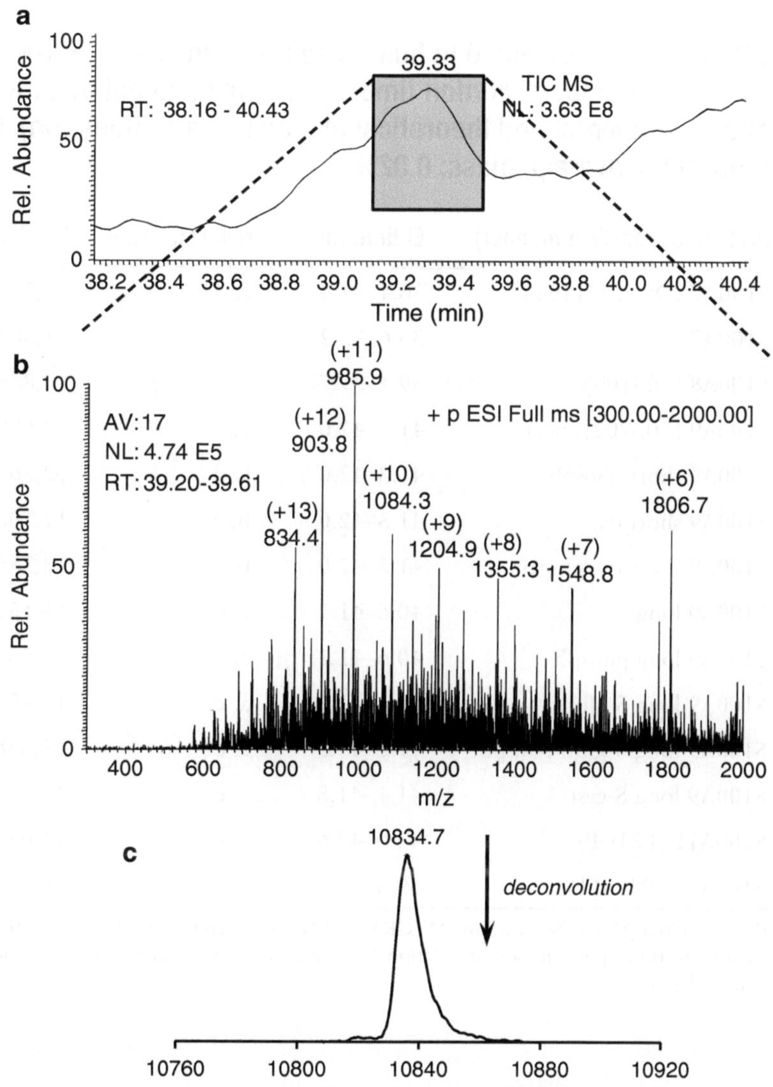

Fig. 1. Enlargement of the total ion current (TIC) chromatographic profile of the acidic soluble fraction of human saliva analyzed by RP-HPLC-ESI-MS according to the conditions described in Subheadings 3.3 and 3.4 (elution range 38.2–40.4 min, panel (**a**)). The ESI mass spectrum registered between 39.20 and 39.61 min is shown in panel (**b**), where the multiply-charged ions belonging to the protein identified as S100A8 are labeled. The deconvolution procedure applied on the ESI mass spectrum resulted in the Mav 10,834.7 Da, which is in good agreement with the molecular mass expected for S100A8 (panel (**c**)). (*RT* retention time, *NL* normalization level, *AV* average).

- S100A9 short phosphorylated. It has lost the MTCKM N-terminal pentapeptide, it is N-terminally acetylated and phosphorylated at the penultimate Thr_{108} residue (6).
- S100A9 short oxidized. It has lost the MTCKM N-terminal pentapeptide, it is N-terminally acetylated and oxidized at the Met_{89} residue (7).

Table 1
S100A proteins detected in human saliva by RP-HPLC-ESI-MS, Swiss-Prot Accession numbers, elution times, post-translational modifications, and experimental and theoretical average mass values (Da). Experimental error of the average mass: 0.02%

Protein (Swiss-Prot number)	El time (min)	N-term. PTMs	Exp Mav (Da)	Theor Mav (Da)
S100A7 (D_{27}) (P31151)	36.4–37.0	a, c	11,368	11,368
S100A7 (E_{27})	36.6–37.2	a, c	11,382	11,382
S100A8 (P05109)	39.1–39.7	–	10,835	10,835
S100A9 (P06702) short	41.3–42.0	b, c	12,689	12,689
S100A9 short phosph.	41.3–42.0	b, c	12,769	12,769
S100A9 short ox.	41.3–42.0	b, c	12,705	12,705
S100A9 short ox. phosph.	41.3–42.0	b, c	12,785	12,785
S100A9 long	40.6–41.2	a, c	13,153	13,153
S100A9 long phosph.	40.6–41.2	a, c	13,233	13,233
S100A9 long S-glut.	41.1–41.8	a, c	13,458	13,458
S100A9 long S-glut. phosph.	41.1–41.8	a, c	13,538	13,538
S100A9 long S-cyst.	41.1–41.8	a, c	13,272	13,272
S100A11 (P31949)	42.0–42.6	a, c	11,651	11,651
S100A12 (P80511)	39.5–40.2	a	10,444	10,444

(a) N-term loss: M; (b) N-term loss: MTCKM; (c) N-term acetylation. Phosphorylation, glutathionylation, and cysteinylation are other possible post-translational modifications: the occurrence is specified in the column reporting the name of the proteins

- S100A9 short oxidized phosphorylated. It has lost the MTCKM N-terminal pentapeptide, it is N-terminally acetylated, oxidized at the Met_{89} residue and phosphorylated at the penultimate Thr_{108} residue (7).
- S100A9 long. It has lost the N-terminal Met residue and it is N-terminally acetylated (6).
- S100A9 long phosphorylated. It has lost the N-terminal Met residue, it is N-terminally acetylated and phosphorylated at the penultimate Thr_{112} residue (6).
- S100A9 long S-glutathionylated. It has lost the N-terminal Met residue, it is N-terminally acetylated and S-glutathionylated at the Cys_2 residue.

- S100A9 long S-glutathionylated phosphorylated. It has lost the N-terminal Met residue, it is N-terminally acetylated, phosphorylated at the penultimate Thr_{112} residue and S-glutathionylated at the Cys_2 residue.
- S100A9 long S-cysteinylated. It has lost the N-terminal Met residue, it is N-terminally acetylated and S-cysteinylated at the Cys_2 residue.
- S100A11 (Accession Number P31949). It has lost the N-terminal Met residue and it is N-terminally acetylated (6).
- S100A12 (Accession Number P80511). It has lost the N-terminal Met residue and it is not N-terminally acetylated (6).

It is relevant to outline that (1) some of the post-translational modifications above described are not reported in Swiss-Prot data bank (6); (2) acidic hydro-organic solutions are commonly used in the HPLC-ESI-MS analysis in order to facilitate the ionization process. Many Ca^{2+}-binding S100 proteins exert their function throughout a subtle interplay between monomeric and homo- and hetero-oligomeric not-covalent assemblies, which are unstable under the acidic conditions. Thereby, only monomeric S100 proteins, soluble in acidic solutions, are detectable by this method.

2. Materials

Prepare all solutions using ultrapure water (prepared by purifying deionized water with a Milli-Q Ultrapure Water purification system to attain a sensitivity of 18 M Ω cm at 25°C). Acetonitrile and 2,2,2-trifluoroacetic acid (TFA) are of LC-MS reagent grade (Mallinckrodt Baker, Deventer, The Netherlands). Collect saliva using sterile soft plastic (see Note 1) disposable serological pipette and store in safe-lock tube. For HPLC separation use a chromatographic C_8 column (column dimensions (150×2.1 mm, 5 μm particle size; Vydac, Hesperia, CA, USA)).

2.1. Eluents and Sample Solution

1. Eluent A: 0.056% TFA in water.

   Prepare a stock solution of 1%TFA adding at 1 mL of pure TFA 999 mL of water. Add 28 mL of 1% TFA to 472 mL of water (final volume 500 mL, see Note 2).
2. Eluent B: 0.050% TFA in acetonitrile-water 80/20 (v/v).

   Add 75 mL of water and 25 mL of 1%TFA to 400 mL of acetonitrile (final volume 500 mL, see Note 2).
3. Sample solution: 0.2% TFA in water.

   Add 20 mL of 1%TFA to 80 mL of water (final volume 100 mL, see Note 2).

2.2. Apparatus

The HPLC-ESI-IT-MS apparatus is a Surveyor HPLC system (ThermoFinnigan, San Jose, CA, USA) connected to an LCQ Advantage mass spectrometer equipped with an ESI source and version 2.0.7 of Excalibur software. For the deconvolution of the ESI spectrum the freeware MagTran 1.0 software (8) can be used.

3. Methods

3.1. Collection and Treatment of Salivary Samples

Collect resting whole saliva using a sterile plastic soft disposable serological pipette aspirator between 11 and 12 a.m. in a safe-lock tube. Immediately after collection, add to the salivary sample the sample solution exactly in 1:1 v/v ratio at 4°C (see Note 3).

After stirring, centrifuge the solution at 8,000 × *g* for 5 min (4°C) and then separate the acidic supernatant from the precipitate.

Analyze the supernatant immediately by HPLC-MS or store at −80°C.

3.2. Setting of the MS Apparatus

Optimize electrospray ion source and ion trap parameters before starting with the analysis by following steps 1–6. The ion signal abundances strictly depend on the instrumental characteristics.

1. Use purified histatin 1 (4,928.15 Da) as standard peptide to optimize the tune parameters in LC-ESI MS mode.
2. Dissolve the peptide at concentration of 0.1 mM in 70/30 eluent A/B (v/v) and fill a 500 μL syringe with the peptide solution.
3. Set up the instrument to introduce sample by syringe pump into solvent flow from the HPLC by using a tee union, that is connected to the fused-silica infusion line (capillary i.d. 100 μm). Set the flow at 100 μL/min.
4. Manually change the following parameters, in order to obtain the best signal to noise ratio for peptide multiply-charged ions: capillary temperature, spray voltage, capillary voltage, sheath gas flow rate, auxiliary gas flow rate. Optimize these parameters to obtain a mass spectrum where the most abundant ion is the penta-charged ion at m/z 986.6 Th.
5. Optimize ion trap, tube lens, and multipole lens parameters by performing the auto tuning, setting the most abundant peptide ion in the mass spectra.
6. Save the tune method file containing the optimized parameters and use it for chromatograms and mass spectra recording.
7. Set mass spectra acquisition in positive ion and full scan mode in a range 300–2,000 m/z and mass spectra collection every 3 ms.
8. Save all these instrumental settings in the final mass spectrometry analysis method file.

3.3. Reversed Phase HPLC Separation

1. Carry out HPLC analysis of salivary proteins by using a C_8 chromatographic column (150×2.1 mm, 5 μm particle size; Vydac, Hesperia, CA, USA) equipped with C_8 guard cartridge (7.5×2.1 mm, 5 μm particle size; Vydac, Hesperia, CA, USA, see Note 4).
2. Set column temperature at 25°C.
3. Let conditioning the column for 5 min with 100% of mobile phase A at a flow rate of 100 μl/min.
4. Inject 30 μl of sample solution.
5. Carry out analyte elution with a binary high-pressure gradient at flow rate of 100 μl/min.
6. Perform analyte elution increasing the percentage of solvent B through the column in a multi-step mode: from 0% to 54% in 39 min, from 54% to 100% in 10 min, then to 100% in 1 min.
7. Do not address the elution solution to the mass spectrometer during the first 5 min of separation, to avoid instrument damage due to the high salt concentration.
8. Utilize, for the acquisition of the HPLC-ESI-MS profile, the parameters optimized during the tuning procedure (see Section 3.2).

3.4. Protein Detection and Quantification by the XIC Procedure

To detect S100 proteins in your samples and to quantify them by XIC procedure use the m/z values reported in Table 2.

1. Select in the total ion chromatogram (TIC) the region corresponding to the elution range (considering a possible variability of ±0.7 min) of each S100 protein reported in Table 1, and evidence the ESI spectrum.
2. Perform the deconvolution of the ESI spectrum to obtain the average protein mass value. To perform deconvolution see Note 5.
3. The difference between the experimental and theoretical average mass value must be within 2 Da.
4. Select in the experimental ESI spectrum the ions to be used to extract the ion current peak according the following criteria:

   Select multiply-charged ions only if the S/N ratio is at least 5.

   Select the most intense ions (relative abundance ≥ 25%).

   Exclude the ions that are common with co-eluting proteins.
5. Table 2 reports the m/z values of the multiply-charged ions selected to reveal S100A protein peaks in the chromatographic profiles registered by using the apparatus reported in Subheading 3 according the above reported criteria. A variability of ±0.7 Th may be accepted.
6. If the intensity of the ions of the ESI spectrum registered with the apparatus reported in Subheading 2.2 is largely different

Table 2
Multiply-charged ions utilized for the determination of the XIC peak area of the S100A proteins of Table 1, and their relative abundance usually registered under the experimental conditions described and by using the apparatus reported in Subheading 3

Protein name	Multiply-charged ions used for XIC search m/z Th (charge) [Relative Abundance ± 6%]
S100A7 (D_{27})	1,034.4 (+11) [25%], 1,137.8 (+10) [40%], 1,264.1 (+9) [52%], 1,422.0 (+8) [64%]
S100A7 (E_{27})	1,035.7 (+11) [19%], 1,139.2 (+10) [37%], 1,265.6 (+9) [60%], 1,423.7 (+8) [55%]
S100A8	986.0 (+11) [100%], 1,084.5 (+10) [60%], 1,204.8 (+9) [46%], 1,355.3 (+8) [58%]
S100A9 short	977.1 (+13) [84%], 1,058.4 (+12) [98%], 1,154.6 (+11) [85%], 1,269.9 (+10) [64%], 1,410.9 (+9) [42%]
S100A9 short phosph.	983.3(+13) [83%], 1,065.1 (+12) [100%], 1,161.8 (+11) [63%], 1,277.9 (+10) [53%], 1,419.8 (+9) [55%]
S100A9 short ox.	978.3 (+13) [81%], 1,059.8 (+12) [100%], 1,156.0 (+11) [74%], 1,271.5 (+10) [68%], 1,412.7 (+9) [44%]
S100A9 short ox. phosph.	984.5 (+13) [100%], 1,066.4 (+12) [79%], 1,163.3 (+11) [62%], 1,279.5 (+10) [54%], 1,421.6 (+9) [45%]
S100A9 long	940.5 (+14) [90%], 1,012.8 (+13) [100%], 1,097.1 (+12) [80%], 1,196.7 (+11) [55%], 1,316.3 (+10) [60%]
S100A9 long phosph.	946.2 (+14) [80%], 1,018.9 (+13) [100%], 1,103.7 (+12) [85%], 1,204.0 (+11) [60%], 1,324.3 (+10) [50%]
S100A9 long S-glut.	962.3 (+14) [93%], 1,036.2 (+13) [100%], 1,122.5 (+12) [82%], 1,224.5 (+11) [51%], 1,346.8 (+10) [46%]
S100A9 long S-glut. phosph.	968.0 (+14) [84%], 1,042.4 (+13) [100%], 1,129.2 (+12) [85%], 1,231.7 (+11) [67%], 1,354.8 (+10) [38%]
S100A9 long S-cyst.	949.0 (+14) [41%], 1,021.9 (+13) [100%], 1,107.0 (+12) [56%], 1,207.5 (+11) [40%], 1,328.2 (+10) [30%]
S100A11	1,060.2 (+11) [50%], 1,166.1 (+10) [100%], 1,295.6 (+9) [63%], 1,457.4 (+8) [35%]
S100A12	950.4 (+11) [61%], 1,045.4 (+10) [77%], 1,161.4 (+9) [90%], 1,306.5 (+8) [100%]

from that reported in Table 2 (more than 20%), change the tuning parameters of the ESI source by a new optimization (see Subheading 3.2) and verify again the gaussian distribution of the ion intensities.

7. To extract the ion current peak apply the proper list of selected multiply-charged ions (layout), as follows: open the

"chromatogram ranges" window in the Xcalibur software and insert the selected m/z values on the "mass range" plot type separating the values with a comma or a plus sign.

8. Extract the ion current (XIC) peak and integrate the area of the peak, which is proportional to the concentration of the protein and may be used to perform its relative quantification. Figures 2–7 show the XIC procedures utilized to detect the S100A proteins reported in the introduction and Table 1 in typical samples of adult human saliva.

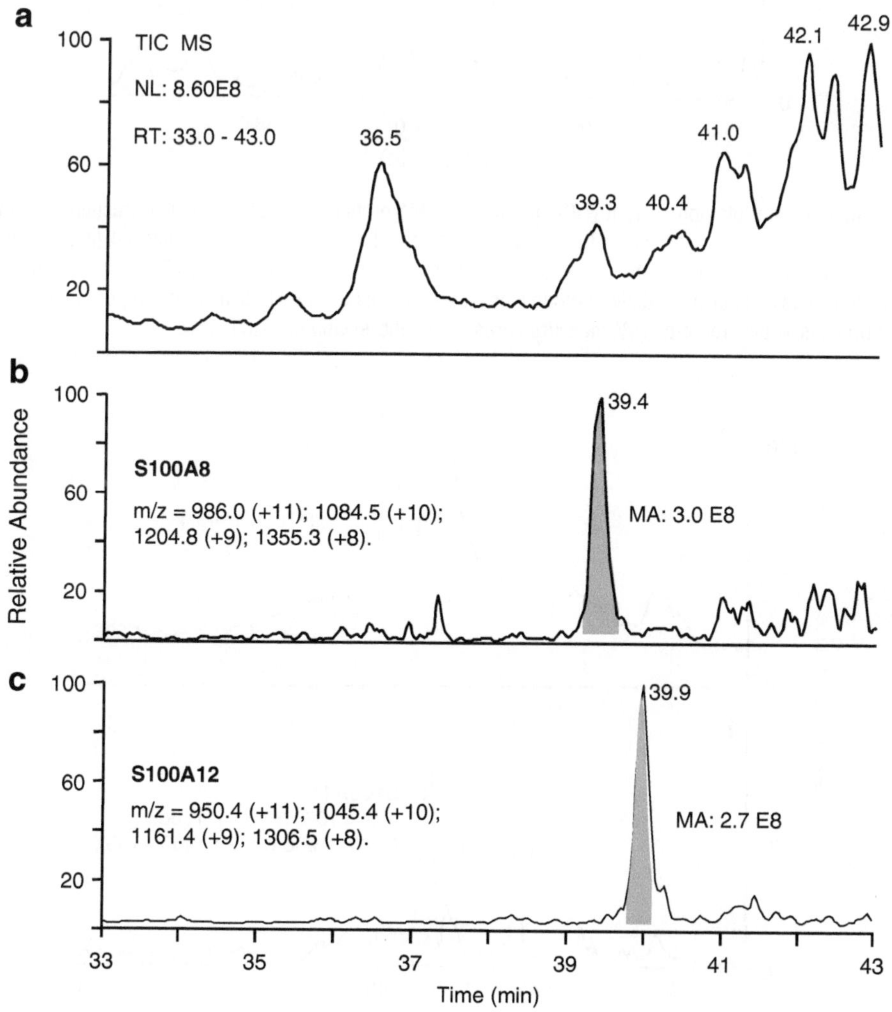

Fig. 2. Enlargement of the total ion current (TIC) chromatographic profile of the acidic soluble fraction of human saliva analyzed by RP-HPLC-ESI-MS according to the conditions described in Subheadings 3.3 and 3.4 (elution range 33–43 min, panel (**a**)). Panels (**b**) and (**c**) show the extracted ion current (XIC) peaks of S100A8 and S100A12 obtained by selective extraction of some of their distinctive multiply-charged ions (see also Table 2). The XIC peak area, proportional to the concentration of the proteins, is also reported (*MA* measured area; other abbreviations as in Fig. 1).

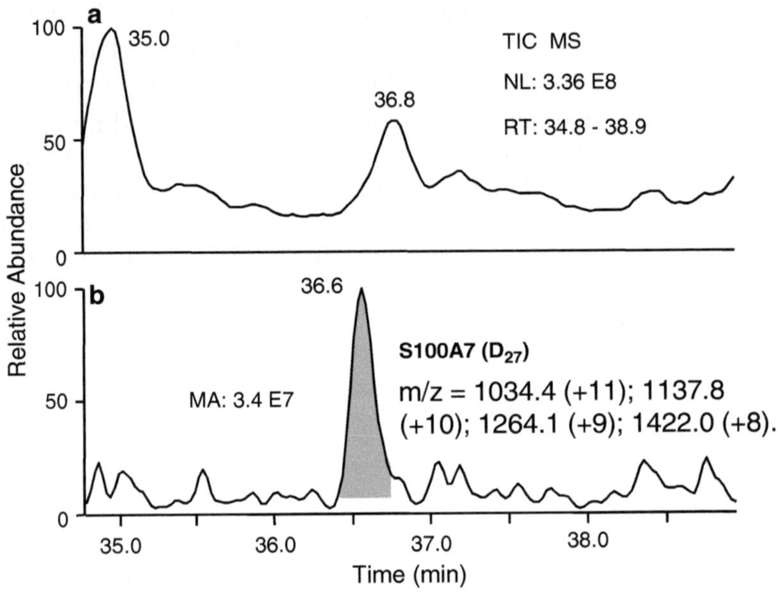

Fig. 3. Enlargement of the total ion current (TIC) chromatographic profile of the acidic soluble fraction of human saliva analyzed by RP-HPLC-ESI-MS according to the conditions described in Subheadings 3.3 and 3.4 (elution range 34.8–38.9 min, panel (**a**)). Panel (**b**) shows the extracted ion current (XIC) peaks of S100A7 (D27 var.) obtained by selective extraction of some of its distinctive multiply-charged ions (see also Table 2). The XIC peak area, proportional to the concentration of the proteins, is also reported (*MA* measured area; other abbreviations as in Fig. 1).

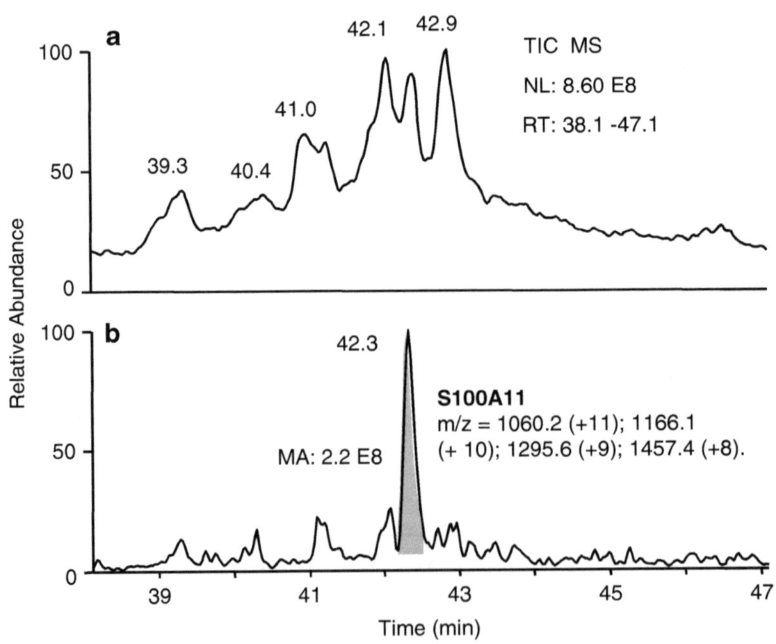

Fig. 4. Enlargement of the total ion current (TIC) chromatographic profile of the acidic soluble fraction of human saliva analyzed by RP-HPLC-ESI-MS according to the conditions described in Subheadings 3.3 and 3.4 (elution range 38.1–47.1 min, panel (**a**)). Panel (**b**) shows the extracted ion current (XIC) peaks of S100A11 obtained by selective extraction of some of its distinctive multiply-charged ions (*see* also Table 2). The XIC peak area, proportional to the concentration of the proteins, is also reported (*MA* measured area; other abbreviations as in Fig. 1).

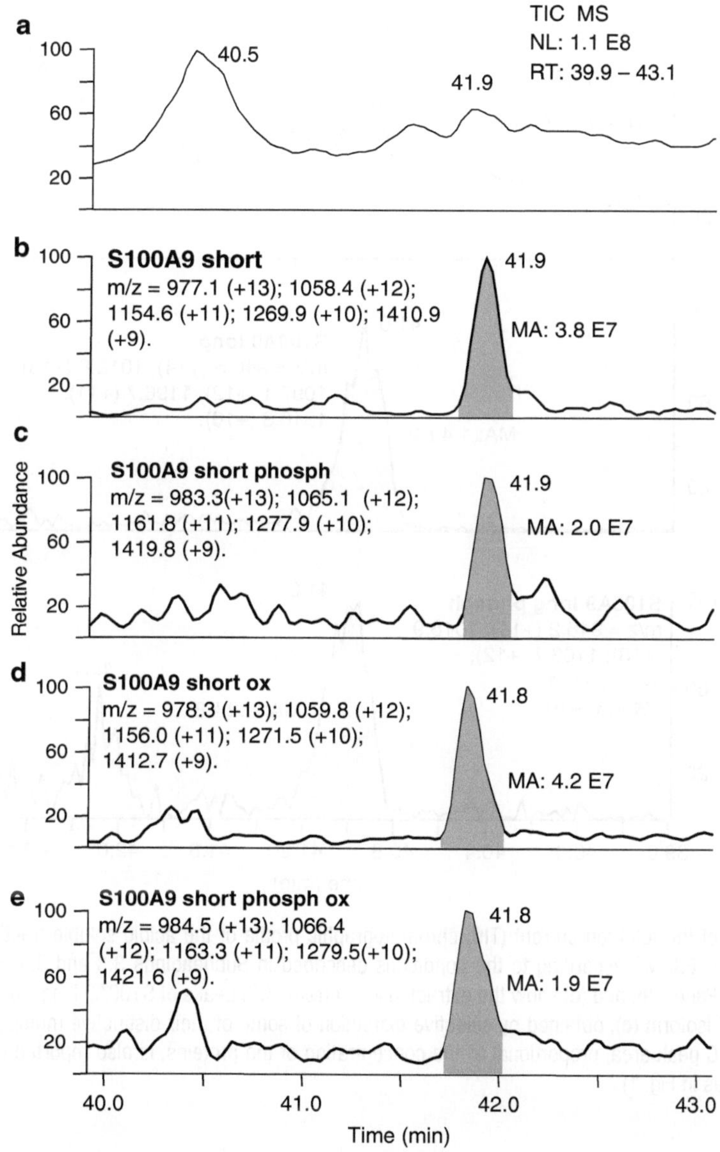

Fig. 5. Enlargement of the total ion current (TIC) chromatographic profile of the acidic soluble fraction of human saliva analyzed by RP-HPLC-ESI-MS according to the conditions described in Subheadings 3.3 and 3.4 (elution range 39.9–43.1 min, panel (**a**)). Panels from (**b**) to (**e**) show the extracted ion current (XIC) peaks of S100A9 short isoform (**b**), S100A9 short phosphorylated isoform (**c**), S100A9 short oxidized isoform (**d**), and S100A9 short phosphorylated oxidized isoform (**e**) obtained by selective extraction of some of their distinctive multiply-charged ions (see also Table 2). The XIC peak area, proportional to the concentration of the proteins, is also reported (*MA* measured area; other abbreviations as in Fig. 1).

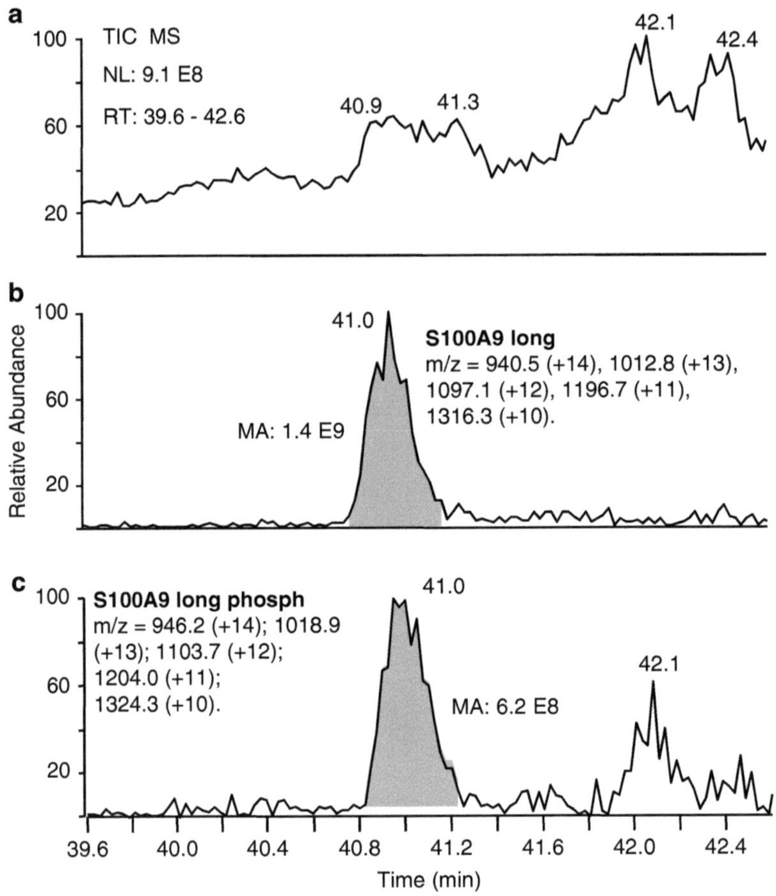

Fig. 6. Enlargement of the total ion current (TIC) chromatographic profile of the acidic soluble fraction of human saliva analyzed by RP-HPLC-ESI-MS according to the conditions described in Subheadings 3.3 and 3.4 (elution range 39.6–42.6 min, panel (**a**)). Panels (**b**) and (**c**) show the extracted ion current (XIC) peaks of S100A9 long isoform (**b**) and S100A9 long phosphorylated isoform (**c**), obtained by selective extraction of some of their distinctive multiply-charged ions (see also Table 2). The XIC peak area, proportional to the concentration of the proteins, is also reported (*MA* measured area; other abbreviations as in Fig. 1).

9. Integration of the XIC peak area may be done by using appropriate Excalibur tools.
10. Save the layout in a file to extract the peak of the same protein in any other chromatographic profile.
11. By applying the layout of each specific S100 protein, the extracted ion current peak will appear on the chromatographic profile in the proper elution range. The integration of the peak area must be manually repeated for any sample.
12. In any new HPLC-ESI-MS chromatographic profile check the ESI spectrum of the S100 protein and its average mass after deconvolution.

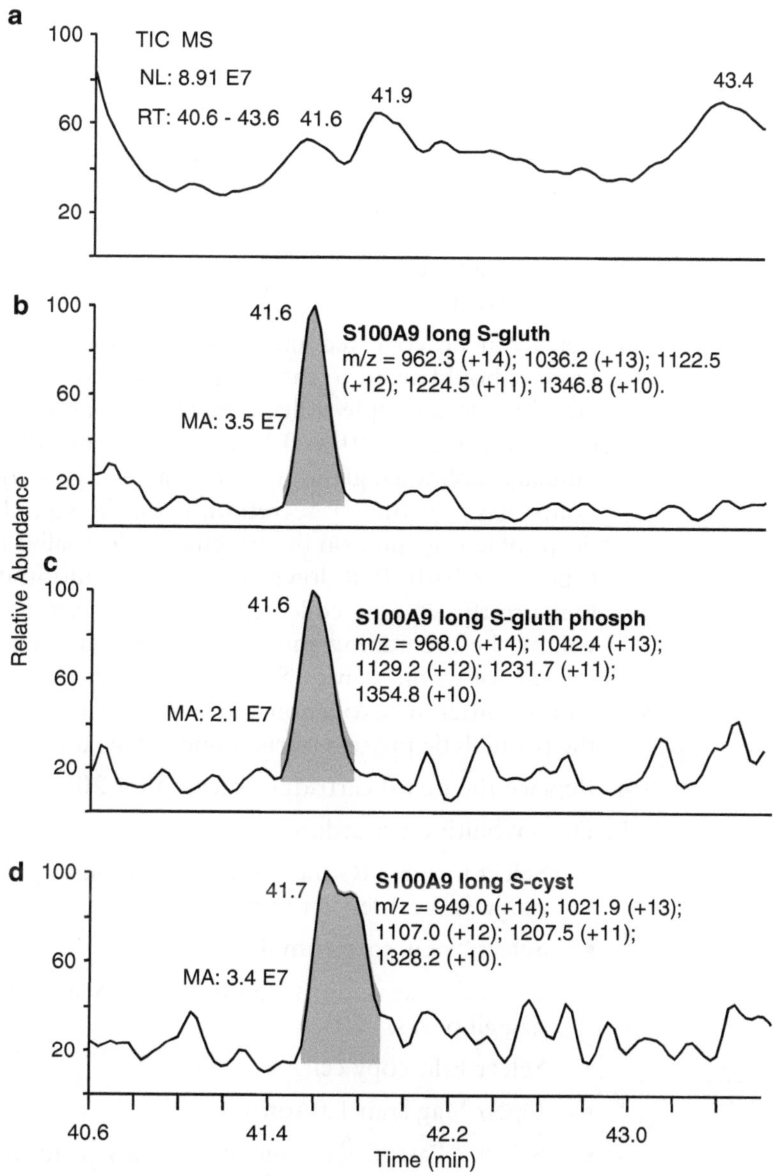

Fig. 7. Enlargement of the total ion current (TIC) chromatographic profile of the acidic soluble fraction of human saliva analyzed by RP-HPLC-ESI-MS according to the conditions described in Subheadings 3.3 and 3.4 (elution range 40.6–43.6 min, panel (**a**)). Panels (**b**)–(**d**) show the extracted ion current (XIC) peaks of S100A9 long S-glutathionylated (Cys_2) isoform (**b**), S100A9 long S-glutathionylated phosphorylated isoform (**c**) and S100A9 long S-cysteinylated (Cys_2) isoform (**d**) obtained by selective extraction of some of their distinctive multiply-charged ions (*see* also Table 2). The XIC peak area, proportional to the concentration of the proteins, is also reported (*MA* measured area; other abbreviations as in Fig. 1).

4. Notes

1. Use only pipettes and safe-lock tubes (i.e., certified Eppendorf tubes) with low plastic release. Plastic contamination can derive from polymer degradation products and surfactants. They can generate interference in the ESI spectrum.
2. It is advisable to prepare eluents and sample solutions at least every 2 weeks.
3. Saliva can be collected under the tongue, near to caruncle. Donor should avoid to eat and to drink (except water) at least 30 min before sample collection. The best period for specimen collection is from 10 to 12 a.m. when parotid and submandibular/sublingual gland secretions are roughly comparable. If donor has low salivary secretion, it can be stimulated by two drops of lemon juice on the tongue. Collect saliva in a safe-lock tube in ice bath. Pull down the tube by hand in order to well measure the volume collected. Do not accumulate too much volume before adding the sample solution (i.e., from 0.2 to 0.5 mL of whole saliva). Saliva is a dynamic bodily fluid containing different exogenous and endogenous proteinase and the proteolytic process starts immediately after secretion.
4. Replace the guard cartridge at least after 20 runs.
5. Deconvolution procedure:
   - Select in the TIC the region corresponding to the elution range and evidence the ESI spectrum.
   - Select "view spectrum list".
   - Open display options window in the Xcalibur software and flag all peaks.
   - Select Edit copy cell.
   - Open MagTran 1.0 software.
   - Select "Past Data" button to copy your ESI spectrum inside the MagTran software.
   - In Magtran→Option→Transform options set:
     - Mass range from 1 to 40,000 Da
     - Min. Peak-Width (Da) 1 and Max. Peak-Width 30.
     - Mass unit/point: 0
     - S/N Threshold: 2
     - Max. No. of species: 8
     - Mass accuracy: 0.05
     - Charge Determined by: Charge envelop only

– Select "Start Transform" button. If the original spectrum is too noisy, select "Smooth original spectrum" button to reduce the noise and then select "Start Transform" button again.

Acknowledgments

We acknowledge the financial support of the Università di Cagliari, Università Cattolica, MIUR, the Italian National Research Council (CNR) and Regione Sardegna. We thank them for their programs on the promotion of scientific research and diffusion.

References

1. Covey T (1996) Analytical characteristics of the electrospray ionization process. Chapter 2. In: Snyder AP (ed) Biochemical and biotechnological applications of electrospray ionization mass spectrometry. ACS Symposium Series, vol 619, Oxford University Press, Oxford, pp 21–59
2. Fenn JB, Mann M, Meng CK, Wong SF, Whitehouse CM (1989) Electrospray ionization for mass spectrometry of large biomolecules. Science 246:64–71
3. March RE (2000) Quadrupole ion trap mass spectrometry: a view at the turn of the century. Int J Mass Spectrom 200:285–312
4. Cottrell JS, Greathead RJ (1986) Extending the mass range of a sector mass spectrometer. Mass Spectrom Rev 5:215–247
5. Marshall AG, Hendrickson CL, Jackson GS (1998) Fourier transform ion cyclotron resonance mass spectrometry: a primer. Mass Spectrom Rev 17:1–35
6. Castagnola M, Inzitari R, Fanali C, Iavarone F, Vitali A, Desiderio C et al (2011) The surprising composition of the salivary proteome of preterm human newborn. Mol Cell Proteomics 10:M110.003467
7. Cabras T, Pisano E, Mastinu A, Denotti G, Pusceddu PP, Inzitari R et al (2010) Alterations of the salivary secretory peptidome profile in children affected by type 1 diabetes. Mol Cell Proteomics 9:2099–2108
8. Zhang Z, Marshall AG (1998) A universal algorithm for fast and automated charge state deconvolution of electrospray mass-to-charge ratio spectra. J Am Soc Mass Spectrom 9:225–233

Chapter 23

Clinical Use of the Calcium-Binding S100B Protein

Ramona Astrand, Johan Undén, and Bertil Romner

Abstract

S100B is a calcium-binding protein most abundant in neuronal tissue. It is expressed in glia cells and Schwann cells and exerts both intra- and extracellular effects. Depending on the concentration, secreted S100B exerts either trophic or toxic effects. Its functions have been extensively studied but are still not fully understood. It can be measured in cerebrospinal fluid and blood, and increased S100B level in blood can be seen after, e.g., traumatic brain injury, certain neurodegenerative disorders and malignant melanoma.

This chapter provides a short background of protein S100B, commercially available methods of analysis, and its clinical use.

Key words: S100B, EF-hand, Calcium, Traumatic brain injury, Minor head injury, Malignant melanoma, Alzheimer's disease

1. Introduction

1.1. Biochemistry of S100 Proteins

The first member of the S100 family was isolated by Moore in 1965. It was given its name S100 due to solubility in 100% saturated ammonium sulfate at neutral pH. The protein S100 was initially purified from bovine brain and thought to be specific to neuronal tissue since it was mainly expressed in glial and Schwann cells (1). Subsequent studies studies showed that the fraction contained two polypeptides, S100A1 and S100B; the two S100 proteins most abundant in the nervous system (2).

The S100 family is a multigenic family membering at least 25 proteins characterized by the presence of two calcium-binding, EF-hand motifs (helix-loop-helix structure). The proteins S100A1–S100A16 are coded by genes clustered at human chromosome 1q21. The other proteins S100B, S100P, S100Z, and S100G are

Claus W. Heizmann (ed.), *Calcium-Binding Proteins and RAGE: From Structural Basics to Clinical Applications*, Methods in Molecular Biology, vol. 963, DOI 10.1007/978-1-62703-230-8_23, © Springer Science+Business Media New York 2013

coded by genes located at different chromosome 21q22, 4p16, 5q14, and Xp22, respectively (3–5).

The S100 proteins are small acidic proteins, with molecular weights between 9 and 13 kDa. They exist within cells as homodimers (e.g., S100BB), with the exception of S100G (calbindin D9k) which is monomeric. In some cases, heterodimers can form, as in the case with S100A1/S100B, and even oligomers in some specific cells. Dimerization and oligomerisation seem to be important for the biological activity, and upon calcium binding, conformational changes on the protein occur in order to be able to recognize its target proteins/receptors (e.g. RAGE) (6). Several S100 proteins not only bind calcium but also zinc and copper.

1.2. Functional Roles of S100B

The S100 proteins relay signals from receptors on the cell surface to target molecules inside the cell, in the cytoplasm or nucleus. Extracellularly, several S100 proteins bind to RAGE (receptor for advanced glycation end products), which activates several intracellular signaling pathways (see Chapters 15 and 16). In general, there are five major intracellular functions for S100 proteins: (1) regulations of phosphorylation mediated by protein kinases, (2) modulation of enzymatic activity, (3) maintenance of cell shape and motility, (4) influence of some signal-transduction pathways, and (5) promotion of calcium homeostasis (7). The function of S100B is, among others, regulation of calcium homeostasis and regulation of cell motility. Depending on the concentration, secreted S100B exerts trophic as well as toxic effects. At lower concentrations (nanomolar) S100B is thought to stimulate neuronal growth and enhancement of neuronal survival during development, while at higher concentrations (micromolar) S100B may have deleterious effects by increasing expression of the pro-inflammatory cytokine IL-6 and inducing apoptotic neuronal death (8).

1.3. Sources of S100B

Several of the S100 proteins are abundant in the central nervous system. S100B is expressed in astrocytes, oligodendrocytes, and Schwann cells, but it can also be found outside the nervous system in chondrocytes, adipocytes, and melanocytes (8, 9). The extracranial concentration of S100B is much lower than that found in neuronal tissue (10).

S100B can be measured in both cerebrospinal fluid and in blood serum. The median concentration in blood among healthy adults is 0.05 μg/L and is independent of both age and gender (11). There are still some uncertainties about the release mechanism of S100B into the blood. The most accepted theory is that of extracellular S100B passing into the blood due to a disrupted blood–brain barrier (BBB) (12). Hence, serum protein S100B is considered an early marker of the disruption of BBB after a blow to the head, thus not always as a sign of neuronal damage. It is metabolized and excreted via the kidneys and its half-life is estimated to 30–130 min (13, 14). The stability of protein S100B in serum has been investigated by Raabe and colleagues, who found no effects of storage time and temperature prior to analysis (15). S100B has also been found in other extracellular fluids, such as saliva and urine. The stability of urine S100B seems to be less stable (16). Studies comparing jugular vein S100B to arterial concentrations have found contradicting results, one finding a significantly higher concentration in jugular blood (17) and the other with almost equivalent levels (18).

Reference levels for S100B have been found to be age- and gender independent (11), and a cut-off value ≤0.10 μg/L has been considered normal (within 95th percentile) (19). Reference levels for children are slightly increased compared to adults (20).

2. Clinical Use of S100B

2.1. Head Injury

Traumatic brain injury is a significant cause of mortality and morbidity in adults and a leading cause of death in childhood. The diagnostic process includes clinical examination and, in more severe cases, neuroimaging, such as computed tomography of the head (CT) or magnetic resonance imaging (MRI). The majority of head injuries are so-called minor head injuries (85–90%), whereas only 5–10% are moderate to severe head injuries. The majority of the minor head injured are fully awake (Glasgow coma scale—GCS 15), without any loss of consciousness or amnesia post-injury and have a very low risk of developing intracranial complications. These are according to the Scandinavian Neurotrauma Committee classified as minimal head injuries (21). Those defined as being "mild head injuries" are ones who appear fully or nearly fully awake (GCS 14–15) but have had or might have had a brief loss of consciousness and/or amnesia. The mild head injured have an approximately 1% risk of developing intracranial complications in need of neurosurgical intervention; less than 10% have pathological intracranial findings on CT (22).

2.1.1. Sensitivity and Specificity

Serum S100B increases almost immediately after a relevant brain injury, due to a brief breach of the blood–brain barrier (BBB) (12). It has been discussed that since epidural hematomas may initially not be associated with an actual brain damage, the biomarker might not be increased. However, epidural hematomas reported in the literature have all shown elevated S100B > 0.10 μg/L; although only slightly elevated (23). Thus, S100B could be a marker of BBB disruption as well as brain cell damage (24). The sensitivity of the protein is high, making S100B a better marker for identifying patients with a negligible risk for intracranial complications

(25–29). In a recent meta-analysis by Undén et al., the sensitivity for S100B (cut-off 0.10 μg/L and sample time within 3 h after minor head trauma) was found to be 98% for any type of intracranial pathology on CT. For clinically relevant findings/neurosurgical intervention, the sensitivity was 100% (30).

Protein S100B is not specific to the brain. Its presence is also found in low concentrations in adipose tissue, melanocytes in the skin, bone marrow, and heart muscle (31). Undén et al. also showed that serum S100B is increased in patients with especially long bone fractures (32), hence the extracerebral sources may be a limiting factor in the immediate assessment of multi-trauma patients (33). S100B is not influenced by blood-alcohol levels (28, 34).

2.1.2. Minor Head Injury

The best results for clinical use of S100B is in the field of minor head injuries. Due to the high specificity and high negative predictive value for neurosurgically relevant intracranial complications, it has been estimated that S100B could reduce the frequency of head CT by 30%, and still maintain patient safety. This was formerly proposed by Biberthaler et al., and further studied in a meta-analysis by Undén et al., strengthening the theory (30, 35). Recently, Zongo et al. published a prospective study on the evaluation of S100B as a screening tool in minor head injury (GCS 13–15), which confirms the potential benefit of implementing S100B into clinical use (36). A total of 1,646 patients were included in the study. The cut-off value was chosen to be 0.12 μg/L and blood was collected within 6 h after trauma. None of the patients were considered false negative. The sensitivity was 99.1%, and negative predictive value 99.7. The potential percentage of avoided CT scans were calculated to 19%. These studies only apply on adult patients so far.

The Scandinavian head injury guidelines are currently being revised, and serum S100B will be recommended as an option in mild head injury, as a decision tool to perform a CT or not.

2.1.3. Severe Traumatic Brain Injury

Research on the potential use of neuromarkers in severe head injury has primarily been directed at severity and outcome prediction. The clinical presentation of an unconscious patient, who is often sedated and/or with external mechanical ventilation, makes traditional clinical evaluation difficult. Initial level of consciousness (according to GCS), CT classifications (Marshall), pupil response, and events of intracranial hypertension or hypoperfusion/hypoxia are some of the traditional clinical tools for estimating clinical progress and outcome. Thus a biochemical brain marker could be most valuable in the prediction of secondary complications in a neurointensive care patient.

Previous studies have found a correlation between serum S100B levels and clinical outcome according to Glasgow Outcome Scale (GOS) after severe head injury (37, 38). Patients with poor outcome (vegetative state or death) had significantly higher S100B levels than patients with favourable to severely disabled outcome (39, 40). One study found a 2.1-fold increase in serum S100B in those with unfavorable vs. favorable outcome (41). In earlier studies, Raabe et al. found a strong association between a cut-off level of 2.5 μg/L (any time) and unfavorable outcome (37, 42) and Woertgen et al. found that serum levels above 2 μg/L (drawn within 6 h after trauma) predicted unfavorable outcome (positive predictive value 87%; negative predictive value, 77%) (43). Nylén et al. reported a cut-off of 0.55 μg/L (at admission) with a 100% specificity for predicting unfavorable outcome 1 year postinjury (44). Böhmer et al. collected daily CSF samples in 20 patients with severe TBI and found that early elevations of S100B (up to 3 days) predicted deterioration to brain death (45). Dimopoulou et al. has reported elevated median serum S100B levels of 2.32 μg/L in those progressing to brain death, while survivors had significantly lower median values, 1.04 μg/L (46).

Contrary to previous studies, some of the more recent studies have shown no significant correlation between early CSF-S100B or serum S100B and GOS or other outcome scales (47), nor between dichotomized GOS (48, 49). They conclude there is no clinically significant value of the marker as predictor of clinical outcome (49). In a prospective, double-blind, randomized study, Olivecrona et al. investigated the prognostic value of biomarkers and concluded that serum is a poor predictor of secondary insults (49).

2.1.4. Multi-trauma

One drawback of S100B is the presence of extracranial sources, especially bone marrow and adipose tissues (32, 33). Pelinka et al. found that all patients studied with multiple organ trauma demonstrated raised S100B levels whether or not TBI was present, and they concluded that serum levels drawn during the first 24 h did not reliably predict clinical outcome in these patients (50). Daily measurements were advocated. However, another study by da Rocha et al. did not find any correlation between higher S100B values and the presence of multitrauma (39).

2.1.5. S100B Protein in Children

There are, by far, fewer studies performed on pediatric TBI and biomarkers, but similar to the studies on adults, most studies have focused on the ability of S100B to predict outcome after TBI. Bechtel and colleagues described use of serum S100B as a screening tool to detect intracranial injuries in children admitted to the emergency department after closed head trauma. Although children with intracranial injuries had significantly higher S100B concentrations than children without intracranial injuries, the ability of S100B levels to detect such injuries was found to be poor (51). Child abuse and inflicted TBI are both serious and non-negligible in this patient group. The presenting symptoms are diffuse, such as headache, abdominal pain, and excessive vomiting, and can be difficult to differentiate between other medical conditions, especially among infants. Neither CSF nor serum biomarkers have shown sufficient discriminative abilities to be used as a screening tool for inflicted TBI (52).

Castellani et al. recently investigated the correlation of S100B to CT results after mild TBI (GCS 13–15). Serum S100B concentrations were significantly higher in patients with abnormal CT (mean 0.64 μg/L) compared with patients with normal CT (mean 0.50 μg/L) and even though specificity was poor (0.42), the sensitivity and negative predictive value was 1.00. The authors concluded that normal S100B levels (<0.16 μg/L, Elecsys) safely rules out CT pathologies in minor head injured children (53). Two independent studies have recently been performed to investigate the reference values of S100B in children, and conclude that healthy children have increased S100B levels compared to adults, and especially among infants (20, 54). These studies encourage further research on this topic to pursue optimizing routines and management in pediatric head injury.

2.2. Neurodegenerative Diseases and Cerebrovascular Insults

The levels of S100B in blood and CSF has also been correlated to a number of other diseases, such as Alzheimers disease (AD) (58), autoimmune disorders such as multiple sclerosis (MS) (59), psychiatric disorders such as schizofrenia (60), and acute cerebrovascular insult (61).

Alzheimer's disease (AD) is so far the most extensively studied neurodegenerative disorder coupled with S100B levels. S100B in CSF has previously been shown to be increased in mild to moderate stage of AD, but nearly equal or decreased in severe AD compared to controls (62). A more recent study reported positive correlations of the severity of dementia in AD and serum S100B levels (63). This result differs from a previous study by Gruden et al. 2007, who reported lower levels in severe dementia compared to moderate, and a peak of S100B immune response during moderate dementia (64).

The S100B findings in MS and schizophrenia supports the theory that there are neurodegenerative mechanisms involved in the pathogenesis of these diseases. CSF and/or serum levels are increased in the active phase of MS (59, 65) and in the acute schizophrenic phase (66). However, the role of the S100B protein in these diseases is still quite poorly understood, and further studies are warranted.

2.3. Malignant Melanoma

Immunohistochemical procedures have improved the correct classification of tumors, although conventional histology is still considered the golden standard. Melanocytes, as opposed to nonmelanocytic tumors, express S100B, which can be detected with monoclonal or polyclonal antibodies. Protein S100 has been used to improve melanoma staging; the more advanced tumor stage, the higher the blood levels. The sensitivity of serum S100B in patients with stage I and II malignant melanoma has been reported to be < 15%, for stage III varying from 10 to 50%, and for stage IV malignant melanoma 60–85% (9, 55). S100B is so far considered as the most sensitive serum marker for detecting metastatic disease in malignant melanoma.

The prognostic factors for survival in malignant melanoma are tumor thickness and lymph node status. However, S100B has been shown to be a fairly good prognostic marker for survival. Patients with low levels of S100B had on average 29.6 months longer survival than patients with elevated levels (56).

Serum S100B is not specific to malignant melanoma. It is elevated in several malignant diseases and increased levels can also be found in liver and renal diseases and in inflammatory and infectious diseases (57).

3. Diagnostic Tests

There are a number of commercially available immunoassays for serum S100B. The two most common are the immunoluminometric assay Liaison® Sangtec 100 (DiaSorin AB, Bromma, Sweden) and the electrochemiluminescence immunoassay Elecsys S100 (Roche Diagnostics, Mannheim, Germany). Both assays detect the heterodimers S100A1B and the homodimer S100BB; both are designated as S100B protein.

Other assays commonly used in recent studies are: the immunoluminometric assay LIA-mat® Sangtec 100 (DiaSorin AB, Bromma, Sweden), the ELISA Nexus DX™ S100 (SynX Pharma Inc., York, UK.), the enzyme-linked immunosorbent assay (Nanogen) and the enzyme-linked immunoassay Elecsys® S100 (Roche, Mannheim, Germany).

CanAg S100A1B EIA and CanAg S100BB EIA (Fujirebio Diagnostics AB, Gothenburg, Sweden) are so far the only assays to specifically analyze the S100A1B (focusing on the A1 dimer) and S100BB, respectively. Although commercially available, the analysis of S100BB in former studies has been limited. However, Nylén et al. investigated the relation of S100B, S100A1B, and S100BB in serum after severe TBI to outcome, with the hypothesis that S100BB dimer should be better related to outcome after severe TBI than either S100A1B or S100B (sum concentration), due to the greater abundancy of S100BB in the brain. They concluded that there most probably was no advantage of analysing S100BB instead of S100B alone (44).

There are a few studies performed to compare some of the commercially available assays. Muller et al. compared the Liaison® Sangtec 100 (DiaSorin AB) and the Elecsys® S100 (Roche) and found a mean difference of 0.14 μg/L between the measured values, where Liaison® Sangtec 100 was consistently higher than the Elecsys® S100 method. Similar results were found by Hallén et al., with a mean difference of 0.12 μg/L. In conclusion, differences between the analytical methods should be considered when interpreting results from different studies (16, 67, 68).

Some of the assays have also been used for measuring S100B in CSF, saliva and other biological fluids, although their clinical implications (apart fron CSF) have yet been quite limited.

4. Notes

Below we give a proposal for the management of minimal to moderate head injuries with the use of protein S100B (see flow-chart 1).

The assay currently used for measuring S100B in the event of brain injuries, in the Scandinavian countries, is the electrochemiluminescence Elecsys S100 assay (Roche Diagnostics, Mannheim, Germany), mainly due to its availability. The majority of hospitals already have the laboratory setup and analyzers Elecsys 2010, Modular analytics E170 or Cobas, which are all compatible with the Elecsys S100 assay. The flow-chart for the management of minimal, mild, and moderate head injuries gives a cut-off level of S100B $\leq$10μg/L, which is based on measurements made with the Elecsys S100 assay on Modular E170.

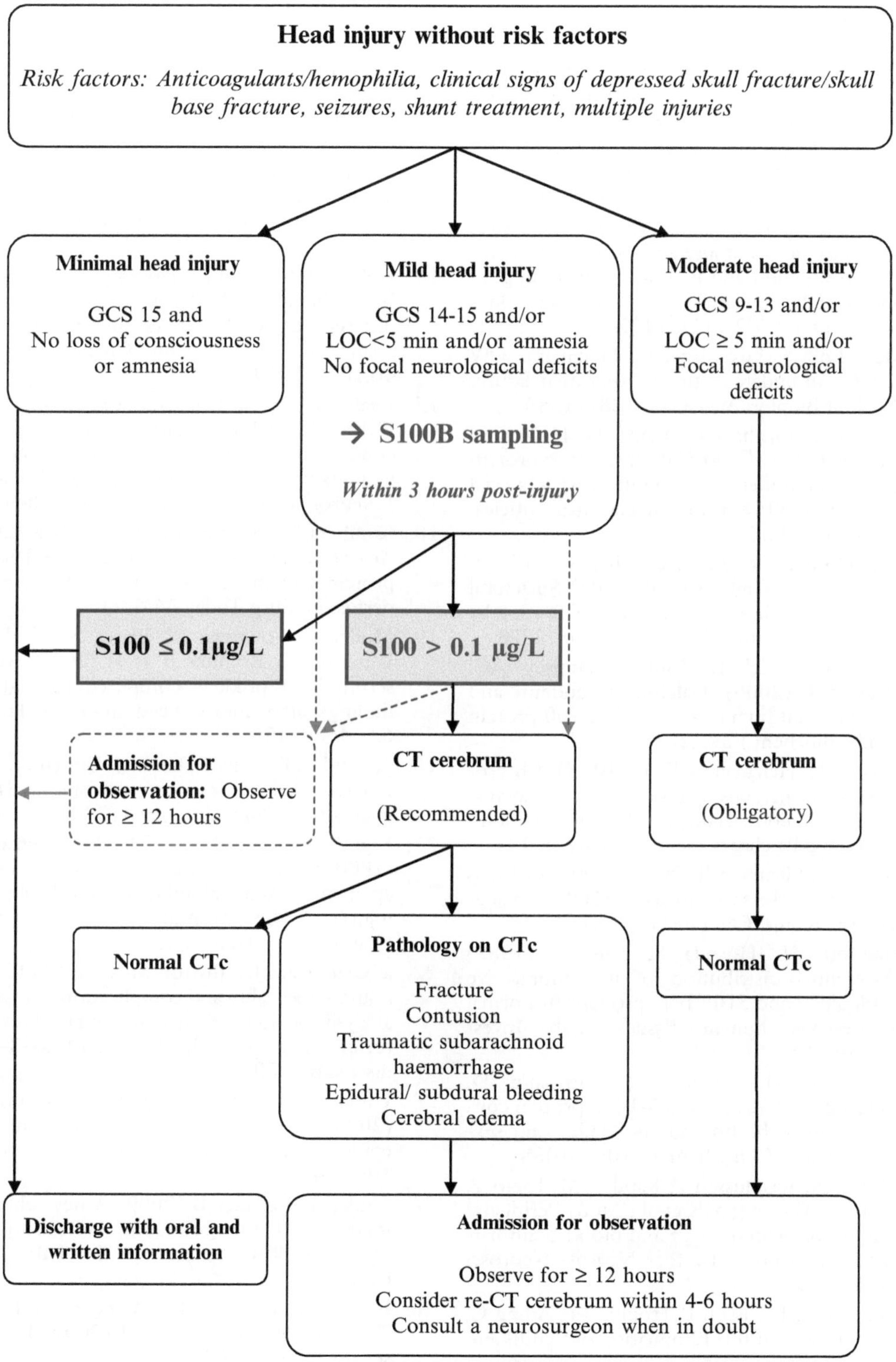
Head injury without risk factors
Risk factors: Anticoagulants/hemophilia, clinical signs of depressed skull fracture/skull base fracture, seizures, shunt treatment, multiple injuries
Minimal head injury
GCS 15 and
No loss of consciousness or amnesia
Mild head injury
GCS 14-15 and/or
LOC<5 min and/or amnesia
No focal neurological deficits
→ S100B sampling
Within 3 hours post-injury
Moderate head injury
GCS 9-13 and/or
LOC ≥ 5 min and/or
Focal neurological deficits
S100 ≤ 0.1μg/L
S100 > 0.1 μg/L
Admission for observation: Observe for ≥ 12 hours
CT cerebrum
(Recommended)
CT cerebrum
(Obligatory)
Normal CTc
Pathology on CTc
Fracture
Contusion
Traumatic subarachnoid haemorrhage
Epidural/ subdural bleeding
Cerebral edema
Normal CTc
Discharge with oral and written information
Admission for observation
Observe for ≥ 12 hours
Consider re-CT cerebrum within 4-6 hours
Consult a neurosurgeon when in doubt

References

1. Moore BW (1965) A soluble protein characteristic of the nervous system. Biochem Biophys Res Commun 19:739–744
2. Isobe T, Okuyama T (1981) The amino-acid sequence of the alpha subunit in bovine brain S-100a protein. Eur J Biochem 116:79–86
3. Marenholz I, Heizmann CW, Fritz G (2004) S100 proteins in mouse and man: from evolution to function and pathology (including an update of the nomenclature). Biochem Biophys Res Commun 322:1111–1122
4. Marenholz I, Lovering RC, Heizmann CW (2006) An update of the S100 nomenclature. Biochim Biophys Acta 1763:1282–1283
5. Kizawa K, Takahara H, Unno M, Heizmann CW (2011) S100 and S100 fused-type protein families in epidermal maturation with special focus on S100A3 in mammalian hair cuticles. Biochimie 93:2038–2047
6. Ostendorp T, Leclerc E, Galichet A, Koch M, Demling N, Weigle B et al (2007) Structural and functional insights into RAGE activation by multimeric S100B. EMBO J 26:3868–3878
7. Santamaria-Kisiel L, Rintala-Dempsey AC, Shaw GS (2006) Calcium-dependent and -independent interactions of the S100 protein family. Biochem J 396:201–214
8. Donato R, Heizmann CW (2010) S100B protein in the nervous system and cardiovascular apparatus in normal and pathological conditions. Cardiovasc Psychiatry Neurol 2010:929712
9. Harpio R, Einarsson R (2004) S100 proteins as cancer biomarkers with focus on S100B in malignant melanoma. Clin Biochem 37:512–518
10. Haimoto H, Hosoda S, Kato K (1987) Differential distribution of immunoreactive S100-alpha and S100-beta proteins in normal nonnervous human tissues. Lab Invest 57:489–498
11. Wiesmann M, Missler U, Gottmann D, Gehring S (1998) Plasma S-100b protein concentration in healthy adults is age- and sex-independent. Clin Chem 44:1056–1058
12. Marchi N, Rasmussen P, Kapural M, Fazio V, Kight K, Mayberg MR et al (2003) Peripheral markers of brain damage and blood-brain barrier dysfunction. Restor Neurol Neurosci 21:109–121
13. Jonsson H, Johnsson P, Hoglund P, Alling C, Blomquist S (2000) Elimination of S100B and renal function after cardiac surgery. J Cardiothorac Vasc Anesth 14:698–701
14. Blomquist S, Johnsson P, Luhrs C, Malmkvist G, Solem JO, Alling C et al (1997) The appearance of S-100 protein in serum during and immediately after cardiopulmonary bypass surgery: a possible marker for cerebral injury. J Cardiothorac Vasc Anesth 11:699–703
15. Raabe A, Kopetsch O, Gross U, Zimmermann M, Gebhart P (2003) Measurements of serum S-100B protein: effects of storage time and temperature on pre-analytical stability. Clin Chem Lab Med 41:700–703
16. Hallen M, Carlhed R, Karlsson M, Hallgren T, Bergenheim M (2008) A comparison of two different assays for determining S-100B in serum and urine. Clin Chem Lab Med 46:1025–1029
17. Raabe A, Menon DK, Gupta S, Czosnyka M, Pickard JD (1998) Jugular venous and arterial concentrations of serum S-100B protein in patients with severe head injury: a pilot study. J Neurol Neurosurg Psychiatry 65:930–932
18. Kunihara T, Shiiya N, Bin L, Yasuda K (2006) Arterio-jugular differences in serum S-100beta proteins in patients receiving selective cerebral perfusion. Surg Today 36:6–11
19. Muller K, Townend W, Biasca N, Unden J, Waterloo K, Romner B et al (2007) S100B serum level predicts computed tomography findings after minor head injury. J Trauma 62:1452–1456
20. Astrand R, Romner B, Lanke J, Unden J (2011) Reference values for venous and capillary S100B in children. Clin Chim Acta 412:2190–2193
21. Ingebrigtsen T, Romner B, Kock-Jensen C (2000) Scandinavian guidelines for initial management of minimal, mild, and moderate head injuries. The Scandinavian Neurotrauma Committee. J Trauma 48:760–766
22. af Geijerstam JL, Britton M (2003) Mild head injury—mortality and complication rate: meta-analysis of findings in a systematic literature review. Acta Neurochir (Wien) 145:843–850, discussion 850
23. Unden J, Bellner J, Astrand R, Romner B (2005) Serum S100B levels in patients with epidural haematomas. Br J Neurosurg 19:43–45
24. Unden J, Romner B (2009) A new objective method for CT triage after minor head injury—serum S100B. Scand J Clin Lab Invest 69:13–17
25. Biberthaler P, Mussack T, Wiedemann E, Gilg T, Soyka M, Koller G et al (2001) Elevated serum levels of S-100B reflect the extent of brain injury in alcohol intoxicated patients after mild head trauma. Shock 16:97–101
26. Biberthaler P, Mussack T, Wiedemann E, Kanz KG, Koelsch M, Gippner-Steppert C et al

(2001) Evaluation of S-100b as a specific marker for neuronal damage due to minor head trauma. World J Surg 25:93–97

27. Ingebrigtsen T, Romner B, Marup-Jensen S, Dons M, Lundqvist C, Bellner J et al (2000) The clinical value of serum S-100 protein measurements in minor head injury: a Scandinavian multicentre study. Brain Inj 14:1047–1055
28. Mussack T, Biberthaler P, Kanz KG, Heckl U, Gruber R, Linsenmaier U et al (2002) Immediate S-100B and neuron-specific enolase plasma measurements for rapid evaluation of primary brain damage in alcohol-intoxicated, minor head-injured patients. Shock 18:395–400
29. Poli-de-Figueiredo LF, Biberthaler P, Simao Filho C, Hauser C, Mutschler W, Jochum M (2006) Measurement of S-100B for risk classification of victims sustaining minor head injury—first pilot study in Brazil. Clinics (Sao Paulo) 61:41–46
30. Unden J, Romner B (2010) Can low serum levels of S100B predict normal CT findings after minor head injury in adults?: an evidence-based review and meta-analysis. J Head Trauma Rehabil 25:228–240
31. Kleine TO, Benes L, Zofel P (2003) Studies of the brain specificity of S100B and neuron-specific enolase (NSE) in blood serum of acute care patients. Brain Res Bull 61:265–279
32. Unden J, Bellner J, Eneroth M, Alling C, Ingebrigtsen T, Romner B (2005) Raised serum S100B levels after acute bone fractures without cerebral injury. J Trauma 58:59–61
33. Anderson RE, Hansson LO, Nilsson O, Dijlai-Merzoug R, Settergren G (2001) High serum S100B levels for trauma patients without head injuries. Neurosurgery 48:1255–1258, discussion 1258–1260
34. Biberthaler P, Mussack T, Wiedemann E, Kanz KG, Gilg T, Gippner-Steppert C et al (2000) Influence of alcohol exposure on S-100b serum levels. Acta Neurochir Suppl 76:177–179
35. Biberthaler P, Linsenmeier U, Pfeifer KJ, Kroetz M, Mussack T, Kanz KG et al (2006) Serum S-100B concentration provides additional information fot the indication of computed tomography in patients after minor head injury: a prospective multicenter study. Shock 25:446–453
36. Zongo D, Ribereau-Gayon R, Masson F, Laborey M, Contrand B, Salmi LR et al (2011) S100-B protein as a screening tool for the early assessment of minor head injury. Ann Emerg Med 59:209–218
37. Raabe A, Grolms C, Seifert V (1999) Serum markers of brain damage and outcome prediction in patients after severe head injury. Br J Neurosurg 13:56–59
38. Nylen K, Ost M, Csajbok LZ, Nilsson I, Blennow K, Nellgard B et al (2006) Increased serum-GFAP in patients with severe traumatic brain injury is related to outcome. J Neurol Sci 240:85–91
39. da Rocha AB, Schneider RF, de Freitas GR, Andre C, Grivicich I, Zanoni C et al (2006) Role of serum S100B as a predictive marker of fatal outcome following isolated severe head injury or multitrauma in males. Clin Chem Lab Med 44:1234–1242
40. Wiesmann M, Steinmeier E, Magerkurth O, Linn J, Gottmann D, Missler U (2010) Outcome prediction in traumatic brain injury: comparison of neurological status, CT findings, and blood levels of S100B and GFAP. Acta Neurol Scand 121:178–185
41. Vos PE, Jacobs B, Andriessen TM, Lamers KJ, Borm GF, Beems T et al (2010) GFAP and S100B are biomarkers of traumatic brain injury: an observational cohort study. Neurology 75:1786–1793
42. Raabe A, Grolms C, Sorge O, Zimmermann M, Seifert V (1999) Serum S-100B protein in severe head injury. Neurosurgery 45:477–483
43. Woertgen C, Rothoerl RD, Metz C, Brawanski A (1999) Comparison of clinical, radiologic, and serum marker as prognostic factors after severe head injury. J Trauma 47:1126–1130
44. Nylen K, Ost M, Csajbok LZ, Nilsson I, Hall C, Blennow K et al (2008) Serum levels of S100B, S100A1B and S100BB are all related to outcome after severe traumatic brain injury. Acta Neurochir (Wien) 150:221–227, discussion 227
45. Bohmer AE, Oses JP, Schmidt AP, Peron CS, Krebs CL, Oppitz PP et al (2011) Neuron-specific enolase, S100B, and glial fibrillary acidic protein levels as outcome predictors in patients with severe traumatic brain injury. Neurosurgery 68:1624–1631
46. Dimopoulou I, Korfias S, Dafni U, Anthi A, Psachoulia C, Jullien G et al (2003) Protein S-100b serum levels in trauma-induced brain death. Neurology 60:947–951
47. Bellander BM, Olafsson IH, Ghatan PH, Bro Skejo HP, Hansson LO, Wanecek M et al (2011) Secondary insults following traumatic brain injury enhance complement activation in the human brain and release of the tissue damage marker S100B. Acta Neurochir (Wien) 153:90–100
48. Unden J, Astrand R, Waterloo K, Ingebrigtsen T, Bellner J, Reinstrup P et al (2007) Clinical significance of serum S100B levels in neurointensive care. Neurocrit Care 6:94–99

49. Olivecrona M, Rodling-Wahlstrom M, Naredi S, Koskinen LO (2009) S-100B and neuron specific enolase are poor outcome predictors in severe traumatic brain injury treated by an intracranial pressure targeted therapy. J Neurol Neurosurg Psychiatry 80:1241–1247
50. Pelinka LE, Toegel E, Mauritz W, Redl H (2003) Serum S 100 B: a marker of brain damage in traumatic brain injury with and without multiple trauma. Shock 19:195–200
51. Bechtel K, Frasure S, Marshall C, Dziura J, Simpson C (2009) Relationship of serum S100B levels and intracranial injury in children with closed head trauma. Pediatrics 124:e697–e704
52. Berger RP, Adelson PD, Pierce MC, Dulani T, Cassidy LD, Kochanek PM (2005) Serum neuron-specific enolase, S100B, and myelin basic protein concentrations after inflicted and noninflicted traumatic brain injury in children. J Neurosurg 103:61–68
53. Castellani C, Bimbashi P, Ruttenstock E, Sacherer P, Stojakovic T, Weinberg AM (2009) Neuroprotein s-100B—a useful parameter in paediatric patients with mild traumatic brain injury? Acta Paediatr 98:1607–1612
54. Castellani C, Stojakovic T, Cichocki M, Scharnagl H, Erwa W, Gutmann A et al (2008) Reference ranges for neuroprotein S-100B: from infants to adolescents. Clin Chem Lab Med 46:1296–1299
55. Hauschild A, Engel G, Brenner W, Glaser R, Monig H, Henze E et al (1999) S100B protein detection in serum is a significant prognostic factor in metastatic melanoma. Oncology 56:338–344
56. Hauschild A, Michaelsen J, Brenner W, Rudolph P, Glaser R, Henze E et al (1999) Prognostic significance of serum S100B detection compared with routine blood parameters in advanced metastatic melanoma patients. Melanoma Res 9:155–161
57. Molina R, Navarro J, Filella X, Castel T, Ballesta AM (2002) S-100 protein serum levels in patients with benign and malignant diseases: false-positive results related to liver and renal function. Tumour Biol 23:39–44
58. Petzold A, Jenkins R, Watt HC, Green AJ, Thompson EJ, Keir G et al (2003) Cerebrospinal fluid S100B correlates with brain atrophy in Alzheimer's disease. Neurosci Lett 336:167–170
59. Rothermundt M, Peters M, Prehn JH, Arolt V (2003) S100B in brain damage and neurodegeneration. Microsc Res Tech 60:614–632
60. Schroeter ML, Abdul-Khaliq H, Krebs M, Diefenbacher A, Blasig IE (2009) Neuron-specific enolase is unaltered whereas S100B is elevated in serum of patients with schizophrenia–original research and meta-analysis. Psychiatry Res 167:66–72
61. Beer C, Blacker D, Bynevelt M, Hankey GJ, Puddey IB (2010) Systemic markers of inflammation are independently associated with S100B concentration: results of an observational study in subjects with acute ischaemic stroke. J Neuroinflammation 7:71
62. Peskind ER, Griffin WS, Akama KT, Raskind MA, Van Eldik LJ (2001) Cerebrospinal fluid S100B is elevated in the earlier stages of Alzheimer's disease. Neurochem Int 39:409–413
63. Chaves ML, Camozzato AL, Ferreira ED, Piazenski I, Kochhann R, Dall'Igna O et al (2010) Serum levels of S100B and NSE proteins in Alzheimer's disease patients. J Neuroinflammation 7:6
64. Gruden MA, Davidova TB, Malisauskas M, Sewell RD, Voskresenskaya NI, Wilhelm K et al (2007) Differential neuroimmune markers to the onset of Alzheimer's disease neurodegeneration and dementia: autoantibodies to Abeta((25-35)) oligomers, S100b and neurotransmitters. J Neuroimmunol 186:181–192
65. Mitosek-Szewczyk K, Gordon-Krajcer W, Flis D, Stelmasiak Z (2011) Some markers of neuronal damage in cerebrospinal fluid of multiple sclerosis patients in relapse. Folia Neuropathol 49:191–196
66. Rothermundt M, Ahn JN, Jorgens S (2009) S100B in schizophrenia: an update. Gen Physiol Biophys 28 Spec No Focus:F76–F81
67. Muller K, Elverland A, Romner B, Waterloo K, Langbakk B, Unden J et al (2006) Analysis of protein S-100B in serum: a methodological study. Clin Chem Lab Med 44:1111–1114
68. Alber B, Hein R, Garbe C, Caroli U, Luppa PB (2005) Multicenter evaluation of the analytical and clinical performance of the Elecsys S100 immunoassay in patients with malignant melanoma. Clin Chem Lab Med 43:557–563

Chapter 24

Sensible Use of High-Sensitivity Troponin Assays

Danielle Hof, Roland Klingenberg, and Arnold von Eckardstein

Abstract

The first intracellular Ca^{2+}-sensor protein to be described was the troponin complex. Only later it was discovered that cardiac-specific isoforms of troponin I (cTnI) and troponin T (cTnT) exist, and nowadays, measurement of cardiac troponins is a corner stone in the diagnosis of patients with acute coronary syndrome (ACS). High-sensitivity (hs-) assays have been developed that can record slightly elevated plasma concentrations of cardiac troponins as early as 3 h after onset of clinical symptoms. International guidelines defined a diagnostic cut-off at cardiac troponin levels corresponding to the 99th percentile of a healthy reference population and require that hs-assays measure this concentration with an interassay coefficient of variation ≤10%. This review provides an overview of the diagnostic and prognostic use of cardiac troponins.

Key words: Acute coronary syndrome, Cardiovascular disease, High-sensitivity assay, NSTEMI, STEMI, Cardiac troponin, Laboratory medicine, Calcium, EF-hand

1. Introduction

In the 1960s, troponin was discovered as the first intracellular Ca^{2+}-sensor protein, when Ebashi performed a series of experiments on Ca^{2+}-sensitivity of crude and reconstituted actomyosin (1). Upon purification of the protein complex in the 1970s, it was found out that troponin actually consists of three subunits and these subunits were named after their function: the calcium-sensing troponin C (TnC), the actomyosin ATPase inhibitory troponin I (TnI), and the tropomyosin-binding troponin T (TnT). It was only in 1987 that Cummins and colleagues reported the use of cardiac TnI (cTnI) for the diagnosis of acute myocardial infarction (AMI) (2), and the first commercial assay was marketed by Dade Behring for use on their Stratus I analyzer in 1996.

Claus W. Heizmann (ed.), *Calcium-Binding Proteins and RAGE: From Structural Basics to Clinical Applications*, Methods in Molecular Biology, vol. 963, DOI 10.1007/978-1-62703-230-8_24, © Springer Science+Business Media New York 2013

The troponin complex (see Fig. 1) is found in the thin filaments of striated skeletal and cardiac muscle, where it functions as a regulator of muscle contraction. The thin filaments are composed of two twisted fibrous actin strands that each consist of around 300–400 globular actin molecules. In a resting state, the binding sites on globular actin for the myosin heads of the thick filaments are covered by tropomyosins, which are associated with troponin complexes. Upon a transient rise of intracellular Ca^{2+} and the subsequent binding of Ca^{2+} to TnC, TnI moves away from its blocking position. Consequently, tropomyosin changes into an "open" state, the myosin heads of the thick filaments can bind to the actin subunits, the actomyosin ATPase can generate force and muscle contraction can take place (see review ref. 3). Recent molecular dynamics simulations suggest that calcium-dependent changes actually occur throughout the entire troponin complex (4).

The approximately 37 kDa TnT exists as multiple isoforms, which occur due to the expression from different genes, transcriptional control, alternative splicing, and posttranslational modifications (5, 6). Three TnT genetic isoforms exist that are encoded by three homologous genes, namely the TNNT1 gene for "slow skeletal" TnT, the TNNT2 gene for "cardiac" TnT, and the TNNT3 gene for "fast skeletal" TnT. They are specifically expressed by transcriptional regulation in slow skeletal muscle, heart muscle, and fast skeletal muscle, respectively. In addition, alternative RNA splicing—mainly of the N-terminal variable region—leads to multiple protein isoforms of the three TnT isoforms that differ in length and charge (5). For example, in cardiac TnT (cTnT) three of the 17 exons can be alternatively spliced, leading to four cTnT isoforms that are designated $cTnT_1$ to $cTnT_4$. TnI is structurally related to TnT and these two proteins have been hypothesized to derive from a common ancestral gene (6). The 21–24 kDa TnI also exists as three genetic isoforms, where the TNNI3 gene encodes the cardiac-specific TnI protein. So far, no alternative splicing processes have been described for TnI. Both cTnT and cTnI are posttranslationally modified by phosphorylation through various kinases (5, 7, 8), leading to conformational changes of the troponin complex, and by restrictive proteolytic cleavage (5, 9). Furthermore, cTnI but not cTnT contains cysteine residues that can lead to intramolecular disulfide bond formation. The 18 kDa TnC protein is a member of the calmodulin family and contains four EF-hands, with two of them located in the N-terminal domain (1). In mammalian TnC only one of these two EF hands can bind Ca^{2+} and thus act as calcium sensors. In contrast to TnT and TnI, TnC exists as two instead of three isoforms, namely "fast skeletal" TnC encoded by TNNC2 and "slow skeletal-cardiac" TnC encoded by TNNC1 (5). Accordingly, no cardiac-specific TnC isoform exists.

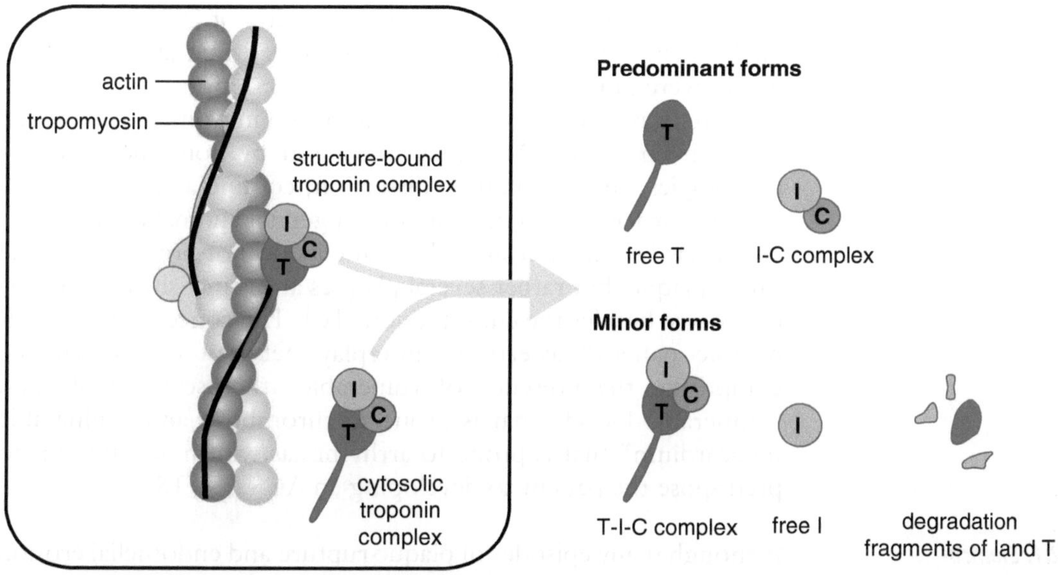

Fig. 1. Schematic overview of the troponin complex in cardiomyocytes. The majority of troponin complexes are bound to the thin filaments, whereas around 8% are present in the cytoplasm. The predominant forms detected in blood after cardiomyocyte necrosis are free cTnT and the binary cTnI–cTnC complex.

Taken together, although many different forms of TnT, TnI, and TnC exist, distinct TnT and TnI isoforms are specifically expressed in the adult heart tissue, and as the consequence can be applied as organ-specific markers of cardiac necrosis.

2. Pathophysiology of the Acute Coronary Syndrome

Acute coronary syndrome (ACS) results from the partial or total occlusion of coronary arteries with atherothrombosis as the hallmark of plaque rupture or erosion, leading to ischemia of the heart and eventually to necrosis of the heart tissue. The underlying pathogenic cause is atherosclerosis, which progresses from early lesions—so-called fatty streaks that are already present at young age—to complicated, late-stage atherosclerotic plaques. The sequential changes that occur during the progression of atherosclerosis have been classified by Stary et al. into six stages (10, 11). Late-stage, vulnerable lesions, which are prone to rupture, are characterized by a large necrotic core, in which cholesterol and active inflammatory cells have accumulated, and are covered by a thin fibrous cap. In the past two decades, it has become evident that inflammation is a key hallmark of atherosclerotic plaques (12). In addition to thrombus formation during plaque rupture, thrombi can also be formed by endothelial erosion (13). A study of 291 patients that died of acute myocardial infarction observed that around 75% of deaths were caused by plaque rupture and 25% by erosion, and that the latter was a more prevalent phenomenon in women than in men (14). In rare cases, ACS may be caused by non-atherosclerotic pathologies, such as trauma, dissection, cocaine abuse, complications of cardiac catheterization, or congenital anomalies (15).

It has been appreciated that patients do not present only one single plaque, but rather several plaques at distinct sites in the coronary tree besides the culprit lesion (16). The concept of the "vulnerable patient" asserts an interplay between diverse factors, comprising the presence of vulnerable atherosclerotic plaques, "vulnerable blood" that is prone to thrombosis and "vulnerable myocardium" that is prone to arrhythmias, which are thought to predispose the patient to developing an ACS (17, 18).

2.1. Electrocardiography

Although many episodes of plaque rupture and endothelial erosion may go unnoticed without clinical symptoms, once a patient does present with symptoms of ACS, this is a critical and potentially life-threatening situation that requires prompt medical attention. Ischemia caused by coronary artery occlusion may—but not necessarily—lead to structural changes in the myocardial tissue that can be observed as ST-segment changes in electrocardiograms (ECGs). As depicted in Fig. 2, interpretation of ECG is the first step in the management of patients with acute chest pain and should be made within 10 min after first medical contact (15). The ECG will classify the patients into two categories, namely patients with persistent (>20 min) ST-elevation (STE-ACS) and patients without persistent ST-elevations (NSTE-ACS) (15), with NSTE-ACS being more frequent than STE-ACS. NSTE-ACS patients may have abnormal (e.g., ST-depression or T-wave inversion) or normal ECG findings and need monitoring with serial ECGs and serial measurement of blood biomarkers indicating heart tissue damage. Patients may subsequently be categorized into non-ST-elevation myocardial infarction (NSTEMI) or unstable angina (UA) patients, based on the presence/absence of circulating biomarkers of cardiac necrosis (cardiac troponins in particular). Rapid coronary revascularization, by percutaneous coronary intervention (PCI) or if not available fibrinolysis, is the mainstay of therapy in both STE-ACS and NSTE-ACS, complemented by medications aimed at secondary prevention, as summarized in current guidelines (15, 19).

2.2. Release of Markers of Cardiac Necrosis

Upon and at designated time intervals during hospital admission, blood is sampled to measure markers of myocardial necrosis, such as myoglobin, creatine kinase (CK), CK-MB, and cardiac troponins. CK-MB and myoglobin are mostly present in the cytoplasm (91 and 98%, respectively) (20) and are released very quickly upon ACS, with myoglobin showing the fastest release (see Fig. 3).

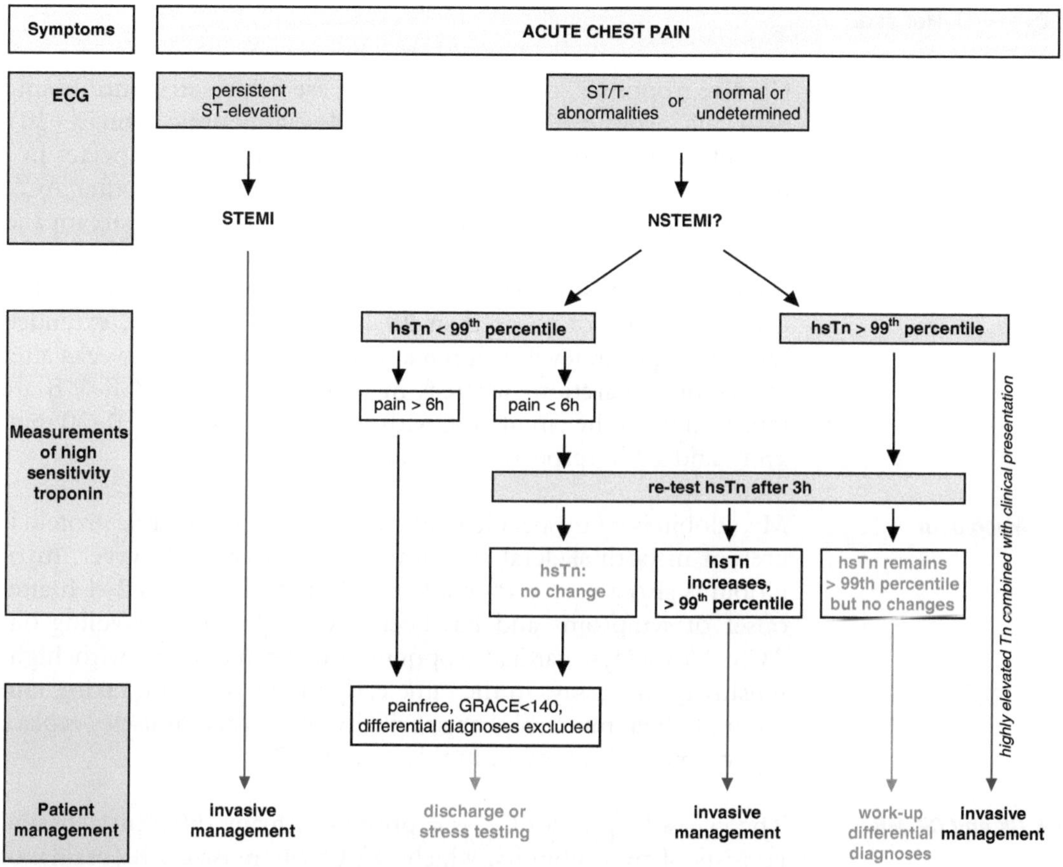

Fig. 2. Guideline for the use of troponin (hsTn) measurements in the management of patients with acute chest pain. GRACE = Global Registry of Acute Coronary Events.

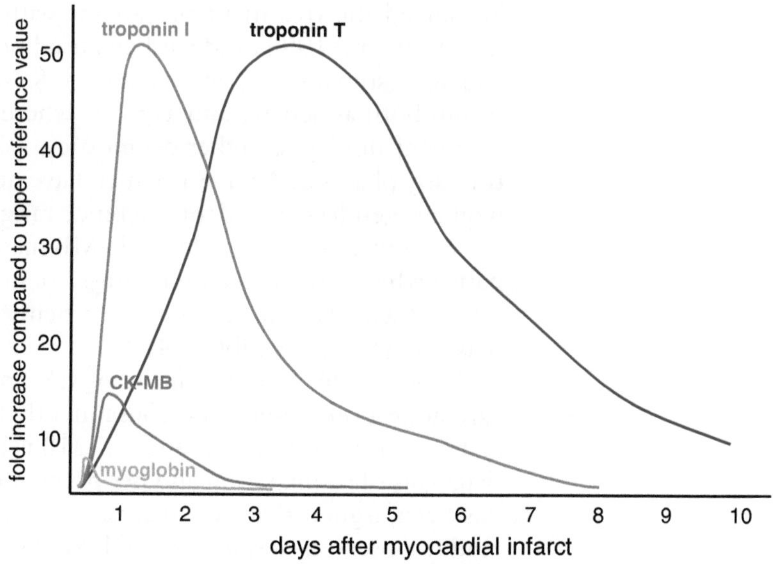

Fig. 3. Rise and fall of markers of myocardial damage in plasma after onset of symptoms. Shown are myoglobin (*solid line*), CK-MB (*broken line*), troponin I (*broken/dotted line*), and troponin T (*dotted line*).

Cardiac troponins, on the other hand, are mostly structure-bound with only around 8% present as soluble cytoplasmic protein (20). Therefore, troponins are released from dying cardiomyocytes in a two-step manner (21). Initially, during the first hours after ACS, the soluble, cytoplasmic pool is released into the blood stream and has a half-life of around 2 h. In a second phase, also structure-bound troponins are released, at much slower rates and with a clinical half-life of more than 20 h. Due to this slow, extended release, troponin levels can remain elevated for days or weeks after myocardial infarction, whereas myoglobin, CK, and CK-MB are removed from the circulation with half-lives of around 10–20 min, 15 h, and 12 h, respectively (22, 23).

2.3. Myoglobin

Myoglobin is an unspecific marker, since this cytoplasmic protein is present in both skeletal and heart muscle tissue. However, myoglobin is also a very early marker with increased levels 2–4 h after onset of symptoms and has been used especially for ruling-out ACS. Nowadays, cardiac troponins can be measured with high-sensitivity (hs) assays in the same early time frame and, taking into account their much higher specificity for cardiac muscle, replace myoglobin measurement in this regard (22).

2.4. CK and CK-MB

The cytosolic protein CK is expressed in many different cells and consists of two subunits, which can be M (muscle), B (brain), or Mi (mitochondrial) type. The CK-MB heterodimer is relatively specific for heart tissue, although it can also derive from skeletal muscle under certain pathological conditions or after extreme physical exercises. The measurement of CK and CK-MB has revolutionized the treatment of patients with acute cardiac events in the 1970s and 1980s and has been used both for diagnosis of AMI and for estimation of infarct size. CK-MB can be measured in serum both as activity and as mass, where the CK-MB mass measurement displays a higher diagnostic specificity and specificity for the early phases of AMI. Since men have higher baseline levels than women, gender-dependent reference ranges are recommended by international guidelines (24). CK-MB mass can be used as an alternative when cardiac troponin assays are not available, and CK and CK-MB activity can be used when neither troponin nor CK-MB mass assays are accessible (24, 25).

Due to the shorter half-life of CK and CK-MB compared to cardiac troponins and their consequently faster clearance from the system, monitoring of CK and CK-MB is believed to be helpful for diagnosis of re-infarction and periprocedural MI (26). However, it has been argued that new increases in cardiac troponin levels are readily observed despite elevated baseline levels, and that cardiac troponins can be used for diagnosing re-infarction with the same or even higher diagnostic sensitivity and specificity (27, 28).

2.5. Cardiac Troponins

Among the different markers, cardiac troponins have the highest specificity for myocardial necrosis and are therefore the gold standard in the diagnosis of ACS. Indeed, cardiac troponins, measured by hs-assays, are the only blood biomarkers described in the current European Society of Cardiology (ESC) guidelines for the diagnosis of NSTEMI (15). In agreement with these guidelines, many hospitals have removed CK-MB from their cardiac biomarker panels (27). On the other hand, many laboratories still use cardiac troponins in combination with other markers of myocardial damage, as was demonstrated by questionnaire inquiries made by the European Federation of Clinical Chemistry and Laboratory Medicine (EFCC) in 2008 and 2011. In 2008, the EFCC analyzed the use of cardiac necrosis markers for the diagnosis of ACS in 220 laboratories in eight countries, in the so-called Cardiac Marker Guideline Uptake in Europe (CARMAGUE) study (29). It was found that 94% of the laboratories used cardiac troponin (51% using cTnT and 49% using cTnI) as the preferred test. In 34% of the laboratories, cardiac troponins were combined with a second blood biomarker, such as CK, CK-MB, or (in only two laboratories) myoglobin. Recently, the CARMAGUE study was repeated in 303 laboratories in 28 European countries (30). Around 95% of the laboratories used cardiac troponin as the preferred test, with 50% using cTnT and 45% using cTnI assays. Cardiac troponins were used as the sole marker of cardiac necrosis in only 31% of the laboratories. In 69% of the laboratories, troponins were offered in combination with other markers, such as CK (59%), lactate dehydrogenase (LDH) (30%), aspartate transaminase (AST) (34%), and CK-MB (8.3%). It must be noted that international guidelines strongly advise against the use of CK, LDH, and AST for diagnostics of ACS (15, 22, 24).

3. Diagnostic Tests

3.1. Cardiac Troponin I and Troponin T Immunoassays

The cardiac troponins cTnI and cTnT are highly specific for myocardial necrosis and both can be detected in the blood after AMI. Depending on the analytical performance of the test, cardiac troponins can be measured as early as 3 h after AMI.

3.2. Quick Tests

Several different quick tests and point-of-care tests for the measurement of cardiac troponins, also in combination with myoglobin and CK-MB, are commercially available in the form of strip tests. They are recommended only for use in settings where laboratories cannot deliver troponin results within 60 min (15). Since the analytical sensitivity of these tests is low, a negative result during the first 10 h after onset of symptoms does not exclude ACS and measurements should be repeated at a later time point. In general, positive results can be reliably interpreted as indicative of ACS. Because the visual read-out of the quick tests does not give quantitative results, an increase in troponin concentration cannot be measured. Recently, a study by Than et al. comprising around 3,500 patients with acute chest pain but without ST-segment elevations, provided a fast (within 2 h after admission) rule-out protocol (31). By combining the Thrombolysis in Myocardial Infarction (TIMI) score, ECG and point-of-care tests measuring cTnI, CK-MB, and myoglobin (TRIAGE and CardioProfilER from Alere, USA), a negative predictive value of 99.1% could be demonstrated. Fast and safe rule-out protocols reduce the length of stay at the emergency department and are therefore economically attractive.

3.3. Automated Assays

Whereas patent rights protect the measurement of cTnT by only one manufacturer (Roche Diagnostics, Switzerland), many different automated assays are commercially available for the measurement of cTnI in plasma (see Table 1). Cardiac troponins are predominantly released as free cTnT and binary cTnI–cTnC complexes (Fig. 1), and to a much lesser extent as ternary cTnT–cTnI–cTnC and binary cTnT–cTnI complexes, free cTnI and degraded cTnI and cTnT fragments, with the relative release pattern changing over time and demonstrating interindividual differences (32–35). In addition, extensive modification of cTnI in necrotic heart tissue (36–38) as well as in serum (36, 39) by proteolysis and phosphorylation leads to the occurrence of multiple smaller (and some larger) fragments in blood after AMI, as detected by Western blotting (36, 39). Gel filtration chromatography (32) did not detect fragments of cTnI, which could be due to a lower sensitivity of this method compared to Western blotting. Due to these various forms of cTnI, large interassay differences in measured cTnI plasma concentrations have been noticed that result from differences in the antibody designs and their corresponding epitopes. Consequently, difficulties arise in comparing data from clinical studies, as well as in comparing cTnI levels in patients that are transferred between hospitals.

3.4. Standardization of cTnI Measurement

The International Federation of Clinical Chemistry and Laboratory Medicine (IFCC) has addressed the problem of large differences in cTnI results by appointing a working group on the standardization of cTnI measurement. Amongst others, an overview of peer-reviewed publications on the analytical evaluation of commercial cTnI assays is provided on the IFCC website. To minimize the differences that occur due to differences in antibody specificities, it has been postulated to avoid the use of antibodies against the rapidly removed N- and C-termini of cTnI and only use antibodies against the central, most stable region of cTnI (amino acids 30–110) (40). An overview of the epitopes recognized in the

Table 1
Overview of different commercial assays for troponin I and troponin T

Assay	LoB# (μg/l)	LoD* (μg/l)	99th % (μg/l)	% CV at 99th %	10% CV (μg/l)	Risk stratification	Epitopes recognized by antibodies	Detection antibody tag
cTnI assays								
Abbott AxSYM ADV	0.02		0.04	14.0	0.16	Yes	C: 87–91, 41–49; D: 24–40	ALP
Abbott ARCHITECT	<0.01		0.028	14.0	0.032	Yes (no in USA)	C: 87–91, 24–40; D: 41–49	Acridinium
Abbott i-STAT	0.02		0.08	16.5	0.10	Yes	C: 41–49, 88–91; D: 28–39, 62–78	ALP
Alere Triage SOB	0.05		NAD	NA	NA	No	C: NA; D: 27–40	Fluorophor
Alere Triage Cardio 3 (r)	0.01		NAD	17.0 (at 0.02)	NA	No	NA	Fluorophor
Beckman Coulter Access Accu	0.01		0.04	14.0	0.06	Yes	C: 41–49, D: 24–40	ALP
bioMerieux Vidas Ultra	0.01		0.01**	27.7	0.11	No	C: 41–49, 22–29; D: 87–91, 7B9	ALP
Mitsubishi Chemical PATHFAST	0.008		0.029	5.0	0.014	No	C: 41–49; D: 71–116, 163–209	ALP
Ortho VITROS Troponin I ES	0.007	0.012	0.034	10.0	0.034	Yes	C: 24–40, 41–49; D: 87–91	HRP
Radiometer AQT90 FLEX TnI		0.0095	0.023	17.7	0.039	NA	C: 41–49, 190–196; D: 137–149	Europium
Response Biomedical RAMP	0.03		NAD	18.5 (at 0.05)	0.21	No	C: 85–92; D: 26–38	Fluorophor
Roche E2010/cobase e411/E170/cobas e601/602 cTnI		0.16	0.16**	NA	0.3	No	C: 87–91, 190–196; D: 23–29, 27–43	Ruthenium

(continued)

Table 1 (continued)

Assay	LoB[#] (μg/l)	LoD[*] (μg/l)	99th % (μg/l)	% CV at 99th %	10% CV (μg/l)	Risk stratification	Epitopes recognized by antibodies	Detection antibody tag
Siemens Centaur Ultra	0.006		0.04	8.8	0.03	Yes	C: 41–49, 87–91 D: 27–40	Acridinium
Siemens Dimension RxL	0.04		0.07	20.0	0.14	Yes	C: 27–32; D: 41–56	ALP
Siemens Dimension EXL	0.017		0.056	10.0	0.05	Yes	C: 27–32; D: 41–56	Chemiluminescence
Siemens Immulite 2500 STAT	0.1		0.2	NA	0.42	No	C: 87–91; D: 27–40	ALP
Siemens Immulite 1000 Turbo	0.15		NA	NA	0.64	No	C: 87–91; D: 27–40	ALP
Siemens Stratus CS	0.03		0.07	10.0	0.06	Yes	C: 27–32; D: 41–56	ALP
Siemens VISTA	0.015		0.045	10.0	0.04	Yes	C: 27–32; D: 41–56	Chemiluminescence
Tosoh ST AIA-PACK	0.06		0.06[**]	8.5	NA	No	C: 41–49; D: 87–91	ALP
Beckman Coulter Access hs-cTnI	0.002		0.0086	10.0	0.0086	NA	C: 41–49; D: 24–40	ALP
Nanosphere VeriSens hs-cTnI	0.0002		0.0028	9.5	0.0005	NA	C: 136–147; D: 49–52, 70–73, 88, 169	Gold nanoparticles
Singulex hs-cTnI	0.00009		0.0101	9.0	0.00088	NA	C: 41–49; D: 27–41	Capillary flow fluorescence
cTnT assays								
Radiometer AQT90 FLEX TnT		0.010	0.017	15.2	0.025	NA	C: 125–131; D: 136–147	Europium
Roche Cardiac Reader cTnT	<0.05		NAD	NA	NA	No	C: 125–131; D: 136–147	Gold particles

Roche E2010/cobase e411/E170/cobas e601/602 TnT (4th generation)	0.01		NAD	NA	0.03	Yes	C: 125–131; D:136–147	Ruthenium
Roche E2010/cobase e411/E170/cobas e601/602 hs-TnT		0.005	0.014	10.0	0.013	NA	C: 125–131; D:136–147	Ruthenium

Reproduced with permission from the International Federation of Clinical Chemistry and Laboratory Medicine (IFCC). Version: December 2010. # LoB = limit of blank; *LoD = limit of detection; NAD = the 99th percentile concentration of the value distribution of a reference population is indeterminate; NA = no data available; ** = a 99th percentile concentration equal to the assay's LoD is unlikely to have acceptable imprecision for reliable troponin measurement; C = capture antibodies; D = detection antibodies; ALP = alkaline phosphatase; HRP = horseradish peroxidase; hs = high-sensitivity different commercial assays for cTnI, as well as for cTnT, is given in Table 1. In addition, poor agreement between assays is explained by the lack of a suitable reference material for calibration of the cTnI assays. The IFCC working group proposes a metrologically based reference measurement system with primary, secondary, and third reference materials that can subsequently be linked to methods and calibrators in an unbroken traceability chain (40).

3.5. High-Sensitivity Assays

Since conventional, early-generation troponin assays measured low troponin serum concentrations with very high imprecision, manufacturers have tried to improve the analytical sensitivity and precision of their tests in new-generation assays. As recommended by international cardiology and laboratory medicine guidelines, hs-troponin assays must not have an interassay imprecision greater than 10% at the cut-off for the diagnosis of AMI, which is the 99th percentile of a healthy reference population, (24, 25, 28) (see also Table 1 for the 99th percentile concentrations and the corresponding coefficients of variation for different commercially available assays). According to guidelines from the Clinical Laboratory Standards Institute (CLSI) a group of at least 120 healthy individuals should be analyzed to be able to appropriately calculate the 97.5^{th} percentile. For valid establishment of the 99th percentile cut-off, troponin concentrations are recommended to be measured in 300–500 healthy individuals. It must be kept in mind that the choice of the reference population (e.g., age, ethnicity), exclusion criteria, sample size, and specimen type will influence the cut-off. Ideally, the 99th percentile value should be established by each individual laboratory for the assay that will be used. It is clear that, due to failing resources, not all laboratories have the ability to do this evaluation and for these laboratories it is advised to refer to peer-reviewed cut-off levels (24).

3.6. Analytical Precautions

In general, measurements of cardiac troponins are not interfered by hemolysis, icterus or high triglyceride levels. Cardiac TnT is relatively stable, with a decrease of less than 5% in 24 h when the sample is stored at 4°C. The stability of cTnI depends on the assay that is used. For immunoassays employing antibodies against the stable central region of cTnI, samples are stable for 2 days at room temperature and 3 days at 4°C. At −20°C or lower temperatures, cTnT and cTnI are stable for at least 12 and 2 months, respectively, although it is recommended to freeze–thaw the sample only once. Falsely elevated troponin measurements may be assay-dependently caused by heterophilic antibodies, human anti-mouse antibodies (HAMA), rheumatoid factor (RF), and fibrin clots. When suspicion of false-high troponin concentration exists, the measurement should be repeated with a fresh blood sample to exclude mix-up of samples. Subsequently, the sample can be assayed with another cardiac troponin test that uses different antibodies against other epitopes or a test that measures the other cardiac troponin (e.g., cTnI when cTnT was assessed first), or interferences can be reduced by treatment of the sample with heterophilic blocking reagents prior to measurement of cardiac troponins. In addition to the removal of heterophilic antibodies, some of these blocking reagents also diminish interferences by HAMA (41) and RF (42). False-low results can derive from the presence of biotin in patients under high-dose biotin therapy. Autoantibodies against cTnI have been described to produce false-negative results in some patients with chest pain, as well as in patients with noncardiac symptoms (43, 44), which could be explained by blocking of the epitope that is recognized in the immunoassay. Furthermore, autoantibodies against cTnI (12.7%) and against cTnT (9.9%) have been described in healthy blood donors (45, 46). On the other hand, autoantibodies could also stabilize troponins, thereby prolonging the half-life and leading to falsely elevated levels. If false-negative cTnI is suspected, the sample could be analyzed for cTnT, since autoantibodies against cTnI may not cross-react with cTnT. However, a recent study found a rather high prevalence of coinciding autoantibodies against cTnI and cTnT, where 55 of 345 sera were found to contain autoantibodies against cTnI (n = 27), cTnT (n = 22), or both (n = 6) (47).

4. Clinical Applications of Troponins

4.1. Diagnostics of Acute Coronary Syndrome

The guidelines for diagnosis and clinical management of patients with acute chest pain (15) are schematically summarized in Fig. 2, where cardiac troponins are especially useful to discriminate between NSTEMI and UA. Cardiac troponins are measured, ideally with hs-tests, upon admission to the hospital and subsequently at one or more time point(s) during a few hours of follow-up. The most powerful characteristic of troponin measurement with hs-assays is that undetectable troponin levels at admission to the hospital exclude AMI with very high certainty (48, 49), which is reflected by a high negative predictive value of around 95% (15). In combination with a repeated measurement 3 h after hospitalization, the diagnostic sensitivity increases to almost 100% (15). However, several cardiovascular diseases other than AMI and also non-cardiovascular diseases may cause elevated cTn plasma concentration, especially in the lower concentration range of sensitive cTn assays (see Subheading 4.4 and Table 2). As a consequence, high-sensitive troponin assays have a lower specificity and positive predictive value than conventional troponin assays. If not resolved by clinical judgment, for example, by integrating data from patient history, clinical presentation, and ECG using TIMI or Global

Table 2
Overview of circumstances other than ACS that are associated with elevated troponin levels

Cardiac pathologies	Noncardiac pathologies	Non-pathological circumstances	Assay-dependent false results: False-positive results	Assay-dependent false results: False-negative results
Acute pericarditis	Acute pulmonary embolism	High-intensity exercise (e.g., marathon running)	Human anti-mouse antibodies (HAMA) after therapy with biologicals	Biotin treatment (assay dependent)
Acute or severe heart failure	Sepsis	Prolonged, moderate-intensity exercise (e.g., long distance walking)	Human anti-animal antibodies	Autoantibodies against cardiac troponins (blocking the epitope on cTn)
Myocarditis	Critical illness		Heterophilic antibodies	
Aortic dissection	End stage renal disease		Rheumatoid factor	
Cardiac ablation	Stroke		Autoantibodies against cardiac troponins (increasing the half-life of cTn)	
Heart transplantation	Intracranial hemorrhage		Fibrin clots in the sample	
Cardiac trauma	Epileptic seizures			
Postoperative cardiac surgery	Postoperative noncardiac surgery			
Cardiopulmonary resuscitation	Diabetes mellitus			
Arrhythmias	Acute rheumatic fever			
Left ventricular hypertrophy	Infiltrative disorders (e.g., amyloidosis, sarcoidosis)			
Cardio-toxic chemotherapy	Rhabdomyolysis			
	Extensive burns			
	Inflammatory muscle diseases (poly-/dermatomyositis)			

Registry of Acute Coronary Events (GRACE) scores (50–52), the problem of reduced specificity can be partially solved by repeated measurements at (a) defined time interval(s), and potentially by a combination of hs-cTn assays with additional biomarkers. According to international guidelines, the diagnosis of AMI is based not solely on the elevation cTn plasma concentrations but by their subsequent rise or fall (53). However, the magnitude of change in troponin levels (the so-called delta criterion) as compared to baseline values at admission is still under debate.

Not infrequently, ACS occurs in patients who underwent coronary interventions, i.e., PCI or coronary artery bypass surgery (CABG). In these situations the diagnostic cut-off for the diagnosis of AMI is threefold and fivefold, respectively, higher than the concentration at the 99th percentile (15).

4.2. Risk Stratification of Patients with ACS

Risk stratification of patients with ACS is of utmost clinical importance. In several studies that compared sensitive versus conventional troponin assays, it was demonstrated that hs-cTn assays improve risk prediction of event-free survival for 30 days as well as for 1 year (54–58). Nonetheless, the prognostic benefits of high-sensitive over conventional troponin assays must be valued in the light of clinical scores that are recommended by current guidelines (15, 59), as e.g. the TIMI (50, 52) and GRACE (51) risk scores, which combine clinical parameters, ECG findings, and basic laboratory parameters, including conventional cTn concentration. A recent multi-marker study analyzed the effect of adding different biomarkers to the clinical TIMI risk score in 4,352 NSTEMI patients that participated in the Metabolic Efficiency With Ranolazine for Less Ischemia in Non-ST Elevation Acute Coronary-Thrombolysis in Myocardial Infarction 36 (MERLIN-TIMI 36) trial. For hs-cTnI a net improvement over the clinical TIMI risk score (excluding troponin) was found, with a c-index increasing from 0.784 to 0.805 ($p=0.005$) and an NRI of 0.389 ($p<0.001$) for cardiovascular death. Similar results were demonstrated for MI, hospitalization for congestive heart failure and the composite endpoint cardiovascular death and hospitalization for congestive heart failure (60). In another recent study, which enrolled 1,159 patients with acute chest pain and of whom 30% were diagnosed with ACS, it was found that the addition of hs-cTnT instead of conventional cTnT to the TIMI risk score improved the prediction of death, but not subsequent AMI, with an improved reclassification of patients (net reclassification improvement (NRI) of 0.91) (61). On the other hand, in a recent study that comprised 50 STEMI, 147 NSTEMI, and 173 UA patients, addition of hs-cTn concentration to the GRACE risk score (including conventional cTn) did not give incremental prognostic value for prediction of in-hospital mortality, 1-year mortality, and combined death/AMI at 1 year (62).

It was also recently demonstrated that hs-cTnT, instead of conventional cTnT, reclassified patients into NSTEMI from UA (63) and improved long-term risk prediction (55, 63). In conclusion, if combined with clinical scores hs-troponin assays appear to improve risk prediction in NSTEMI patients but not, or less so, in STEMI patients.

4.3. Clinical Management of Patients with ACS

ST-elevations upon ECG and detectable cardiac troponins, measured by conventional assays, help to stratify ACS patients into subgroups who benefit from acute PCI (53). As yet, only little is known about the potential benefits of introducing hs-Tn assays into clinical practice on patient management and outcome after therapeutic intervention. A recent study addressed this question by lowering the diagnostic threshold of cTnI from 0.20 to 0.05 μg/L by the introduction of a sensitive troponin assay (1,038 patients before and 1,054 after implementation). Changing the clinical protocol did not only increase the rate of AMI diagnosed in patients with suspected ACS, but most importantly led to major reductions in morbidity and mortality (64). Future trials are required to examine the benefit of invasive therapy in AMI patients with elevated levels as detected by hs-cTn assays. Quantitative cut-offs for hs-cTn and changes over time (delta criterion) will become more important as guidance for therapeutic interventions.

4.4. Elevation of Troponins in Other Conditions than ACS

Troponins can be elevated in other diseases that affect the heart, such as stable CAD, acute heart failure, or myocarditis as well as in noncardiac diseases, like pulmonary embolism and sepsis, and in non-pathological circumstances, such as after high-intensity sports (see reviews refs. 65, 66). An overview of known circumstances, other than ACS, that give rise to elevated levels of troponins is given in Table 2.

In conventional assays, cardiac troponins were below the assay's detection limit in most patients with stable CAD. Recently, Omland et al. (67) measured cTnT levels with a hs-assay, having a tenfold increased sensitivity in comparison with the conventional cTnT assay, in serum samples of over 8,000 patients with stable CAD that were collected in the Prevention of Events with Angiotensin Converting Enzyme Inhibition (PEACE) trial. It was found that 97.7% of these patients had cTnT levels above the detection limit, with 11.1% displaying cTnT serum levels at or above the 99th percentile. Furthermore, these low cTnT levels significantly associated with the incidence of future cardiovascular death and heart failure but not with AMI. The fact that more than 10% of CAD patients have cardiac troponin levels at or above the 99th percentile in current hs-assays could reduce the diagnostic specificity of cardiac troponins for AMI in these patients. Recently, Reiter et al. (68) analyzed serum concentrations of cardiac troponins in 1,098 patients with acute chest pain, of which 401 had preexisting CAD. Again, it was shown that a significant number of patients with CAD, but without AMI, had troponin levels above the 99th percentile. Furthermore, it was demonstrated that the hs-cTnT and the two sensitive cTnI assays that were used in this clinical study maintain their prognostic value, although the cut-off levels to discriminate AMI from non-AMI tended to be higher in patients with preexisting CAD than in patients without CAD.

Although cardiac troponins are cleared via the kidney and elevated cardiac troponin levels are observed in patients with renal dysfunction, they are predictive of AMI and death in patients with renal dysfunction that display acute chest pain (69). Elevated troponin levels in end-stage renal disease, which could be caused by cardiac dysfunction, left ventricular hypertrophy or subclinical MI, predict mortality in these patients (70–72). Upon employing hs-cTnT or sensitive cTnI assays, slight elevations above the 99th percentile are commonly observed in elderly patients without AMI, as was recently shown by Reiter et al. (73). The cut-off values to discriminate AMI from non-AMI varied significantly between elderly (> 70 years) and younger patients (< 70 years), highlighting the need for age-optimized cut-off values.

Not in all instances is it understood why troponin levels are elevated without cardiac necrosis. The half-life of the troponins measured after, e.g., marathon running is much shorter than the troponin half-life observed after AMI, pointing at a cytoplasmic rather than a structural origin. Recently, Hickman et al. proposed a mechanism, by which cytoplasmic troponins are released in membranous blebs into the blood stream upon ischemia (74).

4.5. Troponins in the Healthy Population: Risk Markers of Future Coronary Artery Disease Events?

To study the clinical significance of cTnT detection in the general population, Saunders et al. (75) recently measured cTnT with a hs-assay in over 9,000 subjects, aged 54–74, of the Atherosclerotic Risk in Communities (ARIC) study. In these apparently healthy, unselected individuals from the general population, 66.5% displayed cTnT serum levels above the detection limit and 7.4% had cTnT serum levels at or higher than the 99th percentile. Associations with CAD (especially with fatal CAD), all-cause mortality and above all heart failure were found. Furthermore, the addition of cTnT to risk prediction models improved risk prediction better than hs-C-reactive protein (CRP) and in a similar way as N-terminal pro-B-type natriuretic peptide (NT-proBNP). Similar results were reported by de Lemos et al. (76) in a study comprising around 3,500 individuals, aged 30–65 years, that participated in the Dallas Heart Study. Again, cTnT, as measured by a hs-assay, associated with heart disease events and all-cause mortality, and improved risk prediction models. In a multimarker study of 7,915 Finnish men and women and 2,551 Irish men, sensitive cTnI together with NT-proBNP and hs-CRP were the only out of 30 parameters which allowed to improve the prediction of cardiovascular events beyond classical risk prediction rules (77). Taken together, these data indicate that cardiac troponins measured by hs-assays may become part of cardiovascular risk prediction in asymptomatic populations to identify high-risk individuals for intensive risk factor intervention, for example by lowering of cholesterol or blood pressure.

5. Conclusions

The discovery of the cardiac-specific troponins cTnI and cTnT and the subsequent introduction of cardiac troponins tests into clinical laboratory medicine has greatly improved the management of patients presenting to the emergency department with acute chest pain in the past two decades. The most recent introduction of hs-cTn assays has helped to further shorten the diagnostic time interval. However, the trade-off for the increased sensitivity and high negative predictive value of hs-cTn tests, allowing a quick "rule-out" of AMI, is the diminished specificity with a lower positive predictive value, making the diagnosis of AMI ("rule-in") more demanding. To address this problem, absolute and relative increases in hs-cTn levels over time are followed for the diagnosis AMI. Furthermore, combining hs-cTn test with another disease-specific and complementary biomarker may improve the diagnostic efficacy for AMI at a single time point in the future. It is expected that the diagnostic specificity of cardiac troponins will even further decrease with novel cTn tests that possess even higher analytical sensitivities, since low levels of cardiac troponins will be detected in more patients that do not suffer from ACS. Each new high- or ultra-sensitive cardiac troponin assay will require thorough clinical evaluation, and establishing cut-off levels for specific groups, e.g., stratifying for age, ethnicity, preexisting, and concomitant diseases, will become more and more important. At least in NSTEMI patients, more and more evidence is gathered, indicating that hs-cTn assays improve short- and long-term risk prediction beyond clinical risk prediction rules, including cTn concentrations measured by conventional assays. In addition, the role of hs-cTn testing on the choice of therapeutic intervention and clinical outcome is an important topic of ongoing research. Finally, hs-cTn assays may open new diagnostic areas for the use of cardiac troponins, namely cardiovascular risk prediction in asymptomatic individuals.

References

1. Schaub MC, Heizmann CW (2008) Calcium, troponin, calmodulin, S100 proteins: from myocardial basics to new therapeutic strategies. Biochem Biophys Res Commun 369:247–264
2. Cummins B, Auckland ML, Cummins P (1987) Cardiac-specific troponin-I radioimmunoassay in the diagnosis of acute myocardial infarction. Am Heart J 113:1333–1344
3. Gordon AM, Homsher E, Regnier M (2000) Regulation of contraction in striated muscle. Physiol Rev 80:853–924
4. Manning EP, Tardiff JC, Schwartz SD (2011) A model of calcium activation of the cardiac thin filament. Biochemistry 50:7405–7413
5. Jin JP, Zhang Z, Bautista JA (2008) Isoform diversity, regulation, and functional adaptation of troponin and calponin. Crit Rev Eukaryot Gene Expr 18:93–124
6. Wei B, Jin JP (2011) Troponin T isoforms and posttranscriptional modifications: evolution, regulation and function. Arch Biochem Biophys 505:144–154
7. Solaro RJ, van der Velden J (2010) Why does troponin I have so many phosphorylation sites? Fact and fancy. J Mol Cell Cardiol 48:810–816
8. Zhang J et al (2011) Phosphorylation, but not alternative splicing or proteolytic degradation, is conserved in human and mouse cardiac troponin T. Biochemistry 50:6081–6092
9. Di Lisa F et al (1995) Specific degradation of troponin T and I by mu-calpain and its modulation by substrate phosphorylation. Biochem J 308:57–61
10. Stary HC et al (1995) A definition of advanced types of atherosclerotic lesions and a histological classification of atherosclerosis. A report from the Committee on Vascular Lesions of the Council on Arteriosclerosis, American Heart Association. Circulation 92:1355–1374
11. Stary HC et al (1994) A definition of initial, fatty streak, and intermediate lesions of atherosclerosis. A report from the Committee on Vascular Lesions of the Council on Arteriosclerosis, American Heart Association. Circulation 89:2462–2478
12. Libby P, Ridker PM, Hansson GK (2011) Progress and challenges in translating the biology of atherosclerosis. Nature 473:317–325
13. Libby P (2009) Molecular and cellular mechanisms of the thrombotic complications of atherosclerosis. J Lipid Res 50(Suppl):S352–357
14. Arbustini E et al (1999) Plaque erosion is a major substrate for coronary thrombosis in acute myocardial infarction. Heart 82:269–272
15. Hamm CW et al (2011) ESC Guidelines for the management of acute coronary syndromes in patients presenting without persistent ST-segment elevation: the Task Force for the management of acute coronary syndromes (ACS) in patients presenting without persistent ST-segment elevation of the European Society of Cardiology (ESC). Eur Heart J 32:2999–3054
16. Rioufol G et al (2002) Multiple atherosclerotic plaque rupture in acute coronary syndrome: a three-vessel intravascular ultrasound study. Circulation 106:804–808
17. Naghavi M et al (2003) From vulnerable plaque to vulnerable patient: a call for new definitions and risk assessment strategies: part II. Circulation 108:1772–1778
18. Naghavi M et al (2003) From vulnerable plaque to vulnerable patient: a call for new definitions and risk assessment strategies: part I. Circulation 108:1664–1672
19. Van de Werf F et al (2008) Management of acute myocardial infarction in patients presenting with persistent ST-segment elevation: the Task Force on the Management of ST-Segment Elevation Acute Myocardial Infarction of the European Society of Cardiology. Eur Heart J 29:2909–2945
20. Bleier J et al (1998) Different intracellular compartmentations of cardiac troponins and myosin heavy chains: a causal connection to their different early release after myocardial damage. Clin Chem 44:1912–1918
21. Gerhardt W et al (1991) S-troponin T in suspected ischemic myocardial injury compared with mass and catalytic concentrations of S-creatine kinase isoenzyme MB. Clin Chem 37:1405–1411
22. Morrow DA et al (2007) National Academy of Clinical Biochemistry Laboratory Medicine Practice Guidelines: clinical characteristics and utilization of biochemical markers in acute coronary syndromes. Circulation 115:e356–375
23. Thomas L (2005) Labor und Diagnose, 6th edn. TH-Books Verlagsgesellschaft, Frankfurt
24. Apple FS et al (2007) National Academy of Clinical Biochemistry and IFCC Committee for Standardization of Markers of Cardiac Damage Laboratory Medicine Practice Guidelines: analytical issues for biochemical markers of acute coronary syndromes. Clin Chem 53:547–551
25. Morrow DA et al (2007) National Academy of Clinical Biochemistry Laboratory Medicine Practice Guidelines: clinical characteristics and utilization of biochemical markers in acute coronary syndromes. Clin Chem 53:552–574

26. Anderson JL et al (2007) ACC/AHA 2007 guidelines for the management of patients with unstable angina/non ST-elevation myocardial infarction: a report of the American College of Cardiology/American Heart Association Task Force on Practice Guidelines (Writing Committee to Revise the 2002 Guidelines for the Management of Patients With Unstable Angina/Non ST-Elevation Myocardial Infarction): developed in collaboration with the American College of Emergency Physicians, the Society for Cardiovascular Angiography and Interventions, and the Society of Thoracic Surgeons: endorsed by the American Association of Cardiovascular and Pulmonary Rehabilitation and the Society for Academic Emergency Medicine. Circulation 116:e148–304
27. Saenger AK, Jaffe AS (2008) Requiem for a heavyweight: the demise of creatine kinase-MB. Circulation 118:2200–2206
28. Thygesen K et al (2007) Universal definition of myocardial infarction. Circulation 116:2634–2653
29. Collinson P et al (2008) How well do laboratories follow guidelines on cardiac markers? The cardiac marker guideline uptake in Europe study. Clin Chem 54:448–449
30. Collinson P et al (2011) Evidence-based laboratory medicine: how well do laboratories follow recommendations and guidelines? The Cardiac Marker Guideline Uptake in Europe (CARMAGUE) study. Clin Chem 58:305–306
31. Than M et al (2011) A 2-h diagnostic protocol to assess patients with chest pain symptoms in the Asia-Pacific region (ASPECT): a prospective observational validation study. Lancet 377:1077–1084
32. Bates KJ et al (2010) Circulating immunoreactive cardiac troponin forms determined by gel filtration chromatography after acute myocardial infarction. Clin Chem 56:952–958
33. Giuliani I et al (1999) Determination of cardiac troponin I forms in the blood of patients with acute myocardial infarction and patients receiving crystalloid or cold blood cardioplegia. Clin Chem 45:213–222
34. Katrukha AG et al (1997) Troponin I is released in bloodstream of patients with acute myocardial infarction not in free form but as complex. Clin Chem 43:1379–1385
35. Wu AH et al (1998) Characterization of cardiac troponin subunit release into serum after acute myocardial infarction and comparison of assays for troponin T and I. American Association for Clinical Chemistry Subcommittee on cTnI Standardization. Clin Chem 44:1198–1208
36. Katrukha AG et al (1998) Degradation of cardiac troponin I: implication for reliable immunodetection. Clin Chem 44:2433–2440
37. McDonough JL, Arrell DK, Van Eyk JE (1999) Troponin I degradation and covalent complex formation accompanies myocardial ischemia/reperfusion injury. Circ Res 84:9–20
38. McDonough JL et al (2001) Cardiac troponin I is modified in the myocardium of bypass patients. Circulation 103:58–64
39. Labugger R et al (2000) Extensive troponin I and T modification detected in serum from patients with acute myocardial infarction. Circulation 102:1221–1226
40. Tate JR et al (2010) Standardisation of cardiac troponin I measurement: past and present. Pathology 42:402–408
41. Koshida S et al (2010) Prevalence of human anti-mouse antibodies (HAMAs) in routine examinations. Clin Chim Acta 411:391–394
42. Todd DJ et al (2011) Erroneous augmentation of multiplex assay measurements in patients with rheumatoid arthritis due to heterophilic binding by serum rheumatoid factor. Arthritis Rheum 63:894–903
43. Eriksson S et al (2005) Negative interference in cardiac troponin I immunoassays by circulating troponin autoantibodies. Clin Chem 51:839–847
44. Eriksson S et al (2005) Comparison of cardiac troponin I immunoassays variably affected by circulating autoantibodies. Clin Chem 51:848–855
45. Adamczyk M, Brashear RJ, Mattingly PG (2009) Circulating cardiac troponin-I autoantibodies in human plasma and serum. Ann N Y Acad Sci 1173:67–74
46. Adamczyk M, Brashear RJ, Mattingly PG (2009) Prevalence of autoantibodies to cardiac troponin T in healthy blood donors. Clin Chem 55:1592–1593
47. Adamczyk M, Brashear RJ, Mattingly PG (2010) Coprevalence of autoantibodies to cardiac troponin I and T in normal blood donors. Clin Chem 56:676–677
48. Body R et al (2011) Rapid exclusion of acute myocardial infarction in patients with undetectable troponin using a high-sensitivity assay. J Am Coll Cardiol 58:1332–1339
49. Januzzi JL et al (2010) High-sensitivity troponin T concentrations in acute chest pain

patients evaluated with cardiac computed tomography. Circulation 121:1227–1234
50. Antman EM et al (2000) The TIMI risk score for unstable angina/non-ST elevation MI: a method for prognostication and therapeutic decision making. JAMA 284:835–842
51. Eagle KA et al (2004) A validated prediction model for all forms of acute coronary syndrome: estimating the risk of 6-month postdischarge death in an international registry. JAMA 291:2727–2733
52. Morrow DA et al (2000) TIMI risk score for ST-elevation myocardial infarction: a convenient, bedside, clinical score for risk assessment at presentation: an intravenous nPA for treatment of infarcting myocardium early II trial substudy. Circulation 102:2031–2037
53. Morrow DA et al (2001) Ability of minor elevations of troponins I and T to predict benefit from an early invasive strategy in patients with unstable angina and non-ST elevation myocardial infarction: results from a randomized trial. JAMA 286:2405–2412
54. Apple FS et al (2009) Role of monitoring changes in sensitive cardiac troponin I assay results for early diagnosis of myocardial infarction and prediction of risk of adverse events. Clin Chem 55:930–937
55. Celik S et al (2011) Cardiac troponin T concentrations above the 99th percentile value as measured by a new high-sensitivity assay predict long-term prognosis in patients with acute coronary syndromes undergoing routine early invasive strategy. Clin Res Cardiol 100:1077–1085
56. Kavsak PA et al (2009) Short- and long-term risk stratification using a next-generation, high-sensitivity research cardiac troponin I (hs-cTnI) assay in an emergency department chest pain population. Clin Chem 55:1809–1815
57. Kurz K et al (2011) Comparison of the new high sensitive cardiac troponin T with myoglobin, h-FABP and cTnT for early identification of myocardial necrosis in the acute coronary syndrome. Clin Res Cardiol 100:209–215
58. Lindahl B, Venge P, James S (2010) The new high-sensitivity cardiac troponin T assay improves risk assessment in acute coronary syndromes. Am Heart J 160:224–229
59. O'Connor RE et al (2010) Part 10: acute coronary syndromes: 2010 American Heart Association Guidelines for Cardiopulmonary Resuscitation and Emergency Cardiovascular Care. Circulation 122:S787–817
60. Scirica BM et al (2011) Assessment of multiple cardiac biomarkers in non-ST-segment elevation acute coronary syndromes: observations from the MERLIN-TIMI 36 trial. Eur Heart J 32:697–705
61. Hochholzer W et al (2011) Incremental value of high-sensitivity cardiac troponin T for risk prediction in patients with suspected acute myocardial infarction. Clin Chem 57:1318–1326
62. Meune C et al (2011) The GRACE score's performance in predicting in-hospital and 1-year outcome in the era of high-sensitivity cardiac troponin assays and B-type natriuretic peptide. Heart 97:1479–1483
63. Ndrepepa G et al. (2011) Comparison of prognostic value of high-sensitivity and conventional troponin T in patients with non-ST-segment elevation acute coronary syndromes. Clin Chim Acta 412: 1350–1356
64. Mills NL et al (2011) Implementation of a sensitive troponin I assay and risk of recurrent myocardial infarction and death in patients with suspected acute coronary syndrome. JAMA 305:1210–1216
65. Korff S, Katus HA, Giannitsis E (2006) Differential diagnosis of elevated troponins. Heart 92:987–993
66. Roongsritong C, Warraich I, Bradley C (2004) Common causes of troponin elevations in the absence of acute myocardial infarction: incidence and clinical significance. Chest 125:1877–1884
67. Omland T et al (2009) A sensitive cardiac troponin T assay in stable coronary artery disease. N Engl J Med 361:2538–2547
68. Reiter M et al (2011) Early diagnosis of acute myocardial infarction in patients with pre-existing coronary artery disease using more sensitive cardiac troponin assays. Eur Heart J 33:988–997. doi:10.1093/eurheartj/ehr376, E-pub ahead of print
69. Aviles RJ et al (2002) Troponin T levels in patients with acute coronary syndromes, with or without renal dysfunction. N Engl J Med 346:2047–2052
70. Apple FS et al (2002) Predictive value of cardiac troponin I and T for subsequent death in end-stage renal disease. Circulation 106:2941–2945
71. Dierkes J et al (2000) Cardiac troponin T predicts mortality in patients with end-stage renal disease. Circulation 102:1964–1969
72. Ooi DS et al (1999) Increased mortality in hemodialyzed patients with elevated serum troponin T: a one-year outcome study. Clin Biochem 32:647–652
73. Reiter M et al (2011) Early diagnosis of acute myocardial infarction in the elderly using more sensitive cardiac troponin assays. Eur Heart J 32:1379–1389

74. Hickman PE et al (2010) Cardiac troponin may be released by ischemia alone, without necrosis. Clin Chim Acta 411:318–323
75. Saunders JT et al (2011) Cardiac troponin T measured by a highly sensitive assay predicts coronary heart disease, heart failure, and mortality in the Atherosclerosis Risk in Communities Study. Circulation 123: 1367–1376
76. de Lemos JA et al (2010) Association of troponin T detected with a highly sensitive assay and cardiac structure and mortality risk in the general population. JAMA 304:2503–2512
77. Blankenberg S et al (2010) Contribution of 30 biomarkers to 10-year cardiovascular risk estimation in 2 population cohorts: the MONICA, risk, genetics, archiving, and monograph (MORGAM) biomarker project. Circulation 121:2388–2397

Chapter 25

S100A1 Gene Therapy in Small and Large Animals

Patrick Most, Philip Raake, Christophe Weber, Hugo A. Katus, and Sven T. Pleger

Abstract

Myocardial in vivo gene delivery is a valuable technique to investigate the relevance of a protein of interest on cardiac contractile function, hypertrophy, and energy state in healthy animals as well as in a variety of models of cardiovascular disease. Rodent models are used to screen effects and to investigate molecular mechanisms, while large animal models, more closely reflecting human anatomy, physiology, and function, are inevitable for translational therapeutic approaches. The gene of interest, whose expression is driven by a non-cardioselective or cardioselective promotor is cloned into a viral vector. This vehicle is then delivered using an appropriate administration route to target the heart and to achieve efficient protein expression in myocardium.

Here we describe myocardial gene therapy in small and large animal models of postischemic heart failure used to reveal the positive inotrope, antihypertrophic, and pro-energetic action of the small calcium sensor protein S100A1.

Key words: S100A1, Calcium, EF-hand, Myocardial gene therapy, Retroinfusion, Antegrade delivery, Adenovirus, Adeno-associated virus, Efficacy, Cardioselective

1. Introduction

There was tremendous progress in the field of in vivo genetic engineering during the last decade contributing significantly to reveal molecular mechanisms and to explore molecular-targeted treatment strategies in cardiovascular disease (1–6). Myocardial gene delivery is an add-on to transgenic animal models offering the possibility to initiate protein (over)expression at a certain time point in healthy animals, and thus circumventing potential side effects such as the possibility of a lethal embryonic phenotype or an evolutionary switch of the mode of action of the protein of interest. Myocardial gene delivery can also be used as a treatment option (gene therapy)

Claus W. Heizmann (ed.), *Calcium-Binding Proteins and RAGE: From Structural Basics to Clinical Applications*, Methods in Molecular Biology, vol. 963, DOI 10.1007/978-1-62703-230-8_25, © Springer Science+Business Media New York 2013

in animal models of cardiovascular disease. The choice of species depends on suitable or existing animal models as well as on the intention to either screen effects or perform translational research and is of utmost importance for the technique of cardiac gene delivery, the administration route, and the viral vector (3, 6–9).

A significant aspect of myocardial gene therapy is analyses of potential detrimental effects of (a) the applied vector, (b) the gene of interest, and (c) the delivery procedure itself. Especially dosing of the vector, the route of vector administration, vector purification, choice of vector, choice of promotor, and mechanical aspects such as injection pressure may be of major importance interpreting observed effects prior clinical translation (10–12).

Here we demonstrate efficient cardioselective S100A1 gene therapy in rat and pig models causing both, increased global cardiac function and cardiac reverse-remodeling in postischemic heart failure (13, 14). Positive inotrope, antihypertrophic and pro-energetic effects of S100A1 gene therapy in heart failure were mediated by calcium dependent direct binding to distinct intracellular target structures: the ryanodine receptor (RyR2), the sarcoplasmic reticulum calcium ATPase (SERCA2a), the sarcomeric titin, and mitochondrial enzymes such as the F1-ATPase (8, 13–26). The described protocols are based on techniques established by Walter Koch et al., Roger Hajjar et al., Kenneth Chien et al., and Peter Boekstegers et al. which were modified to work best in our hands for efficient and reproducible cardiac gene delivery (27–30). To achieve myocardial S100A1 overexpression in different species, distinct AAV serotypes, cardioselective promotors, and administration routes were used (13, 14). Thus, we describe methods to genetically target the whole heart in a rat model, while for anatomy reasons, in pig the target area of gene delivery is the supply zone of the anterior descending coronary artery (LAD) (13, 14).

Adeno-associated virus (AAV) vectors were produced, purified, and quantitated in core facilities of the "Deutsches Krebsforschungszentrum (DKFZ) and the Center for Translational Medicine at Thomas Jefferson University in order to achieve high-titer AAV vector solutions without toxic contamination. The described procedures should enable to reproduce efficient cardioselective gene therapy in small and large animal models and highlight potential pitfalls.

2. Materials

2.1. Myocardial Gene Delivery in Rats

1. Isoflurane condenser (e.g., Dräger or Harvard Apparatus).
2. Surgical cold light source (Harvard Apparatus).
3. Self-made rat laryngoscope (see Note 1).

4. Standard rubber band (~10 cm total length).
5. Styrofoam board.
6. Styrofoam shields to protect the hindlimbs of the rat from freezing.
7. Blunt nose tracheal cannula (Havard Apparatus, size depends on rat's weight).
8. Respirator suitable for rodents (Harvard Apparatus).
9. Thermometer.
10. Ice packs ($18 \times 10 \times 1$ cm).
11. Heating pad (40×20 cm) (Harvard Apparatus).
12. Standard surgical set of sterile drapes and autoclavable scissors, forceps, surgical clamps, and rib stretcher (Hereaus).
13. P-50 tubing (Becton Dickinson).
14. 4-0 and 2-0 silk (Becton Dickinson).
15. Millar pressure transducer system; 2F SPR-320 catheter (Millar instruments).
16. Waterproofed marker.
17. Chapstick (Labello).
18. Adenosine (Glaxo).
19. Substance P (Sigma chemical).
20. Dobutamine Liquid (Fresenius).
21. Syringes (50 ml, 3 ml, 1 ml) and needles (30½ gauge; Becton Dickinson).
22. 2.5×10^{11} total viral particles of an AAV6-transgene.
23. Timer.
24. Defibrillator suitable for kids (Medtronic).
25. Small electrical cauter (Harvard Apparatus).

2.2. Myocardial Gene Delivery in Pigs

1. Respirator suitable for pigs (Dräger).
2. Ketamine, Xylazine, Atropine, Buprenorphine, Heparine, Nitroglycerine, 0.9% saline, and Isoflurane.
3. Tracheal cannula suitable for pigs (Mallinckrodt).
4. ECG monitoring, pulse oximetry monitoring (Dräger).
5. Standard surgical set of sterile drapes and autoclavable scissors, forceps, surgical clamps, and rib stretcher (Hereaus).
6. Percutaneous transluminal coronary artery (PTCA) angioplasty set (Smiths).
7. Contrast dye (Bracco).
8. 3-0 and 4-0 silk (Becton Dickinson).

9. 6F and 9F vascular sheath (Cordis).
10. Peripheral venous line, peripheral venous catheter (20 gauge; Cordis).
11. Coronary sinus catheter (Attain Command 6250 MB or 6250 MPA2; Medtronic).
12. Coronary sinus balloon catheter (Attain 6215-80 cm; Medtronic).
13. 0.0035′ wire (Medex).
14. 6F Judkins Right catheter (Cordis).
15. Angioplasty balloon (3.5×12 mm; Maverick).
16. 1.5×10^{13} total viral particles of AAV6- or AAV9-transgene.
17. Cathlab angiopump (Philips).
18. X-ray C-arm (Philips).
19. Timer.
20. Defibrillator (Medtronic).
21. Balance Middleweight PTCA wire (Abbott).

3. Methods

3.1. Myocardial Gene Delivery in Rats

3.1.1. Intubation of the Rat

1. Place the anesthetized rat (250–300 g) on the table (legs upright), on a flat styrofoam board (~2 cm thick), with the rats head towards you. Tape down the fore- and hindlimbs of the rat on the styrofoam board.
2. To continue anesthesia place the isoflurane line (1.2%) into a 50 ml syringe and place the large open end of the syringe over the rat's head.
3. Move a strong cold light source above the rat's upper thorax/lower neck. Turn off the anesthesia to intubate.
4. Open the rat's mouth and lift up the tongue using a self-made laryngoscope and keep the upper incisors down with a rubber band attached to the Styrofoam box. Intubate the rat by advancing a blunt nose cannula into the trachea (see Note 1).
5. Turn on the respirator; 8–9 ml at 45 resparations/min. When the rat is intubated the chest will go up and down in rhythm with the movements of the respirator, otherwise the stomach will swell. Tape-down the respirator hose to fix it.
6. Shave the animal around the larynx and vacuum. The surgical field will be wiped with alcohol and prepared with Chlorhexadine solution

3.1.2. Preparation Prior Gene Delivery in Rat

1. Place a thermometer in the anus of the rat to monitor body temperature; fix it by taping down.
2. Place ice packs (18 × 10 × 1 cm) on the hindlimbs of the rat. Use two layers of Styrofoam (3 mm thick) to protect the legs from freezing and necrosis (see Note 2).
3. Cool down the rat to below 30°C (between 28 and 30°C) while preparing the carotid artery (see Note 3).
4. Turn on a heating pad and lay aside (will be used after the virus is injected).
5. Using a scalpel blade make an incision from just below the forelimbs anterior to the upper (proximal) third of the sternum along the midline.
6. Grab the tissue with forceps and clamps and expand, do not cut because of blood vessels. There are muscles on either side of the neck that help the head turn, the Sternomastoids, and muscles that lie on top of the trachea, the Sternohyoids, cut the sternohyoids down the center line exposing the ridged trachea.
7. Lift (!) the sternum and cut the upper (proximal) third of the sternum (~ 1 cm) just in the very midline of the sternum to avoid major bleeding by injuring the Aa. Mammaria internae. Cut the sternum with scissors in short snips, significant force is required to cut through, go slow. If some bleeding occurs that's ok, use the electric cauter to stop it. Remember the heart is underneath the sternum.
8. Insert a rib stretcher between the severed bones of the upper third of the sternum. Carefully stretch the upper (proximal) third of the sternum.
9. Open the pericardium and avoid opening of the pleura (!) (see Note 4).
10. Clamp the thymus and deflect to the left side (see Note 5).
11. Identify the right carotid artery and separate it from connecting tissue and the N. Vagus, which runs alongside it.
12. Take a loop of a 4-0 silk suture and slip it underneath the right carotid artery, then split the suture and move one strand towards the heart and the other towards the neck. Make a loop of the suture closest to the heart, but do not tighten.
13. Place a loop of a 2-0 silk suture under the ascending aortic. Loop the aorta ascendens. Don't loop the pulmonary artery or the vena cava superior (see Note 6).

3.1.3. Myocardial Gene Delivery in Rat

Once the body temperature of the rat is below 30°C:

1. Knot the loop on the carotid artery situated towards the head of the rat. Put tension on the silk by taping to the Styrofoam to stretch the carotid artery.

2. Clamp the carotid below the second loop situated towards the heart.
3. Cut a tiny hole (a third of the diameter of the vessel) in the upper carotid artery between the two loops using fine scissors.
4. Insert a 2F Millar pressure transducer (SPR-320) into the right carotid artery and advance it via the ascending aorta and the aortic valve into the left ventricle. Afterwards pull back the pressure transducer just above the aortic valve using the shape of the pressure curve and define the distance between the hole in the carotid artery and the aortic valve. Mark the distance on the pressure transducer and pull out the catheter (see Note 7).
5. Mark the exact distance between the hole of the carotid artery and the aortic valve on the P-50 tubing.
6. Insert the P-50 tubing into the right carotid artery. Advance just above the aortic valve (not in the left ventricular chamber) by using the mark (see Note 8).
7. Reduce the isoflurane to 0.5%.
8. Inject 1.2 mg of adenosine in 400 μl directly into the right ventricle chamber using a 30 ½ gauge needle. This will induce a temporary atrioventricular blockage and stop the heart beating.
9. Immediately, stop blood flow from the ascending aorta proximal of the tip of the P-50 tubing (which is still between the clamp and the aortic valve) by tightening the 2-0 suture and fix the suture using a small but strong clamp. Once you clamp the aorta start the timer.
10. Rapidly inject half of the prepared virus solution: (2.5×10^{11} total viral particles in a volume of 250 μl, 8 μg of substance P in 80 μl, and 1.6 ml of saline totaling to ~1.9 ml volume in a 3 ml syringe) as a bolus via the P-50 tubing. We were using AAV serotype 6. Expression of our transgene was driven by a cardioselective promotor (elongation factor 1α promotor and the α-cardiac actin enhancer). The heart is now resting and the color of the myocardium turns white/yellow (see Note 9).
11. Wait 1-min.
12. Inject the remaining half of the virus solution and flush the P-50 tubing with saline using an appropriate volume (depends on the length of your tubing) to get the virus solution out of the P-50 tubing.
13. Wait 1-min (the heart may eventually start beating).
14. Inject a bolus of 30 μg of dobutamine via the P-50 tubing to restart the heart.
15. Remove the P-50 tubing and clamp from the ascending aorta.

16. Warm the rat to 37°C using the heating pad and removing the ice packs (see Note 10).
17. Make a knot in the carotid artery close to the incision before closing.
18. Suture the fascia and the muscles
19. Suture the skin.
20. Turn off the oxygen and isoflurane start on room air.
21. When the rat starts breathing spontaneously extubate and place in a cage (care for appropriate pain reliever) (see Note 11).

3.2. Myocardial Gene Delivery in Pig

3.2.1. Preparation Prior Myocardial Gene Delivery in Pig

1. Animals (30–35 kg) will not receive food in the morning of procedure.
2. The animals will be anesthetized with an intramuscular injection of a cocktail containing Ketamine (20 mg/kg), Xylazine (2.0 mg/kg), and Atropine (0.04 mg/kg). Buprenorphine (0.05 mg/kg) will be given before the surgical procedure starts.
3. Upon loss of responsiveness and spontaneous movement, the animal will be intubated and maintained on isoflurane (1–4%) and oxygen. If necessary, pigs may be ventilated using an intermittent positive pressure ventilator. Heart rate, EKG, and oxygenation (pulse oximetry) will be monitored throughout the duration of the procedure.
4. The neck area of each animal will be shaved; the surgical field will be wiped with alcohol and prepared with Chlorhexadine solution. A marginal ear vein will be catheterized for fluid administration and any necessary drug delivery during procedure.

3.2.2. Myocardial Gene Delivery in Pig (Target Area Is the Supply Zone of the LAD)

1. Surgically expose the right carotid artery and the right jugular vein. Carefully prepare the carotid artery and the jugular vein from connecting tissue and the N. vagus.
2. Take a loop of a 3-0 or 4-0 silk suture and slip it underneath the right carotid artery, then split the suture and move one strand towards the heart and the other towards the neck. Make a loop of the suture closest to the heart, but do not tighten. Perform the same procedure to prepare the right jugular vein (see Note 12).
3. Ligate the right carotid artery using the suture towards the head. Carefully stretch the vessel towards the head using a clamp.
4. Ligate the right carotid artery using the suture towards the head. Carefully stretch the vessel towards the head using a clamp.
5. Cut a small hole, both in the upper right carotid artery and the upper right jugular vein between the two loops using fine scissors and insert a vascular sheath (6F or larger for carotid artery and 9F for jugular vein).

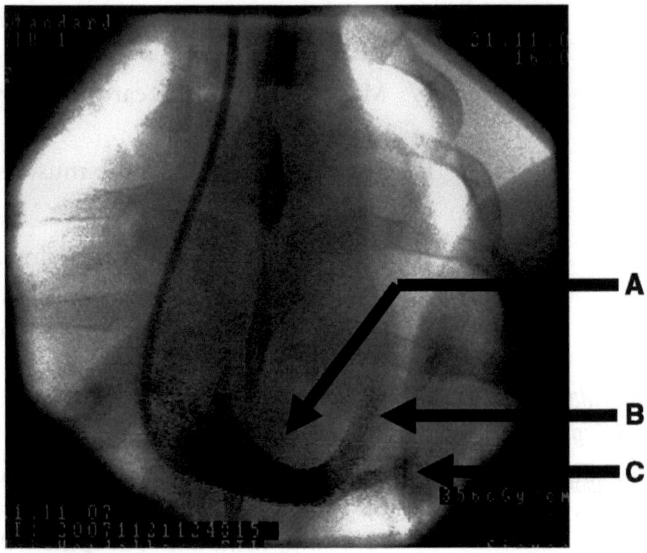

Fig. 1. Visualization of the coronary sinus in German farm pigs. Radioscopic image in anterior posterior projection showing (**A**) the coronary sinus, (**B**) the V. azygos, and (**C**) the vein draining blood from the coronary arteries. Contrast dye is primarily injected into the coronary sinus.

6. Administer heparin (2,500 IU) intravenously using a peripheral line via an ear vein.
7. Advance a coronary sinus catheter (Attain Command 6250 MB2 or 6250 MP A2) into the coronary sinus (Fig. 1; arrow A) and next into the vein draining blood from the coronary arteries (Fig. 1; arrow C) (see Note 13).
8. After the catheter intubates the vein draining blood from the coronary arteries (Fig. 1, arrow C) insert a 0.014′ Balance Middleweight Wire (BMW) or Balance Heavyweight Wire (BHW) into anterior interventricular cardiac vein (AICV) (Figs. 2 and 3) which runs parallel to the left ascending coronary artery (LAD).
9. Advance a coronary sinus balloon catheter (Attain 6215-80 cm or Arrow A1-7F) to the entrance of the AICV (Fig. 2) (see Note 14).
10. Introduce the guiding catheter (6F Judkins right) via the right carotid artery into the ostium of the left coronary artery (LCA) using an LAD 0.035′ standard wire.
11. Via the guiding catheter introduce a 0.014′ BMW wire in the LAD (see Note 15).
12. Introduce a 3.5 mm × 12 mm PTCA balloon into the LAD distal of the first diagonal branch using contrast dye.
13. Occlude LAD distal to the first diagonal branch (see Note 16).

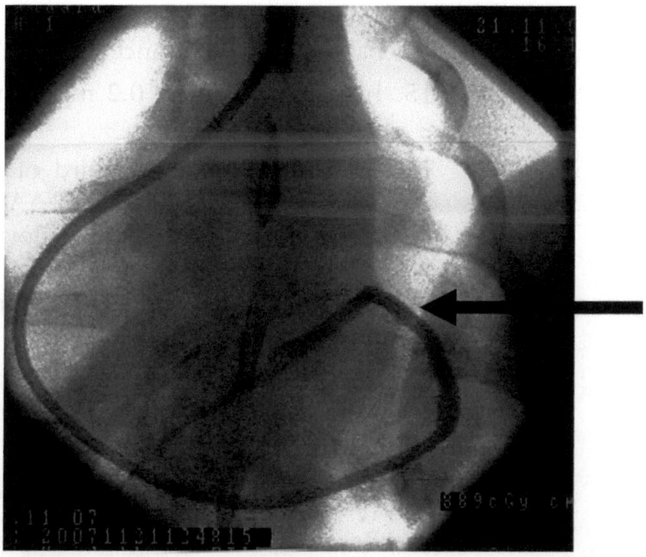

Fig. 2. Visualization of the anterior interventricular cardiac vein (AICV). Radioscopic image in anterior posterior projection. The *arrow* marks the entrance of the AICV draining blood from the left anterior descending coronary artery (LAD). At the entrance of the AICV the coronary sinus balloon catheter should be situated and virus solution should be retrogradely injected in the AICV. The AICV runs towards the *lower left corner* of the image.

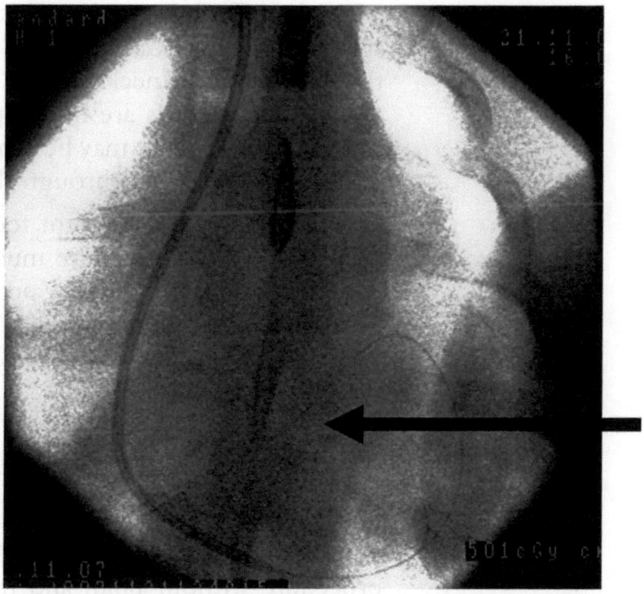

Fig. 3. Visualization of the anterior interventricular cardiac vein (AICV). Radioscopic image in anterior posterior projection. The *arrow* marks a wire situated in the AICV running parallel to the LAD.

14. Immediately, occlude the AICV using the coronary sinus balloon at the entrance of the AICV (Fig. 2).
15. Immediately, inject 0.2 mg of nitroglycerine (NTG) into the AICV (see Note 17).
16. Immediately, inject a third of the virus solution containing 1.5 × 10*13 tvp of AAV6 or AAV9-transgene using a cardioselective promotor (cytomegalovirus/myosin light chain 0.26 (CMV/MLC0.26)), 0.4 mg of NTG and 0.9% saline totaling to a volume of 50 ml using a 50 ml syringe (see Note 18). Start timer.
17. Wait 45 s from the start of virus delivery. Then deflate the balloons in the LAD and in the AICV (see Note 19).
18. Wait 3 min.
19. Repeat steps 14–17 twice to inject the whole virus solution.
20. Remove catheters and balloons from the LAD and the AICV. Perform a follow-up coronary angiography of the LCA to ensure integrity of the vessel.
21. Close the incision using three layers of suture (fascia, muscle, and skin). Care for appropriate pain reliever.

4. Notes

1. This is a difficult procedure that must be done quickly. Use a strong cold light source placed over the neck, not the top of the throat, but the neck near the top of the thorax. The trachea and the vocal cords are quite apparent with the light. Don't push too hard. There may be some resistance, but it should not feel like your pushing through the body.

   It is important to use bright focused surgical light. The light used during this procedure must be a cold light source since the brain will not tolerate the procedure at a temperature above 30°C.

   An iron or aluminum sheet sized 7 cm (length) and 6 mm (wide) can be used to construct a rat laryngoscope (create a flattened S-shape of the sheet).
2. Without Styrofoam to protect the legs you will find necrosis at the hindlimbs of the rat eventually causing loss of toes.
3. A body temperature below 30°C is required to perform the procedure without brain and myocardial damage due to the cardiac arrest for ~2 min. Further reduction of the body temperature (below 28°C) is possible but will cause a significant longer period of time for the rat to wake up and recover.

4. Open the pericardium only at the basis of heart. Do not open the pleura since otherwise the lungs collapse.
5. The thymus may be a large organ especially in young rats. Cutting the thymus will cause significant bleeding. Avoid cutting, mobilize the thymus by using electrical cauterization on the right side just enough to get access to the ascending aorta and deflect the thymus to the left side.
6. A 90°C needle holder with a thin but blunt proximal tip is useful for this procedure.
7. This procedure is necessary since the virus solution need to be injected between the aortic valve and the silk-loop around the ascending aorta to allow perfusion of the coronary arteries. Injection of the virus solution into the left ventricular chamber (having a resting heart) will not cause myocardial gene delivery.
8. This is not easy, take your time, carotid is flexible. Slide the tubing into the carotid without rupturing the carotid artery and advance the catheter past the clamp. Determine the position of the catheter in the ascending aorta using forceps. Advance just above the aortic valve (not in the left ventricle) by using the mark. Use chapstick on the outer side of the P-50 tubing to reduce friction.
9. Myocardial gene delivery will only be efficient in a resting heart, some beats may occur.
10. Because substance P causes the rat to secrete fluid you eventually need to dry the mouth and nose using cotton tips during the following 10 min.
11. During the first hour after the procedure some rats will develop ventricular fibrillation. The animal might suddenly cramp and faint. Use the defibrillator to convert into sinus rhythm.
12. Do not rotate the jugular vein. Otherwise you may occlude the vessel.
13. Carefully turn the catheter to the left side in the proximal third of the right atrium to intubate the coronary sinus (Fig. 1; arrow A). Pigs have a prominent V. azygos (Fig. 1; arrow B) draining into the coronary sinus (Fig. 1; arrow A). To intubate the vein draining blood from the coronary arteries (Fig. 1, arrow C) slightly rotate and advance the catheter clockwise. Sometimes there is a valve in front of the vein draining blood from the coronary arteries (Fig. 1, arrow C). Try to visualize by using contrast dye. Overall aim is to intubate the anterior intraventricular coronary vein (AICV) (Fig. 2; arrow) draining venous blood from the supply zone of the anterior descending coronary artery (LAD). Retrograde delivery of the virus solution via

the AICV will cause myocardial gene delivery to the anterior/ septal myocardium (supply zone of the LAD) which is the target zone for cardiac gene delivery described in this protocol.

14. Alternatively (to steps 7–9) use a multipurpose catheter (7F) to intubate the vein draining blood from the coronary arteries (Fig. 1; arrow C); advance a 260 cm 0.014′ or 0.018′ wire into the AICV. Using the wire in place remove the multipurpose catheter and introduce the coronary sinus balloon catheter (Attain 6215-80 cm or Arrow A1 -7F) into the entrance of the AICV (Fig. 2).
15. The wires in the LAD and the AICV now run in parallel. Of note, there is considerable variation in the anatomy of the AICV depending on the breed of the pigs. In German farm pigs, we always observed an anatomy as described here and also shown in Figs. 1, 2, and 3. However, Yucatan pigs and cross-breeding pigs between Yorkshire pigs and Yucatan pigs sometimes show draining of the AICV directly into the V. cava superior making retrograde myocardial gene delivery as described in this protocol impossible.
16. This will block antegrade perfusion of the target area and facilitate myocardial gene delivery. Further proximal temporary occlusion of the LAD will cause a significant higher rate of ventricular fibrillation and eventually myocardial damage.
17. To reduce the endothelial barrier.
18. Use an angiopump which is standard in every cathlab to control injection velocity and maximal pressure within the vessel. Use a ramp to avoid a pressure peak and thus damage of the AICV. Possible injection velocity may vary between 16 ml within 15–45 s and allowing a maximum injection pressure of 100 mmHg.
19. Injection of the virus solution is divided into three portions to reduce the time of myocardial ischemia but to allow a prolonged time of incubation of the virus within the vessel.

Acknowledgments

This work was supported by NIH grants: R01HL92130 and R01HL92130-02S1 (P.M.); Deutsche Forschungsgemeinschaft: 562/1-1 (P.M. and S.T.P.); and the Bundesministerium für Bildung und Forschung: 01GU0527 (P.M., H.A.K.); the Pennsylvania-Delaware Affiliate of the American Heart Association (S.T.P.); the Lilly-Stipendium of the Deutsche Gesellschaft für Kardiologie (S.T.P.).

References

1. del Monte F, Hajjar R (2003) Targeting calium cycling proteins in heart failure through gene transfer. J Physiol 546:49–61
2. Wehrens XH, Marks AR (2004) Novel therapeutic approaches for heart failure by normalizing calcium cycling. Nat Rev Drug Discov 3:565–573
3. Wasala NB, Shin JH, Duan D (2011) The evolution of heart gene delivery vectors. J Gene Med 13:557–565
4. Raake PW, Tscheschner H, Reinkober J, Ritterhoff J, Katus HA, Koch WJ, Most P (2011) Gene therapy targets in heart failure: the path to translation. Clin Pharmacol Ther 90:542–553
5. Heine HL, Leong HS, Rossi FM, McManus BM, Podor TJ (2005) Strategies of conditional gene expression in myocardium: an overview. Methods Mol Med 112:109–154
6. Ishikawa K, Tilemann L, Fish K, Hajjar RJ (2011) Gene delivery methods in cardiac gene therapy. J Gene Med 13:566–572
7. Dixon JA, Spinale FG (2009) Large animal models of heart failure: a critical link in the translation of basic science to clinical practice. Circ Heart Fail 2:262–271
8. Pleger ST, Boucher M, Most P, Koch WJ (2007) Targeting myocardial beta-adrenergic receptor signaling and calcium cycling for heart failure gene therapy. J Card Fail 13:401–414
9. Raake PW, Hinkel R, Müller S, Delker S, Kreuzpointner R, Kupatt C, Katus HA, Kleinschmidt JA, Boekstegers P, Müller OJ (2008) Cardio-specific long-term gene expression in a porcine model after selective pressure-regulated retroinfusion of adeno-associated viral (AAV) vectors. Gene Ther 15:12–17
10. Emani SM, Shah AS, Bowman MK, Emani S, Wilson K, Glower DD, Koch WJ (2003) Catheter-based intracoronary myocardial adenoviral gene delivery: importance of intraluminal seal and infusion flow rate. Mol Ther 8:306–313
11. Wang J, Faust SM, Rabinowitz JE (2011) The next step in gene delivery: molecular engineering of adeno-associated virus serotypes. J Mol Cell Cardiol 50:793–802
12. Raake P, von Degenfeld G, Hinkel R, Vachenauer R, Sandner T, Beller S, Andrees M, Kupatt C, Schuler G, Boekstegers P (2004) Myocardial gene transfer by selective pressure-regulated retroinfusion of coronary veins: comparison with surgical and percutaneous intramyocardial gene delivery. J Am Coll Cardiol 44:1124–1129
13. Pleger ST, Shan C, Ksienzyk J, Bekeredjian R, Boekstegers P, Hinkel R, Schinkel S, Leuchs B, Ludwig J, Qiu G, Weber C, Raake P, Koch WJ, Katus HA, Müller OJ, Most P (2011) Cardiac AAV9-S100A1 gene therapy rescues post-ischemic heart failure in a preclinical large animal model. Sci Transl Med 20:92ra64
14. Pleger ST, Most P, Boucher M, Soltys S, Chuprun JK, Pleger W, Gao E, Dasgupta A, Rengo G, Remppis A, Katus HA, Eckhart AD, Rabinowitz JE, Koch WJ (2007) Stable myocardial-specific AAV6-S100A1 gene therapy results in chronic functional heart failure rescue. Circulation 115:2506–2515
15. Kraus C, Rohde D, Weidenhammer C, Qiu G, Pleger ST, Voelkers M, Boerries M, Remppis A, Katus HA, Most P (2009) S100A1 in cardiovascular health and disease: closing the gap between basic science and clinical therapy. J Mol Cell Cardiol 47:445–455
16. Boerries M, Most P, Gledhill JR, Walker JE, Katus HA, Koch WJ, Aebi U, Schoenenberger CA (2007) Ca2+ -dependent interaction of S100A1 with F1-ATPase leads to an increased ATP content in cardiomyocytes. Mol Cell Biol 27:4365–4373
17. Most P, Bernotat J, Ehlermann P, Pleger ST, Reppel M, Börries M, Niroomand F, Pieske B, Janssen PM, Eschenhagen T, Karczewski P, Smith GL, Koch WJ, Katus HA, Remppis A (2001) S100A1: a regulator of myocardial contractility. Proc Natl Acad Sci USA 98:13889–13894
18. Most P, Remppis A, Pleger ST, Löffler E, Ehlermann P, Bernotat J, Kleuss C, Heierhorst J, Ruiz P, Witt H, Karczewski P, Mao L, Rockman HA, Duncan SJ, Katus HA, Koch WJ (2003) Transgenic overexpression of the Ca2+ binding protein S100A1 in the heart leads to increased in vivo myocardial contractile performance. J Biol Chem 278:33809–33817
19. Most P, Seifert H, Gao E, Funakoshi H, Völkers M, Heierhorst J, Remppis A, Pleger ST, DeGeorge BR Jr, Eckhart AD, Feldman AM, Koch WJ (2006) Cardiac S100A1 protein levels determine contractile performance and propensity toward heart failure after myocardial infarction. Circulation 114:1258–1268
20. Tsoporis JN, Marks A, Zimmer DB, McMahon C, Parker TG (2003) The myocardial protein S100A1 plays a role in the maintenance of normal gene expression in the adult heart. Mol Cell Biochem 242:27–33

21. Völkers M, Loughrey CM, Macquaide N, Remppis A, DeGeorge BR Jr, Wegner FV, Friedrich O, Fink RH, Koch WJ, Smith GL, Most P (2007) S100A1 decreases calcium spark frequency and alters their spatial characteristics in permeabilized adult ventricular cardiomyocytes. Cell Calcium 41:135–143
22. Most P, Pleger ST, Völkers M, Heidt B, Boerries M, Weichenhan D, Löffler E, Janssen PM, Eckhart AD, Martini J, Williams ML, Katus HA, Remppis A, Koch WJ (2004) Cardiac adenoviral S100A1 gene transfer rescues failing myocardium. J Clin Invest 114:1550–1563
23. Pleger ST, Remppis A, Heidt B, Völkers M, Chuprun JK, Kuhn M, Zhou RH, Gao E, Szabo G, Weichenhan D, Müller OJ, Eckhart AD, Katus HA, Koch WJ, Most P (2005) S100A1 gene therapy preserves in vivo cardiac function after myocardial infarction. Mol Ther 12:1120–1129
24. Rohde D, Ritterhoff J, Voelkers M, Katus HA, Parker TG, Most P (2010) S100A1: a multifaceted therapeutic target in cardiovascular disease. J Cardiovasc Transl Res 3:525–537, Review
25. Völkers M, Rohde D, Goodman C, Most P (2010) S100A1: a regulator of striated muscle sarcoplasmic reticulum Ca2+ handling, sarcomeric, and mitochondrial function. J Biomed Biotechnol 2010:178614
26. Brinks H, Rohde D, Voelkers M, Qiu G, Pleger ST, Herzog N, Rabinowitz J, Ruhparwar A, Silvestry S, Lerchenmüller C, Mather PJ, Eckhart AD, Katus HA, Carrel T, Koch WJ, Most P (2011) S100A1 genetically targeted therapy reverses dysfunction of human failing cardiomyocytes. J Am Coll Cardiol 58:966–973
27. Maurice JP, Hata JA, Shah AS, White DC, McDonald PH, Dolber PC, Wilson KH, Lefkowitz RJ, Glower DD, Koch WJ (1999) Enhancement of cardiac function after adenoviral-mediated in vivo intracoronary beta2-adrenergic receptor gene delivery. J Clin Invest 104:21–29
28. Hajjar RJ, Schmidt U, Matsui T, Guerrero JL, Lee KH, Gwathmey JK, Dec GW, Semigran MJ, Rosenzweig A (1998) Modulation of ventricular function through gene transfer in vivo. Proc Natl Acad Sci USA 95:5251–5256
29. Hoshijima M, Ikeda Y, Iwanaga Y, Minamisawa S, Date MO, Gu Y, Iwatate M, Li M, Wang L, Wilson JM, Wang Y, Ross J Jr, Chien KR (2002) Chronic suppression of heart-failure progression by a pseudophosphorylated mutant of phospholamban via in vivo cardiac rAAV gene delivery. Nat Med 8:864–871
30. Boekstegers P, von Degenfeld G, Giehrl W, Heinrich D, Hullin R, Kupatt C, Steinbeck G, Baretton G, Middeler G, Katus HA, Franz WM (2000) Myocardial gene transfer by selective pressure-regulated retroinfusion of coronary veins. Gene Ther 7:232–240

Index

Claus W. Heizmann (ed.), *Calcium-Binding Proteins and RAGE: From Structural Basics to Clinical Applications*, Methods in Molecular Biology, vol. 963, DOI 10.1007/978-1-62703-230-8,

MIX
Papier aus verantwortungsvollen Quellen
Paper from responsible sources
FSC® C105338

If you have any concerns about our products,
you can contact us on
ProductSafety@springernature.com

In case Publisher is established outside the EU,
the EU authorized representative is:
Springer Nature Customer Service Center GmbH
Europaplatz 3, 69115 Heidelberg, Germany

Printed by Libri Plureos GmbH
in Hamburg, Germany